Lecture Notes in Computer Science 7086

Commenced Publication in 1973
Founding and Former Series Editors:
Gerhard Goos, Juris Hartmanis, and Jan van Leeuwen

Jean-Pierre Jouannaud   Zhong Shao (Eds.)

Certified Programs and Proofs

First International Conference, CPP 2011
Kenting, Taiwan, December 7-9, 2011
Proceedings

Springer

Volume Editors

Jean-Pierre Jouannaud
Tsinghua University
FIT 3-603, Beijing 100084, China
E-mail: jeanpierre.jouannaud@inria.fr

Zhong Shao
Yale University, Department of Computer Science
51 Prospect Street, New Haven, CT 06520-8285, USA
E-mail: zhong.shao@yale.edu

ISSN 0302-9743 e-ISSN 1611-3349
ISBN 978-3-642-25378-2 e-ISBN 978-3-642-25379-9
DOI 10.1007/978-3-642-25379-9
Springer Heidelberg Dordrecht London New York

Library of Congress Control Number: 2011940955

CR Subject Classification (1998): F.3.1, F.4.1, D.3.3, I.2.3, D.2.4, D.2

LNCS Sublibrary: SL 1 – Theoretical Computer Science and General Issues

Typesetting: Camera-ready by author, data conversion by Scientific Publishing Services, Chennai, India

Printed on acid-free paper

Springer is part of Springer Science+Business Media (www.springer.com)

Preface

This volume contains the papers presented at CPP 2011, the First International Conference on Certified Programs and Proofs held during December 7–9, 2011 in Kenting, Taiwan.

CPP is a new international forum on theoretical and practical topics in all areas, including computer science, mathematics and education, that consider certification as an essential paradigm for their work. The aims of CPP are best described in its Manifesto, which comes right after this text.

CPP 2011 was organized in Kenting, a beautiful resort in the south of Taiwan, together with APLAS. APLAS started on Monday, December 5 and ended on Wednesday, December 7, while CPP started on Wednesday, December 7 and ended on Friday, December 9. APLAS and CPP had therefore a full common day on December 9, with two jointly invited keynote speakers. CPP had one invited speaker for each of the two remaining days. Besides its four invited speakers, Andrew Appel, Nicolaj Bjønear, Peter O'Hearn, and Vladimir Voevodsky, CPP 2011 also had two lively panels, on Certificates and on Teaching with Proof Assistants, introduced by Dale Miller for the first, and by Tobias Nipkow for the second. While these introductions and debates could not be part of the proceedings for obvious reasons, they appear on CPP's website http://formes.asia/cpp/.

The CPP Program Committee was intentionally very diverse. We felt that this diversity would act as a strength in both attracting and evaluating submissions, and we think that our goal was met. There were 49 submissions. Each submitted extended abstract was reviewed by at least four program committee members. In some cases, the committee chose to consult additional reviewers whose names are listed at the end of this preface. The Program Committee did not meet in person, but carried out extensive electronic discussions over a period of almost three weeks, and accepted 24 papers. The program also included four invited talks by the invited speakers mentioned above.

CPP is a brand-new conference, with a rather large spectrum for a new conference, since the importance of formal proofs is now widely recognized as an important trend, both in computer science and mathematics, and even beyond. CPP being a new conference, we were prepared to handle a small number of submissions. We were actually quite surprised to have 53 abstracts submitted, resulting in 49 final submissions, which we read as a confirmation that our initiative fulfilled a real need.

Most, but not all, submissions came with a formal proof development, using one of the Proof Assistants on the market, mostly Coq and HOL as it appears from the proceedings. We strongly believe that the presence of these developments influenced very positively the overall quality of submissions. Although we had decided beforehand to clearly privilege quality over quantity, we ended up with a strong program containing almost half of the submissions.

CPP's call for papers announced a best paper award. Unfortunately, we were not able to raise funds for such an award, and have not decided yet, at the time the proceedings go to print, whether we shall give a best paper award without reward.

We are grateful to the Chair of the APLAS/CPP Organizing Committee, Tyng-Ruey Chuang, the CPP 2011 General Chair, Yih-Kuen Tsay, and the Publicity Chair, Bow-Yaw Wang, for their considerable effort in planning and organizing the conference itself and making the conference possible and enjoyable. We would like to thank our fellow Program Committee members for their hard work selecting a high-quality and stimulating program of contributed papers, and advising on invited speakers. On behalf of the emerging CPP community, we thank all authors who submitted papers for consideration. It is the quality of your submissions which makes this program attractive. We also wish to thank the invited speakers who all delivered on their promises. We would also like to thank all those, outside the Program Committee, who contributed with reviews which were sometimes decisive.

Finally, CPP 2011 was sponsored by Academia Sinica (Taiwan), National Taiwan University, Tsinghua University (Institute of Advanced Study and Software Chair), and the National Science Council (Taiwan). We are also grateful for the administrative support from the Institute of Information Science, Academia Sinica (Taiwan) and the Department of Information Management and Yen Tjing Ling Industrial Research Institute at National Taiwan University. Without their help, organizing the first CPP would have simply been impossible.

September 2011

Jean-Pierre Jouannaud
Zhong Shao

Conference Organization

Program Co-chairs

Jean-Pierre Jouannaud	INRIA and Tsinghua University, France/China
Zhong Shao	Yale University, USA

Program Committee

Andrea Asperti	University of Bologna, Italy
Gilles Barthe	IMDEA Software Institute, Spain
Xiao-Shan Gao	Chinese Academy of Sciences, China
Georges Gonthier	Microsoft Research Cambridge, UK
John Harrison	Intel Corporation, USA
Chris Hawblitzel	Microsoft Research Redmond, USA
Akash Lal	Microsoft Research, India
Xavier Leroy	INRIA Paris-Rocquencourt, France
Yasuhiko Minamide	University of Tsukuba, Japan
Shin-Cheng Mu	Academia Sinica, Taiwan
Michael Norrish	NICTA, Australia
Brigitte Pientka	McGill University, Canada
Sandip Ray	University of Texas at Austin, USA
Natarajan Shankar	SRI International, USA
Christian Urban	TU Munich, Germany
Viktor Vafeiadis	MPI-SWS, Germany
Stephanie Weirich	University of Pennsylvania, USA
Kwangkeun Yi	Seoul National University, Korea

General Chair

Yih-Kuen Tsay	National Taiwan University

Publicity Chair

Bow-Yaw Wang	Academia Sinica, Taiwan

Organizing Committee

Tyng-Ruey Chuang	Academia Sinica, Taiwan
Shin-Cheng Mu	National Taiwan University
Yih-Kuen Tsay	Academia Sinica, Taiwan

External Reviewers

Reynald Affeldt
Carlos Areces
Mathieu Boespflug
Chris Casinghino
Ilaria Castellani
Andrew Cave
Juergen Christ
Tyng-Ruey Chuang
Ian Clement
Juan Manuel Crespo
Pierre-Louis Curien
Varacca Daniele
Joshua Dunfield
Stephan Falke
Elena Giachino
Alexey Gotsman
Daniel Hirschkoff
Florent Jacquemard
Yungbum Jung
Cezary Kaliszyk
Iksoon Kim
Cesar Kunz
Gyesik Lee
Wonchan Lee
Daniel Licata
Marco Maggesi
Gregory Malecha
Claude Marché
Marino Miculan
Jean-Francois Monin
David Nowak
Bruno Oliveira
Sam Owre
Rex Page
Sungwoo Park
Andrei Popescu
Donald Porter
Nicolas Pouillard
Wilmer Ricciotti
Michael Rusinowitch
Claudio Sacerdoti Coen
Julien Schmaltz
Robert Simmons
Vilhelm Sjoeberg
Matthieu Sozeau
Antonis Stampoulis
Pierre-Yves Strub
Aaron Stump
Enrico Tassi
Zachary Tatlock
Aditya Thakur
Tjark Weber
Ian Wehrman
S. Zanella Béguelin
Steve Zdancewic
Xingyuan Zhang
Jianzhao Zhao

Sponsoring Institutions

Academia Sinica (Taiwan)
National Taiwan University
Tsinghua University (Institute of Advanced Study and Software Chair)
National Science Council (Taiwan)

CPP Manifesto

In this manifesto, we advocate for the creation of a new international conference in the area of formal methods and programming languages, called *Certified Programs and Proofs (CPP).* Certification here means formal, mechanized verification of some sort, preferably with the production of independently checkable certificates. CPP would target any research promoting formal development of certified software and proofs, that is:

- The development of certified or certifying programs
- The development of certified mathematical theories
- The development of new languages and tools for certified programming
- New program logics, type systems, and semantics for certified code
- New automated or interactive tools and provers for certification
- Results assessed by an original open source formal development
- Original teaching material based on a proof assistant

Software today is still developed without precise specification. A developer often starts the programming task with a rather informal specification. After careful engineering, the developer delivers a program that may not fully satisfy the specification. Extensive testing and debugging may shrink the gap between the two, but there is no assurance that the program accurately follows the specification. Such inaccuracy may not always be significant, but when a developer links a large number of such modules together, these "noises" may multiply, leading to a system that nobody can understand and manage. System software built this way often contains hard-to-find "zero-day vulnerabilities" that become easy targets for Stuxnet-like attacks. CPP aims to promote the development of new languages and tools for building certified programs and for making programming precise.

Certified software consists of an executable program plus a formal proof that the software is free of bugs with respect to a particular dependability claim. With certified software, the dependability of a software system is measured by the actual formal claim that it is able to certify. Because the claim comes with a mechanized proof, the dependability can be checked independently and automatically in an extremely reliable way. The formal dependability claim can range from making almost no guarantee, to simple type safety property, or all the way to deep liveness, security, and correctness properties. It provides a great metric for comparing different techniques and making steady progress in constructing dependable software.

The conventional wisdom is that certified software will never be practical because any real software must also rely on the underlying runtime system which is too low-level and complex to be verifiable. In recent years, however, there have been many advances in the theory and engineering of mechanized proof systems applied to verification of low-level code, including proof-carrying code, certified assembly programming, local reasoning and separation logic, certified linking of heterogeneous components, certified protocols, certified garbage collectors, certified or certifying compilation, and certified OS-kernels. CPP intends to be a driving force that would facilitate the rapid development of this exciting new area, and be a natural international forum for such work.

The recent development in several areas of modern mathematics requires mathematical proofs containing enormous computation that cannot be verified by mathematicians in an entire lifetime. Such development has puzzled the mathematical community and prompted some of our colleagues in mathematics and computer science to start developing a new paradigm, *formal mathematics*, which requires proofs to be verified by a reliable theorem prover. As particular examples, such an effort has been made for the four-color theorem and has started for the sphere packing problem and the classification of finite groups. We believe that this emerging paradigm is the beginning of a new era. No essential existing theorem in computer science has yet been considered worth a similar effort, but it could well happen in the very near future. For example, existing results in security would often benefit from a formal development allowing us to exhibit the essential hypotheses under which the result really holds. CPP would again be a natural international forum for this kind of work, either in mathematics or in computer science, and would participate strongly in the emergence of this paradigm.

On the other hand, there is a recent trend in computer science to formally prove new results in highly technical subjects such as computational logic, at least in part. In whichever scientific area, formal proofs have three major advantages: no assumption can be missing, as is sometimes the case; the result cannot be disputed by a wrong counterexample, as sometimes happens; and more importantly, a formal development often results in a better understanding of the proof or program, and hence results in easier and better implementation. This new trend is becoming strong in computer science work, but is not recognized yet as it should be by traditional conferences. CPP would be a natural forum promoting this trend.

There are not many proof assistants around. There should be more, because progress benefits from competition. On the other hand, there is much theoretical work that could be implemented in the form of a proof assistant, but this does not really happen. One reason is that it is hard to publish a development work, especially when this requires a long-term effort as is the case for a proof assistant. It is even harder to publish work about libraries which, we all know, are fundamental for the success of a proof assistant. CPP would pay particular attention in publishing, publicizing, and promoting this kind of work.

Finally, CPP also aims to be a publication arena for innovative teaching experiences, in computer science or mathematics, using proof assistants in an essential way. These experiences could be submitted in an innovative format to be defined.

CPP would be an international conference initially based in Asia. Formal methods in Asia based on model checking have been boosted by ATVA. An Asian community in formal methods based on formal proofs is now emerging, in China, South Korea, Taiwan, and Japan (where the use of such formal methods is recent despite a strong logical tradition), but is still very scattered and lacks a forum where researchers can easily meet on a regular basis. CPP is intended to nurse such a forum, and help boost this community in Asia as ATVA did for the model checking community.

In the long run, we would target a three-year rotating schema among Asia, Europe, and North America, and favor colocations with other conferences on each continent.

November 2010

Jean-Pierre Jouannaud
Zhong Shao

Table of Contents

APLAS/CPP Invited Talks

Session 1: Logic and Types

Session 2: Certificates

Session 3: Invited Talk

Session 4: Formalization

Session 5: Proof Assistants

Session 6: Teaching

Session 7: Invited Talk

Session 8: Programming Languages

Session 9: Hardware Certification

Session 10: Miscellaneous

Session 11: Proof Pearls

Engineering Theories with Z3

Nikolaj Bjørner

Microsoft Research
nbjorner@microsoft.com

Abstract. Modern Satisfiability Modulo Theories (SMT) solvers are fundamental to many program analysis, verification, design and testing tools. They are a good fit for the domain of software and hardware engineering because they support many domains that are commonly used by the tools. The meaning of domains are captured by theories that can be axiomatized or supported by efficient *theory solvers*. Nevertheless, not all domains are handled by all solvers and many domains and theories will never be native to any solver. We here explore different theories that extend Microsoft Research's SMT solver Z3's basic support. Some can be directly encoded or axiomatized, others make use of user theory plug-ins. Plug-ins are a powerful way for tools to supply their custom domains.

1 Introduction

This talk surveys a selection of theories that have appeared in applications of Z3 [4] and also in recent literature on automated deduction. In each case we show how the theories can be supported using either existing built-in theories in Z3, or by adding a custom decision procedure, or calling Z3 as a black box and adding axioms between each call. The theme is not new. On the contrary, it is very central to research on either encoding (reducing) theories into a simpler basis or developing special solvers for theories. Propositional logic is the most basic such basis e.g., [6]. In the context of SMT (Satisfiability Modulo Theories), the basis is much richer. It comes with built-in support for the theory of equality, uninterpreted functions, arithmetic, arrays, bit-vectors, and even first-order quantification. The problem space is rich, and new applications that require new solutions keep appearing. We don't offer a silver bullet solution, but the "exercise" of examining different applications may give ideas how to tackle new domains.

Z3 contains an interface for plugging in custom theory solvers. We exemplify this interface on two theories: MaxSMT and partial orders. This interface is powerful, but also requires thoughtful interfacing. To date it has been used in a few projects that we are aware of [8,1,7]. Some of our own work can also be seen as an instance of a theory solver. The quantifier-elimination procedures for linear arithmetic and algebraic data-types available in Z3 acts as a special decision procedure [2]. The OpenSMT solver also supports an interface for pluggable theories [3]. We feel that the potential is much bigger.

Z3 also allows interfacing theories in simpler ways. The simplest is by encoding a theory using simpler theories and often also first-order quantification.

J.-P. Jouannaud and Z. Shao (Eds.): CPP 2011, LNCS 7086, pp. 1–2, 2011.

We discuss two encodings for a theory of object graphs. Something between encoding and a user theory, is by calling Z3 repeatedly. Whenever Z3 returns a satisfiable state, then add new axioms that are not satisfied by the current candidate model for the existing formulas. A theory of Higher-Order Logic, HOL, can be solved using this method.

The case studies discussed in this talk are available as F# code samples. An extended version of this abstract is available in the proceedings of APLAS 2011.

References

1. Banerjee, A., Naumann, D., Rosenberg, S.: Decision Procedures for Region Logic. In: Submission (August 2011), `http://www.cs.stevens.edu/naumann/publications/dprlSubm.pdf`
2. Bjørner, N.: Linear quantifier elimination as an abstract decision procedure. In: Giesl, J., Hähnle, R. (eds.) [5], pp. 316–330
3. Bruttomesso, R., Pek, E., Sharygina, N., Tsitovich, A.: The OpenSmt Solver. In: Esparza, J., Majumdar, R. (eds.) TACAS 2010. LNCS, vol. 6015, pp. 150–153. Springer, Heidelberg (2010)
4. de Moura, L., Bjørner, N.S.: Z3: An Efficient SMT Solver. In: Ramakrishnan, C.R., Rehof, J. (eds.) TACAS 2008. LNCS, vol. 4963, pp. 337–340. Springer, Heidelberg (2008)
5. Giesl, J., Hähnle, R. (eds.): IJCAR 2010. LNCS, vol. 6173. Springer, Heidelberg (2010)
6. Lahiri, S.K., Seshia, S.A., Bryant, R.E.: Modeling and Verification of Out-of-Order Microprocessors in Uclid. In: Aagaard, M.D., O'Leary, J.W. (eds.) FMCAD 2002. LNCS, vol. 2517, pp. 142–159. Springer, Heidelberg (2002)
7. Rümmer, P., Wintersteiger, C.: Floating-point support for the Z3 SMT Solver, `http://www.cprover.org/SMT-LIB-Float`
8. Suter, P., Steiger, R., Kuncak, V.: Sets with Cardinality Constraints in Satisfiability Modulo Theories. In: Jhala, R., Schmidt, D. (eds.) VMCAI 2011. LNCS, vol. 6538, pp. 403–418. Springer, Heidelberg (2011)

Algebra, Logic, Locality, Concurrency

Peter W. O'Hearn

Queen Mary University of London

This talk reports on ongoing work – with Tony Hoare, Akbar Hussain, Bernhard Möller, Rasmus Petersen, Georg Struth, Ian Wehrman, and others – on models and logics for concurrent processes [10,6,5]. The approach we are taking abstracts from syntax or particular models. Message passing and shared memory process interaction, and strong (interleaving) and weak (partial order) approaches to sequencing, are accomodated as different models of the same core axioms. Rules of program logic, related to Hoare and Separation logics, flow at once from the algebraic axioms. So, one gets a generic program logic from the algebra, which holds for a range of concrete models.

The most notable amongst the algebra laws is an ordered cousin of the exchange law of 2-categories or bicategories, which here links primitives for sequential and parallel composition

$$(p \parallel r); (q \parallel s) \sqsubseteq (p; q) \parallel (r; s).$$

This law was was noticed in work on pomsets and traces in the 1980s and 1990s [4,1], and emphasized recently in the formulation of Concurrent Kleene Algebra [5]. An important observation of [5] is that by viewing the pre/post spec $\{p\}\, c\, \{q\}$ as a certain relation in the algebra – there are actually two such, $p; c \sqsubseteq q$ and $c; q \sqsubseteq p$ – one obtains a number of rules for program logic. The use of ; to separate the precondition and program, or program and postcondition, has an interesting consequence: if the sequential composition is a 'weak' one that allows statement re-ordering (as in weak or relaxed memory models that do not guarantee sequentially consistent behaviour, or more generally as available in partial order models such as pomsets or event structures [11,9]) then we still obtain rules of sequential Hoare logic. And when combined with $\parallel$ using the exchange law, it results in very general versions of the rules

$$\frac{\{P_1\}\, C_1\, \{Q_1\} \quad \{P_2\}\, C_2\, \{Q_2\}}{\{P_1 * P_2\}\, C_1 \| C_2\, \{Q_1 * Q_2\}}\ \texttt{Concurrency} \qquad \frac{\{P\}\, C\, \{Q\}}{\{P * F\}\, C\, \{Q * F\}}\ \texttt{Frame}$$

which in Concurrent Separation Logic support modular reasoning about concurrent processes [7], where $*$ is the separating conjunction (which holds when its conjuncts holds of separate resources).

A remarkable fact is that the initial conception of these rules from Concurrent Separation Logic is strongly based on an idea of 'locality of resource access' [8,2,3], where such intuitions do not seem to be present in the algebraic theory. For instance, in the frame rule we understand that $\{P\}\, C\, \{Q\}$ implies that command C only accesses those resources described by precondition P, and this justifies tacking on a description of separate resources that will thus not be altered (the $*F$ part). Similarly, in the concurrency rule we understand that

J.-P. Jouannaud and Z. Shao (Eds.): CPP 2011, LNCS 7086, pp. 3–4, 2011.

processes started in separate states will not trample on one another's resources, because of locality. The notion of 'locality of resource access' is a semantic notion that underlies the semantics of Separation Logic: the soundness of the Frame and Concurrency has been proven by validating properties of the semantics of programs that express locality of resource access (properties which incidentally are independent of the syntax of the logic) [12,3]. However, such forms of justification are not needed at all in the algebra.

The understanding of this point – how locality and the algebra are related – is a particular focus of the talk. We start from a standard model of resources, and construct an algebra from it, making a link between the intuitions concerning locality of resource access and the axioms in the algebra. Perhaps surprisingly, the algebra is seen to contain a general account of locality, which strictly generalizes the modular reasoning of Concurrent Separation Logic [5].

On the other hand, the algebra has as instances concrete models that are far removed conceptually from the resource models at the basis of Separation Logic (e.g., models based on interleaving and independence of events), and this leads to the question of whether it is possible to uniformly obtain effective modular reasoning techniques for a wide range of models of concurrency.

References

1. Bloom, S.L., Ésik, Z.: Free shuffle algebras in language varieties. Theor. Comput. Sci. 163(1&2), 55–98 (1996)
2. Brookes, S.D.: A semantics of concurrent separation logic. Theoretical Computer Science 375(1-3), 227–270 (2007); Prelim. version appeared in CONCUR 2004
3. Calcagno, C., O'Hearn, P.W., Yang, H.: Local action and abstract separation logic. In: LICS, pp. 366–378. IEEE Computer Society (2007)
4. Gischer, J.L.: The equational theory of pomsets. Theor. Comput. Sci. 61, 199–224 (1988)
5. Hoare, C.A.R., Hussain, A., Möller, B., O'Hearn, P.W., Petersen, R.L., Struth, G.: On Locality and the Exchange Law for Concurrent Processes. In: Katoen, J.-P., König, B. (eds.) CONCUR 2011 – Concurrency Theory. LNCS, vol. 6901, pp. 250–264. Springer, Heidelberg (2011)
6. Hoare, T., Möller, B., Struth, G., Wehrman, I.: Concurrent Kleene algebra and its foundations. J. Log. Algebr. Program (2011); Preliminary verson in CONCUR 2009
7. O'Hearn, P.W.: Resources, concurrency and local reasoning. Theoretical Computer Science 375(1-3), 271–307 (2007); Prelim. version appeared in CONCUR 2004
8. O'Hearn, P.W., Reynolds, J.C., Yang, H.: Local Reasoning about Programs that Alter Data Structures. In: Fribourg, L. (ed.) CSL 2001 and EACSL 2001. LNCS, vol. 2142, pp. 1–9. Springer, Heidelberg (2001)
9. Pratt, V.: Modelling concurrency with partial orders. International Journal of Parallel Programming 15(1), 33–71 (1986)
10. Wehrman, I., Hoare, C.A.R., O'Hearn, P.W.: Graphical models of separation logic. Inf. Process. Lett. 109(17), 1001–1004 (2009)
11. Winskel, G.: Events in Computation. Ph.D. thesis, University of Edinburgh (1980)
12. Yang, H., O'Hearn, P.W.: A Semantic Basis for Local Reasoning. In: Nielsen, M., Engberg, U. (eds.) FOSSACS 2002. LNCS, vol. 2303, pp. 402–416. Springer, Heidelberg (2002)

Constructive Formalization of Hybrid Logic with Eventualities

Christian Doczkal and Gert Smolka

Saarland University, Saarbrücken, Germany
{doczkal,smolka}@ps.uni-saarland.de

Abstract. This paper reports on the formalization of classical hybrid logic with eventualities in the constructive type theory of the proof assistant Coq. We represent formulas and models and define satisfiability, validity, and equivalence of formulas. The representation yields the classical equivalences and does not require axioms. Our main results are an algorithmic proof of a small model theorem and the computational decidability of satisfiability, validity, and equivalence of formulas. We present our work in three steps: propositional logic, modal logic, and finally hybrid logic.

Keywords: hybrid logic, eventualities, small model theorem, decision procedures, Coq, Ssreflect.

1 Introduction

We are interested in the formalization of decidable logics in constructive type theory. Of particular interest are logics for reasoning about programs, as exemplified by PDL [6] and CTL [4]. Given that these logics enjoy the small model property, one would hope that they can be formalized in constructive type theory without using classical assumptions. In this paper, we report about the constructive formalization of H^* [12], a hybrid logic [1] with eventualities (iteration in PDL, "exists finally" in CTL). We employ the proof assistant Coq [15] with the Ssreflect extension [9].

Our formalization represents formulas and models and defines a two-valued function evaluating formulas in models. Our main result is an algorithmic proof of a small model theorem, from which we obtain the computational decidability of satisfiability, validity, and equivalence of formulas. We do not require axioms and rely on the native notion of computability that comes with constructive type theory.

Hybrid logics [1] extend modal logics with nominals. The models of a modal logic can be seen as transition systems. The formulas of a modal logic describe predicates on the states of a model. Nominals are primitive predicates that hold for exactly one state. Since we formalize a classical modal logic in constructive type theory, we require that the formulas denote boolean state predicates. To make this possible, we employ models that come with localized modal operations mapping boolean state predicates to boolean state predicates. While localized

J.-P. Jouannaud and Z. Shao (Eds.): CPP 2011, LNCS 7086, pp. 5–20, 2011.

modal operations are essential to our constructive development, they are superfluous in a conventional mathematical development (since their existence is obvious). In the constructive setting, the localized modal operations constitute localized decidability assumptions, which eliminate the need for global decidability assumptions.

A conventional proof of a small model theorem starts from a formula and a model satisfying it. From the formula one obtains a finite syntactic closure (roughly the subformulas) and projects the states of the model to Hintikka sets contained in the closure. One then shows that the finitely many Hintikka sets obtained this way constitute a model of the formula.

The conventional proof does not work in our constructive setting since the Hintikka projection cannot be obtained from the model. However, there is an algorithmic technique known as pruning that originated with Pratt [14] that obtains from the system of all Hintikka sets contained in the finite syntactic closure a subsystem that constitutes a small model of all satisfiable formulas in the closure. As we show in this paper, the correctness of pruning can be shown constructively and provides for a constructive proof of the small model theorem. Interestingly, the pruning technique results in a worst-case optimal decision procedure (exponential complexity) while a naive search based on the small model theorem results in a double exponential decision procedure.

The formalization presented in this paper is based on the mathematical development presented in [11]. Small model theorems and pruning-based decision procedures originated in the work of Fischer and Ladner [6], Pratt [14], and Emerson and Halpern [5].

There appears to be no formalized decidability result for classical modal logic in the literature. Formalizing decidability results in classical logics like HOL requires an explicit formalization of computability. While there are formalizations of computability theory in HOL [13,17], we are not aware of any decidability results based on these. However, there is work on the verification of decision procedures in constructive logic, often with an emphasis on code extraction. An early example is a decision procedure for classical propositional logic verified by Caldwell [3] in the Nuprl system. Another example is Buchberger's algorithm for polynomial rings verified by Théry [16] in Coq. Also, there is recent work on the constructive metatheory of classical logics. Ilik et. al. [10] give a constructive completeness proof for a classical first order sequent calculus with respect to a certain class of Kripke models. In contrast to [10], we work with a notion of model that closely resembles the usual mathematical definition. We can do this since our proofs only require the construction of finite models.

Given that we work with finite structures and finite sets, we could profit much from Coq's Ssreflect extension. In particular, we make use of Ssreflect's support for boolean propositions, finite types, and finite sets [8,7].

The paper presents our work in three cumulative steps: Propositional logic, modal logic with eventualities, and finally modal logic with eventualities and nominals. For each logic, we present the mathematical theory underlying the formalization and comment on its realization in Coq. In each case, we work with a finite formula closure and prove a small model theorem based on Hintikka sets. The Coq formalizations of the three logics appear in separate files that can be found at `http://www.ps.uni-saarland.de/extras/cpp11/`.

2 Propositional Logic

We start with the theory of classical propositional logic we have formalized. We call this theory P. Theory P is arranged such that it fits a constructive formalization that scales to modal logic. We first outline the mathematical theory and then sketch its formalization.

2.1 Mathematical Development

We assume a countable alphabet of names called *variables* and declare the letters p and q as ranging over variables. *Formulas* are defined by the grammar

$$s, t ::= p \mid \neg p \mid s \wedge t \mid s \vee t$$

A *model* $\mathcal{M}$ is a set of variables. The *satisfaction relation* $\mathcal{M} \models s$ between models and formulas is defined by induction on formulas.

$$\mathcal{M} \models p \iff p \in \mathcal{M} \qquad \mathcal{M} \models s \wedge t \iff \mathcal{M} \models s \text{ and } \mathcal{M} \models t$$
$$\mathcal{M} \models \neg p \iff p \notin \mathcal{M} \qquad \mathcal{M} \models s \vee t \iff \mathcal{M} \models s \text{ or } \mathcal{M} \models t$$

Satisfiability, *validity*, and *equivalence* of formulas are defined as follows.

- s is *satisfiable* if $\mathcal{M} \models s$ for some model $\mathcal{M}$.
- s is *valid* if $\mathcal{M} \models s$ for all models $\mathcal{M}$.
- s and t are *equivalent* ($s \equiv t$) if $\mathcal{M} \models s$ iff $\mathcal{M} \models t$ for all models $\mathcal{M}$.

To express general negation we define a *negation operator* $\sim$ by induction on formulas:

$$\sim p = \neg p \qquad \sim(s \wedge t) = \sim s \vee \sim t$$
$$\sim(\neg p) = p \qquad \sim(s \vee t) = \sim s \wedge \sim t$$

Proposition 2.1. *Let s and t be formulas.*

1. $\sim\sim s = s$
2. $\mathcal{M} \models \sim s$ *iff* $\mathcal{M} \not\models s$
3. *s is valid iff $\sim s$ is unsatisfiable.*
4. *$s \equiv t$ iff $(s \wedge t) \vee (\sim s \wedge \sim t)$ is valid.*

The *syntactic closure* $\mathcal{C}s$ of a formula s is the set of all subformulas of s. We define $\mathcal{C}s$ inductively.

$$\mathcal{C}p = \{p\} \qquad \mathcal{C}(\neg p) = \{\neg p\}$$
$$\mathcal{C}(s \wedge t) = \{s \wedge t\} \cup \mathcal{C}s \cup \mathcal{C}t \qquad \mathcal{C}(s \vee t) = \{s \vee t\} \cup \mathcal{C}s \cup \mathcal{C}t$$

We fix some formula s_0. A *Hintikka set* is a set $H \subseteq \mathcal{C}s_0$ satisfying:

H1. If $\neg p \in H$, then $p \notin H$.
H2. If $s \wedge t \in H$, then $s \in H$ and $t \in H$.
H3. If $s \vee t \in H$, then $s \in H$ or $t \in H$.

Proposition 2.2. *Let H be a Hintikka set. Then $\{ p \mid p \in H \}$ is a model that satisfies every formula $s \in H$.*

Theorem 2.3. *A formula $s \in \mathcal{C}s_0$ is satisfiable if and only if there exists a Hintikka set H such that $s \in H$.*

Proof. Let $\mathcal{M} \models s$. Then $\{ t \in \mathcal{C}s_0 \mid \mathcal{M} \models t \}$ is a Hintikka set containing s. The other direction follows from Proposition 2.2. □

We now have a decision procedure for satisfiability. Given a formula s, the procedure checks whether the finite set $\mathcal{C}s$ has a subset that contains s and is a Hintikka set.

Corollary 2.4. *Satisfiability, validity, and equivalence of formulas are decidable.*

Proof. Follows from Theorem 2.3 and Proposition 2.1. □

2.2 Decidability, Finite Types and Finite Sets

We formalize our results in the proof assistant Coq, a system implementing the Calculus of Inductive Constructions [15]. All functions definable in Coq (without axioms) are total and computable. Hence, to show that a predicate `P : X -> Prop` is `decidable` we define a decision function of type

```
forall x:X , { P x } + { ~ P x }
```

returning for every `x:X` either a proof of `P x` or a proof of `~ P x`.

Our formal proofs rely heavily on the Ssreflect extension to Coq, so we briefly describe the most important features we use. For technical details refer to [7,8].

Ssreflect defines an implicit coercion from `bool` to `Prop`, allowing booleans to appear in place of Propositions. The type of boolean predicates over a type `T` (i.e., `T -> bool`) is abbreviated `pred T`.

In Ssreflect, a finite type is a type together with an explicit enumeration of its elements. Finite types can be constructed from finite sequences using `seq_sub` and finiteness is preserved by many type constructors. For a sequence `xs:seq T` the finite type `X := seq_sub xs` comes with a generic injection `val` from `X` into `T`. Finite types come with boolean quantifiers `forallb` and `existsb` taking boolean predicates and returning booleans.

If `X` is a finite type, the type `{set X}` is the type of sets over `X`, which is itself a finite type. Ssreflect provides the usual set theoretic operations on `{set X}` including membership, written `x \in X`, and set comprehensions `[set x:X | p]`.

Ssreflect also provides a choice operator for boolean predicates over finite types. We use choice and boolean quantifiers to specify decision procedures in a declarative way.

2.3 Formalization of Propositional Logic

We now outline the formalization of P in Coq with Ssreflect. We start with the definition of types for variables, formulas, and models.

```
Definition var := nat.
Inductive form := Var : var -> form | ...
Definition model := var -> bool.
```

For convenience, we choose `nat` to be the type of variables. To obtain a representation that is faithful to classical logic, we represent models as boolean predicates. The satisfaction relation is then obtained with a recursive evaluation function:

```
Fixpoint eval (M : model) (s : form) : bool := ...
```

The definitions of satisfiability, validity, and equivalence are straightforward.

```
Definition sat s : Prop := exists M, eval M s.
Definition valid s : Prop := forall M, eval M s.
Definition equiv s t : Prop := forall M, eval M s = eval M t.
```

The proof of Proposition 2.1 can be carried out constructively since formulas evaluate to booleans. For (3) the de Morgan law for the existential quantifier is needed, which is intuitionistically provable. An analogous proof of the statement

$$s \text{ satisfiable iff } {\sim}s \text{ is not valid}$$

is not possible at this point since it would require the de Morgan law for the universal quantifier, which is not provable intuitionistically. As is, we can prove that decidability of satisfiability implies decidability of validity and equivalence.

```
Lemma dec_sat2valid : decidable sat -> decidable valid.
Lemma dec_valid2equiv : decidable valid -> forall s, decidable (equiv s).
```

We define the syntactic closure operator $\mathcal{C}$ as a recursive function from formulas to lists of formulas.

```
Fixpoint synclos (s : form) : seq form := ...
```

Given a formula s_0, we obtain $\mathcal{C}s_0$ as a finite type `F`.

```
Variable s0 : form.
Definition F : finType := seq_sub (synclos s0).
```

We identify Hintikka sets by a boolean predicate: [1]

```
Definition Hcond (t : F) (H : {set F}) :=
match val t with
  | NegVar v => ~~ (Var v \in' H)
  | And s t => s \in' H && t \in' H
  | Or s t  => s \in' H || t \in' H
  | _   => true
end.
Definition hintikka (H : {set F}) : bool :=
  forallb t, (t \in H) ==> Hcond t H.
```

[1] The operators `~~`, `&&`, and `||`, denote boolean negation, conjunction, and disjunction.

Our alternative membership `\in'` extends membership in `{set F}` from `F` to `form`, separating the definition of Hintikka sets and the membership proofs for `synclos s0` associated with `F`. Defining Hintikka sets only for sets over `F` allows us to make use of Ssreflect's extensive library on finite sets.

We then prove Proposition 2.2 for Hintikka sets in `{set F}` and Theorem 2.3 for formulas in `F`.

```
Theorem decidability (t:F) :
  sat (val t) <-> existsb H, hintikka H && (t \in H).
```

From this, we obtain Corollary 2.4. See the theory file `P.v` for full details.

3 Modal Logic

We now present the mathematical theory of modal logic with eventualities we have formalized. We call this theory K*. As before, we first outline the mathematical theory and then turn to formalization aspects.

3.1 Mathematical Development

We assume that the reader has seen modal logic before. We see the models of modal logic as transition systems where the states are labeled with variables. Formulas are evaluated at a state of a transition system. A primitive formula p holds at a state w if w is labeled with p, a formula $\Box s$ holds at w if s holds at all successors of w, and a formula $\Diamond s$ holds at w if s holds at some successor of w. A formula $\Box^* s$ ($\Diamond^* s$) holds at a state w if all (some) states reachable from w satisfy s. We call formulas of the form $\Diamond^* s$ *eventualities.*

We assume a countable alphabet $\mathcal{V}$ of names called *variables* and declare the letters p and q as ranging over variables. *Formulas* are defined by the grammar

$$s, t ::= p \mid \neg p \mid s \wedge t \mid s \vee t \mid \Box s \mid \Diamond s \mid \Box^* s \mid \Diamond^* s$$

A *model* $\mathcal{M}$ is a triple consisting of the following components:

- A *carrier* set $|\mathcal{M}|$ whose elements are called *states.*
- A relation $\to_{\mathcal{M}} \subseteq |\mathcal{M}| \times |\mathcal{M}|$ called *transition relation.*
- A function $\Lambda_{\mathcal{M}} \colon \mathcal{V} \to 2^{|\mathcal{M}|}$ called *labeling function.*

We deviate from the standard definition by admitting models with an empty set of states. This does not make a difference as it comes to satisfiability and validity of formulas. We write $\to^*_{\mathcal{M}}$ for the reflexive transitive closure of $\to_{\mathcal{M}}$. The *satisfaction relation* $\mathcal{M}, w \models s$ between models, states, and formulas is defined by induction on formulas.

$$\mathcal{M}, w \models p \iff w \in \Lambda_{\mathcal{M}} p \qquad \mathcal{M}, w \models s \wedge t \iff \mathcal{M}, w \models s \text{ and } \mathcal{M}, w \models t$$

$$\mathcal{M}, w \models \neg p \iff w \notin \Lambda_{\mathcal{M}} p \qquad \mathcal{M}, w \models s \vee t \iff \mathcal{M}, w \models s \text{ or } \mathcal{M}, w \models t$$

$$\mathcal{M}, w \models \Box s \iff \mathcal{M}, v \models s \text{ for all } v \text{ such that } w \to_{\mathcal{M}} v$$

$$\mathcal{M}, w \models \Diamond s \iff \mathcal{M}, v \models s \text{ for some } v \text{ such that } w \to_{\mathcal{M}} v$$

$$\mathcal{M}, w \models \Box^* s \iff \mathcal{M}, v \models s \text{ for all } v \text{ such that } w \to^*_{\mathcal{M}} v$$

$$\mathcal{M}, w \models \Diamond^* s \iff \mathcal{M}, v \models s \text{ for some } v \text{ such that } w \to^*_{\mathcal{M}} v$$

Satisfiability, *validity*, and *equivalence* of formulas are defined as follows.

- s is *satisfiable* if $\mathcal{M}, w \models s$ for some model $\mathcal{M}$ and some state $w \in |\mathcal{M}|$.
- s is *valid* if $\mathcal{M}, w \models s$ for all models $\mathcal{M}$ and all states $w \in |\mathcal{M}|$.
- s and t are *equivalent* ($s \equiv t$) if $\mathcal{M}, w \models s$ iff $\mathcal{M}, w \models t$ for all models $\mathcal{M}$ and all states $w \in |\mathcal{M}|$.

For a set of formulas A, we write $\mathcal{M} \models A$ if there exists some $w \in |\mathcal{M}|$ such that $\mathcal{M}, w \models t$ for all $t \in A$. We call a set of formulas A *satisfiable* if there is some model $\mathcal{M}$ such that $\mathcal{M} \models A$. We extend the *negation operator* to modal formulas:

$$\sim(\Box s) = \Diamond(\sim s) \qquad \sim(\Diamond s) = \Box(\sim s)$$
$$\sim(\Diamond^* s) = \Box^*(\sim s) \qquad \sim(\Box^* s) = \Diamond^*(\sim s)$$

Proposition 3.1. *Let s and t be formulas.*

1. $\sim(\sim s) = s$
2. $\mathcal{M}, w \models \sim s$ *iff not* $\mathcal{M}, w \models s$
3. s *is valid iff* $\sim s$ *is unsatisfiable.*
4. $s \equiv t$ *iff* $(s \wedge t) \vee (\sim s \wedge \sim t)$ *is valid.*
5. $\Box^* s \equiv s \wedge \Box\Box^* s$ *and* $\Diamond^* s \equiv s \vee \Diamond\Diamond^* s$.

We also extend the *syntactic closure*:

$$\mathcal{C}(\Box s) = \{\Box s\} \cup \mathcal{C}s \qquad \mathcal{C}(\Diamond s) = \{\Diamond s\} \cup \mathcal{C}s$$
$$\mathcal{C}(\Box^* s) = \{\Box^* s, \Box\Box^* s\} \cup \mathcal{C}s \qquad \mathcal{C}(\Diamond^* s) = \{\Diamond^* s, \Diamond\Diamond^* s\} \cup \mathcal{C}s$$

We again fix a formula s_0. A *Hintikka set* is a set $H \subseteq \mathcal{C}s_0$ satisfying (H1) to (H3) as defined for P and the following conditions (cf. Proposition 3.1(5)):

H4. If $\Box^* s \in H$, then $s \in H$ and $\Box\Box^* s \in H$.
H5. If $\Diamond^* s \in H$, then $s \in H$ or $\Diamond\Diamond^* s \in H$.

A *Hintikka system* is a set of Hintikka sets. The *transition relation* $\rightarrow_{\mathcal{S}}$ of a Hintikka system $\mathcal{S}$ is defined as follows: $H \rightarrow_{\mathcal{S}} H'$ iff $H \in \mathcal{S}$, $H' \in \mathcal{S}$, and $t \in H'$ whenever $\Box t \in H$. We define the *model* $\mathcal{M}_{\mathcal{S}}$ described by a Hintikka system $\mathcal{S}$ as follows: $|\mathcal{M}_{\mathcal{S}}| = \mathcal{S}$, $\rightarrow_{\mathcal{M}_{\mathcal{S}}} = \rightarrow_{\mathcal{S}}$, and $\Lambda_{\mathcal{M}_{\mathcal{S}}} p = \{\, H \in S \mid p \in H \,\}$. A *demo* is a Hintikka system $\mathcal{D}$ such that the following conditions are satisfied:

(D$\Diamond$) If $\Diamond s \in H \in \mathcal{D}$, then $H \rightarrow_{\mathcal{D}} H'$ and $s \in H'$ for some $H' \in \mathcal{D}$.
(D$\Diamond^*$) If $\Diamond^* s \in H \in \mathcal{D}$, then $H \rightarrow^*_{\mathcal{D}} H'$ and $s \in H'$ for some $H' \in \mathcal{D}$.

Proposition 3.2. *Let $\mathcal{D}$ be a demo and $s \in H \in \mathcal{D}$. Then $\mathcal{M}_{\mathcal{D}}, H \models s$.*

3.2 Demo Theorem

By Proposition 3.2, demos can be seen as syntactic models. We now show that every satisfiable formula $t \in \mathcal{C}s_0$ is satisfied by a demo. Note that, given s_0, there are only finitely many demos. The *Hintikka universe* $\mathcal{H}$ is the (finite) set of all Hintikka sets. For models $\mathcal{M}$ and states $v \in |\mathcal{M}|$, we define $H_v := \{t \in \mathcal{C}s_0 \mid \mathcal{M}, v \models t\}$.

Proposition 3.3. *Let $\mathcal{M}$ be a model and $v \in |\mathcal{M}|$. Then H_v is a Hintikka set.*

Demos are closed under union. Hence, there exists a largest demo contained in $\mathcal{H}$. Starting from $\mathcal{H}$, we construct this demo by successively pruning sets that violate the demo conditions. The pruning technique originated with Pratt [14].

Proposition 3.4. *Let S be a Hintikka system containing all satisfiable Hintikka sets. Then:*

1. *If $\Diamond t \in H \in S$ and $\forall H'.\ H \to_S H' \Rightarrow t \notin H'$, then H is unsatisfiable.*
2. *If $\Diamond^* t \in H \in S$ and $\forall H'.\ H \to_S^* H' \Rightarrow t \notin H'$, then H is unsatisfiable.*

Proof. 1. Assume $\mathcal{M}, w \models H$. Hence, there exists a state v such that $w \to_M v$ and $\mathcal{M}, v \models t$. Thus, we have $t \in H_v$. This leads to a contradiction since $H \to_S H_v$. (H_v is satisfiable and therefore in S).
2. Assume $\mathcal{M}, w \models H$. Since H is a Hintikka set, we have $\Diamond\Diamond^* t \in H$. Hence, there exists a state v such that $\mathcal{M}, v \models \Diamond^* t$ and $H \to_S H_v$. To obtain a contradiction, it suffices to show that there exists a u such that $H_v \to_S^* H_u$ and $t \in H_u$. This follows easily by induction on $v \to_{\mathcal{M}}^* u$ and the fact that $v \to_{\mathcal{M}} u$ implies $H_v \to_S H_u$. □

We define a relation on Hintikka systems representing a single pruning action: $S \xrightarrow{p} S'$ iff $S' = S \setminus \{H\}$ for some H violating (D$\Diamond$) or (D$\Diamond^*$). We extend this to the *pruning relation* on Hintikka systems: $S \overset{p}{\rightsquigarrow} S'$ iff $S \xrightarrow{p}{}^* S'$ and S' is terminal for $\xrightarrow{p}$.

Proposition 3.5. *Let S and S' be Hintikka systems such that $S \overset{p}{\rightsquigarrow} S'$. Then:*

1. *S' satisfies (D$\Diamond$) and (D$\Diamond^*$).*
2. *If S contains all satisfiable Hintikka sets, so does S'.*

Let Δ be the set such that $\mathcal{H} \overset{p}{\rightsquigarrow} \Delta$. By Propositions 3.2 and 3.5, Δ is the demo containing exactly the satisfiable Hintikka sets and is thus uniquely determined.

Theorem 3.6 (Demo Theorem). *A formula $t \in \mathcal{C}s_0$ is satisfiable if and only if there exists a Hintikka set $H \in \Delta$ such that $t \in H$.*

Proof. The direction form right to left follows from Proposition 3.2. For the other direction, assume $\mathcal{M}, v \models t$. Then $t \in H_v \in \Delta$. □

We now have a decision procedure for satisfiability. Given an input formula s, the procedure constructs the set of all Hintikka sets contained in $\mathcal{C}s$. It then removes Hintikka sets violating (D$\Diamond$) or (D$\Diamond^*$) until no such sets remain and returns satisfiable iff the resulting demo contains some H such that $s \in H$.

Corollary 3.7. *Satisfiability, validity, and equivalence of formulas are decidable.*

3.3 Formalization of Modal Logic

The most important design decision in formalizing modal logic is the representation of models. We require that formulas evaluate to boolean state predicates, i.e., functions of type `state -> bool`. To meet this requirement, we need boolean versions of the logical operations. For instance, for the $\Diamond$-modality we need an operation

```
EXb : pred state -> pred state
```

satisfying

```
forall p w : EXb p w <-> exists v, trans w v /\ p v
```

Since the boolean versions of the logical operations do not automatically exist in a constructive setting, we require that they are provided by the model. As it turns out, it suffices that a model comes with a boolean labeling function and the boolean operations for the existential modalities (i.e., $\Diamond$ and $\Diamond^*$). This leads to the definition of models appearing in Fig. 1. The boolean operations for $\Box$ and $\Box^*$ can be defined from their duals `EXb` and `EFb`. For $\Box^*$ we have:

```
CoInductive AG X (R : X -> X -> Prop) (P : X -> Prop) (w : X) : Prop :=
  | AGs : P w -> (forall v, R w v -> AG R P v) -> AG R P w.
Definition AGb p w := ~~ EFb (fun v => ~~ p v) w.
Lemma AXbP p w : AGb p q <-> AG trans p w.
```

Note that the (co)inductive definitions of `AG` and `EF` are provably equivalent to more conventional definitions employing the reflexive transitive closure of the transition relation.

We can now define a boolean evaluation function:

```
Fixpoint eval M s :=
  match s with Var v => label v |...| Dia s => EXb (eval M s) |... end.
```

We have now arrived at a faithful representation of classical modal logic providing the usual equivalences between formulas.

On the syntactic side we proceed similarly as we did for P. Given a formula s_0, we again represent the syntactic closure $\mathcal{C}s_0$ as a finite type `F`. The definition of Hintikka sets is adapted to cover conditions (H4) and (H5). Hintikka systems are represented as elements of `{set {set F}}`. The transition relation $\rightarrow_S$ and the demo conditions (D$\Diamond$) and (D$\Diamond^*$) are easily expressed as boolean predicates.

Proposition 3.2 and Proposition 3.4 can be shown as one would expect from the mathematical proofs. Proposition 3.2 requires the construction of a finite

```
Definition EX X (R : X -> X -> Prop) (P : X -> Prop) (w : X) : Prop :=
  exists v, R w v /\ P v.
Inductive EF X (R : X -> X -> Prop) (P : X -> Prop) (w : X) : Prop :=
  | EF0 : P w -> EF R P w
  | EFs v : R w v -> EF R P v -> EF R P w.
Record model := Model {
  state :> Type;
  trans : state -> state -> Prop;
  label : var -> pred state;
  EXb : pred state -> pred state;
  EXbP p w : EXb p w <-> EX trans p w ;
  EFb : pred state -> pred state;
  EFbP p w : EFb p w <-> EF trans p w }.
```

Fig. 1. Definition of modal models

model from a demo. Since the carrier of the constructed model is finite, `label`, `EXb`, and `EFb` are easily defined using Ssreflect's `fintype` and `fingraph` libraries.

To constructively prove the demo theorem, we require some implementation of the pruning relation $\overset{p}{\rightsquigarrow}$. For this, we define a function

```
pick_dia : {set {set F}} -> option {set F}
```

selecting, if possible, in a Hintikka system S some $H \in S$ violating (D$\Diamond$). Likewise, we define a function `pick_dstar` for (D$\Diamond^*$). Both functions are defined using the choice operator provided by Ssreflect. From this, it is easy to define a pruning function:

```
Definition step S := if pick_dia S is Some H then S :\ H else
  if pick_dstar S is Some H then S :\ H else S.
Function prune (S : {set {set F}}) {measure (fun S => #|S|) S}
  : {set {set F}} := if step S == S then S else prune (step S).
```

It is easy to show that the result of pruning satisfies (D$\Diamond$) and (D$\Diamond^*$). To obtain Proposition 3.5, we have to show that the precondition of Proposition 3.4 is an invariant of the pruning algorithm.

```
Definition HU := [set H | hintikka H].
Definition invariant (S: {set {set F}}) :=
  S \subset HU /\ forall H, H \in HU -> satF H -> H \in S.
Lemma invariant_prune S : invariant S -> invariant (prune S).
```

Finally, we obtain:

```
Theorem demo_theorem (t : F) :
  sat (val t) <-> existsb H, (H \in Delta) && (t \in H).
```

4 Hybrid Logic

Hybrid logic [2] extends modal logic with special variables called nominals that must label exactly one state. We extend K* with nominals and call the resulting logic H*.

4.1 Mathematical Development

We assume a countable set $\mathcal{N}$ of nominals and let x and y range over $\mathcal{N}$. The grammar of formulas is extended accordingly:

$$s,t ::= p \mid \neg p \mid s \wedge t \mid s \vee t \mid \Box s \mid \Diamond s \mid \Box^* s \mid \Diamond^* s \mid x \mid \neg x$$

We extend the definition of models with a *nominal labeling* $\Phi_{\mathcal{M}} : \mathcal{N} \to 2^{|\mathcal{M}|}$ and require $|\Phi_{\mathcal{M}} x| = 1$ for all x.

We extend the *syntactic closure* to cover nominals:

$$\mathcal{C}x = \{x\} \qquad\qquad \mathcal{C}(\neg x) = \{\neg x, x\}$$

As before, we fix a formula s_0 and define Hintikka sets as subsets of $\mathcal{C}s_0$. The Hintikka condition for nominals is identical to the condition for variables.

Constructing models $\mathcal{M}_S$ from arbitrary Hintikka systems S, does not work for H*. To extend Proposition 3.2 to H*, we adapt the notion of demo. A *demo* is a nonempty Hintikka system satisfying (D$\Diamond$) and (D$\Diamond^*$) as well as

(Dx) For every nominal $x \in \mathcal{C}s_0$, there exists exactly one $H \in \mathcal{D}$ such that $x \in H$.

We define the *model* $\mathcal{M}_{\mathcal{D}}$ described by a demo $\mathcal{D}$ as follows: $|\mathcal{M}_{\mathcal{D}}|$, $\to_{\mathcal{M}_{\mathcal{D}}}$, and $\Lambda_{\mathcal{M}}$ are defined as for $\mathcal{M}_S$; for $\Phi_{\mathcal{M}_{\mathcal{D}}}$ we choose some $H_0 \in \mathcal{D}$ and define

$$\Phi_{\mathcal{M}_{\mathcal{D}}} x = \begin{cases} \{H_0\} & x \notin \mathcal{C}s_0 \\ \{H \in \mathcal{D} \mid x \in H\} & \text{otherwise} \end{cases}$$

Due to condition (Dx), every nominal is mapped to a singleton and we obtain:

Proposition 4.1. *If $\mathcal{D}$ is a demo and $t \in H \in \mathcal{D}$, then $\mathcal{M}_{\mathcal{D}}, H \models t$.*

4.2 Demo Theorem for Hybrid Logic

We now show that every satisfiable formula $t \in \mathcal{C}s_0$ is satisfied by a demo. We call a Hintikka system

- *nominally coherent* if it satisfies (Dx)
- *maximal*, if it is nominally coherent and contains all Hintikka sets not containing nominals.

Due to condition (Dx), demos for H* are not closed under union. Hence, there is no largest demo and the pruning technique from Section 3.2 is not directly applicable. However, demos contained in a maximal Hintikka system are closed under union. This allows the search for a demo to be separated into two parts: guessing a suitable maximal Hintikka system and pruning it.

This two stage approach first appeared in [11], where it is used to obtain a complexity optimal decision procedure for hybrid PDL. In contrast to [11], where correctness is argued after establishing the small model property, we use the procedure as the basis for our algorithmic proof of the demo theorem.

Pruning a maximal Hintikka system may remove satisfiable Hintikka sets. To account for this, we refine the pruning invariant. Instead of requiring all satisfiable sets to be present, we state the invariant with respect to a model $\mathcal{M}$ and only require the sets H_w with $w \in |\mathcal{M}|$ to be present. We adapt Proposition 3.4 as follows:

Proposition 4.2. *Let $\mathcal{M}$ be a model and S be a Hintikka system such that for all $w \in |\mathcal{M}|$, we have $H_w \in S$. Then:*

1. *If $\Diamond t \in H \in S$ and $\forall H'.\ H \rightarrow_S H' \Rightarrow t \notin H'$, then $\mathcal{M} \not\models H$.*
2. *If $\Diamond^* t \in H \in S$ and $\forall H'.\ H \rightarrow_S^* H' \Rightarrow t \notin H'$, then $\mathcal{M} \not\models H$.*

We also need to adapt Proposition 3.5.

Proposition 4.3. *Let $\mathcal{M}$ be a model and S be a maximal Hintikka system such that for all $w \in |\mathcal{M}|$, $H_w \in S$. If $S \overset{p}{\rightsquigarrow} S'$, then S' is nominally coherent and for all $w \in |\mathcal{M}|$, $H_w \in S'$.*

Proposition 4.4. *For every model $\mathcal{M}$, there exists a maximal Hintikka system S such that for all $w \in |\mathcal{M}|$, $H_w \in S$.*

We fix a function Δ returning for a Hintikka system S some S' such that $S \overset{p}{\rightsquigarrow} S'$.

Theorem 4.5. *A formula $t \in \mathcal{C}s_0$ is satisfiable iff there exists a maximal Hintikka system S such that $\Delta(S)$ is nominally coherent and contains some H such that $t \in H$.*

Proof. "$\Rightarrow$" Let $\mathcal{M}, w \models t$ and S be some maximal Hintikka system such that $H_w \in S$ for all $w \in |\mathcal{M}|$ (Proposition 4.4). Then $t \in H_w \in \Delta(S)$ and $\Delta(S)$ nominally coherent by Proposition 4.3.

"$\Leftarrow$" Satisfiability of t follows from Proposition 4.1, since $\Delta(S)$ is a demo by Proposition 3.5(1). □

We now have a decision procedure for satisfiability. Given an input formula s, the procedure guesses for every nominal $x \in \mathcal{C}s$ a Hintikka set H such that $x \in H \subseteq \mathcal{C}s$. It then adds all Hintikka sets contained in $\mathcal{C}s$ that do not contain nominals and prunes the resulting Hintikka system. It returns satisfiable iff the pruned Hintikka system contains for every $x \in \mathcal{C}s$ some H such that $x \in H$ and some H' such that $s \in H'$.

4.3 Formalization of Hybrid Logic

To formalize H*, we first need to adapt the formal representation of models accordingly.

```
Record model := Model {
  ...
  nlabel : nvar -> pred state;
  nlabelP : forall x : nvar, exists! w, w \in nlabel x }.
```

This representation gives us all the required properties of nominals without having to assume that equality on `state` is decidable.

We define `N` to be the finite type of nominals occurring in `F`. We separate (Dx) into a nominal consistency condition `Dxc` requiring at most one occurrence of every nominal in `N` and a nominal existence condition `Dxe` requiring at least one occurrence. Condition `Dxc` is trivially preserved by pruning, while `Dxe` follows from the refined pruning invariant:

```
Definition invariant M (S : {set {set F}}) :=
  S \subset HU /\ forall v:M, H_at v \in S.
Lemma invariant_prune S : invariant S -> invariant (prune S).
Lemma invariant_xe S : invariant S -> Dxe S.
```

To prove Proposition 4.4 for a model $\mathcal{M}$, it is sufficient to prove the existence of a function assigning to every nominal in $x \in \mathcal{C}s_0$ the Hintikka set H_w, where $w \in |\mathcal{M}|$ is the unique w such that $\mathcal{M}, w \models x$

```
Lemma guess :
  exists f : N -> {set F}, forall x,
    exists2 w : M, eval M (val x) w & f x = H_at w.
```

This easily follows from the following choice principle

```
Lemma finite_choice (X : finType) Y (R : X -> Y -> Prop) :
  (forall x : X, exists y , R x y) -> exists f, forall x, R x (f x).
```

which is provable by induction on the enumeration of `X`. Finally we obtain:

```
Theorem demo_theorem (t : F) :
  sat (val t) <-> existsb S, maximal S &&
    let D := prune S in Dxe D && existsb H, (H \in D) && (t \in H).
```

Note that it is sufficient to check `Dxe` after pruning.

5 Conclusions

We have formalized propositional logic, modal logic with eventualities, and modal logic with eventualities and nominals in constructive type theory. Our main results are algorithmic proofs of small model theorems and the computational decidability of satisfability, validity, and equivalence of formulas.

We represent models such that we can define a boolean evaluation function for formulas. This allows us to formalize classical modal logic. We do not assume axioms and employ the notion of computational decidability that comes with constructive type theory. This is possible since we localize the required classical assumptions to the models.

Representation of Models. The most important design decision in our formalization is the representation of models. The reason for this is that in the constructive logic of Coq, the naive representation of models

```
Record naive_model : Type := Model {
  state : Type ;
  trans : state -> state -> Prop ;
  label : var -> state -> Prop }.
```

does not allow the definition of an evaluation function satisfying the classical equivalences of modal logic. This problem would disappear if we were to assume informative excluded middle

```
Axiom IXM : forall P:Prop, { P } + { ~ P }
```

But then our definition of decidability would no longer imply computational decidability. Hence, we have localized specific instances of `IXM` to the models. [2] Regarding the exact form of these instances, there is room for variation, provided that the following conditions are met:

1. The class of models must admit an evaluation function for formulas satisfying the classical dualities.
2. The asserted functions need to be definable for finite carrier types.

We mention a second class of models for K*.

```
Record model := Model {
  state : Type;
  trans : state -> state -> bool;
  label : state -> var -> bool;
  exs : pred state -> bool;
  exsP p : exs p <-> exists w, p w ;
  trans_star : state -> state -> bool;
  trans_starP w v : trans_star w v <-> clos_refl_trans trans w v }.
```

For the purpose of this discussion, we call these models *strong models* and refer to the models defined in Section 3.3 as *weak models*.[3] The assumptions `exs` and `exsP` give us a decidable existential quantifier for states and boolean state predicates. This way one can define a boolean evaluation function directly following the mathematical presentation. The decidable existential quantifier also provides for a direct definition of a demo from a model:

```
Definition D (M:model) := [set H | exs (fun (w : M) => H == H_at w)]
```

[2] `EXb`, `EXbP`, ... are easily definable from `IXM`.

[3] For every strong model, one can define a corresponding weak model. The converse does not seem to be true (consider a model $\mathcal{M}$ where $|\mathcal{M}| = \mathbb{N}$ and $n \to_{\mathcal{M}} m$ iff $n = m + 1$).

This allows the formalization of the usual, non-algorithmic proof of the small model theorem.

Proposition 5.1. *A formula $t \in \mathcal{C}s_0$ is satisfiable iff there exists a demo containing some H, such that $t \in H$.*

Proof. Let $\mathcal{M}, w \models t$. The set $\{ H_w \mid w \in |\mathcal{M}| \}$ is a demo as required. The other direction follows as before. □

The algorithmic proof we have given for weak models provides a more informative small model theorem and shows that the additional strength of boolean existential quantification (i.e. `exs` and `exsP`) is not required to prove the small model theorem.

The file `Kstar_strong.v` contains a formal proof of a small model theorem for K* using the strong representation of models. The non-algorithmic formalization is not significantly shorter than the algorithmic formalization presented in Section 3.3.

Extension to Temporal Logics. The particular representation of models we use in this paper is motivated by the wish to find a design that extends, in a uniform way, to temporal logics like CTL [4]. Temporal logics employ models with a total transition relation and define the semantics of their modal operators using infinite paths. For the modal operators AF and EG one typically has the definitions

$$\mathcal{M}, w \models \mathsf{AF}\, s \iff \mathcal{M}, \sigma_n \models s \text{ for some } n, \text{ for all } \sigma \in \mathcal{M}^\omega \text{ such that } \sigma_0 = w$$
$$\mathcal{M}, w \models \mathsf{EG}\, s \iff \mathcal{M}, \sigma_n \models s \text{ for all } n, \text{ for some } \sigma \in \mathcal{M}^\omega \text{ such that } \sigma_0 = w$$

where $\mathcal{M}$ is a model, $\mathcal{M}^\omega$ is the set of all infinite paths in $\mathcal{M}$, and σ_n is the n-th state of an infinite path σ. The infinite path semantics does not seem to be feasible in constructive logic. However, inductive and coinductive definitions for AF and EG as we have used them in this paper for EF and AG seem to work fine:

```
Inductive AF (p : X -> Prop) (w : X) : Prop :=
| AF0 : p w -> AF p w
| AFs : (forall v, e w v -> AF p v) -> AF p w.
CoInductive EG (p : X -> Prop) (w : X) : Prop :=
| EGs v : p w -> e w v -> EG p v -> EG p w.
```

To support AF and EG, models would come with a boolean operator `AFb` and a proof `AFbP` that `AFb` agrees with `AF` for boolean predicates on the states of the model. With `AFb` and `AFbP` one can define `EGb` and a proof that `EGb` agrees with `EG`. With `AFb` and `EGb` one can then define an evaluation function satisfying the classical dualities. Moreover, given a finite type of states, one can define `AFb` and `AFbP`.

Acknowledgements. We thank Chad Brown for many inspiring discussions concerning the research presented in this paper. We also thank the people from the Coq and Ssreflect mailing lists (in particular Georges Gonthier) for their helpful answers.

References

1. Areces, C., ten Cate, B.: Hybrid logics. In: Blackburn, P., et al. (eds.) [2], pp. 821–868
2. Blackburn, P., van Benthem, J., Wolter, F. (eds.): Handbook of Modal Logic, Studies in Logic and Practical Reasoning, vol. 3. Elsevier (2007)
3. Caldwell, J.L.: Classical Propositional Decidability Via Nuprl Proof Extraction. In: Grundy, J., Newey, M. (eds.) TPHOLs 1998. LNCS, vol. 1479, pp. 105–122. Springer, Heidelberg (1998)
4. Emerson, E.A., Clarke, E.M.: Using branching time temporal logic to synthesize synchronization skeletons. Sci. Comput. Programming 2(3), 241–266 (1982)
5. Emerson, E.A., Halpern, J.Y.: Decision procedures and expressiveness in the temporal logic of branching time. J. Comput. System Sci. 30(1), 1–24 (1985)
6. Fischer, M.J., Ladner, R.E.: Propositional dynamic logic of regular programs. J. Comput. System Sci., 194–211 (1979)
7. Garillot, F., Gonthier, G., Mahboubi, A., Rideau, L.: Packaging Mathematical Structures. In: Berghofer, S., Nipkow, T., Urban, C., Wenzel, M. (eds.) TPHOLs 2009. LNCS, vol. 5674, pp. 327–342. Springer, Heidelberg (2009)
8. Gonthier, G., Mahboubi, A., Rideau, L., Tassi, E., Théry, L.: A Modular Formalisation of Finite Group Theory. In: Schneider, K., Brandt, J. (eds.) TPHOLs 2007. LNCS, vol. 4732, pp. 86–101. Springer, Heidelberg (2007)
9. Gonthier, G., Mahboubi, A., Tassi, E.: A Small Scale Reflection Extension for the Coq system. Research Report RR-6455, INRIA (2008), `http://hal.inria.fr/inria-00258384/en/`
10. Ilik, D., Lee, G., Herbelin, H.: Kripke models for classical logic. Ann. Pure Appl. Logic 161(11), 1367–1378 (2010)
11. Kaminski, M., Schneider, T., Smolka, G.: Correctness and Worst-Case Optimality of Pratt-Style Decision Procedures for Modal and Hybrid Logics. In: Brünnler, K., Metcalfe, G. (eds.) TABLEAUX 2011. LNCS (LNAI), vol. 6793, pp. 196–210. Springer, Heidelberg (2011)
12. Kaminski, M., Smolka, G.: Terminating Tableaux for Hybrid Logic with Eventualities. In: Giesl, J., Hähnle, R. (eds.) IJCAR 2010. LNCS, vol. 6173, pp. 240–254. Springer, Heidelberg (2010)
13. Norrish, M.: Mechanised Computability Theory. In: van Eekelen, M., Geuvers, H., Schmaltz, J., Wiedijk, F. (eds.) ITP 2011. LNCS, vol. 6898, pp. 297–311. Springer, Heidelberg (2011)
14. Pratt, V.R.: Models of program logics. In: Proc. 20th Annual Symp. on Foundations of Computer Science (FOCS 1979), pp. 115–122. IEEE Computer Society Press (1979)
15. The Coq Development Team: The Coq Proof Assistant Reference Manual - Version 8.3. INRIA, France (2011), `http://coq.inria.fr`
16. Théry, L.: A machine-checked implementation of Buchberger's algorithm. J. Autom. Reasoning 26(2), 107–137 (2001)
17. Zammit, V.: A Mechanization of Computability Theory in HOL. In: von Wright, J., Harrison, J., Grundy, J. (eds.) TPHOLs 1996. LNCS, vol. 1125, pp. 431–446. Springer, Heidelberg (1996)

Proof-Carrying Code in a Session-Typed Process Calculus

Frank Pfenning[1], Luis Caires[2], and Bernardo Toninho[1,2]

[1] Computer Science Department, Carnegie Mellon University, Pittsburgh, PA, USA

[2] Faculdade de Ciencias e Tecnologia, Universidade Nova de Lisboa, Lisboa, Portugal

Abstract. Dependent session types allow us to describe not only properties of the I/O behavior of processes but also of the exchanged data. In this paper we show how to exploit dependent session types to express proof-carrying communication. We further introduce two modal operators into the type theory to provide detailed control about how much information is communicated: one based on traditional proof irrelevance and one integrating digital signatures.

Keywords: Process calculus, session types, proof irrelevance, proof-carrying code.

1 Introduction

Session types [10] provide high-level specifications for the communication behavior of interacting processes along bidirectional channels. Recently, logical foundations for session types have been established via Curry-Howard correspondences with linear logic [5,11]. Besides clarifying and unifying concepts in session types, such logical underpinnings provide simple means for generalization. One such extension to *dependent session types* [4,18] allows us to express and enforce complex properties of data transmitted during sessions.

In this paper we build upon dependent session types to model various aspects of systems employing certified code. Already, just dependent session types can model basic proof-carrying code since a dependent type theory uniformly integrates proofs and programs. Briefly, a process implementing a session type $\forall x{:}\tau.A(x)$ will input a data value M of type τ and then behave as $A(M)$, while $\exists x{:}\tau.A(x)$ will output a data value M of type τ and then behave as $A(M)$. The data values are taken from an underlying functional layer which is dependently typed. The session type $\mathbf{1}$ indicates termination of the session. For example, the following is the specification of a session that accepts the code for a function on natural numbers, a proof that the function is decreasing, and emits a fixed point of that function.

$$\forall f{:}\mathsf{nat} \rightarrow \mathsf{nat}.\, \forall p{:}(\Pi x{:}\mathsf{nat}.\, f(x) \leq x).\, \exists y{:}\mathsf{nat}.\, \exists q{:}(y = f(y)).\, \mathbf{1}$$

J.-P. Jouannaud and Z. Shao (Eds.): CPP 2011, LNCS 7086, pp. 21–36, 2011.

In a session of the type above, two proof objects will be transmitted: one (labeled p) showing that the function f is decreasing will be passed to the server, and a second one (labeled q), that the returned value y is indeed a fixed point, will be passed back to the client. Note that the propositions $n \leq m$ and $n = m$ act as types of their proofs, according to the usual Curry-Howard correspondence.

The client may easily check that the returned value y is indeed a fixed point by computing $f(y)$ itself, so we would like to avoid transmitting a proof q of $y = f(y)$. But we do not want to erase this requirement entirely, of course, just avoid sending a proof term. We can do this by using the type-theoretic concept of *proof irrelevance* [15,14,2]. Generally, a type $[A]$ (pronounced "*bracket* A") is the type inhabited by proofs of A, all of which are identified. This is only meaningful if such proofs play no computational role, so there is some subtlety to the type system presented in Section 3. The revised specification would be:

$$\forall f{:}\mathsf{nat} \rightarrow \mathsf{nat}.\, \forall p{:}(\Pi x{:}\mathsf{nat}.\, f(x) \leq x).\, \exists y{:}\mathsf{nat}.\, \exists q{:}[y = f(y)].\, \mathbf{1}$$

Irrelevant proofs terms are eliminated in the operational semantics of our type theory, so just a unit element would be communicated instead of a proof. The residual communication overhead can also be optimized away using two different techniques (see Section 3).

The proof that a given function is decreasing may be complex, so the server may try to avoid checking this proof, delegating it instead to a trusted verifier. This verifier would sign a digital certificate to the effect that there is a proof that the function is decreasing. We integrate this into our type theory with a type constructor $\Diamond_K A$ (read "K *says* A"), where K is a principal and A is a proposition. We want this certificate not to contain the proof, so the proof itself is marked as irrelevant. We obtain:

$$\forall f{:}\mathsf{nat} \rightarrow \mathsf{nat}.\, \forall p{:}\Diamond_{\mathsf{verif}}[\Pi x{:}\mathsf{nat}.\, f(x) \leq x].\, \exists y{:}\mathsf{nat}.\, \exists q{:}[y = f(y)].\, \mathbf{1}$$

In the implementation, we assume a public key infrastructure so that the verifier can sign a certificate containing the proposition $[\Pi x{:}\mathsf{nat}.\, f(x) \leq x]$ and the server can reliably and efficiently check the signature. Our experience with a proof-carrying file system [8] shows that digitally signed certificates are much more compact and can be checked much more quickly than proofs themselves and are one of the cornerstones to make the architecture practical. In this paper we show that they can be accommodated elegantly within session types, based on logical grounds.

We begin in Section 2 with an overview of dependent session types in a term passing variant of the π-calculus, as formulated in previous work by the authors. In Section 3 we define proof irrelevance and how it is used in our operational model, followed by a discussion of affirmation as a way of integrating digitally signed certificates into sessions in Section 4. We sketch some standard meta-theoretic results regarding progress and preservation in Section 5 and conclude in Section 6.

2 Dependent Session Types

In this section we will briefly review dependent session types and the facilities they provide in terms of proof-carrying communication. Dependent session types [4,18] are a conservative extension of session types [10,5,6,9] that allow us to not only describe the behavior of processes in terms of their input and output behavior but also enable us to describe rich properties of the communicated data themselves.

In [18], the authors investigated a natural interpretation of linear type theory as a dependent session typed π-calculus.

Definition 1 (Types). *Types in linear type theory are freely generated by the following grammar, given types τ from a standard dependent type theory:*

$$\begin{array}{rl} A, B ::= & \mathbf{1} \mid A \multimap B \mid A \otimes B \mid A \,\&\, B \mid A \oplus B \mid\, !A \\ & \forall x{:}\tau.A \mid \exists x{:}\tau.A \end{array}$$

A process P offering a service A along a channel z is typed as $P :: z{:}A$ and we obtain an interpretation of the types as follows:

$P :: x : \mathbf{1}$	inaction
$P :: x : A \multimap B$	input a channel of type A along x and continue as B
$P :: x : A \otimes B$	output a fresh channel y of type A along x and continue as B
$P :: x : A \,\&\, B$	offer the choice between A and B along x
$P :: x : A \oplus B$	provide either A or B along x
$P :: x : !A$	provide a persistent (replicating) service A along x
$P :: x : \forall y{:}\tau.\, A$	input a value M of type τ along x and continue as $A\{M/y\}$
$P :: x : \exists y{:}\tau.\, A$	output a value M of type τ along x and continue as $A\{M/y\}$

As an example consider the following type:

$$\mathsf{T} \triangleq \forall n{:}\mathsf{nat}.\, \forall p{:}(n > 0).\, \exists y{:}\mathsf{nat}.\, \exists q{:}(y > 0).\, \mathbf{1}$$

The type T specifies a session that receives a positive number n and sends another positive number y. A process that implements this session (along channel x) is:

$$\mathsf{P} :: x : \mathsf{T} \triangleq x(n).x(p).x\langle n+1\rangle.x\langle \mathsf{incp}\, n\, p\rangle.\, \mathbf{0}$$

where $\mathsf{incp}\, n\, p$ denotes a proof term of type $n + 1 > 0$, computed by a function:

$$\mathsf{incp} : \Pi m : \mathsf{int}.(m > 0) \to (m + 1 > 0)$$

The properties of the communicated data (in this case, the positivity of both numbers) are made explicit by the exchange of terms that act as proof certificates for the properties, by inhabiting the appropriate types.

Our type system arises as a direct interpretation of the rules of linear logic as typing rules for processes, thus our typing judgment is essentially the same as that for a linear logic sequent calculus with a proof term assignment, but singling out a specific channel in which the considered session is being offered. The typing judgment for our system is written as $\Psi; \Gamma; \Delta \Rightarrow P :: z : A$, where Ψ consists of assumptions of the form $x{:}\tau$, Γ consists of persistent assumptions of the form $u{:}A$, and Δ consists of linear assumptions of the form $x{:}A$. We assume all variables in these contexts to be distinct.

The typing judgment above denotes that process P implements session A along channel z, provided it is placed in a process environment that offers the sessions and values specified in contexts Ψ, Γ and Δ. The typing rules for our system are given below in Fig. 1, and are defined modulo structural congruence of processes. Following standard sequent calculus presentations of logic, our system is made up of so-called right and left rules that define the types, and structural rules that denote sound reasoning principles in logic. In our interpretation, right rules define how to implement a session of a particular type, while left rules define how to use such a session. The standard reasoning principles of cut and identity correspond to process composition and channel forwarding (i.e., communication along a channel being replaced by communication on another).

As previously mentioned, our process calculus is a π-calculus where processes can communicate not only channel names as usual, but also terms from a typed functional language, defined by the typing judgment $\Psi \vdash N{:}\tau$, whose proof rules we deliberately leave open.

Definition 2 (Processes). *Processes are defined by the following grammar, where P, Q range over processes, x, y over names and N over terms.*

$$\begin{array}{rl} P, Q ::= & \mathbf{0} \mid P|Q \mid (\boldsymbol{\nu} y)P \mid x\langle y\rangle.P \mid x(y).P \mid x\langle N\rangle.P \\ & !x(y).P \mid x.\mathsf{inl}; P \mid x.\mathsf{inr}; P \mid x.\mathsf{case}(P, Q) \mid [y \leftrightarrow x] \end{array}$$

Most constructs are standard. We highlight the term output construct $x\langle N\rangle.P$, the binary guarded choice constructs $x.\mathsf{inl}; P$ and $x.\mathsf{inr}; P$ with the corresponding case construct; the channel forwarding or renaming construct $[y \leftrightarrow x]$ that links the channels x and y.

Processes are equated up to a structural congruence $\equiv$, defined below.

Definition 3 (Structural Congruence). *Structural congruence is defined as the least congruence relation closed under the following rules:*

$$\begin{array}{ll} P \mid \mathbf{0} \equiv P & P \equiv_\alpha Q \Rightarrow P \equiv Q \\ P \mid (Q \mid R) \equiv (P \mid Q) \mid R & P \mid Q \equiv Q \mid P \\ x \notin \mathit{fn}(P) \Rightarrow P \mid (\boldsymbol{\nu} x)Q \equiv (\boldsymbol{\nu} x)(P \mid Q) & (\boldsymbol{\nu} x)\mathbf{0} \equiv \mathbf{0} \\ (\boldsymbol{\nu} x)(\boldsymbol{\nu} y)P \equiv (\boldsymbol{\nu} y)(\boldsymbol{\nu} x)P & [y \leftrightarrow x] \equiv [x \leftrightarrow y] \end{array}$$

The operational semantics for the process calculus are standard. The semantics for the $[y \leftrightarrow x]$ construct, as informed by the proof theory, consist of channel renaming.

$$\frac{}{\Psi; \Gamma; \cdot \Rightarrow \mathbf{0} :: z : \mathbf{1}}\ \mathsf{1R} \qquad \frac{\Psi; \Gamma; \Delta \Rightarrow P :: z : C}{\Psi; \Gamma; \Delta, x : \mathbf{1} \Rightarrow P :: z : C}\ \mathsf{1L}$$

$$\frac{\Psi; \Gamma; \cdot \Rightarrow P :: y : A}{\Psi; \Gamma; \cdot \Rightarrow !z(y).P :: z : !A}\ \mathsf{!R} \qquad \frac{\Psi; \Gamma, u : A; \Delta \Rightarrow P :: z : C}{\Psi; \Gamma; \Delta, x : !A \Rightarrow P\{x/u\} :: z : C}\ \mathsf{!L}$$

$$\frac{\Psi; \Gamma; \Delta \Rightarrow P :: z : A \quad \Psi; \Gamma; \Delta \Rightarrow Q :: z : B}{\Psi; \Gamma; \Delta \Rightarrow z.\mathsf{case}(P, Q) :: z : A \mathbin{\&} B}\ \mathsf{\&R}$$

$$\frac{\Psi; \Gamma; \Delta, x : A \Rightarrow P :: z : C}{\Psi; \Gamma; \Delta, x : A \mathbin{\&} B \Rightarrow x.\mathsf{inl}; P :: z : C}\ \mathsf{\&L_1}$$

$$\frac{\Psi; \Gamma; \Delta, x : B \Rightarrow P :: z : C}{\Psi; \Gamma; \Delta, x : A \mathbin{\&} B \Rightarrow x.\mathsf{inr}; P :: z : C}\ \mathsf{\&L_2}$$

$$\frac{\Psi; \Gamma; \Delta_1 \Rightarrow P :: y : A \quad \Psi; \Gamma; \Delta_2 \Rightarrow Q :: z : B}{\Psi; \Gamma; \Delta_1, \Delta_2 \Rightarrow (\boldsymbol{\nu} y)z\langle y\rangle.(P \mid Q) :: z : A \otimes B}\ \mathsf{\otimes R}$$

$$\frac{\Psi; \Gamma; \Delta, y : A, x : B \Rightarrow P :: z : C}{\Psi; \Gamma; \Delta, x : A \otimes B \Rightarrow x(y).P :: z : C}\ \mathsf{\otimes L}$$

$$\frac{\Psi; \Gamma; \Delta \Rightarrow P :: z : A}{\Psi; \Gamma; \Delta \Rightarrow z.\mathsf{inl}; P :: z : A \oplus B}\ \mathsf{\oplus R_1} \qquad \frac{\Psi; \Gamma; \Delta \Rightarrow P :: z : B}{\Psi; \Gamma; \Delta \Rightarrow z.\mathsf{inr}; P :: z : A \oplus B}\ \mathsf{\oplus R_2}$$

$$\frac{\Psi; \Gamma; \Delta, x : A \Rightarrow P :: z : C \quad \Psi; \Gamma; \Delta, x : B \Rightarrow Q :: z : C}{\Psi; \Gamma; \Delta, x : A \oplus B \Rightarrow x.\mathsf{case}(P, Q) :: z : C}\ \mathsf{\oplus L}$$

$$\frac{}{\Psi; \Gamma; x : A \Rightarrow [x \leftrightarrow z] :: z : A}\ \mathsf{id} \qquad \frac{\Psi, x : \tau; \Gamma; \Delta \Rightarrow P :: z : A}{\Psi; \Gamma; \Delta \Rightarrow z(x).P :: z : \forall x : \tau.A}\ \mathsf{\forall R}$$

$$\frac{\Psi \vdash N : \tau \quad \Psi; \Gamma; \Delta, x : A\{N/y\} \Rightarrow P :: z : C}{\Psi; \Gamma; \Delta, x : \forall y : \tau.A \Rightarrow x\langle N\rangle.P :: z : C}\ \mathsf{\forall L}$$

$$\frac{\Psi \vdash N : \tau \quad \Psi; \Gamma; \Delta \Rightarrow P : A\{N/x\}}{\Psi; \Gamma; \Delta \Rightarrow z\langle N\rangle.P :: z : \exists x : \tau.A}\ \mathsf{\exists R} \qquad \frac{\Psi, y : \tau; \Gamma; \Delta, x : A \Rightarrow P :: z : C}{\Psi; \Gamma; \Delta, x : \exists y : \tau.A \Rightarrow x(y).P :: z : C}\ \mathsf{\exists L}$$

$$\frac{\Psi; \Gamma, u : A; \Delta, y : A \Rightarrow P :: z : C}{\Psi; \Gamma, u : A; \Delta \Rightarrow (\boldsymbol{\nu} y)u\langle y\rangle.P :: z : C}\ \mathsf{copy}$$

$$\frac{\Psi; \Gamma; \Delta_1 \Rightarrow P :: x : A \quad \Psi; \Gamma; \Delta_2, x : A \Rightarrow Q :: z : C}{\Psi; \Gamma; \Delta_1, \Delta_2 \Rightarrow (\boldsymbol{\nu} x)(P \mid Q) :: z : C}\ \mathsf{cut}$$

$$\frac{\Psi; \Gamma; \cdot \Rightarrow P :: x : A \quad \Psi; \Gamma, u : A; \Delta \Rightarrow Q :: z : C}{\Psi; \Gamma; \Delta \Rightarrow (\nu u)((!u(x).P) \mid Q) :: z : C}\ \mathsf{cut}^{!}$$

Fig. 1. Dependent Session Types

Definition 4 (Reduction). *The reduction relation on processes, $P \rightarrow Q$ is defined by the following rules:*

$$
\begin{array}{l}
x\langle y\rangle.Q \mid x(z).P \rightarrow Q \mid P\{y/z\} \\
x\langle y\rangle.Q \mid !x(z).P \rightarrow Q \mid P\{y/z\} \mid !x(z).P \\
x\langle N\rangle.Q \mid x(z).P \rightarrow Q \mid P\{N/z\} \\
(\boldsymbol{\nu}x)([x \leftrightarrow y] \mid P) \rightarrow P\{y/x\} \\
x.\mathsf{inl};\, P \mid x.\mathsf{case}(Q, R) \rightarrow P \mid Q \\
x.\mathsf{inr};\, P \mid x.\mathsf{case}(Q, R) \rightarrow P \mid R \\
Q \rightarrow Q' \Rightarrow P \mid Q \rightarrow P \mid Q' \\
P \rightarrow Q \Rightarrow (\boldsymbol{\nu}y)P \rightarrow (\boldsymbol{\nu}y)Q \\
P \equiv P',\, P' \rightarrow Q',\, Q' \equiv Q \Rightarrow P \rightarrow Q
\end{array}
$$

A labeled transition system can be defined in a somewhat standard manner, where a label denotes a silent action, an output or input of a (bound) name or of a term (note that terms do not contain channel names, so no issues of scope extrusion arise).

The language of terms is intentionally left open-ended. We only suppose that they contain no (π-calculus) names, and that it satisfies substitution, progress, and preservation properties as we usually suppose for functional languages. In the next section we will postulate some particular constructs that allow us to specify different versions of proof-carrying code protocols.

3 Proof Irrelevance

In a dependent type theory, proofs are represented as terms. Even with basic dependent function types we already have the means to model proof-carrying code, as explained in the introduction and the previous section. This assumes that data values transmitted along channels are type-checked when received before we continue to compute with them in a type-safe way.

Under which circumstances can we avoid type-checking a proof object, or perhaps even avoid transmitting it entirely? One class of examples is provided by cases where the property of the objects we specified is (easily) decidable. Then we can check the property itself without the need to obtain an explicit proof object. However, this only works if the proof object is also of no actual *operational* significance, that is, it is computationally irrelevant. The previous section (e.g., $\forall p{:}(n > 0)$) and the introduction (e.g., $\exists q{:}(y = f(y))$) contain examples of this kind. But we do not want to presuppose or "bake in" any particular analysis or strategy, but formulate the type theory so that we can seamlessly move between different specifications. This is what a modality for *proof irrelevance* [15,14,2] in the type theory allows us to do.

Proof irrelevance is a technique that allows us to selectively hide portions of a proof (and by the proofs-as-programs principle, portions of a program). The idea is that these "irrelevant" proof objects are required to exist for the purpose of type-checking, but they must have no bearing on the computational outcome of the program. This means that typing must ensure that these hidden proofs are never required to compute something that is not itself hidden.

We internalize proof irrelevance in our functional language by requiring a modal type constructor, $[\tau]$ (read *bracket* τ), meaning that there is a term of type τ, but the term is deemed irrelevant from a computational point of view. We give meaning to $[\tau]$ by adding an introduction form for irrelevant terms, written $[M]$, that states that M is not available computationally; and a new class of assumptions $x \div \tau$, meaning that x stands for a term of type τ that is not computationally available; we then define a promotion operation on contexts that transforms computationally irrelevant hypotheses into ordinary ones, to account for type-checking within the bracket operator.

Definition 5 (Promotion)

$$\begin{array}{ll} (\cdot)^{\oplus} & \triangleq \cdot \\ (\Psi, x : \tau)^{\oplus} & \triangleq \Psi^{\oplus}, x : \tau \\ (\Psi, x \div \tau)^{\oplus} & \triangleq \Psi^{\oplus}, x : \tau \end{array}$$

The introduction and elimination forms of proof irrelevant terms are defined by the following rules:

$$\frac{\Psi^{\oplus} \vdash M : \tau}{\Psi \vdash [M] : [\tau]}\,[]I \qquad \frac{\Psi \vdash M : [\tau] \quad \Psi, x \div \tau \vdash N : \sigma}{\Psi \vdash \mathbf{let}\ [x] = M\ \mathbf{in}\ N : \sigma}\,[]E$$

The introduction rule states that any term M (that may use irrelevant hypotheses) of type τ induces a proof irrelevant term $[M]$ of type $[\tau]$. The elimination rule states that we can unwrap the bracket operator only by binding its contents to a variable classified as proof irrelevant. This new class of variables is given meaning by an appropriate substitution principle.

Theorem 1 (Irrelevant substitution). *If* $\Psi^{\oplus} \vdash M : \tau$ *and* $\Psi, x \div \tau, \Psi' \vdash N : \sigma$ *then* $\Psi, \Psi' \vdash N\{M/x\} : \sigma$

Proof. By structural induction on the derivation of $\Psi, x \div \tau, \Psi' \vdash N : \sigma$

We generally prefer a call-by-value operational semantics for the type theory so that we can restrict communication to values without complications. We first extend this to a version that computes explicit evidence for inhabitation of type $[\tau]$, although the intent is to actually erase rather than compute irrelevant objects. The single-step reduction relation would then contain the following congruence and reduction rules (treating irrelevant terms lazily):

$$\frac{M \longrightarrow M'}{\mathbf{let}\ [x] = M\ \mathbf{in}\ N \longrightarrow \mathbf{let}\ [x] = M'\ \mathbf{in}\ N}$$

$$\frac{}{\mathbf{let}\ [x] = [M]\ \mathbf{in}\ N \longrightarrow N\{M/x\}}$$

As motivated above, the next step is to check that irrelevant terms do not need to be computed at the functional level or communicated at the process level. We formalize this through a notion of *erasure* that replaces computationally irrelevant types by a unit type unit and irrelevant terms by corresponding unit elements $\langle\rangle$.

Definition 6 (Erasure). *The erasure operation † is defined on contexts, types, processes and terms. It is compositional everywhere, with the following special cases.*

$$\begin{array}{ll}
(\Psi, x{:}\tau)^\dagger & \triangleq \Psi^\dagger, x{:}\tau^\dagger \\
(\Psi, x{\div}\tau)^\dagger & \triangleq \Psi^\dagger \\
[\tau]^\dagger & \triangleq \mathsf{unit} \\
[M]^\dagger & \triangleq \langle\rangle \\
(\mathbf{let}\ [x] = M\ \mathbf{in}\ N)^\dagger & \triangleq N^\dagger
\end{array}$$

The erasure from the definition above does not affect the process structure. It simply traverses processes down to the functional terms they manipulate and replaces bracketed terms by the unit element as specified above.

Theorem 2 (Erasure correctness)
If $\Psi; \Gamma; \Delta \Rightarrow P :: z : A$ *then* $\Psi^\dagger; \Gamma^\dagger; \Delta^\dagger \Rightarrow P^\dagger :: z : A^\dagger$.

Proof Straightforward, by induction on the typing derivation. Note that in the case for the let-binding for bracket types we rely on the fact that the variable $[x]$ can only occur in a bracketed term (which is itself replaced by $\langle\rangle$ in †).

Note that the lack of computational significance of proof-irrelevant terms ensures that the meanings of programs are preserved. Since erasure does not affect the structure of processes, we need only focus on the functional language itself (which we fix to be well-behaved in terms of the standard properties of progress and preservation). We can establish that erasure and evaluation commute, in the following sense (where $\equiv$ is a standard notion of equality).

Theorem 3 (Erasure Preservation). *If* $\Psi \vdash M : \tau$ *and* $M \longrightarrow N$, *then there exists* N' *such that* $M^\dagger \longrightarrow^* N'$ *and* $N^\dagger \equiv N'$.

Proof. By induction on the operational semantics.

However, the erasure operation is just a step in the optimization mentioned above, since the processes in the image of the erasure still perform some communication (of unit elements) in the same places where proof objects were previously exchanged. To fully remove the potentially unnecessary communication, we consistently appeal to type isomorphisms regarding the interaction of unit with the universal and existential quantifiers:

$$\begin{array}{c}
\forall x{:}\mathsf{unit}.A \cong A \\
\exists x{:}\mathsf{unit}.A \cong A
\end{array}$$

Since we only allow for types of the functional language in the universal and existential quantifiers (and terms in the appropriate process constructs), the isomorphisms above allow us to remove a communication step. For example, if we revisit our initial example of Section 2, we can reformulate the type and process as:

$$\mathsf{T}_1 \triangleq \forall n{:}\mathsf{nat}.\, \forall p{:}[n > 0].\, \exists y{:}\mathsf{nat}.\, \exists q{:}[y > 0].\, \mathbf{1}$$

$$\mathsf{P}_1 :: x : \mathsf{T}_1 \triangleq x(n).x(p).x\langle n+1\rangle.x\langle [\mathsf{incp}\, n\, p)]\rangle.\mathbf{0}$$

By bracketing the types for the universally and existentially quantified variables p and q, we are effectively stating that we only require some proof that p and y are positive, but the content of the proof itself does not matter. Of course, since determining the positivity of an integer is easily decidable, and the form of the proof is irrelevant, we can erase the proofs using †, obtaining the following process (and type):

$$\mathsf{T}_1^\dagger \triangleq \forall n{:}\mathsf{nat}.\, \forall p{:}\mathsf{unit}.\, \exists y{:}\mathsf{nat}.\, \exists q{:}\mathsf{unit}.\, \mathbf{1}$$

$$\mathsf{P}_1^\dagger :: x : \mathsf{T}_1 \triangleq x(n).x(p).x\langle n+1\rangle.x\langle\langle\rangle\rangle.\mathbf{0}$$

By consistently appealing to the type isomorphisms mentioned above, we obtain the process below that simply inputs a number n and outputs its increment:

$$\mathsf{P}_{1\cong}^{\dagger} \triangleq x(n).x\langle n+1\rangle.\mathbf{0}$$

An alternative technique familiar from type theories is to replace sequences of data communications by a single communication of pairs. When proof objects are involved, these become Σ-types which are inhabited by pairs. For example, we can rewrite the example above as

$$\mathsf{T}_2 \triangleq \forall p{:}(\Sigma n{:}\mathsf{nat}.\, [n > 0]).\, \exists q{:}(\Sigma y{:}\mathsf{nat}.\, [y > 0]).\, \mathbf{1}$$

$$\mathsf{P}_2 :: x : \mathsf{T}_2 \triangleq x(\langle n, p\rangle).\, x\langle\, \langle n + 1, [\mathsf{incp}\, n\, p]\rangle\,\rangle.\, 0$$

where we have take the liberty of using pattern matching against $\langle n, p\rangle$ instead of writing first and second projections. Applying erasure here only simplifies the communicated terms without requiring us to change the structure of the communication.

$$\mathsf{T}_2^\dagger \triangleq \forall p{:}(\Sigma n{:}\mathsf{nat}.\, \mathsf{unit}).\, \exists q{:}(\Sigma y{:}\mathsf{nat}.\, \mathsf{unit}).\, \mathbf{1}$$

$$\mathsf{P}_2^\dagger :: x : \mathsf{T}_2 \triangleq x(\langle n, \_\rangle).\, x\langle\, \langle n+1, \langle\rangle\rangle\,\rangle.\, 0$$

This solution is popular in type theory, where $\Sigma x{:}\tau.\, [\sigma]$ is a formulation of a *subset type* [15], $\{x{:}\tau \mid \sigma\}$. Conversely, bracket types $[\sigma]$ can be written as $\{x{:}\mathsf{unit} \mid \sigma\}$, except that the proof object is *always* erased. Under some restrictions on σ, subset types can be seen as predicate-based type refinement as available, for example, in Fine [17] where it used for secure communication in distributed computation.

4 Affirmation

In many distributed communicating systems there are trade-offs between trust and explicit proofs. For example, when we download a large application we may be willing to trust its safety if it is digitally signed by a reputable vendor. On the other hand, if we are downloading and running a piece of Javascript code embedded in a web page, we may insist on some explicit proof that it is safe and adheres to our security policy. The key to making such trade-offs explicit in session types is a notion of *affirmation* (in the sense of [7]) of propositions and proofs by principals. Such affirmations can be realized through explicit digital signatures on proofs by principals, based on some underlying public key infrastructure.

An affirmation judgment, written $\Psi \vdash M :_K \tau$, means that principal K attests a proof M for τ. As in prior work [7], this may be realized by a digitally signed certificate, although in our case it will be both the proof and the propositions that are signed by a principal K, written as $\langle M{:}\tau\rangle_K$.

We add the affirmation judgment to the type system of our functional language through the following rule:

$$\frac{\Psi \vdash M : \tau}{\Psi \vdash \langle M{:}\tau\rangle_K :_K \tau}\ (\mathsf{affirms})$$

The rule states that any principal can affirm the property τ by virtue of a proof M. In the implementation, a process wishing to create such an affirmation must have access to K's private key so it can sign the pair consisting of the term M and its type τ.

Such an affirmation may seem redundant: after all, the certificate contains the term M which can be type-checked. However, checking a digitally signed certificate may be faster than checking the validity of a proof, so we may speed up the system if we trust K's signature. More importantly, if we have proof irrelevance, and some parts of M have been erased, then we have in general no way to reconstruct the proofs. In this case we must trust the signing principal K to accept the τ as true, because we cannot be sure if K played by the rules and did indeed have a proof. Therefore, in general, the affirmation of τ by K is *weaker* than the truth of τ, for which we demand explicit evidence. Conversely, when τ is true K can always sign it and be considered as "playing by the rules", as the inference rule above shows.

Now, to actually be able to use affirmation with the other types in our system, we internalize the judgment as a modal operator. We write $\Diamond_K\tau$ for the type that internalizes the judgment $:_K \tau$ (e.g. in the same way that implication internalizes entailment), and $\mathbf{let}\ \langle x{:}\tau\rangle_K = M\ \mathbf{in}\ N$ for the corresponding destructor.

$$\frac{\Psi \vdash M :_K \tau}{\Psi \vdash M : \Diamond_K\tau}\ \Diamond I \qquad \frac{\Psi \vdash M : \Diamond_K\tau \quad \Psi, x{:}\tau \vdash N :_K \sigma}{\Psi \vdash \mathbf{let}\ \langle x{:}\tau\rangle_K = M\ \mathbf{in}\ N :_K \sigma}\ \Diamond E$$

The introduction rule simply internalizes the affirmation judgment. The elimination rule requires the type we are determining to be an affirmation of the same principal K, adding an assumption of τ – we can assume the property τ from an affirmation made by K only if we are reasoning about affirmations of K. Affirmation in this sense works as a principal-indexed monad. The reduction rules for affirmation are straightforward:

$$\frac{M \longrightarrow M'}{\textbf{let } \langle x{:}\tau\rangle_K = M \textbf{ in } N \longrightarrow \textbf{let } \langle x{:}\tau\rangle_K = M' \textbf{ in } N}$$

$$\frac{}{\textbf{let } \langle x{:}\tau\rangle_K = \langle M{:}\tau\rangle_K \textbf{ in } N \longrightarrow N\{M/x\}}$$

Returning now to the example in the introduction, the type

$$\mathsf{fpt} : \forall f{:}\mathsf{nat} \rightarrow \mathsf{nat}.\, \forall p{:}\Diamond_{\mathsf{verif}}[\Pi x{:}\mathsf{nat}.\, f(x) \le x].\, \exists y{:}\mathsf{nat}.\, \exists q{:}[y = f(y)].\, \mathbf{1}$$

expresses the type of a server that inputs a function f, accepts a verifier's word that it is decreasing, and returns a fixed point of f to the client. A client that passes the identity function to fpt may be written as follows:

$$\mathsf{fpt}\langle\lambda x.\, x\rangle.\, \mathsf{fpt}\langle\langle[\lambda x.\, \mathsf{refl}\ x]{:}[\Pi x{:}\mathsf{nat}.\, f(x) \le x]\rangle_{\mathsf{verif}}\rangle.\, \mathsf{fpt}(y).\, \mathsf{fpt}(q).\, \mathbf{0}.$$

If we want to explicate that the digital signature is supplied by another process associated with access to the private key with the principal verif, we could write a polymorphic process with type

$$\mathsf{v} : \forall \alpha{:}\mathsf{type}.\, \forall x{:}\alpha.\, \exists y{:}\Diamond_{\mathsf{verif}}[\alpha].\, \mathbf{1}$$

which could be

$$\mathsf{v}(\alpha).\, \mathsf{v}(x).\, \mathsf{v}\langle\langle[x]{:}[\alpha]\rangle_{\mathsf{verif}}\rangle :: \mathsf{v} : \forall \alpha{:}\mathsf{type}.\, \forall x{:}\alpha.\, \exists y{:}\Diamond_{\mathsf{verif}}[\alpha].\, \mathbf{1}$$

The client would then call upon this service and pass the signed certificate (without the proof term) on to fpt.

$$\mathsf{fpt}\langle\lambda x.\, x\rangle.\, \mathsf{v}\langle\mathsf{nat} \rightarrow \mathsf{nat}\rangle.\, \mathsf{v}\langle\lambda x.\, x\rangle.\, \mathsf{v}(c).\, \mathsf{fpt}\langle c\rangle.\, \mathsf{fpt}(y).\, \mathsf{fpt}(q).\, \mathbf{0}.$$

In fact, the implementation of the proof-carrying file system [8] (PCFS) provides such a generic trusted service. In PCFS, the access control policy is presented as a logical theory in the access control logic. Access to a file is granted if a proof of a corresponding access theorem can be constructed with the theory in access control logic and is presented to the file system. Such proofs are generally small when compared to proof-carrying code in the sense of Necula and Lee [13,12] in which the type safety and memory safety of binary code is certified, but they are still too big to be transmitted and checked every time a file is accessed. Instead, we call upon the trusted verification service to obtain a digitally signed certificate of type $\langle \mathsf{verif}{:}[\alpha]\rangle$ called a *procap* (for *proven capability*). Procaps are generally very small and fast to verify, leading to an acceptably small overhead when compared to checking access control lists.

As another example, we consider a toy scenario where the customer of a store uses a paying machine to make a purchase. The machine receives the account balance from the bank in order to ensure that the client has enough money for the purchase (realistically the bank would decide if the client has enough money, not the machine, but this suits our illustrative purposes best), if that is not the case it must abort the transaction, otherwise the purchase goes through. We can model this system in our setting by specifying a type for the bank and a type for the machine. We abbreviate $\forall x{:}\tau.\, A$ as $\tau \supset A$ and $\exists x{:}\tau.\, A$ as $\tau \wedge A$ when x is not free in A:

$$\begin{aligned}\mathsf{TBank} \triangleq\ & \forall s{:}\mathsf{string}.\, \Diamond_{\mathsf{M}}[\mathsf{uid}(s)] \supset \\ & (\Sigma n{:}\mathsf{int}.\, \Diamond_{\mathsf{B}}[\mathsf{bal}(s,n)]) \wedge ((\forall m{:}\mathsf{nat}.\, \Diamond_{\mathsf{M}}[\mathsf{charge}(s,m)] \supset \mathbf{1}) \,\&\, \mathbf{1})\end{aligned}$$

The type for the bank describes part of the protocol we wish this system to observe: the bank will receive a string and a signed certificate from the paying machine (we use M and B as the principal identifiers for the machine and for the bank, respectively), that asserts the client's identification data. It then sends back the account balance to the machine, attaching a signed certificate that it is indeed the appropriate balance information. It will wait for the decision of the machine to charge the account or not. This is embodied in the use of the additive conjunction (&), that allows the bank to branch on doing nothing ($\mathbf{1}$) or inputting the appropriate charge information. The type for the interface of the machine with the client is as follows:

$$\mathsf{TMClient} \triangleq \forall s{:}\mathsf{string}.\, ((\Diamond_{\mathsf{M}}[\mathsf{ok}] \wedge \mathbf{1}) \oplus (\Diamond_{\mathsf{M}}[\mathsf{nok}] \wedge \mathbf{1}))$$

The client inputs his pin number in the machine and then simply waits for the machine to inform him if the transaction went through or not. A process implementing the bank session (along channel x) is given below:

$$\begin{aligned}\mathsf{Bank} \triangleq\ & x(s).x(u).x\langle \mathsf{sign}_2(\mathsf{db\_getbal}(s))\rangle. \\ & x.\mathsf{case}(x(m).x(c).\mathbf{0}; \mathbf{0}) :: x : \mathsf{TBank}\end{aligned}$$

We assume a function db_getbal that interacts with the bank database to fetch the appropriate balance information and a generic function sign_2 (making type arguments implicit) which is like the earlier generic verifier and uses the bank's private key.

$$\begin{aligned}\mathsf{db\_getbal} &: \Pi s{:}\mathsf{string}.\, \Sigma n{:}\mathsf{int}.\mathsf{bal}(s,n) \\ \mathsf{sign}_2 &: (\Sigma n{:}\alpha.\, \beta) \to (\Sigma n{:}\alpha.\, \Diamond_{\mathsf{B}}[\beta])\end{aligned}$$

The machine process is typed in an environment containing the bank session along channel x and implementing the interface with the client along channel z, as follows:

$$\mathsf{Machine} \triangleq z(s).\, x\langle s\rangle.\, x\langle\langle[\mathsf{gen\_uid}]{:}[\mathsf{uid}(s)]\rangle_{\mathsf{M}}\rangle.\, x(n).\, x(b).\, P_{\mathsf{decide}}$$

We assume a function gen_uid of type $\Pi s{:}\mathsf{string}.\, \mathsf{uid}(s)$ that takes the clients input and generates the appropriate uid object. We abstract away the details of deciding if the client has enough money for the purchase in process P_{decide}. This process will simply perform the check and then either terminate and send to the client the nok signal, if the client has insufficient balance, or send the charge information to the bank and inform the client that the transaction went through.

As another example, illustrating an application to distributed certified access control, consider the following types

$$\begin{array}{ll}\mathsf{Server} & \triangleq \forall uid{:}\mathsf{string}.\,(\mathbf{1} \oplus (\Diamond_{\mathsf{S}}[\mathsf{perm}(uid)] \wedge \mathsf{Session}(uid)))\\ \mathsf{Session}(uid) & \triangleq (\mathsf{productid} \multimap \Diamond_{\mathsf{S}}[\mathsf{may}(uid, \mathsf{buy})] \supset \mathsf{rcp} \otimes \mathbf{1})\\ & \quad \&\\ & \quad (\mathsf{productid} \multimap \Diamond_{\mathsf{S}}[\mathsf{may}(uid, \mathsf{quote})] \supset \mathsf{ans} \otimes \mathbf{1})\end{array}$$

The type Server specifies a server that receives an user id (of type string), and then either refuses the session ($\mathbf{1}$), or sends back a proof of access permissions granted to the given user, before proceeding. Here, we might have

$$\mathsf{perm}(uid) \triangleq \mathsf{may}(uid, \mathsf{quote}) \vee \mathsf{may}(uid, \mathsf{buy}) \vee \mathsf{may}(uid, \mathsf{all})$$

In order to access an operation (say buy), the client must exhibit a proof of authorization, necessarily computed from the permission proof sent by the server (assuming that only the server can provide such proofs).

The examples above illustrates how proof certificates might be used in our process setting. Recall that, since the proof certificates are always marked as proof irrelevant, we can use the erasure of Section 3 and remove them from the protocol if we so desire.

5 Progress and Preservation

In [18] we established the type safety results of progress and preservation for our dependent session type theory for an unspecified functional language. In fact, we made no mention of when reduction of the functional terms happens. Here, we work under the assumption that processes always evaluate a term to a value *before* communication takes place, and therefore progress and preservation are contingent on the functional layer also being type safe in this sense (which can easily be seen to be the case for the connectives we have presented in this development).

The proof of type preservation then follows the same lines of [18], using a series of reduction lemmas that relate process reductions with parallel composition through an instance of the cut rule and appealing to the type preservation of the functional layer when necessary.

Theorem 4 (Type Preservation). *If* $\Psi; \Gamma; \Delta \Rightarrow P :: z : A$ *and* $P \rightarrow Q$ *then* $\Psi; \Gamma; \Delta \Rightarrow Q :: z : A$

Proof. By induction on the typing derivation. When the last rule is an instance of cut, we appeal to the reduction lemmas mentioned above (and to type preservation of the functional language when the premises of cut are of existential or universal type), which are presented in more detail in [18].

The case for the proof of progress is identical. The result in [18] combined with progress of the functional language establishes progress for the system of this paper. For the purpose of having a self-contained document, we will sketch the proof here as well.

Definition 7 (Live Process)

$$live(P) \triangleq P \equiv (\boldsymbol{\nu}\overline{n})(Q \mid R) \quad \textit{for some } Q, R, \overline{n}$$

where $Q \equiv \pi.Q'$ *(*π *is a non-replicated prefix) or* $Q \equiv [x \leftrightarrow y]$

We begin by defining the form of processes that are live. We then establish a contextual progress theorem from which progress follows (Theorem 5 relies on several inversion lemmas that relate types to action labels). Given an action label α, we denote by $s(\alpha)$ the subject of the action α (i.e., the name through which the action takes place).

Theorem 5 (Contextual Progress). *Let* $\Psi; \Gamma; \Delta \Rightarrow P :: z : C$. *If* $live(P)$ *then there is* Q *such that one of the following holds:*

(a) $P \rightarrow Q$*,*
(b) $P \xrightarrow{\alpha} Q$ *for some* α *where* $s(\alpha) \in z, \Gamma, \Delta$ *and* $s(\alpha) \in \Gamma, \Delta$ *if* $C = !A$*,*
(c) $P \equiv_S [x \leftrightarrow z]$*, for some* $x \in \Delta$*.*

Proof. By induction on typing, following [18].

The theorem above states that live processes are either able to reduce outright, are able to take an action α or are equivalent to a channel forwarding (modulo structural congruence extended with a "garbage collection rule" for replicated processes that are no longer usable).

Theorem 6 (Progress). *If* $\cdot; \cdot; \cdot \Rightarrow P :: x : \mathbf{1}$*, and* $live(P)$*, then there exists a process* Q *such that* $P \rightarrow Q$*.*

Finally, Theorem 6 follows straightforwardly from Theorem 5 since P can never offer an action α along x, due to its type. Note that $\mathbf{1}$ types not just the inactive process but also all closed processes (i.e. processes that consume all ambient sessions).

6 Concluding Remarks

In this paper, we have built upon previous work on dependent session types to account for a flexible notion of proof-carrying code, including digitally signed certificates in lieu of proof objects. To this end, we integrated proof irrelevance and affirmations to the underlying functional language, giving the session type language fine control over which code and data are accompanied by explicit proof, which are supported by digital signature only, and which are trusted outright. We had previously considered proof irrelevance only as a means of optimizing communication in trusted or decidable settings. In a concrete implementation, the operational semantics must be supported by cryptographic infrastructure to digitally sign propositions and proofs and check such signatures as authentic.

Ours is one amongst several Curry-Howard interpretations connecting linear logic to concurrency. Perhaps closest to session types is work by Mazurak and Zdancewic [11] who develop a Curry-Howard interpretation of classical linear logic as a functional programming language with explicit constructs for concurrency. Their system is based on natural deduction and is substantially different from ours, and they consider neither dependent types nor unrestricted sessions.

The work on Fine [17], F7 [3], and more recently F* [16] has explored the integration of dependent and refinement types in a suite of functional programming languages, with the aim of statically checking assertions about data and state, and enforcing security policies. In our line of research, investigating how closely related mechanisms may be essentially extracted from a Curry-Howard interpretation of fragments of linear and affirmation logics, building on proof irrelevance to express a counterpart of the so-called ghost refinements in F*.

The work on PCML5 [1] has some connection to our own in the sense that they also use affirmation in their framework. PCML5, however, is mostly concerned with authorization and access control, while we employ affirmation as a way of obtaining signatures. Furthermore, PCML5 has no concurrency primitives, while our language consists of a process calculus and thus is inherently concurrent. Nevertheless, it would be quite interesting to explore the possibilities of combining PCML5's notion of authorization with our concurrent setting.

For future work, we wish to explore the applications of proof irrelevance and affirmation in the process layer. Proof irrelevance at the process level is not well understood since it interacts with linearity (if a channel is linear, it must be used, but because it is irrelevant it may not) and communication, considered as an effect. The monadic flavor of affirmation seems to enforce a very strong notion of information flow restrictions on processes, where a process that provides a session of type $\Diamond_K A$ is only able to do so using public sessions, or other sessions of type $\Diamond_K\ T$. It would nevertheless be very interesting to investigate how more flexible information flow disciplines might be expressed in our framework, based on modal logic interpretations.

References

1. Avijit, K., Datta, A., Harper, R.: Distributed programming with distributed authorization. In: Proceedings of the 5th Workshop on Types in Language Design and Implementation, TLDI 2010, pp. 27–38. ACM, New York (2010)
2. Awodey, S., Bauer, A.: Propositions as [types]. Journal of Logic and Computation 14(4), 447–471 (2004)
3. Bengtson, J., Bhargavan, K., Fournet, C., Gordon, A.D., Maffeis, S.: Refinement types for secure implementations. In: 21st Computer Security Foundations Symposium, CSF 2008, Pittsburgh, Pennsylvania, pp. 17–32. IEEE Computer Society (June 2008)

4. Bonelli, E., Compagnoni, A., Gunter, E.L.: Correspondence Assertions for Process Synchronization in Concurrent Communications. J. of Func. Prog. 15(2), 219–247 (2005)
5. Caires, L., Pfenning, F.: Session Types as Intuitionistic Linear Propositions. In: Gastin, P., Laroussinie, F. (eds.) CONCUR 2010. LNCS, vol. 6269, pp. 222–236. Springer, Heidelberg (2010)
6. Dezani-Ciancaglini, M., de'Liguoro, U.: Sessions and Session Types: An Overview. In: Laneve, C., Su, J. (eds.) WS-FM 2009. LNCS, vol. 6194, pp. 1–28. Springer, Heidelberg (2010)
7. Garg, D., Bauer, L., Bowers, K.D., Pfenning, F., Reiter, M.K.: A Linear Logic of Authorization and Knowledge. In: Gollmann, D., Meier, J., Sabelfeld, A. (eds.) ESORICS 2006. LNCS, vol. 4189, pp. 297–312. Springer, Heidelberg (2006)
8. Garg, D., Pfenning, F.: A proof-carrying file system. In: Evans, D., Vigna, G. (eds.) Proceedings of the 31st Symposium on Security and Privacy (Oakland 2010), Berkeley, California. IEEE (May 2010); Extended version available as Technical Report CMU-CS-09-123 (June 2009)
9. Honda, K.: Types for Dyadic Interaction. In: Best, E. (ed.) CONCUR 1993. LNCS, vol. 715, pp. 509–523. Springer, Heidelberg (1993)
10. Honda, K., Vasconcelos, V.T., Kubo, M.: Language Primitives and Type Discipline for Structured Communication-Based Programming. In: Hankin, C. (ed.) ESOP 1998. LNCS, vol. 1381, pp. 122–138. Springer, Heidelberg (1998)
11. Mazurak, K., Zdancewic, S.: Lolliproc: To concurrency from classical linear logic via Curry-Howard and control. In: Hudak, P., Weirich, S. (eds.) Proceedings of the 15th International Conference on Functional Programming (ICFP 2010), Baltimore, Maryland, pp. 39–50. ACM (September 2010)
12. Necula, G.C.: Proof-carrying code. In: Jones, N.D. (ed.) Conference Record of the 24th Symposium on Principles of Programming Languages (POPL 1997), Paris, France, pp. 106–119. ACM Press (January 1997)
13. Necula, G.C., Lee, P.: Safe kernel extensions without run-time checking. In: Proceedings of the Second Symposium on Operating System Design and Implementation (OSDI 1996), Seattle, Washington, pp. 229–243 (October 1996)
14. Pfenning, F.: Intensionality, extensionality, and proof irrelevance in modal type theory. In: Halpern, J. (ed.) Proceedings of the 16th Annual Symposium on Logic in Computer Science (LICS 2001), Boston, Massachusetts, pp. 221–230. IEEE (June 2001)
15. Salvesen, A., Smith, J.M.: The strength of the subset type in Martin-Löf's type theory. In: 3rd Annual Symposium on Logic in Computer Science (LICS 1988), Edinburgh, Scotland, pp. 384–391. IEEE (July 1988)
16. Swamy, N., Checn, J., Fournet, C., Strub, P.-Y., Bhargavan, K., Yang, J.: Secure distributed programming with value-dependent types. In: Danvy, O. (ed.) International Conference on Functional Programming (ICFP 2011), Tokyo, Japan. ACM (September 2011) (to appear)
17. Swamy, N., Chen, J., Chugh, R.: Enforcing Stateful Authorization and Information Flow Policies in Fine. In: Gordon, A.D. (ed.) ESOP 2010. LNCS, vol. 6012, pp. 529–549. Springer, Heidelberg (2010)
18. Toninho, B., Caires, L., Pfenning, F.: Dependent session types via intuitionistic linear type theory. In: Proceedings of the 13th International Symposium on Principles and Practice of Declarative Programming (PPDP 2011), pp. 161–172. ACM (July 2011)

Automated Certification of Implicit Induction Proofs

Sorin Stratulat and Vincent Demange

LITA, Paul Verlaine-Metz University, Ile du Saulcy, 57000, Metz, France
{stratulat,demange}@univ-metz.fr

Abstract. Theorem proving is crucial for the formal validation of properties about user specifications. With the help of the Coq proof assistant, we show how to certify properties about conditional specifications that are proved using automated proof techniques like those employed by the Spike prover, a rewrite-based implicit induction proof system. The certification methodology is based on a new representation of the implicit induction proofs for which the underlying induction principle is an instance of Noetherian induction governed by an induction ordering over equalities. We propose improvements of the certification process and show that the certification time is reasonable even for industrial-size applications. As a case study, we automatically prove and certify more than 40% of the lemmas needed for the validation of a conformance algorithm for the ABR protocol.

1 Introduction

Theorem proving is a crucial domain for validating properties about user specifications. The properties formally proved with the help of theorem provers are valid if the proofs are sound. Generally speaking, there are two methods to certify (the soundness of the) proofs: either i) by certifying the implementation of the inference systems; in this way, any generated proof is implicitly sound, or ii) by explicitly checking the soundness of the proofs generated by not-yet certified theorem provers using certified proof environments like Coq [25].

We are interested in certifying properties about conditional specifications using automated proof techniques like those employed by Spike [5,19,3], a rewrite-based implicit induction proof system. The implementation of Spike's inference system is spread over thousands of lines of OCaml [12] code. Its certification, as suggested by method i), would require a tremendous proving effort. For example, [11] reports a cost of 20 person year for the certification of the implementation of another critical software: an OS-kernel comprising about 8,700 lines of C and 600 lines of assembler. For this reason, we followed the method ii), that has been firstly tested manually in [24], then automatically on toy examples in [23]. The method directly translates every step of a Spike proof into Coq scripts, which distinguishes it from previous methods based on proof reconstruction techniques [7,10,13] that mainly transform implicit into explicit induction proofs.

J.-P. Jouannaud and Z. Shao (Eds.): CPP 2011, LNCS 7086, pp. 37–53, 2011.

In this paper, we report improvements in order to certify implicit induction proofs concerning industrial-size applications. The case study of our choice is the validation proof of a conformance algorithm for the ABR protocol [14]. An interactive proof using PVS [18] was firstly presented in [16], then it has been shown in [17] that more than a third of the user interactions can be avoided using implicit induction techniques, Spike succeeding to prove 60% of the user-provided lemmas automatically. Now, a simpler but more restrictive version of the Spike inference system has been shown powerful enough to prove 2/3 out of these lemmas. Moreover, any generated proof has been automatically translated into a Coq script, then automatically certified by Coq. We stress the importance of the automatic feature since the proof scripts are in many cases big and hard to manipulate by the users. The bottom-line is that these improvements allowed us to certify big proof scripts in a reasonable time, 20 times faster than in [23].[1]

The structure of the paper is as follows: after introducing the basic notions and notations in Section 2, we present in Section 3 the restricted inference system and a new representation of the implicit induction proofs for which the underlying induction principle is an instance of Noetherian induction governed by an induction ordering over equalities. The conformance algorithm and its Spike specification are discussed in Section 4. In Section 5, we describe a full implicit induction proof of one of the lemmas used in the ABR proof, then explain in Section 6 its translation into Coq script following the new representation of the implicit induction proofs. We detail the improvements we have made to speed-up the certification process and give statistics about the certification of the proofs of any of the 33 lemmas proved with the restricted inference system. Further improvements are discussed at the end of the section, including the parallelisation of the certification process. The conclusions and directions for future work are given in the last section.

2 Background and Notations

This section briefly introduces the basic notions and notations related to proving properties about conditional specifications by implicit induction techniques. More detailed presentations of them, and about equality reasoning in general, can be found elsewhere, for example in [2].

We assume that $\mathcal{F}$ is an alphabet of arity-fixed function symbols and $\mathcal{V}$ is a set of universally quantified variables. The set of functions symbols is split into *defined* and *constructor* function symbols. We also assume that the function symbols and variables are *sorted* and that for each sort s there is at least one constructor symbol of sort s. The set of *terms* is denoted by $\mathcal{T}(\mathcal{F}, \mathcal{V})$ and the set of variable-free (ground) terms by $\mathcal{T}(\mathcal{F})$. The sort of a non-variable term of the form $f(\ldots)$ is the sort of f, where $f \in \mathcal{F}$. Relations between terms can be established by the means of *equalities*. An *unconditional* equality is denoted by $s = t$, where s and t are two terms of same sort. Unconditional equalities and

[1] The code of the Spike prover and the generated Coq scripts can be downloaded from `http://code.google.com/p/spike-prover/`.

their negations are *literals*. A *clause* is a disjunction of literals. *Horn* clauses, consisting of clauses with at most one unconditional equality, are represented as implications. In its most usual form, $\neg e1 \vee \ldots \vee \neg e_n \vee e$ is a *conditional* equality, denoted by $e_1 \wedge \ldots \wedge e_n \Rightarrow e$, where e_i $(i \in [1..n])$ are *conditions* and e is the *conclusion*. Sometimes, we emphasize a particular condition e_i w.r.t. the other conditions Γ by writing $\Gamma \wedge e_i \Rightarrow e$. We denote by $e_1 \wedge \ldots \wedge e_n \Rightarrow$ an *impossible* set of conditions.

Equality reasoning may require transformations over equalities. A basic such transformation is the *substitution* operation which consists in simultaneous replacements of variables with terms. Formally, a substitution is represented as a finite mapping $\{x_1 \mapsto t_1; \ldots; x_n \mapsto t_n\}$, where $x_i \in \mathcal{V}$ and $t_i \in \mathcal{T}(\mathcal{F}, \mathcal{V})$. If σ is such a substitution, and t a term (resp. e an equality), then $t\sigma$ (resp. $e\sigma$) is an *instance* of t (resp. e). In the following, we assume that the variables from the replacing terms in σ are new w.r.t. the variables of t (resp. e). A term s *matches* a term t if there exists a (matching) substitution σ such that $s\sigma \equiv t$, where $\equiv$ is the *identity* relation. A *unifier* of two terms s and t is a substitution σ such that $s\sigma \equiv t\sigma$. In the rest of the paper, we will consider only the most general unifiers (mgu), and write $\sigma = mgu(s,t)$ whenever $s\sigma \equiv t\sigma$. Another kind of transformation operation is the replacement of a non-variable subterm of a term or equality by another term. The replaced subterm can be uniquely identified by its *position*. If p is a position and e (resp. t) an equality (resp. term), its subterm at position p is denoted by e_p (resp. t_p). Formally, $e_p[s]$ (resp. $t_p[s]$) states that s is a subterm of e (resp. t) at position p.

Any induction principle is based on an (induction) ordering. A *quasi-ordering* $\leq$ is a reflexive and transitive binary relation. The strict part of a quasi-ordering is called *ordering* and is denoted by $<$. We write $x >(\geq)\ y$ iff $y <(\leq)\ x$. An ordering $<$, defined over the elements of a nonempty set A, is *well-founded* if it is impossible to build an infinite strictly descending sequence $x_1 > x_2 > \ldots$ of elements of A. A binary relation R is *stable under substitutions* if whenever $s\,R\,t$ then $(s\sigma)\,R\,(t\sigma)$, for any substitution σ. Induction orderings can be defined over terms as well as equalities. An example of an induction ordering over terms is the recursive path ordering (for short, *rpo*), denoted by $\prec_{rpo}$, recursively defined from a well-founded ordering $<_{\mathcal{F}}$ over $\mathcal{F}$. If $s, t \in \mathcal{T}(\mathcal{F}, \mathcal{V})$, we write $t \prec_{rpo} s$ if $s \equiv f(s_1, \ldots, s_m)$ and i) either $s_i \equiv t$ or $t \prec_{rpo} s_i$ for some s_i, $i \in [1..m]$, or ii) $t \equiv g(t_1, \ldots, t_n), t_i \prec_{rpo} s$ for all i, $i \in [1..n]$, and either a) $g <_{\mathcal{F}} f$, or b) $f \equiv g$ and $(t_1, \ldots, t_n) \prec\!\!\prec_{rpo} (s_1, \ldots, s_n)$. $\prec\!\!\prec_{rpo}$ is the *multiset extension* of $\prec_{rpo}$. Given two multisets of terms A_1 and A_2, we write $A_1 \prec\!\!\prec_{rpo} A_2$ if, after the pairwise elimination of the identical terms from A_1 and A_2, $\forall s \in A_1$, $\exists t \in A_2$ such that $s \prec_{rpo} t$. An induction ordering over equalities, denoted by $\prec$, is defined as follows: $s_1 = t_1 \wedge \ldots \wedge s_n = t_n \Rightarrow s = t \prec s'_1 = t'_1 \wedge \ldots \wedge s'_m = t'_m \Rightarrow s' = t'$ if $\{s_1, t_1, \ldots, s_n, t_n, s, t\} \prec\!\!\prec_{rpo} \{s'_1, t'_1, \ldots, s'_m, t'_m, s', t'\}$. It can be shown that $\prec$ is well-founded and stable under substitutions.

We are interested in proving properties, or *conjectures*, about *conditional specifications* consisting in sets of conditional equalities defining (defined) function symbols, also referred to as *axioms*. An axiom $s_1 = t_1 \wedge \ldots \wedge s_n = t_n \Rightarrow s = t$ can be transformed into the (conditional) *rewrite rule* $s_1 = t_1 \wedge \ldots \wedge s_n = t_n \Rightarrow s \rightarrow t$ if $\{s_1, t_1, \ldots, s_n, t_n, t\} \lll_{rpo} \{s\}$ and the variables of $s_1, t_1, \ldots s_n, t_n, t$ are among the variables of s. Given a substitution σ, a rewrite rule $a = b \Rightarrow l \rightarrow r$ and an equality e such that $e[l\sigma]_u$, a *rewrite operation* replaces $e[l\sigma]_u$ by $e[r\sigma]_u$, then enriches the conditions of $e[r\sigma]_u$ with the new conditions $a\sigma = b\sigma$. A *rewrite system* $\mathcal{R}$ consists of a set of rewrite rules. The rewrite relation $\rightarrow_{\mathcal{R}}$ denotes rewrite operations only with rewrite rules from $\mathcal{R}$ and it can be shown well-founded. The equivalence closure of $\rightarrow_{\mathcal{R}}$ is denoted by $\overset{*}{\leftrightarrow}_{\mathcal{R}}$.

The equality $s_1 = t_1 \wedge \ldots \wedge s_n = t_n \Rightarrow s = t$ is an *inductive theorem* of a set of axioms orientable into a rewrite system $\mathcal{R}$ if, for any of its ground instances $s_1\sigma = t_1\sigma \wedge \ldots \wedge s_n\sigma = t_n\sigma \Rightarrow s\sigma = t\sigma$, whenever $s_i\sigma \overset{*}{\leftrightarrow}_{\mathcal{R}} t_i\sigma$, for each $i \in [1..n]$, we have $s\sigma \overset{*}{\leftrightarrow}_{\mathcal{R}} t\sigma$. *Tautologies* are inductive theorems either of the form $\wedge_i s_i = t_i \Rightarrow t = t$ or equalities for which the conclusion is among the conditions. An equality e_1 is *subsumed* by another equality e_2 if there exists a substitution σ such that $e_2\sigma$ is a *sub-clause* of e_1, i.e. e_1 is of the form $e_2\sigma \vee \Gamma$.

3 Noetherian Induction for Implicit Induction Proofs

Noetherian induction is a widely used induction principle. In its most general form, it allows to prove the validity of a property ϕ for all elements of a potentially infinite poset $(S, <)$, provided that $<$ is a well-founded ordering. The powerful of the principle consists in the possibility to use in the proof of any $\phi(m)$, with $m \in S$, the assumption that $\phi(k)$ is true, for all $k \in S$ smaller than m. Formally,

$$(\forall m \in S, (\forall k \in S, k < m \Rightarrow \phi(k)) \Rightarrow \phi(m)) \Rightarrow \forall m \in S, \phi(m)$$

The soundness of the principle is guaranteed by the well-foundedness property of $<$. The formulas $\phi(k)$, with $k < m$, can safely participate to the proof of $\phi(m)$ as *induction hypotheses*.

The Noetherian induction principle can be used to prove properties about conditional specifications if S is a set of equalities, $\phi(x) = x$ for all $x \in S$, and the well-founded ordering is $\prec$, as defined in the previous section. In this case, the Noetherian induction principle becomes a contrapositive version of the 'Descente Infinie' induction principle we have previously used in [20,21,22,23]:

$$(\forall \psi \in S, (\forall \rho \in S, \rho \prec \psi \Rightarrow \rho) \Rightarrow \psi) \Rightarrow \forall \gamma \in S, \gamma$$

This means that, in the proof of any equality, one can use smaller equalities.

To prove that a set of conditional equalities are inductive theorems w.r.t. a rewrite system $\mathcal{R}$, we will consider a simplified version of the Spike inference system [5,3], represented in Fig. 1, that implements this particular case of Noetherian induction in order to perform implicit induction proofs. The set S will consist of all ground instances of the conjectures encountered in the proof of the inductive theorems, defined as follows.

An implicit induction *proof* represents a finite sequence of *states* $(E^0, \emptyset) \vdash (E^1, H^1) \vdash \ldots \vdash (\emptyset, H^n)$, where E^i $(i \in [0..n-1])$ are multisets of conjectures

GENERATE: $(E \cup \{e[t]_p\},\ H) \vdash (E \cup (\cup_\sigma \{e\sigma[r\sigma]_p$ enriched with cond. $a\sigma = b\sigma\})),\ H \cup \{e\})$
if $a = b \Rightarrow l \rightarrow r \in \mathcal{R}$, $\sigma = mgu(l, t)$.

TOTAL CASE REWRITING: $(E \cup \{e[t]_p\},\ H) \vdash (E \cup (\cup_\sigma \{$cond. $a\sigma = b$ added to $e[r\sigma]_p\})), H)$
if $a = b \Rightarrow l \rightarrow r \in \mathcal{R}$, $l\sigma \equiv t$, b is either $True$ or $False$.

(UNCONDITIONAL) REWRITING: $(E \cup \{e\},\ H) \vdash (E \cup \{e'\},\ H)$
if $e \rightarrow_{\mathcal{R} \cup \mathcal{L} \cup (H \cup E)_{\prec e}} e'$.

AUGMENTATION: $(E \cup \{\Gamma \wedge l \Rightarrow s = t\},\ H) \vdash (E \cup \{\Gamma \wedge \wedge_k l'_k \Rightarrow s = t\},\ H)$
if, for each k, it exists a lemma $u \Rightarrow v$ and a substitution σ such that
$u\sigma \equiv l$, $v\sigma \equiv l'_k$ and $l'_k \prec l$.

TAUTOLOGY: $(E \cup \{e\},\ H) \vdash (E,\ H)$
if e is a tautology.

SUBSUMPTION: $(E \cup \{e\},\ H) \vdash (E,\ H)$
if e is subsumed by some conjecture from $H \cup \mathcal{L}$.

Fig. 1. A simplified version of the Spike inference system

and H^i ($i \in [1..n]$) are multisets of *premises* made of previously treated conjectures. A *step* between two states is performed by the application of one of the Spike *inference rules*. Each rule transforms one of the conjectures, called *current* conjecture, into a potentially empty set of new conjectures. GENERATE applies on equalities incorporating subterms that can unify with some lhs of the rewrite rules from $\mathcal{R}$. Such a subterm should be tested for unification against all the rewrite rules from $\mathcal{R}$, each mgu σ playing the role of *test* substitution. The resulted set of test substitutions is expected to be complete. The method for computing the test substitutions has been presented in [3] and is different from the 'test set' method from [5]. Finally, the current conjecture is added to the set of premises. TOTAL CASE REWRITING can apply on equalities with subterms that are matched by some lhs of rewrite rules from $\mathcal{R}$. As for the unification test, the matching test should be done against all the rewrite rules from $\mathcal{R}$. The rewrite rules have the conditions restricted to the form $a = True$ or $a = False$ in order to simplify the certification process [23]. REWRITING rewrites an equality e with rewrite rules from $\mathcal{R}$, lemmas $\mathcal{L}$ consisting of previously proved conjectures, or smaller instances of premises H and conjectures E from the current state, denoted by $(H \cup E)_{\prec e}$. AUGMENTATION replaces one condition of the current conjecture with smaller conditions. The fact that the replaced condition implies the smaller conditions is stated by lemmas. TAUTOLOGY deletes tautologies. The last rule, SUBSUMPTION, deletes the equalities subsumed by either lemmas or premises. The *proof strategy* indicating the application order of the Spike rules is the following: on a given conjecture, TAUTOLOGY is firstly tried, then REWRITING followed by AUGMENTATION, SUBSUMPTION and TOTAL CASE REWRITING. When none of these rules works, GENERATE is finally applied.

The soundness property of the Spike inference system states that whenever a proof is derived then the initial conjectures are inductive consequences of the axioms. The inference system implements the 'Descente Infinie' principle such that whenever a current conjecture from a proof step can be instantiated to a false ground formula, there is another conjecture in the proof that can be instantiated to a smaller false ground formula, under the following conditions: i) the rewrite system $\mathcal{R}$ respects some syntactic criteria to ensure its *coherence*, i.e. an equality and its negation cannot be simultaneously consequences of the axioms, and ii) *completeness*, i.e. each function is defined in any point of the domain. In [19], we have proposed a methodology for a modular checking of the soundness of implicit induction inference systems. The heart of the methodology is a very general inference system consisting only of two rules, AddPremise and Simplify, which has been shown sound, like any system built from instantiations of its rules. In our case, it can be shown that Generate is an instance of AddPremise, and the other rules are instances of Simplify.

4 Case Study: Validation of a Conformance Algorithm for the ABR Protocol

Available Bit Rate (ABR) [9] is one of the protocols for the ATM networks that fit well to manage the data rates between several applications sharing simultaneously a common physical link. What distinguishes ABR is its flexibility: the provider guarantees a minimum rate to the user, but the rate can fluctuate over time: it can increase if resources are available and should decrease if the network is congestioned. The rate management is a complex task for the provider. Firstly, the user is informed by the new rate values, then the user has to adapt the rates of its applications to the new values. Finally, the provider should test that the user rates are *conform* to a value which may vary in time, called *conformity value*, computed by the conformance algorithm running on a device positioned between the provider and the user. The new rate values arrive from the provider to the device by the means of Resource Management (RM-) cells. The rate and the arrival time of RM-cells are stored in a buffer such that the conformity value, at a given time, will depend on the rate values and the arrival time of the stored RM-cells.

There are several conformance algorithms, but here we will mention only two: Acr and Acr1. Both of them are ideal since they assume that the buffer can store an infinite number of RM-cells in order to give optimal conformity control from the user's point of view, i.e. the users are informed a) as soon as possible when the rate for their applications can augment, and b) as late as possible when the rate should decrease. Acr [4] has been standardized by the ATM Forum and considered as a landmark for the other algorithms. Defined later, Acr1 [14] is more efficient since it provides the conformity value quasi-instantaneously. The idea is to schedule into the future the rate of the RM-cell such that most of the computation is done at the time when an RM-cell arrives to the device and not when the conformity value is really needed. More exactly, at a given time t, the conformity value will be the rate of the first cell from the scheduled cell buffer, assuming that all the cells scheduled in the past (w.r.t. t) can be ignored or simply deleted.

Formally, a (RM- or scheduled) cell is a pair of naturals (t, er), where t is the (arrival or scheduled) time of the cell and er is its rate. A buffer is a list of cells $[(t_1, er_1), (t_2, er_2), \ldots, (t_n, er_n), \ldots]$. The RM-cell buffer is *time-decreasing*, i.e. $t_1 \geq t_2 \geq \ldots \geq t_n \geq \ldots$. In [15,17], Acr1 has been shown equivalent to Acr, i.e. for any configuration of the RM-cell buffer and any time t, both algorithms compute the same conformity value. Before proving in the next subsection one of the lemmas from the 'equivalence' proof, we introduce the following functions: i) $time(c)$ - returns the time value of the cell c, ii) $timel(l)$ - gives the time value of the first cell from the buffer l, iii) $le(n_1, n_2)$ - tests if the natural n_1 is less than or equal to the natural n_2, iv) $sortedT(l)$ - returns true if the buffer l is time-decreasing, and v) $insAt(l, t, e)$ - deletes all the cells from the head of the buffer l having time values greater than t, then adds a new cell with the rate e and time t to the head of the buffer.

4.1 The Spike Specification

We will use the sort NAT to represent naturals, $BOOL$ for booleans, OBJ for cells, and $PLAN$ for lists of cells. The constructor symbols for the sorts NAT, $BOOL$, OBJ and $PLAN$, together with their profiles, are: $0 : NAT$, $S : NAT \rightarrow NAT$, $True : BOOL$, $False : BOOL$, $C : NAT\ NAT \rightarrow OBJ$, $Nil : PLAN$, and $Cons : OBJ\ PLAN \rightarrow PLAN$.

The defined functions and their defining axioms are:

$time : OBJ \rightarrow NAT$
$\langle 105 \rangle\ time(C(u_1, u_2)) = u_1$

$timel : PLAN \rightarrow NAT$
$\langle 98 \rangle\ timel(Nil) = 0$
$\langle 99 \rangle\ timel(Cons(u_1, u_2)) = time(u_1)$

$le : NAT\ NAT \rightarrow BOOL$
$\langle 102 \rangle\ le(0, u_1) = True$
$\langle 103 \rangle\ le(S(u_1), 0) = False$
$\langle 104 \rangle\ le(S(u_1), S(u_2)) = le(u_1, u_2)$

$sortedT : PLAN \rightarrow BOOL$
$\langle 107 \rangle\ sortedT(Nil) = True$
$\langle 108 \rangle\ sortedT(Cons(u_1, Nil)) = True$
$\langle 109 \rangle\ le(u_1, u_2) = True \Rightarrow sortedT(Cons(C(u_2, u_3), Cons(C(u_1, u_4), u_5))) = sortedT(Cons(C(u_1, u_4), u_5))$
$\langle 110 \rangle\ le(u_1, u_2) = False \Rightarrow sortedT(Cons(C(u_2, u_3), Cons(C(u_1, u_4), u_5))) = False$

$insAt : PLAN\ NAT\ NAT \rightarrow PLAN$
$\langle 130 \rangle\ insAt(Nil, u_1, u_2) = Cons(C(u_1, u_2), Nil)$
$\langle 131 \rangle\ le(time(u_1), u_2) = True \Rightarrow insAt(Cons(u_1, u_3), u_2, u_4) = Cons(C(u_2, u_4), Cons(u_1, u_3))$
$\langle 132 \rangle\ le(time(u_1), u_2) = False \Rightarrow insAt(Cons(u_1, u_3), u_2, u_4) = insAt(u_3, u_2, u_4)$

The axioms can be oriented from left to right using the *rpo* ordering based on the following precedence over the function symbols: $0 <_F S <_F \mathit{True} <_F \mathit{False} <_F C <_F \mathit{Nil} <_F \mathit{Cons} <_F \mathit{time} <_F \mathit{timel} <_F \mathit{le} <_F \mathit{sortedT} <_F \mathit{insAt}$. For sake of simplicity, the axioms are referred to by unique identifiers that prefix them, like any conditional equality from the next section.

5 An Example of Implicit Induction Proof

The conjecture $\langle 301 \rangle$ $\mathit{sortedT}(u_1) = \mathit{True} \Rightarrow \mathit{sortedT}(\mathit{insAt}(u_1, u_2, u_3)) = \mathit{True}$, denoted by `sorted_insat1` in [17], will be proved with the inference system from Fig. 1 and the axioms from subsection 4.1, using the lemmas

$$\begin{array}{l} \langle 159 \rangle\ \mathit{le}(u_1, u_2) = \mathit{False}\ \wedge\ \mathit{le}(u_1, u_2) = \mathit{True} \Rightarrow \\ \langle 196 \rangle\ \mathit{sortedT}(\mathit{Cons}(u_1, u_2)) = \mathit{True} \Rightarrow \mathit{sortedT}(u_2) = \mathit{True} \\ \langle 226 \rangle\ \mathit{sortedT}(\mathit{Cons}(u_1, u_2)) = \mathit{True} \Rightarrow \mathit{le}(\mathit{timel}(u_2), \mathit{time}(u_1)) = \mathit{True} \\ \langle 268 \rangle\ \mathit{sortedT}(u_1) = \mathit{True}\ \wedge\ \mathit{le}(\mathit{timel}(u_1), u_2) = \mathit{True} \Rightarrow \\ \hfill \mathit{sortedT}(\mathit{Cons}(C(u_2, u_3), u_1)) = \mathit{True} \end{array}$$

The proof starts with the state $(\{\langle 301 \rangle\}, \emptyset)$. The conjecture $\langle 301 \rangle$ cannot be simplified or deleted, so the only applicable rule is GENERATE. Firstly, the subterm $\mathit{insAt}(u_1, u_2, u_3)$ is unified with the lhs of the axioms $\langle 130 \rangle, \langle 131 \rangle$ and $\langle 132 \rangle$ using the test substitutions $\{u_1 \mapsto \mathit{Nil}\}$ (for $\langle 130 \rangle$) and $\{u_1 \mapsto \mathit{Cons}(u_5, u_6)\}$ (for $\langle 131 \rangle$ and $\langle 132 \rangle$). Since u_5 is an OBJ variable, it can be expanded to $C(u_8, u_9)$. Then, the resulted instances of $\langle 301 \rangle$ are rewritten with the corresponding axioms, respectively, to

$$\begin{array}{l} \langle 328 \rangle\ \mathit{sortedT}(\mathit{Nil}) = \mathit{True} \Rightarrow \mathit{sortedT}(\mathit{Cons}(C(u_2, u_3), \mathit{Nil})) = \mathit{True} \\ \langle 334 \rangle\ \mathit{sortedT}(\mathit{Cons}(C(u_8, u_9), u_6)) = \mathit{True}\ \wedge\ \mathit{le}(\mathit{time}(C(u_8, u_9)), u_2) = \\ \hfill \mathit{True} \Rightarrow \mathit{sortedT}(\mathit{Cons}(C(u_2, u_3), \mathit{Cons}(C(u_8, u_9), u_6))) = \mathit{True} \\ \langle 340 \rangle\ \mathit{sortedT}(\mathit{Cons}(C(u_8, u_9), u_6)) = \mathit{True}\ \wedge\ \mathit{le}(\mathit{time}(C(u_8, u_9)), u_2) = \\ \hfill \mathit{False} \Rightarrow \mathit{sortedT}(\mathit{insAt}(u_6, u_2, u_3)) = \mathit{True} \end{array}$$

The next proof state is $(\{\langle 328 \rangle, \langle 334 \rangle, \langle 340 \rangle\}, \{\langle 301 \rangle\})$. Each of the conjectures are rewritten with REWRITING as follows: firstly, the term $\mathit{sortedT}(\mathit{Cons}(C(u_2, u_3), \mathit{Nil}))$ from $\langle 328 \rangle$ to True by the axiom $\langle 108 \rangle$, then $\mathit{time}(C(u_8, u_9))$ from $\langle 334 \rangle$ and $\langle 340 \rangle$ to u_8 by $\langle 105 \rangle$. The new conjectures are:

$$\begin{array}{l} \langle 343 \rangle\ \mathit{sortedT}(\mathit{Nil}) = \mathit{True} \Rightarrow \mathit{True} = \mathit{True} \\ \langle 346 \rangle\ \mathit{sortedT}(\mathit{Cons}(C(u_8, u_9), u_6)) = \mathit{True}\ \wedge\ \mathit{le}(u_8, u_2) = \mathit{True} \Rightarrow \\ \hfill \mathit{sortedT}(\mathit{Cons}(C(u_2, u_3), \mathit{Cons}(C(u_8, u_9), u_6))) = \mathit{True} \\ \langle 349 \rangle\ \mathit{sortedT}(\mathit{Cons}(C(u_8, u_9), u_6)) = \mathit{True}\ \wedge\ \mathit{le}(u_8, u_2) = \mathit{False} \Rightarrow \\ \hfill \mathit{sortedT}(\mathit{insAt}(u_6, u_2, u_3)) = \mathit{True} \end{array}$$

The conjecture $\langle 343 \rangle$ is deleted by TAUTOLOGY. Notice that each of the remaining conjectures are governed by the condition $\mathit{sortedT}(\mathit{Cons}(C(u_8, u_9), u_6)) =$

True, so Augment can be applied with the lemmas ⟨196⟩ and ⟨226⟩, to yield, respectively

⟨371⟩ $le(u_8, u_2) = True \wedge sortedT(u_6) = True \wedge le(timel(u_6), time(C(u_8, u_9))) = True \Rightarrow sortedT(Cons(C(u_2, u_3), Cons(C(u_8, u_9), u_6))) = True$
⟨396⟩ $le(u_8, u_2) = False \wedge sortedT(u_6) = True \wedge le(timel(u_6), time(C(u_8, u_9))) = True \Rightarrow sortedT(insAt(u_6, u_2, u_3)) = True$

The term $time(C(u_8, u_9))$ from the newly added conditions is again rewritten to u_9 by ⟨105⟩. Now, the new set of conjectures consists of:

⟨374⟩ $le(u_8, u_2) = True \wedge sortedT(u_6) = True \wedge le(timel(u_6), u_8) = True \Rightarrow sortedT(Cons(C(u_2, u_3), Cons(C(u_8, u_9), u_6))) = True$
⟨399⟩ $le(u_8, u_2) = False \wedge sortedT(u_6) = True \wedge le(timel(u_6), u_8) = True \Rightarrow sortedT(insAt(u_6, u_2, u_3)) = True$

The conjecture ⟨399⟩ is deleted by Subsumption with the premise ⟨301⟩ using the substitution $\{u_1 \mapsto u_6\}$. Total Case Rewriting is applied to the remaining conjecture ⟨374⟩ on the term $sortedT(Cons(C(u_2, u_3), Cons(C(u_8, u_9), u_6)))$ using the axioms ⟨109⟩ and ⟨110⟩, to give

⟨441⟩ $le(u_8, u_2) = True \wedge sortedT(u_6) = True \wedge le(timel(u_6), u_8) = True \wedge le(u_8, u_2) = True \Rightarrow sortedT(Cons(C(u_8, u_9), u_6)) = True$
⟨445⟩ $le(u_8, u_2) = True \wedge sortedT(u_6) = True \wedge le(timel(u_6), u_8) = True \wedge le(u_8, u_2) = False \Rightarrow False = True$

⟨441⟩ is subsumed by the lemma ⟨268⟩ using the substitution $\{u_1 \mapsto u_6; u_2 \mapsto u_8; u_3 \mapsto u_9\}$. Similarly, ⟨445⟩ is subsumed by ⟨159⟩ with $\{u_1 \mapsto u_8\}$.

The proof of `sorted_insat1` ends successfully into the state $(\emptyset, \{\langle 301\rangle\})$.

6 Certifying Implicit Induction Proofs

The implicit induction proof from the previous section can be automatically checked for soundness using the certified reasoning environment provided by Coq [25]. The certification approach is based on the Noetherian induction principle from Section 3.

The Spike specification is firstly translated into a Coq script by the Spike prover before dealing with the proof part. The sorts and the function definitions are manually translated by the user, the following code being inlined in the Spike specification. Notice that only the types **OBJ** and **PLAN** have been defined by the user, **nat** and **bool** being predefined data structures in Coq.

```
Inductive OBJ:Set := C: nat → nat → OBJ.
Inductive PLAN:Set := Nil | Cons: OBJ → PLAN → PLAN.
```

```
Fixpoint le (m n:nat): bool :=
match m,n with
| 0,_⇒ true
| S_, 0 ⇒ false
| S x, S y ⇒ le x y
end.

Definition time (o:OBJ) : nat
:=
match o with
| C t e ⇒ t
end.

Fixpoint insAt (p:PLAN) (t e:nat) :
PLAN :=
match p with
| Nil ⇒ Cons(C t e) Nil
| Cons o pg ⇒
  match le (time o) t with
  | true ⇒ Cons (C t e) (Cons o pg)
  | false ⇒ insAt pg t e
  end
end.

Definition timel (o:PLAN): nat :=
match o with
| Nil ⇒ 0
| Cons o p ⇒ time o
end.

Fixpoint sortedT (p:PLAN) : bool
:=
match p with
| Nil ⇒ true
| Cons (C t1 e1) p1 ⇒
  match p1
  with
  | Nil ⇒ true
  | Cons (C t2 e2) p2 ⇒
     match le t2 t1 with
        | true ⇒ sortedT p1
        | false ⇒ false
     end
  end
end.
```

The Spike induction ordering is syntactic and exploits the tree representation of terms. In Coq, the syntactic representation is made explicit by abstracting the Coq terms into terms built from a term algebra provided by COCCINELLE [6]. Some information has to be fed in order to create operational term algebras, for example the set of function symbols. This information can be automatically inferred, as the rest of the transformations: each function symbol f is translated into `id_f`.

```
Inductive symb : Set :=
  | id_O | id_S | id_true | id_false | id_C | id_Nil | id_Cons | id_le
  | id_time | id_timel | id_sortedT | id_insAt.
```

The type of abstracted terms is defined as `Inductive` **term** : `Set` := | `Var` : `variable` → **term** | `Term` : `symb` → `list` **term** → **term**. Any Coq term can be abstracted by *model* functions. They are defined by the user for each type.

```
Fixpoint model_nat (n:nat) : term
:=
match n with
| 0 ⇒ Term id_0 nil
| S n' ⇒ Term id_S ((model_nat n')::nil)
end.

Definition model_bool (b:bool) :
term :=
match b with
| true ⇒ Term id_true nil
| false ⇒ Term id_false nil
end.
```

```
Definition model_OBJ (o:OBJ) : term :=
match o with
| C x y ⇒ Term id_C
  ((model_nat x)::(model_nat y)::nil)
end.
```

```
Fixpoint model_PLAN (p:PLAN) : term :=
match p with
| Nil ⇒ Term id_Nil nil
| Cons o p ⇒ Term id_Cons ((
model_OBJ o)::(model_PLAN p)::nil)
end.
```

A Spike variable x of sort s can be transformed into the Coq variable x of type s', where s' is the Coq type corresponding to s. A non-variable term $f(t_1, \ldots, t_n)$ can be recursively transformed into the Coq term $(f\ t'_1 \ldots t'_n)$, where $t'_1, \ldots, t'_n$ are the transformations of $t_1, \ldots t_n$, respectively. The equality $e_1 \wedge \ldots e_n \Rightarrow e$ can be automatically translated into the Coq formula $\forall \overline{x}, e'_1 \rightarrow \ldots \rightarrow e'_n \rightarrow e'$, where $\overline{x}$ is the vector of all variables from the equality and $e'_1, \ldots, e'_n, e'$ are the equalities issued from the Coq transformations on the lhs and rhs of $e_1, \ldots, e_n, e$, respectively.

The proof part of the translation can be generated completely automatically. In order to perform induction reasoning, we explicitly pair any Coq formula F produced by the above transformations with a comparison weight $W(F)$ such that a formula F_1 is smaller than a formula F_2 if $W(F_1) \prec W(F_2)$, where $\prec$ is a well-founded and stable under substitutions ordering over weights. In our case, the weight of a formula is given by the list of the COCCINELLE terms abstracting the terms the formula is built from, as shown in [23]. The relation between a Coq formula and its weight should be stable under substitutions. This property is achieved if the common variables are factorized using functionals of the form (fun $\overline{x} \Rightarrow (F, W)$), where F is the Coq formula translating the Spike conjecture, W its weight, and $\overline{x}$ the vector of the common variables. The functional's type is associated to a proof and has to be general enough to represent all the conjectures from the proof. The type for the functionals from the proof of sorted_insat1 and labelled as ⟨301⟩ in the previous section, an example of functional corresponding to ⟨301⟩, and all the functionals are:

```
Definition type_LF_301 := PLAN → nat → nat → nat → nat → (Prop × (list term)).
```

```
Definition F_301 : type_LF_301:= (fun u1 u2 u3 _ _ ⇒ ((sortedT u1) = true → (sortedT (insAt u1 u2 u3)) = true, (Term id_sortedT ((model_PLAN u1)::nil))::(Term id_true nil)::(Term id_sortedT ((Term id_insAt ((model_PLAN u1):: (model_nat u2):: (model_nat u3)::nil))::nil))::(Term id_true nil)::nil)).
```

```
Definition LF_301 := [F_301, F_328, F_343, F_334, F_340, F_346, F_371, F_349, F_396, F_374, F_441, F_445].
```

The Noetherian induction principle is represented as a Coq section, parameterized by four variables and two hypotheses. The variables must be specified and the hypotheses proved at each application of the theorem wf_subset. In our case, the variable T will correspond to (Prop × (**list term**)), R to an ordering on pairs, and *wf_R* to the lemma stating the well-foundedness of the ordering.

```
Section wf_subset.

  Variable T : Type.
  Variable R : T → T → Prop.
  Hypothesis wf_R: well_founded R.
  Variable S : T → Prop.
  Variable P : T → Prop.
  Hypothesis S_acc : ∀ x, S x →
 (∀ y, S y → R y x → P y) → P x.

  Theorem wf_subset: ∀ x, S x → P x.
  Proof.
  intro z; elim (wf_R z).
  intros x H1x H2x H3x.
  apply S_acc;
  intros y H1y H2y; apply H2x;
   trivial.
  Qed.
End wf_subset.
```

The main lemma, denoted by main_301, states that all formulas from the functionals of LF_301 are valid, assuming that for each formula one can use any smaller formula as induction hypothesis. The functions fst and snd return the first and the second projections of a pair, respectively, and less is the ordering over weights, presented in [23] and shown well-founded and stable under substitutions.

```
Lemma main_301 : ∀ F, In F LF_301 → ∀ u1, ∀ u2, ∀ u3, ∀ u4, ∀ u5, (∀ F', In F'
LF_301 → ∀ e1, ∀ e2, ∀ e3, ∀ e4, ∀ e5, less (snd (F' e1 e2 e3 e4 e5)) (snd (F
u1 u2 u3 u4 u5)) → fst (F' e1 e2 e3 e4 e5)) → fst (F u1 u2 u3 u4 u5).
```

The heart of its proof consists of a case analysis on the functionals from LF_301. As in [23], the associated formulas are proved using one-to-one translations of the corresponding Spike inference steps from the implicit induction proof presented in Section 5. We will detail only the translations for some applications of AUGMENTATION and SUBSUMPTION rules. Similar translations for the applications of the other rules can be found in [23].

We will firstly consider the application of AUGMENTATION on ⟨346⟩. In the current state of the proof, F is instantiated with F_346, the last condition of the lemma is denoted by *Hind*, and the variables from the conclusion renamed to correspond to the conjecture ⟨346⟩ from the Spike proof. ⟨371⟩ is the result of the augmentation operation. In the first step, we instantiate F' from *Hind* with F_371 and denote it by *H*. The variables of F_371 are renamed in *HFabs0* to correspond to ⟨371⟩ from the Spike proof. This instance can be used in the proof as long as it is smaller than ⟨346⟩, according to *H*. The comparison between their weights is performed automatically by the user-defined strategy *solve_rpo_mul*, once the weights have been normalized by *rewrite_model*. The strategy *trivial_in* tests that F_371 is indeed at the sixth position (starting from 0) in LF_301. In the last step, the instances of the lemmas ⟨226⟩ and ⟨196⟩ are added as conditions and auto finally establishes the logical equivalence between ⟨371⟩ and ⟨346⟩.

```
assert (H := Hind F_371). assert (HFabs0: fst (F_371 u6 u2 u3 u8 u9)).
apply H. trivial_in 6. rewrite_model. abstract solve_rpo_mul.

specialize true_226 with (u2 := u6) (u1 := (C u8 u9)).
specialize true_196 with (u2 := u6) (u1 := (C u8 u9)). auto.
```

The translation of the SUBSUMPTION application on ⟨445⟩ with lemma ⟨159⟩ is simpler since it does not require weight comparisons. The subsuming instance of ⟨159⟩ is contradicted and `auto` tests afterwards if it is a sub-clause of ⟨445⟩. The subsumption test performed by Spike assumes that the equality operator is symmetric, which is not the case for Coq. For example, `auto` fails if we try to subsume $a = b$ with $b = a$. One solution is to apply `symmetry` before `auto`.

specialize true_159 `with` $(u1 := u8)$ $(u2 := u2)$. `intro` L. *contradict* L.
(`auto` || `symmetry`; `auto`).

Once the main proof is finished, we define the set S_301 of all instances of the functionals from LF_301, then prove the theorem all_true_301 stating that the associated formulas are true. A critical step of this proof is the use of wf_subset with the variable S instantiated by S_301. Finally, we prove ⟨301⟩ in true_301.

`Definition` S_301 := `fun` $f \Rightarrow \exists\ F$, In F LF_301 $\wedge \exists\ e1, \exists\ e2, \exists\ e3, \exists\ e4, \exists\ e5$, $f = F\ e1\ e2\ e3\ e4\ e5$.

`Theorem` all_true_301: $\forall\ F$, In F LF_301 $\rightarrow \forall\ u1$: **PLAN**, $\forall\ u2$: **nat**, $\forall\ u3$: **nat**, $\forall\ u4$: **nat**, $\forall\ u5$: **nat**, fst $(F\ u1\ u2\ u3\ u4\ u5)$.

`Theorem` true_301: $\forall$ ($u1$: **PLAN**) ($u2$: **nat**) ($u3$: **nat**), (sortedT $u1$) = true $\rightarrow$ (sortedT (insAt $u1\ u2\ u3$)) = true.

The same inference system and proof strategy involved in the proof of `sorted_insat1` have been applied to prove other Spike conjectures involved in the ABR proof from [17]. In Table 1, we give some statistics about the proofs that have been automatically certified by Coq. We firstly list the name of the conjectures, as denoted in [17]. Then, for the proof of each conjecture, we show how many times each of the Spike inference rules has been applied, the number of needed lemmas, the size of the list of functionals for the main proof, and the global time of the certification process (lemmas + conjecture). For each conjecture, the Spike proof and its Coq translation lasted less than one second. All tests have been done on a MacBook Air featuring a 2.13 GHz Intel Core 2 Duo processor and 4 GB RAM.

6.1 Improvements

We have also tested the toy examples from [23] and found out that the certification time has been considerably reduced. For example, the certification process concerning the proofs about the validity of the sorting algorithm is now 20 times faster. We list the main improvements that allowed us to achieve this performance.

Weaker Conditions for Weight Comparisons. An important part of the certification time concerning the examples from [23] has been spent in doing arithmetic reasoning, mainly using the expensive `omega` tactic. After fruitful

Table 1. Some statistics about the ABR proofs

# Name	Taut.	Rew.	Aug.	Sub.	Case Rew.	Gen.	Lemmas	LF	time (s)
1. firstat_timeat	2	12	0	4	2	2	1	17	3.06
2. firstat_progat	2	13	0	4	2	2	1	18	3.14
3. sorted_sorted	2	1	0	0	0	1	0	6	1.58
4. sorted_insat1	7	15	2	5	1	5	4	12	6.76
5. sorted_insin2	7	22	2	5	1	5	4	20	7.54
6. sorted_e_two	2	1	0	0	0	1	0	6	1.57
7. member_t_insin	1	12	0	15	11	7	2	52	12.89
8. member_t_insat	1	6	0	8	7	4	2	24	5.99
9. member_firstat	2	12	0	8	6	3	2	29	6.39
10. timel_insat	3	7	0	0	0	1	0	11	2.02
11. erl_insin	3	8	0	0	0	1	0	12	2.17
12. erl_insat	3	7	0	0	0	1	0	11	2.02
13. erl_prog	9	29	0	0	0	3	2	18	8.90
14. time_progat_er	2	11	0	2	1	2	1	15	2.50
15. timeat_tcrt	4	6	0	1	2	1	0	16	3.40
16. timel_timeat_max	5	23	2	3	2	3	3	33	7.63
17. null_listat	1	7	0	3	1	2	1	11	2.42
18. null_listat1	2	0	0	0	0	1	0	4	1.38
19. cons_insat	0	1	0	1	0	1	0	4	1.46
20. cons_listat	2	0	0	0	0	1	0	4	1.39
21. progat_timel_erl	7	27	2	3	4	2	3	33	7.50
22. progat_insat	11	85	1	23	26	4	4	134	57.09
23. progat_insat1	8	30	1	7	7	4	3	42	15.31
24. timel_listupto	3	4	0	0	0	1	0	8	1.81
25. sorted_listupto	10	21	3	4	2	6	4	26	9.79
26. time_listat	3	11	0	3	3	2	1	22	5.09
27. sorted_cons_listat	9	24	1	5	9	4	4	42	15.66
28. null_wind2	1	0	0	1	1	2	1	8	3.66
29. timel_insin1	1	9	0	2	1	2	1	12	2.8
30. null_listupto1	1	0	0	0	0	1	0	4	1.38
31. erl_cons	3	8	0	0	0	1	0	12	2.03
32. no_time	2	19	0	6	4	2	2	29	7.56
33. final	4	4	0	8	5	2	3	13	3.90

discussions with the developpers of COCCINELLE, we changed less to avoid the size comparisons for the terms defining the compared weights. The size of the Coq scripts was also dramatically reduced.

Functionals Labeling. We discovered that the membership test of a functional in a list of functionals is costly since it requires complex unification operations. We have avoided this problem by explicitly labeling the functionals, as for the definition of F_301. In this way, the membership test is performed on labels. Lots of comparisons have been avoided by pointing out the exact position of the label in the list.

Other improvements include a better readability of the Coq scripts by the definition of tactics using Ltac [8], the redefinition of the term algebra, and the separation between the specification (static) and proof (dynamic) parts.

Parallelisation of the Certification Process. The one-to-one translations from the main lemma can be proved independently. We have tested for parallelisation the proof of `progat_insat`, the most complex ABR proof from Table 1, as follows. Firstly, we have manually split the list of 134 functionals in n adjacent partitions such that the formulas from each partition i ($i \in [1..n]$) are proved in a process p_i. Another process p_c will prove the main lemma using the previous results. Ideally, under the assumption that there is only one process allotted per processor and the generation of each process is done automatically and instantaneously, the total certification time is $max\{t(p_i) \mid i \in [1..n]\} + t(p_c)$, where $t(p)$ is the execution time of the process p. We have experimented with different split strategies and found out cases where the certification process of the main lemma of `progat_insat` is 5 times faster.

7 Conclusions and Future Work

We have proposed a new methodology for certifying implicit induction proofs. The underlying induction principle is an instance of Noetherian induction based on an ordering over equalities which represents a contrapositive version of the 'Descente Infinie' induction principle from [23]. The soundness arguments are now fully constructive since there is no need to use additional hypotheses like the 'excluded middle' axiom required in the proof-by-contradiction argumentation inherent to the 'Descente Infinie' proofs.

We have also proposed improvements to automatize and speed-up the certification process of implicit induction proofs in order to deal with industrial-size applications. Concerning the ABR proofs, they have been performed by a restricted version of the Spike inference system and their certification was established within seconds for most of them. Compared to the Spike proofs from [17], the arithmetic reasoning was simulated with lemmas instead of using the complex decision procedures integrated into Spike [1]. Generating *complete* translations, i.e. valid Coq scripts from any valid Spike proof steps, is challenging even for simple inference rules. For example, the symmetry problem related to SUBSUMPTION was fixed only for the conclusion and not for the conditions of subsuming equalities. Hopefully, the proof of the main lemma is modular, so the failure of any one-to-one translation does not affect the rest of the proof.

In the near future, we plan to devise complete translations of the Spike inference rules and automatize the parallelisation process. A challenge will be to translate and certify the proofs involved in the JavaCard platform validation [3].

Acknowledgements. We thank the ProVal team from INRIA Saclay - Île-de-France research center for helpful suggestions about the efficient use of COCCINELLE. Special thanks are addressed to the reviewers for their useful remarks.

References

1. Armando, A., Rusinowitch, M., Stratulat, S.: Incorporating decision procedures in implicit induction. J. Symb. Comput. 34(4), 241–258 (2002)
2. Baader, F., Nipkow, T.: Term Rewriting and All That. Cambridge University Press (1998)
3. Barthe, G., Stratulat, S.: Validation of the JavaCard Platform with Implicit Induction Techniques. In: Nieuwenhuis, R. (ed.) RTA 2003. LNCS, vol. 2706, pp. 337–351. Springer, Heidelberg (2003)
4. Berger, A., Bonomi, F., Fendick, K.: Proposed TM baseline text on an ABR conformance definition. Technical Report 95-0212R1, ATM Forum Traffic Management Group (1995)
5. Bouhoula, A., Kounalis, E., Rusinowitch, M.: Automated mathematical induction. Journal of Logic and Computation 5(5), 631–668 (1995)
6. Contejean, E., Courtieu, P., Forest, J., Pons, O., Urbain, X.: Certification of Automated Termination Proofs. In: Konev, B., Wolter, F. (eds.) FroCos 2007. LNCS (LNAI), vol. 4720, pp. 148–162. Springer, Heidelberg (2007)
7. Courant, J.: Proof reconstruction. Research Report RR96-26, LIP (1996); Preliminary version
8. Delahaye, D.: A Tactic Language for the System Coq. In: Parigot, M., Voronkov, A. (eds.) LPAR 2000. LNCS (LNAI), vol. 1955, pp. 85–95. Springer, Heidelberg (2000)
9. ITU-T. Traffic control and congestion control in B ISDN. Recommandation I.371.1 (1997)
10. Kaliszyk, C.: Validation des preuves par récurrence implicite avec des outils basés sur le calcul des constructions inductives. Master's thesis, Université Paul Verlaine - Metz (2005)
11. Klein, G., Andronick, J., Elphinstone, K., Heiser, G., Cock, D., Derrin, P., Elkaduwe, D., Engelhardt, K., Kolanski, R., Norrish, M., Sewell, T., Tuch, H., Winwood, S.: seL4: Formal verification of an operating-system kernel. Communications of the ACM 53(6), 107–115 (2010)
12. Leroy, X., Doligez, D., Frisch, A., Garrigue, J., Rémy, D., Vouillon, J.: The Objective Caml system - release 3.12. Documentation and user's manual. INRIA
13. Nahon, F., Kirchner, C., Kirchner, H., Brauner, P.: Inductive proof search modulo. Annals of Mathematics and Artificial Intelligence 55(1–2), 123–154 (2009)
14. Rabadan, C., Klay, F.: Un nouvel algorithme de contrôle de conformité pour la capacité de transfert 'Available Bit Rate'. Technical Report NT/CNET/5476, CNET (1997)
15. Rusinowitch, M., Stratulat, S., Klay, F.: Mechanical verification of a generic incremental ABR conformance algorithm. Technical Report 3794, INRIA (1999)
16. Rusinowitch, M., Stratulat, S., Klay, F.: Mechanical Verification of an Ideal Incremental ABR Conformance Algorithm. In: Emerson, E.A., Sistla, A.P. (eds.) CAV 2000. LNCS, vol. 1855, pp. 344–357. Springer, Heidelberg (2000)
17. Rusinowitch, M., Stratulat, S., Klay, F.: Mechanical verification of an ideal incremental ABR conformance algorithm. J. Autom. Reasoning 30(2), 53–177 (2003)
18. Shankar, N., Owre, S., Rushby, J.M., Stringer-Calvert, D.W.J.: PVS prover guide - version 2.4. SRI International (November 2001)
19. Stratulat, S.: A general framework to build contextual cover set induction provers. J. Symb. Comput. 32(4), 403–445 (2001)

20. Stratulat, S.: Automatic 'Descente Infinie' Induction Reasoning. In: Beckert, B. (ed.) TABLEAUX 2005. LNCS (LNAI), vol. 3702, pp. 262–276. Springer, Heidelberg (2005)
21. Stratulat, S.: 'Descente Infinie' induction-based saturation procedures. In: SYNASC 2007: Proceedings of the Ninth International Symposium on Symbolic and Numeric Algorithms for Scientific Computing, Washington, DC, USA, pp. 17–24. IEEE Computer Society (2007)
22. Stratulat, S.: Combining Rewriting with Noetherian Induction to Reason on Non-Orientable Equalities. In: Voronkov, A. (ed.) RTA 2008. LNCS, vol. 5117, pp. 351–365. Springer, Heidelberg (2008)
23. Stratulat, S.: Integrating Implicit Induction Proofs into Certified Proof Environments. In: Méry, D., Merz, S. (eds.) IFM 2010. LNCS, vol. 6396, pp. 320–335. Springer, Heidelberg (2010)
24. Stratulat, S., Demange, V.: Validating implicit induction proofs using certified proof environments. In: Poster Session of 2010 Grande Region Security and Reliability Day, Saarbrucken (March 2010)
25. The Coq Development Team. The Coq reference manual - version 8.2 (2009), `http://coq.inria.fr/doc`

A Proposal for Broad Spectrum Proof Certificates

Dale Miller

INRIA & LIX, École Polytechnique

Abstract. Recent developments in the theory of *focused proof systems* provide flexible means for structuring proofs within the sequent calculus. This structuring is organized around the construction of "macro" level inference rules based on the "micro" inference rules which introduce single logical connectives. After presenting focused proof systems for first-order classical logics (one with and one without fixed points and equality) we illustrate several examples of proof certificates formats that are derived naturally from the structure of such focused proof systems. In principle, a proof certificate contains two parts: the first part describes how macro rules are defined in terms of micro rules and the second part describes a particular proof object using the macro rules. The first part, which is based on the vocabulary of focused proof systems, describes a collection of macro rules that can be used to directly present the structure of proof evidence captured by a particular class of computational logic systems. While such proof certificates can capture a wide variety of proof structures, a proof checker can remain simple since it must only understand the micro-rules and the discipline of focusing. Since proofs and proof certificates are often likely to be large, there must be some flexibility in allowing proof certificates to elide subproofs: as a result, proof checkers will necessarily be required to perform (bounded) proof search in order to reconstruct missing subproofs. Thus, proof checkers will need to do unification and restricted backtracking search.

1 Introduction

Most computational logic systems work in isolation in the sense that they are unable to communication among each other documents that encode formal proofs and that they can check and trust. We propose a framework for designing such documents which we will call *proof certificates*. Being based on foundational aspects of proof theory, these proof certificates hold the promise of working as a common communications medium for a broad spectrum of computational logic systems. Since formal proofs are often large, there can be significant time and space costs in producing, communicating, and checking proof certificate. A central aspect of the framework described here is that it provides for flexible trade-offs between these computational resources. In particular, proof certificates can be made smaller by removing subproofs: in that case, the proof checker must do (bounded) proof search to reconstruct elided proofs.

J.-P. Jouannaud and Z. Shao (Eds.): CPP 2011, LNCS 7086, pp. 54–69, 2011.

After presenting the basics of both sequent calculus and focused sequent calculus for first-order classical logic, we present several examples of proof certificates including those based on matrix and non-matrix (e.g., resolution) formats. We then strengthen first-order logic to include both fixed points and equality: these extensions allow focused proof systems to immediately capture (non-deterministic) computation as well as some model-checking primitives. Additional proof certificates are then possible with these extensions to logic. We then conclude with a brief discussion of related and future work.

2 Proof Theory and Proof Certificates

We shall assume that the reader has some familiarity with the sequent calculus. Here we recall some basic definitions and concepts.

2.1 The Basics of Sequent Calculus

Sequents are a pair $\Gamma \vdash \Delta$ of two (possibly empty) collections of formulas. For Gentzen, these collections were lists but we shall assume that these collections are multisets. Such sequents are also called *two-sided* sequents: formulas on the left-hand-side (in Γ) are viewed as assumptions and formulas on the right-hand-side (in Δ) are viewed as possible conclusions: thus, an informal reading of the judgment described by the sequent $\Gamma \vdash \Delta$ is "if all the formulas in Γ are true then some formula in Δ is true."

Sequent calculus proof systems for classical, intuitionistic, and linear logics come with inference rules in which a sequent is the conclusion and zero or more sequents are premises. We break these rules down into three classes of rules. There are two kinds of *structural rules*:

$$\text{Contraction: } \frac{\Gamma, B, B \vdash \Delta}{\Gamma, B \vdash \Delta} \quad \frac{\Gamma \vdash \Delta, B, B}{\Gamma \vdash \Delta, B} \qquad \text{Weakening: } \frac{\Gamma \vdash \Delta}{\Gamma, B \vdash \Delta} \quad \frac{\Gamma \vdash \Delta}{\Gamma \vdash \Delta, B}$$

The *identity rules* are also just two.

$$\frac{}{B \vdash B}\ \textit{Initial} \qquad\qquad \frac{\Gamma \vdash \Delta, B \qquad \Gamma', B \vdash \Delta'}{\Gamma, \Gamma' \vdash \Delta, \Delta'}\ \textit{Cut}$$

The meta-theory of most sequent calculus presentations of logic includes results that say that most instances of these identity rules are, in fact, not necessary. The cut-elimination theorem states that removing the cut-rule does not change the set of provable sequents. Furthermore, the initial rule can usually be eliminated for all cases except when B is an *atomic* formula, that is, a formula in which the top-level constant is a non-logical (predicate) symbol.

The third and final collection of inference rules are the *introduction rules* which describe the role of the logical connectives in proof. In two-sided sequent calculus proofs, these are usually organized as right and left introduction rules for the same connective. For example, here are two pairs of introduction rules for two connectives.

$$\frac{\Gamma, B, B' \vdash \Delta}{\Gamma, B \wedge B' \vdash \Delta} \qquad \frac{\Gamma \vdash \Delta, B \quad \Gamma' \vdash \Delta', B'}{\Gamma, \Gamma' \vdash \Delta, \Delta', B \wedge B'} \qquad \frac{\Gamma, B[t/x] \vdash \Delta}{\Gamma, \forall x B \vdash \Delta} \qquad \frac{\Gamma \vdash \Delta, B[y/x]}{\Gamma \vdash \Delta, \forall x B}$$

The right-introduction rule for $\forall$ has the proviso that the *eigenvariable* y does not have a free occurrence in any formula in the sequent in the conclusion of the rule. Notice that in both of these sets of rules, there is exactly one new occurrence of a logical connective in the conclusion when compared to the premise(s). Some introduction rules are invertible: that is, if their conclusion is provable then all their premises are provable. Of the four introduction rules above, the left introduction rule for $\wedge$ and the right introduction rule for $\forall$ are invertible: the other two introduction rules are not necessarily invertible.

When presenting a sequent calculus proof system for a specific logic, one usually presents the introduction rules for the logical connectives of the logic and usually accepts both identity inference rules (initial and cut). The structural rules are, however, seldom adopted without restriction. For example, intuitionistic logic is a two-sided sequent calculus in which contraction on the right is not allowed. Multiplicative-additive linear logic (MALL) admits neither weakening nor contraction and full linear logic allows those structural rules only for specially marked formulas (formulas marked with the so-called *exponentials* ! and ?). Classical logic, however, generally admits these structural rules unrestricted.

2.2 Encoding Computation with the Sequent Calculus

The sequent calculus can be used to encode computation as follows. A sequent, as a collection of formulas, can be used to encode the state of a computation. As one attempts to build a (cut-free) proof of a given initial sequent, new sequents appear. The dynamics of computation can be encoded by the changes that take place as one moves from conclusion sequent to premise sequents. The cut rule and the cut-elimination theorem are generally used not as part of computation but as a means to reason about computation. This *proof search* approach to specification has been used to formalize the operational semantics of logic programming [16]. Notice that this approach to encoding computation differs significantly from the approach inspired by the "Curry-Howard correspondence" in which natural deduction proofs can be seen as (functional) programs, normalization represent the process of computing, and normal forms represent values [14]. As is argued later in this paper and in [15], the proof search approach to specifying computation is natural for capturing proof checking and proof reconstruction.

If we try to take the construction of proofs literally as a model for performing computation, one is immediately struck by the inappropriateness of sequent calculus for this task: there are just too many ways to build proofs and most of them differ in truly inconsequential ways. While permuting the application of inference rules may yield proofs of different sequents, quite often such permutations yield different proofs of the same sequent. One would wish to have a much tighter correspondence between the application of an inference rule and sometime that might appear as an interesting "action" within a computation or "high-level" inference rule. Such a correspondence is possible but it requires adding more structure to the sequent calculus.

3 Focused Proof Systems

A normal form of sequent calculus that was designed to link steps in computation with steps in deduction can be found in the work on *uniform proofs* and *backchaining* [16] that was used to provide a proof theoretic foundation for logic programming. Andreoli generalized that work to provide a *focused proof system* [1] that allows one to richly restrict and organize the sequent calculus for linear logic. We provide here a high-level outline of the key ideas behind focused proof systems in the context of classical logic.

Focused proofs are divided into two, alternating phases. The first phase groups together all invertible inference rules. The second phase starts by selecting a formula on which to "focus": the inference rule that is applied to this formula is not necessarily invertible. Furthermore, the (reverse) application of that introduction rule will generate one or more (premise) sequents containing zero or more subformulas of the focus formula. If any of those subformulas require a non-invertible inference rule, then this phase continues with that subformula as the new focus. This second phase, also called the *positive* phase, ends when either the proof ends with an instance of the initial rule or when the focus becomes one needing an invertible inference rule. Certain "structural" rules are used to recognize the end of a phase or the switch from one phase to another.

3.1 LKF: A Focused Proof System for Classical Logic

To illustrate these general comments about focusing proof system more concretely, we now present the LKF proof system for first-order classical logic. We shall adopt a presentation of first-order classical logic in which negations are applied only to atomic formulas (*i.e.*, negation normal form) and where the propositional connectives t, f, $\wedge$, and $\vee$ are replaced by two "polarized" versions: t^-, t^+, f^-, f^+, $\wedge^-$, $\wedge^+$, $\vee^-$, $\vee^+$. To complete the definition of polarized formula, we assume that the atomic formulas are assigned positive or negative polarity following some arbitrary and fixed rule. A formula is *negative* if it is a negative atom, the negation of a positive atom, or if its top-level connective is one of t^-, f^-, $\wedge^-$, $\vee^-$, $\forall$. A formula is *positive* if it is a positive atom, the negation of a negative atom, or if its top-level connective is one of t^+, f^+, $\wedge^+$, $\vee^+$, $\exists$. Notice that taking the De Morgan dual of a formula causes its polarity to flip. Finally, a formula is a *literal* if it is an atom or a negated atom.

The LKF focused proof system [12] for classical logic is given in Figure 1. Since we now restrict our attention to classical logic, we can simplify sequents by making them *one-sided*: that is, we can write the sequent $\Gamma \vdash \Delta$ as $\vdash \neg\Gamma, \Delta$ (placing $\neg$ in front of a collection of formulas is taken as the collection of negated

Structural Rules

$$\frac{\vdash \Theta, C \Uparrow \Gamma}{\vdash \Theta \Uparrow \Gamma, C}\ \mathit{Store} \qquad \frac{\vdash \Theta \Uparrow N}{\vdash \Theta \Downarrow N}\ \mathit{Release} \qquad \frac{\vdash P, \Theta \Downarrow P}{\vdash P, \Theta \Uparrow \cdot}\ \mathit{Decide}$$

Identity Rules

$$\frac{\vdash \Theta \Downarrow P \quad \vdash \Theta \Uparrow \neg P}{\vdash \Theta \Uparrow}\ \mathit{Cut} \qquad \frac{}{\vdash \neg P, \Theta \Downarrow P}\ \mathit{Init}\ (\text{literal } P)$$

Introduction of negative connectives

$$\frac{}{\vdash \Theta \Uparrow \Gamma, t^-} \qquad \frac{\vdash \Theta \Uparrow \Gamma, A \quad \vdash \Theta \Uparrow \Gamma, B}{\vdash \Theta \Uparrow \Gamma, A \wedge^- B}$$

$$\frac{\vdash \Theta \Uparrow \Gamma}{\vdash \Theta \Uparrow \Gamma, f^-} \qquad \frac{\vdash \Theta \Uparrow \Gamma, A, B}{\vdash \Theta \Uparrow \Gamma, A \vee^- B} \qquad \frac{\vdash \Theta \Uparrow \Gamma, A}{\vdash \Theta \Uparrow \Gamma, \forall x A}$$

Introduction of positive connectives

$$\frac{}{\vdash \Theta \Downarrow t^+} \qquad \frac{\vdash \Theta \Downarrow A \quad \vdash \Theta \Downarrow B}{\vdash \Theta \Downarrow A \wedge^+ B}$$

$$\frac{\vdash \Theta \Downarrow A_1}{\vdash \Theta \Downarrow A_1 \vee^+ A_2} \qquad \frac{\vdash \Theta \Downarrow A_2}{\vdash \Theta \Downarrow A_1 \vee^+ A_2} \qquad \frac{\vdash \Theta \Downarrow A[t/x]}{\vdash \Theta \Downarrow \exists x A}$$

Fig. 1. The focused proof system LKF for classical logic. Here, P is positive, N is negative, C is a positive formula or a negative literal, Θ consists of positive formulas and negative literals, and x is not free in Θ and Γ. Endsequents have the form $\vdash \cdot \Uparrow \Gamma$.

formulas). In this setting, right and left-introduction rules are now organized around two right introduction rules for a connective and its De Morgan dual. Sequents in LKF are divided into *negative sequents* $\vdash \Theta \Uparrow \Gamma$ and *positive sequents* $\vdash \Theta \Downarrow B$, where Θ and Γ are multisets of formulas and B is a formula. (These sequents are formally one-sided sequents: formulas on the left of $\Uparrow$ and $\Downarrow$ are *not* negated as they are in two-sided sequents.) Notice that in this focused proof system, we have reused the term "structural rule" for a different set of rules. The weakening and contraction rules are each available in exactly one rule in Figure 1, namely, in the *Init* and the *Decide* rules, respectively. Notice also that in any LKF proof that has a conclusion of the form $\vdash \cdot \Uparrow B$, the only formulas occurring to the left of an $\Uparrow$ or $\Downarrow$ within sequents in that proof are positive formulas or negative literals. There are three immediate consequences of this invariant. (*i*) The proviso on the *Init* rule, that P is a literal, is necessarily satisfied. (*ii*) The only formulas that are weakened (in the *Init* rule) are either positive formulas or negative literals. (*iii*) The only formulas contracted (in the *Decide* rule) are positive formulas. Although linear logic is not employed here directly, non-literal negative formulas are treated linearly in the sense that they are never duplicated nor weakened in an LKF proof.

Let B be a formula of first-order logic. By a *polarization* of B we mean a formula, say B', where all the propositional connectives are replaced by *polarized versions* of the same connective and where all atomic formulas are assigned either a positive or negative polarity. Thus, an occurrence of the disjunction $\vee$ is replaced by an occurrence of either $\vee^+$ or $\vee^-$; similarly with $\wedge$ and with the logical constants for true t and false f. For simplicity, we shall assume that polarization for atomic formulas is a global assignment to all atomic formulas. Properly speaking, focused proof systems contain *polarized* formulas and not simply formulas. Notice that if the formula has n occurrences of these four logical connectives then there are 2^n different polarizations of that formula. The following theorem is proved in [12].

Theorem 1 (Soundness and completeness of LKF). *Let B be a first order formula and let B' be a polarization of B. Then B is provable in classical logic if and only if there is a cut-free LKF proof of $\vdash \cdot \Uparrow B'$.*

Notice that polarization does not affect provability but it does affect the shape of possible LKF proofs. To illustrate an application of the correctness of LKF, we show how it provides a direct proof of the following theorem.

Theorem 2 (Herbrand's Theorem). *Let B be a quantifier-free formula and let $\bar{x}$ be a (non-empty) list of variables containing the free variables of B. The formula $\exists \bar{x} B$ is classically provable if and only if there is a list of substitutions $\theta_1, \ldots, \theta_m$ $(m \geq 1)$, all with domain $\bar{x}$, such that the (quantifier-free) disjunction $B\theta_1 \vee \cdots \vee B\theta_m$ is provable (*i.e., *tautologous).*

Proof. Assume that $\exists \bar{x} B$ is provable and let B' be the result of polarizing all occurrences of propositional connectives negatively. By the completeness of LKF, there is a cut-free LKF proof Ξ of $\vdash \exists \bar{x} B' \Uparrow \cdot$. The only sequents of the form $\vdash \Theta \Uparrow \cdot$ in Ξ are such that Θ is equal to $\{\exists \bar{x} B'\} \cup \mathcal{L}$ for $\mathcal{L}$ a multiset of literals. Such a sequent can only be proved by a *Decide* rule that focuses on either a positive literal in $\mathcal{L}$ (in which case the proof is completed by the *Init* rule) or the original formula $\exists \bar{x} B'$: in the latter case, the positive phase above it provides a substitution for all the variables in $\bar{x}$. One only needs to collect all of these substitutions into a list $\theta_1, \ldots, \theta_m$ and then show that the proof Ξ is essentially also a proof of $\vdash B'\theta_1 \vee^+ \cdots \vee^+ B'\theta_m \Uparrow \cdot$ in the sense that the positive and negative phases correspond exactly. □

3.2 Positive and Negative Macro Inference Rules

We shall call individual introduction rules (such as displayed in Figure 1) "micro-rules" (the atoms of inference). An entire phase within a focused proof can be seen as a "macro-rule" (the molecules of inference). In particular, consider the following derivation, where P is a positive formula in Θ.

$$\dfrac{\dfrac{\vdash \Theta \Uparrow N_1 \quad \cdots \quad \vdash \Theta \Uparrow N_n}{\vdash \Theta \Downarrow P}}{\vdash \Theta \Uparrow \cdot}$$

Here, the selection of the formula P for the focus can be seen as selecting among several macro-rules: this derivation illustrates one such macro-rule: the inference rule with conclusion $\vdash \Theta \Uparrow \cdot$ and with $n \geq 0$ premises $\vdash \Theta \Uparrow N_1, \ldots, \vdash \Theta \Uparrow N_n$ (where $N_1, \ldots, N_n$ are negative formulas). We shall say that this macro-rule is positive. Similarly, there is a corresponding negative macro-rule with conclusion, say, $\vdash \Theta \Uparrow N_i$, and with $m \geq 0$ premises of the form $\vdash \Theta, \mathcal{C} \Uparrow \cdot$, where $\mathcal{C}$ is a multiset of positive formulas or negative literals.

In this way, focused proofs allow us to view the construction of proofs from conclusions of the form $\vdash \Theta \Uparrow \cdot$ as first attaching a positive macro rule (by focusing on some formula in Θ) and then attaching negative inference rules to the resulting premises until one is again to sequents of the form $\vdash \Theta' \Uparrow \cdot$. Focused proofs are built by such alternation of positive and negative macro-rules.

Example 3. Assume that Θ contains the formula $a \wedge^+ b \wedge^+ \neg c$, where a, b, and c are positive atomic formulas. A derivation that focuses on that formula must have the following shape.

$$\dfrac{\dfrac{}{\vdash \Theta \Downarrow a}\,\mathit{Init} \quad \dfrac{}{\vdash \Theta \Downarrow b}\,\mathit{Init} \quad \dfrac{\dfrac{\vdash \Theta, \neg c \Uparrow \cdot}{\vdash \Theta \Uparrow \neg c}\,\mathit{Store}}{\vdash \Theta \Downarrow \neg c}\,\mathit{Release}}{\dfrac{\vdash \Theta \Downarrow a \wedge^+ b \wedge^+ \neg c}{\vdash \Theta \Uparrow \cdot}\,\mathit{Decide}}$$

This derivation is possible only if Θ is of the form $\neg a, \neg b, \Theta'$. Thus, the corresponding "macro-rule" is

$$\frac{\vdash \neg a, \neg b, \neg c, \Theta' \Uparrow \cdot}{\vdash \neg a, \neg b, \Theta' \Uparrow \cdot}\ .$$

Thus, selecting this formula corresponds to the "action" of adding the literal $\neg c$ to the context if the two literals $\neg a$ and $\neg b$ are already present.

The *decide depth* of an LKF proof is the maximum number of *Decide* rules along any path starting from the endsequent. We shall often use the decide depth of proofs to help judge their size: as we shall see, such a measurement is more natural than the measurement that counts occurrences of micro rules.

4 Some Examples of Proof Certificates

Let B be a classical propositional formula in negation normal form. Thus, every connective in B can be given either positive or negative polarity. We now consider the two extremes in which all the connectives are made negative and in which all the connectives are made positive.

Roughly speaking, we shall view proof certificates as documents containing two parts. The first part, the *preamble*, uses the language of focusing (*e.g.*, polarization of connectives and literals) to define macro-level connectives. The second part, the *payload*, contains the direct encoding of a proof using those macro-level connectives.

Example 4. Let B^- be the result of polarizing negatively the connectives of B: that is, B^- contains only the connectives $\wedge^-$, $\vee^-$, t^-, and f^-. In this case, an LKF proof of $\vdash\Uparrow B^-$ has a simple structure: in fact, it has a decide depth of exactly 1. The unique negative phase comprises *all* the introduction rules, leaving one premise for every disjunct in the conjunctive normal form of B. Let one such premise be $\vdash L_1, \ldots, L_j \Uparrow$, where $L_1, \ldots, L_j$ are literals. Such a sequent is provable if and only if it has an LKF proof of the form

$$\dfrac{\overline{\vdash L_1, \ldots, L_j \Downarrow P}\ \mathit{Init}}{\vdash L_1, \ldots, L_j \Uparrow}\ \mathit{Decide}$$

where P is a positive literal from the set $\{L_1, \ldots, L_j\}$ and the complement of P is also in that set. Thus a proof certificate for propositional logic can be described as follows. The preamble declares that all propositional connectives are polarized negatively and that all atoms are polarized, say, negatively. Given this preamble, a proof checker will be able to compute the unique negative macro rule. The only information that is missing from the proof is the actual "mating" of complementary literals [2]. Thus, the payload needs to contain one pair of *occurrences* of literals for each disjunct in the conjunctive normal form of B: the first is positive and is used in the *Decide* rule and the second is the complement of the first and provides the information needed for the *Init* rule.

The proof certificate described in Example 4 is potentially large since it must provide a pair of literal occurrences for every one of the (exponentially many) clauses within the formula B. If we allow proof checkers to also do some simple "proof search" then this certificate can be made to have *constant* size. In this case, the proof certificate can simply tell the proof checker that it should search for a proof of *decide depth* 1 for every premise of the negative phase. In this setting, such proof search is trivial. Furthermore, if the proof checker is a logic program, the transition from proof checker to proof searcher can be done with minimal changes. For example, let $\mathcal{L}$ be the term denoting a list encoding of the set of literals $\{L_1, \ldots, L_j\}$ and let P and Q be two literals provided by the proof certificate described in Example 4. A logic programming system would then attempt to prove the query

$$\mathit{memb}(P, \mathcal{L}) \wedge \mathit{positive}(P) \wedge \mathit{complement}(P, Q) \wedge \mathit{memb}(Q, \mathcal{L}),$$

where the predicates *positive*, *memb*, and *complement* are all written as, say, first-order Horn clauses in the expected way. If the proof certificate elides the pair $\langle P, Q\rangle$ then it could be asked to prove the query

$$\exists P \exists Q\ [\mathit{memb}(P, \mathcal{L}) \wedge \mathit{positive}(P) \wedge \mathit{complement}(P, Q) \wedge \mathit{memb}(Q, \mathcal{L})].$$

Proving this query is, of course, straightforward and something that a proof checker with unification and backtracking search can easily be expected to perform.

While the proof certificate for all propositional tautologies can be described in terms of LKF in constant size, it does require the proof checker to spend an exponential amount of time to do the checking. Of course, pushing for very small proof certificates can go too far since sometimes communicating "clever" choices can make proofs much easier to check. Consider the formula $(\neg p \vee C) \vee p$ where C is some propositional formula with a large conjunctive normal form. Using the above proof certificate means that there is no way to describe the obvious reason why this formula is tautologous. To allow for more interesting information to be put into proofs, consider the following use of LKF where all propositional connectives are polarized positively.

Example 5. Let B^+ be the result of polarizing positively the connectives of B and let Ξ be an LKF proof for $\vdash \cdot \Uparrow B^+$. It is easy to show that every $\Uparrow$-sequent in Ξ with an empty right-hand-side is of the form $\vdash B^+, \mathcal{L} \Uparrow \cdot$ where $\mathcal{L}$ is a multiset of negative literals. Furthermore, every positive phase starts (reading proofs bottom up) with the *Decide* rule on B and then continues with a series of selections among disjunctions. Proofs using exclusive positive polarizations will, in general, have large decide depths and require larger proofs that those described in Example 4. The additional proof information can, however, make proofs easier to check.

To illustrate the last claim in Example 5, consider a focused proof for the formula $(\neg p \vee^+ C) \vee^+ p$. This formula has a proof of decide depth 2: the first (closest to the root) positive phase is a series of selections in this disjunction that selects $\neg p$ to add to the left of the $\Uparrow$. The second positive phase makes such selections to pick the formula p, and the proof is complete. Notice that the right sequence of choices steers the proof away from considering the subformula C.

Of course, there are many ways to polarize formulas since one can easily mix positive and negative polarizations: these are choices that someone wanting to communicate a proof certificate can make as seems appropriate for the proof objects that they wish to communicate.

To complete this treatment of proof certificates based on consideration of a formula's "matrix," consider the following example of proof certificates deriving their structure from Herbrand's theorem.

Example 6. Herbrand's theorem (see Section 3.1) can be used to validate proof certificates of formulas of the form $\exists \bar{x} B$ (where B is propositional): such certificates can contain a list of substitutions $\theta_1, \ldots, \theta_m$ ($m \geq 1$), all with domain $\bar{x}$, and then a proof certificate for the propositional formula $B\theta_1 \vee \cdots \vee B\theta_m$. Above we discussed various ways to build proof certificates for tautologies. The additional substitution information can be transmitted in the proof certificate as a series of *Decide* rules followed by a series of $\exists$-introduction rules. In general, a proof checker based on logic programming might be expected to recover actual substitution terms so these might be left out of the proof certificate (of course, the number of substitutions m must be supplied).

5 Non-Matrix Proof Systems

A great many proof structures are not based on the "matrix" of formulas. We consider a couple of such proof systems here. Both of these make use of the *cut* inference rules [11]. There are various cut rules for LKF given in [12]: the cut inference displayed in Figure 1 is an instance of the "key cut" rule while the following inference rule is an instance of the "prime cut" rule:

$$\frac{\vdash \Theta \Uparrow B \qquad \vdash \Theta \Uparrow \neg B}{\vdash \Theta \Uparrow} \; Cut_p$$

Both the "key" and "prime" cut rules can be eliminated in LKF.

Example 7. When a resolution based theorem prover has succeeded in proving a theorem, it has built a resolution dag in which the leaves are clauses, the root is the empty clause (an inconsistency), and the internal nodes are instances of the resolution rule. A *clause* is a closed formula of the form $\forall x_1 \ldots \forall x_n [L_1 \vee \cdots \vee L_m]$ while a *negated clause* is a closed formula of the form $\exists x_1 \ldots \exists x_n [L_1 \wedge \cdots \wedge L_m]$, where $n, m \geq 0$, $\{L_1, \cdots, L_m\}$ a multiset of literals, and $x_1, \ldots, x_n$ is a list of first-order variables. The following predicates are commonly used in building resolution refutations.

1. A clause C is *trivial* if it contains complementary literals.
2. A clause C_1 *subsumes* clause C_2 if there is a substitution instance of the literals in C_1 which is a subset of the literals in C_2.
3. The usual relationship of *resolution* of two clauses C_1 and C_2 to yield C_3 can be characterized by choosing the most general unifier of two complementary literals, one from each of C_1 and C_2. We shall say that C_3 is an *allowed resolvent* if it is constructed by the same rule except that we allow *some* unifier to be used instead of the most general one.

By polarizing clauses using $\vee^-$ (and negated clauses using $\wedge^+$) it is possible to use small LKF proofs to check each of these properties. In particular, it is easy to show that C is trivial if and only if $\vdash \cdot \Uparrow C$ has a proof of decide depth 1. Similarly, by polarizing literals appropriately, C_1 subsumes a non-trivial clause C_2 if and only if $\vdash \neg C_1 \Uparrow C_2$ has a proof of decide depth 1. Finally, C_3 is an allowed resolvent of C_1 and C_2 if and only if $\vdash \neg C_1, \neg C_2 \Uparrow C_3$ has a proof of decide depth 2. It is now a simple matter to take a resolution refutation (also including checks for trivial clauses and subsumption) of the clauses $C_1, \ldots, C_n$ and describe an LKF proof of the sequent $\vdash \neg C_1, \ldots, \neg C_n \Uparrow \cdot$. For example, the following shows how to incorporate into a full proof certificate the fact that C_1 and C_2 yield the resolvent C_{n+1}.

$$\frac{\vdash \neg C_1, \neg C_2 \Uparrow C_{n+1} \qquad \dfrac{\vdash \neg C_1, \ldots, \neg C_n, \neg C_{n+1} \Uparrow \cdot}{\vdash \neg C_1, \ldots, \neg C_n \Uparrow \neg C_{n+1}} \; Store}{\vdash \neg C_1, \ldots, \neg C_n \Uparrow \cdot} \; Cut_p$$

By repeating this process, an entire refutation can be converted into an LKF proof with cuts. In all cases, the left premises of all occurrences of the cut rule have small proofs that can be replaced in the final proof certificate with a notation that asks the proof checker to search for proofs up to decide depth 2. In this way, the resulting proof certificate is essentially a direct translation of the

$$\frac{\vdash \Theta\sigma \Uparrow \Gamma\sigma}{\vdash \Theta \Uparrow \Gamma, s \neq t}\ \dagger \qquad \frac{}{\vdash \Theta \Uparrow \Gamma, s \neq t}\ \ddagger \qquad \frac{}{\vdash \Theta \Downarrow t = t}$$

$$\frac{\vdash \Theta \Uparrow \Gamma, B(\nu B)\bar{t}}{\vdash \Theta \Uparrow \Gamma, \nu B\bar{t}} \qquad \frac{\vdash \Theta \Downarrow B(\mu B)\bar{t}}{\vdash \Theta \Downarrow \mu B\bar{t}}$$

Fig. 2. Focused inference rules for $=$ and μ and their duals. The proviso † requires the terms s and t to be unifiable and for σ to be their most general unifier. The proviso ‡ requires that the terms s and t are not unifiable.

refutation and yields a proof certificate that an LKF proof checker can easily check. Of course, the proof checker will have to do some (bounded) proof search to reconstruct the proofs of the left premises of the cut rule.

Example 8. For another example of a proof certificate, we briefly mention the encoding of tabled deduction described in [17]. It is illustrated there that focused proofs (this time using the focused intuitionistic proof system LJF [12]) can be used to capture tabled deduction. There are two important ingredients in the use of tables. First, items that have already been proved need to be available for subsequent items that are still to be proved: that is easily captured using the cut rule in the fashion described above. Second, one must enforce that items in a table must be *reused* and not *reproved.* It was shown in [17] how it is possible to identify negative polarity with atoms that are not in the table and positive polarity with atoms in the table. One also must allow a cut-inference rule that permits the polarity of atoms to switch from negative (on the left premise) to positive (on the right premise). In this way, a table can be translated into an LJF proof with cuts such that every left-premise of a cut has a proof of decide depth 1: one should be able to elide such proofs in a proof certificate.

6 Fixed Points and Equality

We now extend the first-order logic underlying LKF by adding equality and fixed points and by giving in Figure 2 the introduction rules for these and their duals. Equality $=$ is a positive connective and its De Morgan dual $\neq$ is negative. Similarly, the two fixed points μ and ν are De Morgan duals in which μ is positive and ν is negative. Given that the rules for μ and ν are simply unfoldings of the fixed point, these operators do not yield any particular fixed points. It is possible to have a more expressive proof theory for fixed points that provides also for least and greatest fixed points (see, for example, [4,5]): in that case, the De Morgan dual of the least fixed point is the greatest fixed point.

Example 9. The following simple logic program defines two predicates on natural numbers, assuming that such numbers are built from zero 0 and successor s.

$$\begin{array}{ll} nat\ 0 \subset true. & leq\ 0\ Y \subset true. \\ nat\ (s\ X) \subset nat\ X. & leq\ (s\ X)\ (s\ Y) \subset leq\ X\ Y. \end{array}$$

The predicate *nat* can be written as the fixed point expression

$$\mu(\lambda p \lambda x.(x = 0) \vee^{+} \exists y.(s\ y) = x \wedge^{+} p\ y)$$

and binary predicate *leq* (less-than-or-equal) can be written as the expression

$$\mu(\lambda q \lambda x \lambda y.(x = 0) \vee^{+} \exists u \exists v.(s\ u) = x \wedge^{+} (s\ v) = y \wedge^{+} q\ u\ v).$$

In a similar fashion, any Horn clause specification can be made into fixed point specifications (mutual recursions requires standard encoding techniques) that contain *only positive connectives.*

Example 10. Consider proving the sequent

$$\vdash \Theta \Downarrow (leq\ m\ n \wedge^{+} N_1) \vee^{+} (leq\ n\ m \wedge^{+} N_2),$$

where m and n are natural numbers and *leq* is the fixed point expression displayed above. If both N_1 and N_2 are negative formulas, then there are exactly two possible macro rules: one with premise $\vdash \Theta \Uparrow N_1$ when $m \leq n$ and one with premise $\vdash \Theta \Uparrow N_2$ when $n \leq m$ (thus, if $m = n$, both premises are possible). In this way, a macro inference rule can contain an entire Prolog-style computation.

7 Computation and Model Checking

A traditional approach to leaving out details within a proof is to identify some aspects of the proof as *computation.* An expression to be computed must be communicated in the certificate but the computation trace and the final value do not need to be communicated. When computation is *determinate* (*i.e.*, when expressions have at most one value) this observation has been called the *Poincaré principle* [8,10]. In rich type systems, such as those found in functional Pure Type System [7], computation is dominated by β-reductions: communicating a proof in that setting does not need to communicate the trace of such β-reductions.

Computation is related to proof certificates in at least three ways. First, the negative phase relates its conclusion to its premises in a determinate fashion: a proof checker simply needs to compute that phase. In Example 4, this phase was determined by the computation of a conjunctive normal form. Second, computation can be inserted *within* an inference rule as is illustrated above in Example 10. In that way, one step in a proof can include arbitrary amounts of computation (all described as a Prolog-like fixed point computation). Third, elided proof details must be reconstructed by a proof-search-style computation and this will also involve computation in the style of logic programming: unification and backtracking can be used to reconstruct elided information.

Besides (determinate) computation, some of the primitives of *model checking* are also naturally captured in this setting of focused proofs. For example, consider the model checking problem of determining if the positive formula $B(x)$ holds for every x that is a member of the set $A = \{a_1, \ldots, a_n\}$. Membership in this set can be encoded as $x = a_1 \vee^{+} \cdots \vee^{+} x = a_n$ (abbreviated as $A(x)$).

An attempt to prove the sequent $\forall x.A(x) \supset B(x)$ yields the following negative macro rule in LKF. (Here, the implication $C \supset D$ is rendered as $\neg C \vee^- D$.)

$$\dfrac{\dfrac{\vdash B(a_1) \Uparrow \cdot}{\vdash B(x) \Uparrow x \neq a_1} \quad \cdots \quad \dfrac{\vdash B(a_n) \Uparrow \cdot}{\vdash B(x) \Uparrow x \neq a_n}}{\vdash \cdot \Uparrow \forall x.[x \neq a_1 \wedge^- \cdots \wedge^- x \neq a_n] \vee^- B(x)}$$

In this way, quantification over a finite set is captured precisely as one macro-level inference rules within LKF. The following example illustrates a typical model checking problem.

Example 11. Assume that a label transition system is described by a recursive fixed point expression named $P \xrightarrow{a} P'$ (consider, for example, writing the operational semantics of CCS as a Prolog-like fixed point expression). Simulation in process calculi can be defined as the (greatest) fixed point of the following recursive definition:

$$sim\ P\ Q \equiv \forall P' \forall a[P \xrightarrow{a} P' \supset \exists Q'[Q \xrightarrow{a} Q' \wedge sim\ P'\ Q']].$$

The right-hand side of this definition is composed of *exactly two* macro-rules. The expression $\forall P' \forall a[P \xrightarrow{a} P' \supset \cdot\ ]$ is a negative macro rule since $P \xrightarrow{a} P'$ is positive. The expression $\exists Q'[Q \xrightarrow{a} Q' \wedge^+ \cdot\ ]$ yields a positive macro rule. In this way, the focused proof system is aligned directly with the structure of the actual (model-checking) problem. Notice that if one wishes to communicate a proof of a simulation to a proof checker, no information regarding the use of the negative macro rule needs to be communicated since the proof checker can also perform the computation behind that inference rule (*i.e.*, enumerating all possible transitions of a given process P). Furthermore, eliding proofs of fixed point expressions such as $P \xrightarrow{a} P'$ might also be sensible since the proof checker might well be able to enumerate possible values for some of the values of P, a, and P' when the other values are known [17]. The resulting proof certificate is essentially just the collection of all expressions of the form $sim\ P\ Q$ that are present in the full proof (such a set is also called a *simulation*).

The fact that an entire computation can fit within a macro rule (using purely positive fixed point expressions) provides great flexibility in designing inference rules. Such flexibility allows inference rules to be designed so that they correspond to an "action" within a given computational system. One should note that placing arbitrary computation within an inference rule means that we can have (macro) rules for which their validity is not decidable. Given our interest in proof certificates, however, this does not seem to be a problem in and of itself. A checker may fail to terminate on a given certificate, in which case, we may chose to reject the certificate after waiting some period of time. The engineer of the certificate failed, in this case, to successfully communicate a proof. The certificate structure might then need to be redesigned.

Example 12. The Lucas-Lehmer theorem states that n is prime if and only if there exists an integer a such that $1 < a < n$ and $a^{(n-1)} \equiv 1 \pmod{n}$ and for all prime factors q of $n-1$, it is the case that $a^{(n-1)/q} \not\equiv 1 \pmod{n}$. This theorem can be used to build certificates of primality [20]: proof certificates of the same claim are also easy to develop. Assume that the Lucas-Lehmer theorem has already been proved and placed into a (trusted) library. In principle, the certificate proving primality of n involves producing the witness a and the prime factors of $n-1$. The rest of this certificate requires various straightforward computations (via fixed points) as well as proof certificates that show that the factors q of $n-1$ are, indeed, primes.

8 Related Work

Proof carrying code [19] was an earlier attempt at producing, communicating, and checking proof objects. Much of that effort was focused on theorems involving assertions about mobile and imperative programs in a setting where proof objects were often highly optimized in order to be used in resource-limited systems. Shankar uses proof certificates as a part of a prover architecture in which the claims of untrusted inference procedures are validated by communicating certificates that are then checked by checkers that have been verified relative to a small kernel checker [21]. We are concerned here, however, with all manner of formal proof objects from a much wider range of applications with no a priori restriction on resources.

The Dedukti [9] proof checker shares some characteristics with our proposal for proof checking. Instead of using proof theory and focusing, Dedukti employs *deduction modulo* [10] as a framework for building large scale inference rules from theories and smaller inferences. This system also separates computation steps from deduction steps: in the case of Dedukti, computation is, however, determinant and is based on functional programming-style computation. Proofs are not permitted to contain holes and proof search for proof reconstruction is not available.

9 Future Work

This proposal can be developed along a number of directions.

Proof reconstruction when equality is a logical connective. Proof reconstruction in first-order logics requires unification. When one introduces equality as a logical connective, as we did in Figure 2, proof reconstruction must deal with unification for elided terms ("logic variables") as well as eigenvariables. Standard procedures for (higher-order) unification work well when only one such class of variables is present: unification with these two classes of variables is still to be developed.

Induction and co-induction. Baelde has investigated focused proof system that incorporate both induction and co-induction [4,5]. An experimental prover [6] has demonstrated that small focused proofs involving induction can be proved completely automatically, leaving open the possibility of proof reconstruction even for proof certificates that leave out subproofs of simple inductive lemmas.

Combining intuitionistic and classical logics. One does not want to have to deal with two sets of different proof theories: one each for classical and intuitionistic logics. Ideally, these should be combined into one logic and (focused) proof system: see [13] for an initial proposal for such a logic.

Counterexamples and partial proofs. Structural proof theory deals with complete proofs. In a setting where proofs are developed in a distributed setting and employs an array of theorem proving technologies, partial proofs (proofs with unproved premises) become important objects that should be studied properly. Similarly, counterexamples are extremely valuable documents that should be formally included in a comprehensive approach to proof certificates. These two concepts should also be tied together with techniques similar to those used to eliminate cuts. For example, when someone finds a counterexample to an open premise of a partial proof, one would like to systematically explore how much of the partial proof needs to be rewound in order to avoid that counterexample.

Building and trusting proof checkers. Significant computational resources and flexibility are need for proof checking and proof reconstruction. The Dedukti proof checker [9] places Haskell into its trusted code base. Our framework here is better served by logic programming. Of course, such logic programming systems must be logically sound in the strongest senses. Since there are bindings within formulas (quantifiers) and bindings within proofs (eigenvariables), the λProlog programming language, which treats bindings entirely declaratively, might make a good implementation language for proof checkers. Of course, this means that a λProlog implementation, such as Teyjus [18], must enter the trusted core. Accepting higher-order logic programming languages into the "trusted base of code" is, in fact, a familiar theme: both λProlog and Twelf have been proposed as part of the trusted code base for supporting proof-carrying code [3].

10 Conclusion

We have overviewed a *foundational* approach to designing proof certificates that satisfies the following four desiderata (described in more depth in [15]): they should be (*i*) checkable by simple proof checkers, (*ii*) flexible enough that existing provers can conveniently produce such certificates from their internal evidence of proof, (*iii*) directly related to proof formalisms used within the structural proof theory literature, and (*iv*) permit certificates to elide some proof information with the expectation that a proof checker can reconstruct the missing information using bounded and structured proof search. Central to our design of such proof certificates is the proof-theoretic notion of focused proof system.

Acknowledgments. I wish to thank Alberto Momigliano, Alwen Tiu, and the anonymous referees for their comments on an earlier draft of this paper.

References

1. Andreoli, J.-M.: Logic programming with focusing proofs in linear logic. J. of Logic and Computation 2(3), 297–347 (1992)
2. Andrews, P.B.: Theorem-proving via general matings. J. ACM 28, 193–214 (1981)

3. Appel, A.W., Felty, A.P.: Polymorphic lemmas and definitions in λProlog and Twelf. Theory and Practice of Logic Programming 4(1-2), 1–39 (2004)
4. Baelde, D.: A linear approach to the proof-theory of least and greatest fixed points. PhD thesis, Ecole Polytechnique (December 2008)
5. Baelde, D.: Least and greatest fixed points in linear logic. Accepted to the ACM Transactions on Computational Logic (September 2010)
6. Baelde, D., Miller, D., Snow, Z.: Focused Inductive Theorem Proving. In: Giesl, J., Hähnle, R. (eds.) IJCAR 2010. LNCS, vol. 6173, pp. 278–292. Springer, Heidelberg (2010)
7. Barendregt, H.: Lambda calculus with types. In: Abramsky, S., Gabbay, D.M., Maibaum, T.S.E. (eds.) Handbook of Logic in Computer Science, vol. 2, pp. 117–309. Oxford University Press (1992)
8. Barendregt, H., Barendsen, E.: Autarkic computations in formal proofs. J. of Automated Reasoning 28(3), 321–336 (2002)
9. Boespflug, M.: Conception d'un noyau de vérification de preuves pour le $\lambda\Pi$-calcul modulo. PhD thesis, Ecole Polytechnique (2011)
10. Dowek, G., Hardin, T., Kirchner, C.: Theorem proving modulo. J. of Automated Reasoning 31(1), 31–72 (2003)
11. Gentzen, G.: Investigations into logical deductions. In: Szabo, M.E. (ed.) The Collected Papers of Gerhard Gentzen, pp. 68–131. North-Holland, Amsterdam (1969); Translation of articles that appeared in 1934-1935
12. Liang, C., Miller, D.: Focusing and polarization in linear, intuitionistic, and classical logics. Theoretical Computer Science 410(46), 4747–4768 (2009)
13. Liang, C., Miller, D.: Kripke semantics and proof systems for combining intuitionistic logic and classical logic (September 2011) (submitted)
14. Martin-Löf, P.: Constructive mathematics and computer programming. In: Sixth International Congress for Logic, Methodology, and Philosophy of Science, Amsterdam, pp. 153–175. North-Holland (1982)
15. Miller, D.: Communicating and trusting proofs: The case for broad spectrum proof certificates (June 2011); Available from author's website
16. Miller, D., Nadathur, G., Pfenning, F., Scedrov, A.: Uniform proofs as a foundation for logic programming. Annals of Pure and Applied Logic 51, 125–157 (1991)
17. Miller, D., Nigam, V.: Incorporating Tables into Proofs. In: Duparc, J., Henzinger, T.A. (eds.) CSL 2007. LNCS, vol. 4646, pp. 466–480. Springer, Heidelberg (2007)
18. Nadathur, G., Mitchell, D.J.: System Description: Teyjus - A Compiler and Abstract Machine Based Implementation of λProlog. In: Ganzinger, H. (ed.) CADE 1999. LNCS (LNAI), vol. 1632, pp. 287–291. Springer, Heidelberg (1999)
19. Necula, G.C.: Proof-carrying code. In: Conference Record of the 24th Symposium on Principles of Programming Languages 1997, Paris, France, pp. 106–119. ACM Press (1997)
20. Pratt, V.R.: Every prime has a succinct certificate. SIAM Journal on Computing 4(3), 214–220 (1975)
21. Shankar, N.: Trust and Automation in Verification Tools. In: Cha, S(S.), Choi, J.-Y., Kim, M., Lee, I., Viswanathan, M. (eds.) ATVA 2008. LNCS, vol. 5311, pp. 4–17. Springer, Heidelberg (2008)

Univalent Semantics of Constructive Type Theories

Vladimir Voevodsky

Institute for Advanced Study, Princeton NJ 08540, USA

Abstract. In this talk I will outline a new semantics for dependent polymorphic type theories with Martin-Lof identity types. It is based on a class of models which interpret types as simplicial sets or topological spaces defined up to homotopy equivalence. The intuition based on the univalent semantics leads to new answers to some long standing questions of type theory providing in particular well-behaved type theoretic definitions of sets and set quotients. So far the main application of these ideas has been to the development of "native" type-theoretic foundations of mathematics which are implemented in a growing library of mathematics for proof assistant Coq. On the other hand the computational issues raised by the univalent semantics may lead in the future to a new class of programming languages.

J.-P. Jouannaud and Z. Shao (Eds.): CPP 2011, LNCS 7086, p. 70, 2011.

Formalization of Wu's Simple Method in Coq*

Jean-David Génevaux, Julien Narboux, and Pascal Schreck

LSIIT UMR 7005 CNRS - Université de Strasbourg

Abstract. We present in this paper the integration within the Coq proof assistant, of a method for automatic theorem proving in geometry. We use an approach based on the validation of a certificate. The certificate is generated by an implementation in Ocaml of a simple version of Wu's method.

Keywords: formalization, automation, geometry, Wu's method, proof assistant, Coq.

1 Introduction

Bringing new automation techniques to interactive theorem provers while preserving safety is an important goal in order to spread the use of proof assistants.

In this paper, we focus on one of the most successful methods for automated theorem proving in geometry, namely Wu's method. We integrate this method into the Coq proof assistant [Coq10]. Indeed, Wu's method can prove hundreds of geometry theorems [Wu78] and some conjectures were first proved using this method [Cho88, Wan89].

One could formally prove correct the implementation of this decision procedure within a proof assistant such as Coq or Isabelle, but this would require a tedious work. Another approach is to modify the decision procedure to make it produce a witness of its correctness (the certificate) which can be checked by an external tool called a validator. The certificate is such that the validator is simpler to prove than the original decision procedure. Using such an approach, the completeness of the method is not guaranteed as the certifying algorithm may generate an erroneous certificate or even fail to generate one. But if we prove formally the correctness of the validator, when the validator confirms a result we get the same level of confidence as if we had proven the correctness of the decision procedure. This approach has several advantages: first the algorithm which generates the certificate is independent of the validator and hence can be written by different people using different languages, second the decision procedure does not need to be proved formally, hence the implementation can be optimized more easily. The difficulty is to generate a certificate which can be checked in a reasonable amount of time.

Gröbner basis is another well known tool for automated deduction in geometry [Kap86]. Gröbner basis have already been integrated into Coq by Benjamin

* This work is partially supported by the ANR project Galapagos.

J.-P. Jouannaud and Z. Shao (Eds.): CPP 2011, LNCS 7086, pp. 71–86, 2011.

Grégoire, Loïc Pottier and Laurent Théry [Pot08, GPT11]. Formalizing Wu's method brings us a tool to decide ideal membership as the Gröbner basis method does. But, in the context of automatic deduction in geometry, Wu's method and its variants developed by Shang-Ching Chou, Xiao-Shan Gao and Dongming Wang have been shown to be generally more efficient [CG90, Wan01, Wan04]. Moreover, in geometry, the statements as given by a user are usually false in some special cases, such as when three points are collinear for example. These conditions under which a statement is true are called non degeneracy conditions (ndgs). An advantage of Wu's method is that it can *generate* the ndgs. This is crucial for the applications we aim at in the context of education. In the future, we would like to integrate this method in a tool to prove geometry theorems within a dynamic geometry software following the work presented in [Pha10, PBN11].

1.1 Related Work

Wu's method has been implemented by Chou, he used his implementation to prove hundreds of geometric problems [Cho88]. Xiao-Shan Gao created GEOMETRY EXPERT (GEX) [GL02, Gao00] and Zheng Ye produced JAVA GEOMETRY EXPERT (Java GEX) [YCG11]. Dongming Wang developed further elimination algorithms based on pseudo-division and implemented them as a Maple library called GEOTHER [Wan02]. Judit Robu implemented Wu's method in THEOREMA [Rob02]. Predrag Janičić implemented the method within GCLC [JQ06].

We are not aware of any formalization of Wu's method inside a proof assistant. But other methods for automatic deduction in geometry have been integrated into proof assistants: Gröbner basis can be used in COQ, HOL-LIGHT and ISABELLE [GPT11, Har07, CW07], the area method for non oriented euclidean geometry has been formalized in COQ as a tactic [CGZ94, Nar04, JNQ10] and geometric algebras are also available in COQ and can be used to prove automatically theorems in projective geometry [FT11].

The closest work to ours is the implementation of a Gröbner basis method within Coq by Benjamin Grégoire, Loïc Pottier and Laurent Théry [GPT11]. In this paper, we present an extension of their work to deal with Wu's method.

This paper is organized as follows. We first give a quick overview of Wu's method and we highlight the facts we need to prove to show the correctness of the method. Then we describe our formalization within the COQ proof assistant.

2 Overview of Wu's Method

2.1 Cartesian Geometry

We are interested in formally proving theorems in geometry, but there are many geometries! Wu's method focuses on geometry with coordinates in a field. To be more precise, we consider here the plane $\mathbb{F}^2$ where $\mathbb{F}$ is a field where all predicates on points, lines, circles, etc. can be expressed under the form $P(X) = 0$, X being a vector of variables corresponding to the coordinates of the involved objects.

Let us give a simple (and very classical) example:

Example 1

In a parallelogram the diagonals intersect in their midpoints:

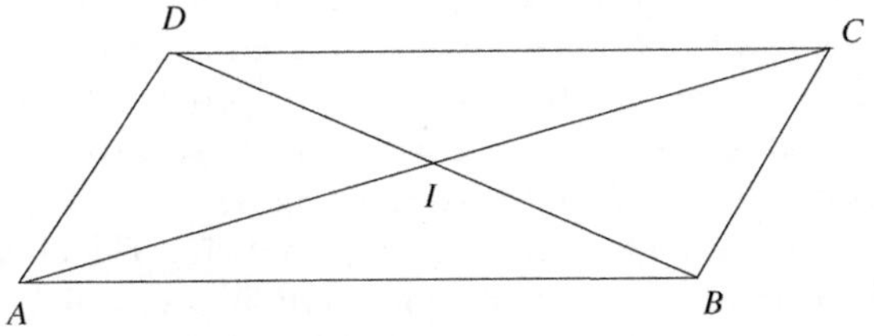

Fig. 1. Classical parallelogram example

In the usual Euclidean geometry, the fact that lines (AB) and (CD) are parallel can be expressed by the polynomial equation[1]
$(x_B - x_A)(y_C - y_D) - (y_B - y_A)(x_C - x_D) = 0$.

The fact that point I is the midpoint of $[AC]$ and the midpoint of $[BD]$ can be expressed by the polynomial equations $(x_B + x_D) - (x_A + x_C) = 0$ and $(y_B + y_D) - (y_A + y_C) = 0$.

The theorem stating that in a parallelogram the diagonals intersect in their midpoints is then directly stated by

$$\begin{cases} (x_B - x_A)(y_C - y_D) - (y_B - y_A)(x_C - x_D) = 0 \\ (x_C - x_B)(y_D - y_A) - (y_C - y_B)(x_D - x_A) = 0 \end{cases}$$

$$\Rightarrow$$

$$(x_B + x_D) - (x_A + x_C) = 0 \ \wedge \ (y_B + y_D) - (y_A + y_C) = 0$$

where all the variables are implicitly universally quantified. □

The goal of Wu's method is precisely to prove geometric theorems which can be put under this algebraic form.

2.2 Rings and Ideals

The general question is the following:

(1) Given k polynomials $H_1, \ldots, H_k$ and G in $\mathbb{F}(x_1, \ldots x_m)$, is it true that

$$\forall x_1, \ldots x_m \in \mathbb{F} : \bigwedge_{i=1}^{k} H_i(x_1, \ldots x_m) = 0 \Rightarrow G(x_1, \ldots, x_m) = 0\ ?$$

[1] Note that this equation is actually equivalent to $A = B$ or $C = D$ or (AB) and (CD) are parallel. See the discussion on ndgs (Sec.3).

This question gets obviously a positive answer when G belongs to the radical of the ideal $I =< H_1, \ldots, H_k >$ generated by polynomials $H_1, \ldots, H_k$, that is, when there exists an integer r and k polynomials $Q_1, \ldots, Q_k$ such that $G^r = \Sigma_i Q_i H_i$. The famous Hilbert's *Nullstellensatz* theorem states that if $\mathbb{F}$ is algebraically closed, then the converse is also true. That is, we can always find such polynomials.

Then, in this framework, proving a geometric theorem consists in proving that the polynomial belongs to an ideal for which a set of generator is known. Note that the *Nullstellensatz* has to be used only to prove that a polynomial does not belong to an ideal: if we want to prove a known theorem, producing a linear combination of polynomials H_i equal to G is enough.

It would be easy to test ideal membership by using Euclidean division. But, unfortunately, the ring of multivariate polynomials over a field is not Euclidean when more than one variable is involved. Nonetheless, the two most famous methods use a kind of Euclidean division to perform such a test. Buchberger's method consists in transforming the generating set into a so-called Gröbner basis in which a division algorithm can be used to give the good result while in the Wu's method a *pseudo-division* is used which closely mimics the Euclidean division.

2.3 Pseudo-Division and Pseudo-Remainder

The idea of pseudo-division consists in multiplying the polynomial to be divided, say P, by the leading coefficient of the dividing polynomial, say Q, to a certain power in order that all the coefficients of this product are divisible by the leading coefficient of Q. More formally, let P and Q be polynomials of $\mathbb{F}[x_1, \ldots, x_m]$, and $I \in \mathbb{F}[x_1, \ldots, x_{m-1}]$ be the leading coefficient of Q in the variable x_m, the pseudo-division of P by Q in the variable x_m yields two polynomials T and R such that

$$I^r P = TQ + R$$

where r is the number of non zero coefficients of P and $R \in \mathbb{F}[x_1, \ldots, x_m]$ with $\deg(x_k, R) < \deg(x_k, Q)$. Establishing the existence of the polynomials T and R and the correction of the pseudo-division algorithm is very similar to the proof of the correction of the Euclidean division algorithm. R is called the pseudo-remainder of P by Q into the variable x_k and it obviously belongs to the ideal $< P, Q >$.

Pseudo-division and pseudo-remainder can be used to perform a triangulation of the system $\bigwedge_i H_i(x_1, \ldots, x_m) = 0$. The triangulation produces an algebraic system $\bigwedge_i T_i(x_1, \ldots, x_m) = 0$ where each T_i belongs to $\mathbb{F}[x_1, \ldots x_i] \cap < H_1, \ldots, H_k >$. This way, the implication

$$\bigwedge_i H_i(x_1, \ldots, x_m) = 0 \Rightarrow \bigwedge_i T_i(x_1, \ldots, x_m) = 0$$

holds, but the converse is true if and only if the leading coefficient considered in the used pseudo-divisions do not vanish. But for the formalization, we do not the need the converse.

Note that, if the statement is given as a ruler-and-compass construction then the triangulation is tractable. Moreover, after choosing a reference, it may even become trivial. For instance, by fixing point A at $(0, 0)$ and point B on Ox axis and observing that $y_C = y_D$, the statement of example 1 becomes :

$$(x_C - x_B - x_D).y_C = 0 \Rightarrow x_B + x_D - x_C = 0$$

where x_B, x_C and y_C are *parameters* of the figure and x_D is a dependent variable. Note that, by making this choice, we implicitly assume that $A \neq B$. The hypotheses are trivially under a triangular form. Next Section explains in details how to fix a *reference*, why this is correct, and how to construct the figure corresponding to the hypotheses.

Eventually, using successive pseudo-divisions of G by $T_k, T_{k-1}, \ldots, T_1$, the method tests if IG belongs to the ideal generated by polynomials H_i where I is a product of the leading coefficients of the triangulated polynomials. Summarizing the calculi, Wu's method, in his simplest form, allows to compute some polynomials I, S_i and R such that:

$$IG = \sum_i (H_i S_i) + R$$

It is clear that if the polynomial R is null, then the theorem is proved under the assumption that $I(x_1, \ldots, x_m) \neq 0$. When the theorem is stated as a construction this assumption corresponds to the non degeneracy conditions [CG92].

In our example, the degenerated case occurs when $y_C = 0$. This matches the case where the four points A, B, C and D are collinear : in this case, the theorem does not hold. The converse is not true: $R \neq 0$ does not mean that the theorem is false. The simple method of Wu is not complete but according to Chou [Cho88] it is powerful enough to prove hundred of classical theorems. A complete method would need to use ascending chains which are considered in the Ritt-Wu principle [Cho88].

The main steps of the Wu's method for geometry theorem proving can be summarized as follows:

1. Transforming the statements into an algebraic form.
2. Choosing the origin and a direction for the system of coordinates.
3. Showing that to prove the statement in general it is sufficient to show that it holds in the given system of coordinates.
4. Simplifying the polynomials thanks to the choice of the reference.
5. Triangulating the list of hypotheses using pseudo-division.
6. Pseudo-dividing successively the goal by the triangulated hypotheses. If the final remainder is the null polynomial, then the statement holds under the condition that some polynomials do not vanish (we call those polynomials the non degeneracy conditions)
7. Re-interpreting non degeneracy conditions as geometric predicates.

3 Formalization in Coq

In this section we describe our formalization of Wu's method in Coq. To formalize a decision procedure within Coq, there are several solutions:

1. We could write a tactic in the implementation language of Coq (OCaml) which generates a proof term.
2. We could write a tactic in Ltac, the tactic language of Coq: a domain specific language which allows pattern matching on the context of the proof and backtracking.
3. We could implement the decision procedure within Coq itself, prove its correctness and use it within Coq using reflection.
4. As explained in the introduction we can implement the decision procedure as an external tool which generates a certificate and then check the certificate by reflection using a procedure written in Coq itself.

The first and third solutions can be seen as special (extreme) cases of the fourth one. For the first solution, the certificate can be considered as the proof itself and the validator is trivial, and for the third solution the certificate contains no information and the validator is the whole decision procedure.

As explained in the introduction for the core of the method we choose an approach based on the validation of a certificate, and we have been able to reuse the machinery developed by Benjamin Grégoire, Loïc Pottier and Laurent Théry for the Gröbner basis. But for some steps of the method we use the tactic language of Coq (Ltac). We use Ltac for the first steps, to put the problem in algebraic form and to choose a system of coordinates. Figure 2 gives an overview of the different parts of the development and their implementation language. The simplification phase consists only perfoming some small simplifications using substitution in Ltac before starting the actual procedure when equations are trivial. The geometrization phase is not yet implemented. The simple procedure of Wu that we use may produce ndgs which are not necessary. We plan to implement geometrization only for the predicates corresponding to ndgs conditions of ruler and compass constructions (collinear, parallel, point equality).

In the next sections, we describe first the algebraization process and then the generation of the certificate. We do not describe fully the validator here because we could reuse the validator of Benjamin Grégoire, Loïc Pottier and Laurent Théry.

3.1 Algebraization

Putting the statement in algebraic form, choosing the right system of coordinates, and simplifying the generated coordinates, is the first step of the method, and it is crucial. In this section we describe the formalization of this step using the tactic language of Coq.

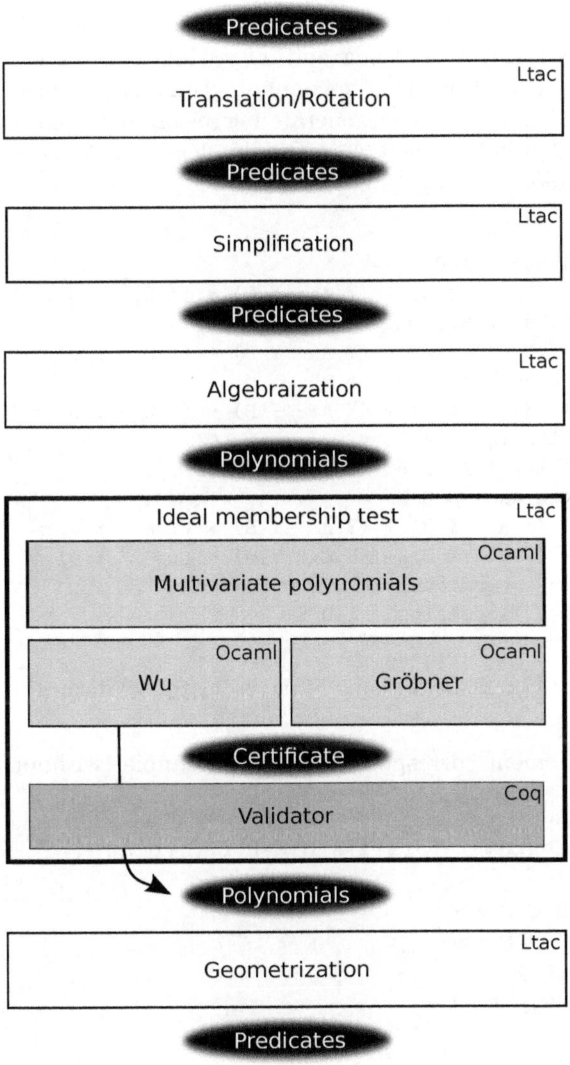

Fig. 2. Overview of the development

Stating the Conjecture. As most automatic theorem provers in geometry, to state a theorem, we will assume that the user provides the assumptions as a list of geometric predicates taking only points as parameters. This has the advantage to simplify the formalization and as shown in [Nar07], it is possible to transform a statement containing points and lines into a statement containing only points. Figure 3 gives the Coq definitions of some common geometric predicates. `X A` and `Y A` denotes respectively the x and y coordinates of point A. Figure 3 lists some of the Coq definitions for the common geometric predicates. These definitions have the advantage that degenerated cases are not excluded. For instance, `parallel A B C D` holds when $A = B$. This leads to statements which are as general as possible.

```
Definition collinear  A B C :=
  (X A - X B) * (Y B - Y C) = (Y A - Y B) * (X B - X C).
Definition parallel A B C D :=
  (X A - X B) * (Y C - Y D) = (Y A - Y B) * (X C - X D).
Definition orthogonal A B C D :=
  (X A - X B) * (X C - X D) + (Y A - Y B) * (Y C - Y D) = 0.
Definition is_midpoint I A B :=
  2 * X I = X A + X B /\ 2 * Y I = Y A + Y B.
Definition length_eq A B C D :=
  (X A - X B) * (X A - X B) + (Y A - Y B) * (Y A - Y B) =
  (X C - X D) * (X C - X D) + (Y C - Y D) * (Y C - Y D).
Definition is_in_intersection A B C D E :=
  collinear A B C /\ collinear A D E.
```

Fig. 3. Definition of some geometric predicates

Then the statement corresponding to our example (without the ndg) is the following:

```
Lemma parallelogram : forall A B C D E F:Point,
  B <> A ->
  parallel A B C D ->
  parallel A D B C ->
  is_midpoint E A C ->
  is_midpoint F B D ->
  equal E F.
```

To put the statement in algebraic form we just need to unfold the definitions of the geometric predicates. But in practice, even for simple examples, if we do not fix a specific coordinate system the computations will take much more time or may even fail due to lack of memory. The classical solution consists in adding additional assumptions to fix the coordinate system, assuming that one point is the origin and another belongs to the x axis:

```
Lemma parallelogram : forall A B C D E F:Point,
  X A = 0 -> Y A = 0 -> Y B = 0 ->
  B <> A ->
  parallel A B C D ->
  parallel A D B C ->
  is_midpoint E A C ->
  is_midpoint F B D ->
  equal E F.
```

This solution is sufficient to perform benchmarks in a context where the user is an expert. As we aim to apply this method in an educational context, we want to provide a procedure which works on the original statement.

We need to show that, *without loss of generality*, we can assume that A is of coordinates $(0, 0)$ and B of coordinates $(x_b, 0)$. Following the idea of John Harrison [Har09], this requires to show that the predicates we use are invariant under translation and rotation. For that purpose, we define a translation function taking as arguments a point and a vector (`trans`) and a rotation function (`rot`) taking as arguments a point and the sine and cosine of the angle of rotation. Then for each predicate (collinear in this example), we prove lemmas of the following form:

```
Lemma collinear_inv_translation: forall A B C V,
 collinear A B C <->
 collinear (trans A V) (trans B V) (trans C V).
Lemma collinear_inv_rotation: forall A B C cos sin,
 cos*cos + sin*sin = 1 ->
 (collinear A B C <->
  collinear (rot A cos sin) (rot B cos sin) (rot C cos sin)).
```

Finally, we have a tactic `Algebraization O I H` which takes as input a point O to be put at the origin, a point I to be put on the x-axis and a proof that $O \neq I$, and which makes use of the above lemmas to perform the required simplifications. The algebraization tactic only work for goals which are stated using function and predicates which take as arguments only points. The following predicates/functions are available: collinear, parallel, orthogonal, midpoint, intersection of lines, square of length, equality of points, angles or lengths. The tactic can not deal with user defined predicates automatically. Adding a new predicate requires to add the lemmas for invariance under translation and rotation and to update the tactic.

We leave for future work, the automatic choice of a reference. Our experiments shows that this choice is crucial : choosing different references can lead to very different computation times. The heuristic proposed by Chou is to choose as an axis a line which contains many points and also a perpendicular line. We noticed that choosing as the origin one of the points of the goal is also often a good choice, because it simplifies the starting polynomial of the successive pseudo-division.

After application of the tactic `Algebraization A B H`, we get the following goal:

```
H : 0 = - X P2 * P3 + X P0 * P3
H0 : - P3 * X P0 + Y P2 * X P0 = X P2 * Y P0
H4 : 2 * X P = X P0
H5 : 2 * Y P = P3 + Y P0
H2 : 2 * X P1 = X P2
H6 : 2 * Y P1 = Y P2
______________________________________(1/1)
   X P = X P0
```

3.2 Certified Implementation

In the section we describe our implementation of a simple version of Wu's method and how it generates certificates.

Implementation of the Method. To integrate Wu's method in Coq, we need to modify the implementation to generate a certificate. Unfortunately, the implementations of Wu's method which are already available are either not open source or rely on proprietary Computer Algebra Systems. Moreover, we hope that in the future our tactic could be distributed with Coq, hence we decided to write our own version of the method in OCaml, the implementation language of Coq. We base our implementation of Wu's method on the OCaml libraries for dealing with multivariate polynomials developed by Loïc Pottier for the integration of Gröbner basis in Coq. He used two different data structures for polynomials : first a list of monomials, each monomial is represented by a coefficient and an array containing the degree of each variable and the total degree of the monomial, second a recursive data structure considering polynomials with several variables as a polynomial with one variable but whose coefficients are polynomials in the other variables. We reused his data-structures, and slightly optimized the one based on list of monomials by changing the representation of monomials: in the arrays of degrees we store only the degrees of the variables until the highest variable which appears in the monomial. This variant reduces both the memory footprint and the number of needed comparisons of degree. This is crucial, because 30% of the computing time is spent in the monomial comparison function which is the core of the multiplication function. We implement the simple algorithm presented by Chou in [Cho88], hence it is incomplete because we do not check polynomials for irreducibility. But in practice many theorems can be proved using this simple method. Our certifying implementation consists in 3000 lines of OCaml half of which are tests and examples.

Certificate Generation. Certificates are pieces of information, which allow to check if the result of a computation is correct. In our case, the certificate we generate is based on the *Nullstellensatz*, hence we need to produce a proof that the polynomial which represents the goal (multiplied by a coefficient representing the ndgs) is in the ideal generated by the hypotheses. We need to show that there are some integer r, polynomials I and S_i such that:

$$I^r G = \sum_i (S_i H_i)$$

As explained in Sec. 2, the core of Wu's method can be decomposed into two main steps: the triangulation and the successive pseudo-divisions.

To generate the certificate we generate three kinds of intermediary certificates: one for the pseudo-divisions, one for the triangulation, and another one for the successive pseudo-division procedure.

Certificates for pseudo-division. The certificates for the pseudo-division will be used both by the triangulation and successive pseudo-division procedure.

Recall that the pseudo-division of A and B return a remainder R[2] such that there exists an integer d and a polynomial Q such that:

$$I^d A = QB + R$$

where I is the leading coefficient of B.

So to certify that R is in the ideal generated by A and B, instead of just returning the remainder R we also return I, d and Q which verify that $R = I^d \times A + (-Q) \times B$.

To ease implementation we number our polynomials. Then the implementation of the pseudo-division is encapsulated into a function which stores in the list of certificates a certificate for each polynomial identified by its number. The certificate for each polynomial is in the form of a list of pairs of a polynomial and the number of a polynomial which we already know to be in the ideal :

```
let pseudo_div_num a b x certif =
    let (r,c,d,s) = pseudo_div (a.p) (b.p) x in
    let new_n = new_num () in
      certif := (new_n, r, [(c^^d, a.n);(p_zero -- s, b.n)])
                ::(!certif);
      {p=r ; n= new_n}
```

Certificates for the triangulation phase. For the triangulation, we do not need to show that the result is triangulated, we just need to show that the polynomials in the triangulation are in the ideal generated by the hypotheses. We notice that the triangulation phase of Wu's method and its variants based on pseudo-division rely on an invariant: the method maintains a list of polynomials which belong to the ideal generated by the hypotheses. At the beginning of the triangulation procedure, the list consists of the original hypotheses themselves, hence they are in the ideal. At each step of the triangulation, the list of polynomials l is replaced

[2] Note that there is a condition on the degree of R but this is not important in this discussion.

by a new list l' such that l and l' differ only in one polynomial $p \in l$ (let's call it $p' \in l'$). This polynomial p' is in the ideal generated by l.

The fact that p' is in the ideal generated by l follows from the fact that p' is computed using a pseudo-division $p' = prem(p, h)$ for some h in l.

Hence, to generate the certificate of the triangulation, we keep track of every intermediate polynomial thanks to its number, and we combine the certificates of the pseudo-divisions.

Certificates for the successive pseudo-division. For the last phase (the successive pseudo-division of the goal by the triangulated polynomials), we just save the S_i, and I as:

$$IG = \sum_i (S_i T_i)$$

where $S_i = q_i \prod_{j=1}^{i-1} c_j^{d_j}$ and $I = \prod_{j=1}^{n} c_j^{d_j}$ and the c_j are the leading coefficients of the polynomials in the triangulation.

But, if during the combination of the different auxiliary certificates we evaluate the S_i, then we get polynomials with thousands of terms. Polynomials of this size cannot be managed by Coq. Grégoire, Pottier and Théry proposed to use certificates involving not simply the list of the S_i but a *straight line program* to evaluate the S_i. This provides the possibility to have some `let ... in` instructions inside the certificates to allow to decompose and to share polynomial expressions. The validator is a function which computes IG and $\sum_i (S_i T_i)$ and which test the equality of these two polynomials.

Using their certificate structure we can define auxiliary polynomials which can then be used to define several other polynomials, *etc.*

To reduce the time required to check certificates, we make use of those `let ... in` instructions as most as possible. But as shown in the Sec. 4 the verification time of the certificates is still significant.

Toward more optimized certificates for ideal membership tests. In this section we present some ideas for improvement (which are not yet implemented) of the structure for certificates.

One limitation of the certificates we use relies on the fact that every polynomial which is defined by a `let ... in` needs to be in the ideal generated by the previous polynomials. This bring no restriction for the polynomials generated by the Gröbner basis method, but in the context of Wu's method, it would be useful to be able to define polynomials using `let ... in` which are not in the ideal generated by the hypotheses. This would allow some sharing between the definition of I and the definition of the S_i. Another idea to reduce drastically the size of the certificates would be to allow to have unevaluated pseudo-divisions in the certificates. This would require to prove formally the pseudo-division. But the pseudo-division is a time consuming task of Wu's method, hence this would imply to do a lot of computations during the validation of the certificate.

4 Benchmark

Table 1 presents our results compared to those of L. Pottier *et al.* based on Gröbner basis. The results show no clear winner, Wu's method is faster in some cases and Gröbner basis is faster in some other cases. We were surprised by the time needed to check the certificate in Coq. To study this, we reimplemented the validator in OCAML and used it with our implementation of the method. The first column provides the computation time to produce the certificate *and* to check it when using an OCAML validator : those timing are encouraging. The percentage of time required to check the certificates using the OCAML validator varies from 1% to 80% but it is about 50% on average.

We did not expect such a difference of computation time between OCAML and Coq. The data structure used by the Coq validator is a a modified Hörner form provided by the ring tactic[GM05]. We tried another implementation based on lists of monomials, but the computation time was similar. The results could be improved in the future by the use of NATIVE-COQ a COQ version using the native OCAML compiler to perform strong reduction [BDG11].

Table 1. Benchmark : Computations are done using an Intel(R) Core(TM) i5 CPU 750 @ 2.67GHz with 4Gb RAM

Theorem	Certificate generated using **Wu** and checking using **Ocaml**	Certificate generated using **Wu** and checking using **Coq**	Certificate generated using **GB** and checking using **Coq**	Relative speed of Wu's method over GB
Pascal_2	0.013	21	-	-
Pascal_1	0.024	22	1652	×75
Ptolemy95	0.010	10	30	×3
Pappus	0.043	3	8	×2.6
Altitudes	0.002	3	7	×2.3
Simson	0.002	5	8	×1.6
Perp-bisect	0.001	2	3	×1.5
Pythagore	0.001	1	1	×1
Feuerbach	0.038	15	15	×1
Isoceles	0.001	1	1	×1
Euler Line	0.063	9	6	×0.6
Medians	0.001	3	2	×0.6
Chords	0.015	4	2	×0.5
Thales	0.003	6	3	×0.5
Bissectors	0.001	6	3	×0.5
Desargues	0.027	99	10	×0.1
Ceva	0.025	98	6	×0.06

5 Conclusion

In this paper, we described the first formalization of Wu's simple method in an interactive proof assistant. We implemented a version of Wu's method which generates certificates to ensure the correctness of the results generated by the method. We reused and extended the framework for certificate validation introduced by Benjamin Grégoire, Loïc Pottier and Laurent Théry, and we present some ideas for improvement. Thanks to the power of Wu's method, the new tactic we obtain does not only prove geometry theorems but also generates non degeneracy conditions necessary to prove a conjecture. Compared to other implementations of the method, as we formalize the invariance by translation and rotation of the statements, we prove the real geometry theorems and not a special case of the algebraic version of the statement given in a specific coordinate system.

In the future, we plan to extend our OCAML implementation with a more optimized version of the pseudo-division (using GCD computations) along with the complete method of Wu-Ritt [Wu78, Cho88, CG90] or its variants developed by Dongming Wang [Wan01, Wan04]. For educational applications our implementation still needs to be polished for example by extending the input language and by choosing automatically an origin and a direction for the system of coordinates.

The formalization of Wu's method in COQ enriches our framework for interactive and automatic theorem proving in geometry. We have now four methods for automated deduction in geometry available in COQ. It opens the door to several directions of research. Among them, it would be interesting to study the combination of Wu's method with other methods inside the same framework, for instance with the cylindrical algebraic decomposition method (CAD) whose formalization in COQ is under way by Assia Mahboubi [Mah06]. We could also study the different applications of Wu's method, such as solving geometric constraint systems [GC98a, GC98b] or finding loop invariants in the context of program verification.

Availability. The current version of the plug-in is available here:
`http://dpt-info.u-strasbg.fr/~narboux/nsatzwu.tar.gz`

Acknowledgments. We wish to thank Loïc Pottier and Laurent Théry for having made their work publicly available and for the discussions we had.

References

[BDG11] Boespflug, M., Dénès, M., Grégoire, B.: Full Reduction at Full Throttle. In: Jouannaud, J.-P., Shao, Z. (eds.) CPP 2011. LNCS, vol. 7086, pp. 357–372. Springer, Heidelberg (2011)

[CG90] Chou, S.-C., Gao, X.-S.: Ritt-Wu's Decomposition Algorithm and Geometry Theorem Proving. In: Stickel, M.E. (ed.) CADE 1990. LNCS, vol. 449, pp. 207–220. Springer, Heidelberg (1990)

[CG92] Chou, S.-C., Gao, X.-S.: A Class of Geometry Statements of Constructive Type and Geometry Theorem Proving. In: Kapur, D. (ed.) CADE 1992. LNCS, vol. 607, pp. 20–34. Springer, Heidelberg (1992)

[CGZ94] Chou, S.-C., Gao, X.-S., Zhang, J.-Z.: Machine Proofs in Geometry. World Scientific, Singapore (1994)

[Cho88] Chou, S.-C.: Mechanical Geometry Theorem Proving. D. Reidel Publishing Company (1988)

[Coq10] Coq development team. The Coq proof assistant reference manual, Version 8.3. LogiCal Project (2010)

[CW07] Chaieb, A., Wenzel, M.: Context Aware Calculation and Deduction. In: Kauers, M., Kerber, M., Miner, R., Windsteiger, W. (eds.) MKM/CALCULEMUS 2007. LNCS (LNAI), vol. 4573, pp. 27–39. Springer, Heidelberg (2007)

[FT11] Fuchs, L., Théry, L.: A Formalization of Grassmann-Cayley Algebra in Coq and its Application to Theorems Proving in Projective Geometry. In: Schreck, P., Narboux, J., Richter-Gebert, J. (eds.) ADG 2010. LNCS (LNAI), vol. 6877, pp. 51–67. Springer, Heidelberg (2011)

[Gao00] Gao, X.-S.: Geometry expert, software package (2000)

[GC98a] Gao, X.-S., Chou, S.-C.: Solving geometric constraint systems. I. A global propagation approach. Computer Aided Design 30(1), 47–54 (1998)

[GC98b] Gao, X.-S., Chou, S.-C.: Solving geometric constraint systems. II. A symbolic approach and decision of Rc-constructibility. Computer Aided Design 30(2), 115–122 (1998)

[GL02] Gao, X.-S., Lin, Q.: MMP/Geometer – A Software Package for Automated Geometric Reasoning. In: Winkler, F. (ed.) ADG 2002. LNCS (LNAI), vol. 2930, pp. 44–66. Springer, Heidelberg (2004)

[GM05] Grégoire, B., Mahboubi, A.: Proving Equalities in a Commutative Ring Done Right in Coq. In: Hurd, J., Melham, T. (eds.) TPHOLs 2005. LNCS, vol. 3603, pp. 98–113. Springer, Heidelberg (2005)

[GPT11] Grégoire, B., Pottier, L., Théry, L.: Proof Certificates for Algebra and Their Application to Automatic Geometry Theorem Proving. In: Sturm, T., Zengler, C. (eds.) ADG 2008. LNCS (LNAI), vol. 6301, pp. 42–59. Springer, Heidelberg (2011)

[Har07] Harrison, J.: Automating Elementary Number-Theoretic Proofs Using Gröbner Bases. In: Pfenning, F. (ed.) CADE 2007. LNCS (LNAI), vol. 4603, pp. 51–66. Springer, Heidelberg (2007)

[Har09] Harrison, J.: Without Loss of Generality. In: Berghofer, S., Nipkow, T., Urban, C., Wenzel, M. (eds.) TPHOLs 2009. LNCS, vol. 5674, pp. 43–59. Springer, Heidelberg (2009)

[JNQ10] Janičić, P., Narboux, J., Quaresma, P.: The Area Method: a Recapitulation. Journal of Automated Reasoning (2010)

[JQ06] Janičić, P., Quaresma, P.: System Description: GCLCprover + Geothms. In: Furbach, U., Shankar, N. (eds.) IJCAR 2006. LNCS (LNAI), vol. 4130, pp. 145–150. Springer, Heidelberg (2006)

[Kap86] Kapur, D.: Geometry Theorem Proving using Hilbert's Nullstellensatz. In: SYMSAC 1986: Proceedings of the fifth ACM Symposium on Symbolic and Algebraic Computation, New York, NY, USA, pp. 202–208. ACM Press (1986)

[Mah06] Mahboubi, A.: Contributions à la certification des calculs dans R: théorie, preuves, programmation. PhD thesis, Université de Nice Sophia-Antipolis (November 2006)

[Nar04] Narboux, J.: A Decision Procedure for Geometry in Coq. In: Slind, K., Bunker, A., Gopalakrishnan, G.C. (eds.) TPHOLs 2004. LNCS, vol. 3223, pp. 225–240. Springer, Heidelberg (2004)

[Nar07] Narboux, J.: A graphical user interface for formal proofs in geometry. J. Autom. Reasoning 39(2), 161–180 (2007)

[PBN11] Pham, T.-M., Bertot, Y., Narboux, J.: A Coq-Based Library for Interactive and Automated Theorem Proving in Plane Geometry. In: Murgante, B., Gervasi, O., Iglesias, A., Taniar, D., Apduhan, B.O. (eds.) ICCSA 2011, Part IV. LNCS, vol. 6785, pp. 368–383. Springer, Heidelberg (2011)

[Pha10] Pham, T.M.: An Additional Tool About the Orientation for Theorem Proving in the Coq Proof Assitant. In: Proceedings of Automated Deduction in Geometry, ADG 2010 (2010)

[Pot08] Pottier, L.: Connecting Gröbner Bases Programs with Coq to do Proofs in Algebra, Geometry and Arithmetics. In: Sutcliffe, G., Rudnicki, P., Schmidt, R., Konev, B., Schulz, S. (eds.) Knowledge Exchange: Automated Provers and Proof Assistants. CEUR Workshop Proceedings, Doha, Qatar p. 418 (2008)

[Rob02] Robu, J.: Geometry Theorem Proving in the Frame of the Theorema Project. PhD thesis, Johannes Kepler Universitt, Linz (September 2002)

[Wan89] Wang, D.: A new theorem discovered by computer prover. Journal of Geometry 36, 173–182 (1989), doi:10.1007/BF01231031

[Wan01] Wang, D.: Elimination Method. Springer, Heidelberg (2001)

[Wan02] Wang, D.: GEOTHER 1.1: Handling and Proving Geometric Theorems Automatically. In: Winkler, F. (ed.) ADG 2002. LNCS (LNAI), vol. 2930, pp. 194–215. Springer, Heidelberg (2004)

[Wan04] Wang, D.: Elimination Practice. Springer, Heidelberg (2004)

[Wu78] Wu, W.-T.: On the Decision Problem and the Mechanization of Theorem Proving in Elementary Geometry. Scientia Sinica 21, 157–179 (1978)

[YCG11] Ye, Z., Chou, S.-C., Gao, X.-S.: An Introduction to Java Geometry Expert. In: Sturm, T., Zengler, C. (eds.) ADG 2008. LNCS, vol. 6301, pp. 189–195. Springer, Heidelberg (2011)

Reasoning about Constants in Nominal Isabelle or How to Formalize the Second Fixed Point Theorem

Cezary Kaliszyk[1] and Henk Barendregt[2]

[1] University of Tsukuba
[2] Radboud University Nijmegen

Abstract. Nominal Isabelle is a framework for reasoning about programming languages with named bound variables (as opposed to de Bruijn indices). It is a definitional extension of the HOL object logic of the Isabelle theorem prover. Nominal Isabelle supports the definition of term languages of calculi with bindings, functions on the terms of these calculi and provides mechanisms that automatically rename binders. Functions defined in Nominal Isabelle can be defined with assumptions: The binders can be assumed fresh for any arguments of the functions. Defining functions is often one of the more complicated part of reasoning with Nominal Isabelle, and together with analysing freshness is the part that differs most from paper proofs. In this paper we show how to define terms from λ-calculus and reason about them without having to carry around the freshness conditions. As a case study we formalize the second fixed point theorem of the λ-calculus.

1 Introduction

Proofs about abstract calculi are often complicated and tedious; which is why they are often done only semi-formally. Nominal Isabelle [13], has been designed to make this kind of proofs easy to formalize. It provides an infrastructure for defining a term language of a calculus (nominal datatypes) together with a reasoning infrastructure about those datatypes. This means a number of lemmas that describe properties of the introduced type, including induction principles that already have the variable convention built in. The framework also provides a number of mechanisms for defining terms and functions over those datatypes. Nominal Isabelle has been used for formalizing proofs about calculi including LF [15], π-calculus [3], and ψ-calculus [2]. Here we will look at formalizing definitions and proofs about terms in the λ-calculus, like pairs, finite sequences, or initial functions [1].

Defining such constants and simple functions can be done in numerous ways. Nominal Isabelle provides *nominal_primrec*, a mechanism for defining functions with primitive recursion [14]. Functions defined with its help can be defined with assumptions: the binders can be assumed fresh for any arguments of the functions. The only proof obligations necessary to fulfill are the FCBs (*fresh*

J.-P. Jouannaud and Z. Shao (Eds.): CPP 2011, LNCS 7086, pp. 87–102, 2011.

condition for binder). This is convenient for functions that analyze term structure, but it does not allow for functions that invent new variables on the right hand side (this is the case for defining constants with λ-abstractions). It is also limited to primitive recursion, but this is not an issue since termination for all the functions that we define here is straightforward.

A different way of defining nominal constants and functions is provided with the *fresh_fun* binder. This construction allows introducing a new variable, that is fresh for the term under the binder. This allows defining more complicated nominal functions, for some CPS translations (the non higher-order ones). Unfortunately reasoning about *fresh_fun* involves reasoning about the term and the freshness obligations *together*, which means that renamings are performed manually. This makes the proofs complicated and very hard to read, especially in the case of iterated applications of *fresh_fun*. In the case of the proofs about the CPS translation, performing them without the help of the Nominal package [9] was less involved than reasoning with *fresh_fun*.

It is also possible to define constants using the Isabelle/HOL function package directly. Given a non-injective datatype, the function package returns a completeness obligation and compatibility obligations. With this approach it is in principle possible to define mutually recursive functions, non-primitive-recursive functions, or even functions on datatypes which abstract multiple binders. In practice fulfilling the proof obligations about the defined function that the function package requires is not feasible with the infrastructure given by the function package (it does not even provide regular induction).

The current implementation of Nominal Isabelle uses quotients [8] to lift the reasoning infrastructure from a raw datatype to a nominal type modulo α-equivalence. One could define constants or functions on the raw datatype and use the same lifting process to obtain a lifted function. This requires showing respectfulness of the raw function, which is true and it is provable. Unfortunately on the raw level we cannot use any of the nominal techniques, and proving such goals does require rederiving the reasoning infrastructure for the raw terms.

This makes defining simple functions or even constants one the more complicated parts of reasoning with Nominal Isabelle. Given that such a function or constant is defined, it normally has freshness obligations for the binders and new variables. When proving properties that involve such obligations with the help of the Nominal package, there is a tendency to leave those freshness obligations to the simplifier. When this does not work, first the user tries to rewrite with a conditional rewrite rule to get from Isabelle the remaining freshness proof obligation. The proof is then changed in such a way that the assumption is available in the goalstate, and the proof can be fully automated with a call to the simplifier. The proof obtained in this way can look similar to a textbook one (with the exception of the step where one specifies the constants to avoid), however this approach has one substantial drawback: the final proof does not reflect the structure in which the proof was written and does not show the reasoning necessary for fulfilling the freshness obligations.

In this paper we look at defining constants. We take an approach different from the ones presented above. Starting with a function that returns different variables, we define constants using those variables and show their convertibility to the forms with freshness obligations. We then look at applied versions of the constants to derive convertibility relations that have no freshness obligations and only add those to the simplifier. This allows reasoning about convertible terms without any freshness assumptions and the formalized proofs resemble textbook proofs. We formalize the second fixed point theorem of the λ-calculus and obtain a formalized proof that resembles a pen and paper one.[1]

The rest of the paper is organized as follows: In Section 2 we introduce the notions from Nominal Logic necessary in later part of the paper. In Section 3 we the introduce the term language and the conversion relation of the λ-calculus, and compare it with classical textbook definition. In Section 4 we show our infrastructure used for defining constants with Nominal Isabelle. Our case-study, the formalization of the second fixed point theorems follows in next two Sections: We show the definitions and the properties of the basic constants in Section 5 and the proof in Section 6. We conclude in Section 7.

2 Nominal Logic Preliminaries

In this section we introduce the core notions of nominal logic, that will be used in the remaining part of the paper. Nominal logic has been introduced by Pitts [12] and has been adapted to the Isabelle/HOL formal setting by Urban. The details of the current version of the adaptation including proofs can be found in [7]. Here we shortly introduce only the main notions that will be used further.

The central notions of nominal logic are sorted atoms and sort-respecting permutations. Atoms are supposed to represent variables (both bound and free ones). Sorts can be used, if a calculus has different kinds of variables, such as names and identifiers in LF. Each atom type has an infinite supply of atoms. Further in this paper we will only talk about one kind of atoms; namely the variables of the classical λ-calculus, thus and we will omit the function *atom* which projects concrete atom types to the type of all atoms.

A permutation is a function that swaps a finite number of atoms. We denote permutations with π and an application of a permutation to an object as $\pi \bullet M$. Permutations applied to an object change only the atoms of the object. The permutation application operation, takes a permutation and an object of the general type-class of objects for which permutations are defined. The identity permutation is written as *0* and the composition of two permutations π_1 and π_2 as $\pi_1 + \pi_2$. We denote the inverse permutation of π as $-\pi$. Application of permutations naturally extends to products, lists, functions and datatypes (nominal datatypes) as described in [16].

The smallest non-trivial permutation is called a swapping. We denote a swapping of atoms a and b as $(a\ b)$, and define by:

[1] The source of the formalization is included in the new version of Nominal Isabelle, available from the repository linked at http://isabelle.in.tum.de/nominal/.

$$(a\ b) = \lambda c.\ if\ a = c\ then\ b\ else\ if\ b = c\ then\ a\ else\ c$$

The most important notion of the nominal logic work is a general definition for the "free variables of an object", called *support*. We will denote it *supp x*. This notion applies not only to terms defined in Nominal Logic, but also to lists, pairs, sets or general functions. For the definition to be applicable, the type only needs to have the permutation operation and equality defined:

$$supp\ x = \{a.\ infinite\ \{b.\ (a\ b) \bullet x \neq x\}\}$$

From the notion of support, the notion of atoms being fresh for an object is derived. Atom a is *fresh* for an object x:

$$a \# x = a \notin supp\ x$$

Nominal logic also defines *equivariance*. An object x is equivariant if applying any permutation to the object leaves it unchanged.

$$\forall \pi \bullet x = x$$

All objects that have no atoms (like natural numbers, booleans, ...) are equivariant. Equivariance is especially important for functions, as this is equivalent to the fact that a permutation applied to the application of the function to its arguments *f a1 a2 .. an* is equal to the application of the permutation to all the arguments. This means that the application of any permutations commutes with the application of the function:

$$\forall \pi.\ \pi \bullet (f\ a_1\ a_2\ \ldots\ a_n) = f\ (\pi \bullet a_1)\ (\pi \bullet a_2)\ \ldots\ (\pi \bullet a_n)$$

Finally, if an object is equivariant, any swappings leave the object unchanged so no atoms are in its support. We will use this property often, to show that a term is a closed term.

3 Lambda Terms and Conversion

The definition of the terms of λ-calculus, as done in textbooks, starts with an alphabet of variables, the λ-abstractor and parentheses and continues with an inductive definition of the set of terms. Below we show the definition according to the second author's book [1].

Definition 1. *(i)* Lambda terms *are words over the following alphabet:*

$v_0, v_1, \ldots$	*variables,*
λ	*abstractor,*
(,)	*parentheses.*

(ii) The set of λ-terms Λ is defined inductively as follows:

1. $x \in \Lambda$;
2. $M \in \Lambda \Rightarrow (\lambda x M) \in \Lambda$;
3. $M, N \in \Lambda \Rightarrow (MN) \in \Lambda$;

where x in 1 or 2 is an arbitrary variable.

A textbook would next add notations that allow writing λ-terms without the brackets that can be unambiguously removed. To make a similar definition in Isabelle one needs to start with defining the set of variables. This is performed with:

atom-decl var

which introduces a new sort *var* that will be used to name the variables. The command from Nominal Isabelle introduces a new set of variables that is fresh for all existing sets and is shown to be infinite. The fact that it is infinite will be necessary to rename variables.

Defining an inductive type in Isabelle requires giving its cases together with the notations. Additionally Nominal Isabelle lets us specify bindings, here we specify that the abstracted variable is bound in the lambda case. This instructs Nominal Isabelle to derive a type where variables in the λ-abstractions can be renamed:

nominal_datatype lam =
  $\overline{v}$
| lam · lam
| λx. lam **bind** x **in** lam

In Isabelle, constructors of inductive types embed particular objects of the type into the type itself. This is why we need a constructor that embeds variables into λ-terms. In the above definition the $\overline{v}$ means a lambda terms that is a variable, as opposed to v which represents a variable itself. Application is denoted with an infix left-associative notation, and λ-abstraction is annotated with a binding. With notations set this way, terms can be written in a way that resembles the textbook notation, but are still an inductive type.

Next, we would like to define convertibility of λ-terms. A formal definition of substitution would often be omitted in textbooks, but Isabelle requires a formal definition of substitution to define β-conversion. Nominal Isabelle allows defining functions where in the clauses the bound variable can be fresh for any other terms. We show this in the standard definition of substitution from Nominal Isabelle:

Definition 2 (Substitution). *Substituting a variable y for a term S in term M is defined by:*

$$T\ [y := S] = \begin{cases} \textit{if } x = y \textit{ then } S \textit{ else } \overline{x} & \textit{if } T = \overline{x} \\ (T_1\ [y := S]) \cdot (T_2\ [y := S]) & \textit{if } T = T_1 \cdot T_2 \\ \lambda x.\ U\ [y := S] & \textit{if } T = \lambda x.\ U \textit{ and } x \# (y,\ S) \end{cases}$$

A regular definition of substitution would require renaming the variable in the lambda case. Nominal Isabelle extends the definition given only for the case where the binder is fresh for the given arguments to a function on the whole domain. Next we show a number of simple properties of substitution, that we want to use to reason about terms involving substitution and freshness later:

Lemma 1. *Substitution is equivariant, substituting a fresh variable does not change the term and substituting a variable for itself does not change the term.*

$$\begin{array}{ll} \pi \bullet a\ [aa := ab] = (\pi \bullet a)\ [(\pi \bullet aa) := (\pi \bullet ab)] & \\ x \# t \Longrightarrow t\ [x := s] = t \qquad M\ [x := \overline{x}] = M & \end{array} \tag{1}$$

Proof. By induction, follows by definitions of substitution and freshness.

Closed terms have an empty support, so any variable will be fresh for one. This implies that substitution does not change closed terms:

$$supp\ t = \varnothing \Longrightarrow t\ [x := s] = t$$

Having defined substitution we can return to the definition of convertibility. Again we try to closely follow [1].

Definition 3 (Convertibility). *Convertibility is an inductively defined relation axiomatized by the following:*

$$\begin{array}{l} (\lambda x.\ M) \cdot N \approx M\ [x := N] \\ M \approx M \\ M \approx N \Longrightarrow N \approx M \\ M \approx N \Longrightarrow N \approx L \Longrightarrow M \approx L \\ M \approx N \Longrightarrow Z \cdot M \approx Z \cdot N \\ M \approx N \Longrightarrow M \cdot Z \approx N \cdot Z \\ M \approx N \Longrightarrow (\lambda x.\ M) \approx (\lambda x.\ N) \end{array} \tag{2}$$

In this paper, we will use two equivalence relations on terms, namely the Isabelle equality = and the convertibility relation ≈ defined above. This is slightly different from the regular textbook approach where convertibility in introduced in addition to syntactic equality. Nominal Isabelle has defined the type of lambda terms in such a way that equality on this type includes α-equivalence, hence it is different from syntactic equality. Nominal Isabelle can automatically derive equivariance for convertibility. Since many of our proofs will talk about chains of convertibilities, we declare convertibility as a transitive relation. This will allow writing $M \approx \ldots \approx N$ in the formal proofs later.

4 Defining λ-Constants with Nominal Isabelle

In order to define all the constants from the next Section, we will need two mechanisms. First a set of variables that will be fresh in the definitions and second a convenient way to show that concepts defined with those fixed variables can be renamed to any fresh variables and therefore applied without freshness obligations.

Nominal Isabelle ensures that there is an infinite amount of atoms in any of the concrete atom sorts. This means that for any finite set of variables we can find a new one, fresh for this set.

Using this property we can define a function $\nu :: \mathbb{N} \Rightarrow var$ which given a number n returns a variable v that is fresh for the set of the results of the function for all smaller numbers:

$$\nu n \sharp \bigcup_{k=0..(n-1)} \{\nu k\} \tag{3}$$

A direct consequence is, that for any two different natural numbers m and n:

$$\nu m \neq \nu n$$

Given a constant defined with the help of a fixed set of νk, we would like to rename the variables to any variables that satisfy the property 3. This can be done by analyzing the equalities between the variable pairs and applying the definition of swapping atoms together with equivariance. With the help of the Sledgehammer SMT linkup [4] this can be performed fully automatically.

Given a definition that depends on freshness obligations, we cannot apply this definition without an additional step in which we obtain the new variables specifying what are they supposed to be fresh for. This is however not going to be the case for constant terms representing functions in the λ-calculus. Since they need to have a λ-abstraction at the top-level, after the abstraction is applied, the newly invented variable can be forgotten. A simplest possible example is:

$$(\lambda x.\ M) \cdot \overline{x} \approx M \tag{4}$$

We can extend this to arbitrary functions, defined by equations that start with a λ-abstraction:

Lemma 2. *Assuming that function L is defined by the equation*

$$\forall a.\ (L = (\lambda a.\ F\ (\overline{a})))$$

and we can substitute variables in F

$$\forall x.\ (x \# A \Longrightarrow F\ (\overline{x})\ [x := A] = F\ A)$$

then

$$L \cdot A \approx F\ A$$

Proof. Obtaining a fresh variable a, such that $a \# A$, using the definition of convertibility (2) and properties of substitution (1) the proof is a simple calculation.

The same mechanism works for functions that require more new variables. The additional assumption that the function makes, is that the variables are distinct, which follows from 3:

Lemma 3. *Assuming that function L is defined by the equation*

$$\forall\, a\ b.\ (a \neq b \Longrightarrow L = (\lambda a.\ \lambda b.\ F\ (\overline{a})\ (\overline{b})))$$

and we can substitute variables in F

$$\forall\, x\ y.\ (x \ \#\ (A,\ B) \wedge y \ \#\ (A,\ B,\ x) \Longrightarrow \\ \Longrightarrow F\ (\overline{x})\ (\overline{y})\ [x := A]\ [y := B] \ = \ F\ A\ B)$$

then

$$L \cdot A \cdot B \approx F\ A\ B$$

Proof. Obtaining fresh variables a, such that $a \ \#\ (A,\ B)$, and b such that $b \ \#\ (A,\ B,\ a)$, using freshness for tuples, the definition of convertibility (2) and properties of substitution (1) the proof is a simple calculation.

5 Basic Constants in λ-Calculus

In this Section we start a presentation of a case-study: formalization of the second fixed point theorem of the λ-calculus. The proof relies on an internal coding of lambda terms as normal lambda terms

$$M \longmapsto \ulcorner M \urcorner.$$

We take the coding used in [5] for its elegance, enabling a short proof, but other codings could have been used as well.

We start with the definition of the initial functions U_n^m. We slightly modify the usual textbook definition, by using a different indexing, starting from 0. We define it for all natural numbers, but intend to use it for $m \leq n$.

Definition 4 (initial functions). *For any natural numbers m and n:*

$$U_n^m = \begin{cases} \lambda\nu\mathit{0}.\ \overline{\nu n} & \textit{provided } m = 0, \\ \lambda\nu m.\ U_n^k & \textit{provided } m = k+1. \end{cases}$$

The terms U_n^m are equivariant and have empty support in the intended domain.

Lemma 4. *Assuming $m \leq n$:*

$$\mathit{supp}\ (U_n^m) = \varnothing \qquad\qquad \pi \bullet U_n^m = U_n^m$$

Proof. By induction on m.

When reasoning about the second fixed point theorems, we are going to use only three cases of U_n^m, for $m = 2$ and $n \leq 2$. By expanding the definition for these three cases we can see that it indeed selects the desired elements. We rename the variables to arbitrary but distinct x, y and z.

Assuming $x \neq y$, $y \neq z$ and $z \neq x$:

$$U^2_0 = \lambda x.\ \lambda y.\ \lambda z.\ \overline{z}$$
$$U^2_1 = \lambda x.\ \lambda y.\ \lambda z.\ \overline{y}$$
$$U^2_2 = \lambda x.\ \lambda y.\ \lambda z.\ \overline{x}$$

Definition 5. *Assuming $x \neq y$, $x \neq e$ and $y \neq e$:*

$$Var = \lambda x.\ \lambda e.\ \overline{e} \cdot U^2_2 \cdot \overline{x} \cdot \overline{e}$$
$$App = \lambda x.\ \lambda y.\ \lambda e.\ \overline{e} \cdot U^2_1 \cdot \overline{x} \cdot \overline{y} \cdot \overline{e}$$
$$Abs = \lambda x.\ \lambda e.\ \overline{e} \cdot U^2_0 \cdot \overline{x} \cdot \overline{e}$$

Using the lemma 3 we can find equations that express convertibility of the applications of *Var*, *App*, *Abs*, that will have no freshness obligations:

$$Var \cdot x \cdot e \approx e \cdot U^2_2 \cdot x \cdot e \tag{5}$$
$$App \cdot x \cdot y \cdot e \approx e \cdot U^2_1 \cdot x \cdot y \cdot e \tag{6}$$
$$Abs \cdot x \cdot e \approx e \cdot U^2_0 \cdot x \cdot e \tag{7}$$

Lemma 5. *The terms Var, App, Abs are closed terms and are equivariant:*

$$supp\ Var = \varnothing \qquad supp\ App = \varnothing \qquad supp\ Abs = \varnothing$$
$$\pi \bullet Var = Var \qquad \pi \bullet App = App \qquad \pi \bullet Abs = Abs$$

Proof. By definitions of the constants, and equivariance rules the terms are equivariant. Hence they have empty support and are closed terms.

We can now define the function that returns the encoding of a given term. The definition proceeds by induction on λ-terms:

Definition 6. *For a given λ-term t, $\ulcorner t \urcorner$ is is defined by:*

$$\ulcorner t \urcorner = \begin{cases} Var \cdot \overline{x} & provided\ t = \overline{x} \\ App \cdot \ulcorner M \urcorner \cdot \ulcorner N \urcorner & provided\ t = M \cdot N \\ Abs \cdot (\lambda x.\ \ulcorner M \urcorner) & provided\ t = (\lambda x.\ M) \end{cases}$$

Lemma 6. *The encoding function $\ulcorner \ldots \urcorner$ is equivariant. It preserves support and freshness:*

$$\pi \bullet \ulcorner a \urcorner = \ulcorner (\pi \bullet a) \urcorner \qquad supp\ \ulcorner x \urcorner = supp\ x \qquad a \# \ulcorner x \urcorner = a \# x$$

Proof. By induction on x, using the equivariance and freshness lemmas.

The next step in the proof will define finite sequences (or tuples). To define it we need a helper definition that given a variable and a list of terms applies all the terms on the list in turn to the variable. We define it by induction on the list l:

Definition 7. *Applying a list of terms l to a variable n is defined by:*

$$app\_lst\ n\ l = \begin{cases} \overline{n} & \text{provided } l \text{ is an empty list} \\ app\_lst\ n\ t \cdot h & \text{provided } l = (h :: t) \end{cases}$$

Lemma 7. *List application is equivariant and preserves support:*

$$\pi \bullet app\_lst\ n\ l = app\_lst\ (\pi \bullet n)\ (\pi \bullet l)$$
$$supp\ (app\_lst\ n\ l) = \{n\} \cup supp\ l$$

Proof. By induction on l, using the equivariance and freshness lemmas.

When defining a finite sequence we can use the nominal function mechanism to write the definition only for the case where the variable is fresh for the list.

Definition 8 (finite sequences). *Given a list of terms l, assuming $x \# l$, we define a finite sequence «l» by:*

$$«l» = \lambda x.\ app\_lst\ x\ (rev\ l)$$

Lemma 8. *Finite sequences are equivariant and preserve support. Finite sequences are equal if their underlying lists are equal. Finite sequences commute with substitution:*

$$\pi \bullet «a» = «\pi \bullet a» \qquad supp\ «t» = supp\ t$$
$$«M» = «N» \text{ if and only if } M = N$$
$$«[M]»\ [x := N] = «[M\ [x := N]]»$$

Proof. To prove equivariance we obtain a variable fresh for both sides of the equation $y \# (t, \pi \bullet t)$, and use the definition with this variable on both sides. To show that equality of finite sequences implies the equality of the underlying term lists we obtain a variable that is fresh for both M and N and use the definition and support equations. To show the support of substitution applied to a sequence, we proceed by induction on the list. For the induction step we use a variable y, fresh for the term, the variable, the substituted term and the result of the substitution: $y \# (M, x, N, M\ [x := N])$. The result follows by a simple computation.

Further in the paper, finite sequences will only refer to lists of one or three elements always written explicitly. Also, we will not use lists for any other purpose. To simplify this notation, we will use the notation $[\ldots]$ in place of «$[\ldots]$» to refer to finite sequences from now in the paper.

Since finite sequences preserve support (lemma 8), using the freshness rules for lists we can write out explicitly the freshness conditions for specific finite sequences:

$$x \# [y] = x \# y \qquad x \# [t, r, s] = x \# (t, r, s)$$

Applying the lemmas 2 and 3 from the previous section we can get terms convertible to the application of a finite sequence. It is indeed the term applied to the elements of the sequence in order:

$$\begin{aligned} [M] \cdot N &\approx N \cdot M \\ [M,\, N,\, P] \cdot R &\approx R \cdot M \cdot N \cdot P \end{aligned} \tag{8}$$

Using the lemmas from Section 4 we can get simpler convertibility rules for finite sequences applied to initial functions. Indeed they return the appropriate elements from the sequences:

$$\begin{aligned} [A,\, B,\, C] \cdot U_2^2 &\approx A \\ [A,\, B,\, C] \cdot U_1^2 &\approx B \\ [A,\, B,\, C] \cdot U_0^2 &\approx C \end{aligned} \tag{9}$$

We can now define the terms F_1, F_2, F_3:

Definition 9. *Assuming $a \neq b$, $b \neq c$ and $c \neq a$:*

$$\begin{aligned} F_1 &= (\lambda a.\ App \cdot \ulcorner Var \urcorner \cdot (Var \cdot \overline{a})) \\ F_2 &= (\lambda a.\ \lambda b.\ \lambda c.\ App \cdot (App \cdot \ulcorner App \urcorner \cdot (\overline{c} \cdot \overline{a})) \cdot (\overline{c} \cdot \overline{b})) \\ F_3 &= (\lambda a.\ \lambda b.\ App \cdot \ulcorner Abs \urcorner \cdot (Abs \cdot (\lambda c.\ \overline{b} \cdot (\overline{a} \cdot \overline{c})))) \end{aligned}$$

Using the lemmas 2 and 3 we can derive the convertibility relations for F_1, F_2, F_3. The first two have no freshness obligations, however in the case of F_3 the internal λ-abstraction is not applied, so the freshness obligations for this variable remain:

$$F_1 \cdot A \approx App \cdot \ulcorner Var \urcorner \cdot (Var \cdot A) \tag{10}$$

$$F_2 \cdot A \cdot B \cdot C \approx App \cdot (App \cdot \ulcorner App \urcorner \cdot (C \cdot A)) \cdot (C \cdot B) \tag{11}$$

Assuming $x \# A$ and $x \# B$:

$$F_3 \cdot A \cdot B \approx App \cdot \ulcorner Abs \urcorner \cdot (Abs \cdot (\lambda x.\ B \cdot (A \cdot \overline{x}))) \tag{12}$$

Lemma 9. *The terms F_1, F_2, F_3 are equivariant and are closed terms:*

$$\begin{array}{lll} \pi \bullet F_1 = F_1 & \pi \bullet F_2 = F_2 & \pi \bullet F_3 = F_3 \\ supp\ F_1 = \varnothing & supp\ F_2 = \varnothing & supp\ F_3 = \varnothing \end{array}$$

Proof. Using the equivariance of the subterms and the fact that equivariance implies empty support.

Next, we define the terms A_1, A_2, A_3:

Definition 10. *Assuming $a \neq b$, $b \neq c$ and $c \neq a$:*

$$\begin{aligned} A_1 &= (\lambda a.\ \lambda b.\ F_1 \cdot \overline{a}) \\ A_2 &= (\lambda a.\ \lambda b.\ \lambda c.\ F_2 \cdot \overline{a} \cdot \overline{b} \cdot [\overline{c}]) \\ A_3 &= (\lambda a.\ \lambda b.\ F_3 \cdot \overline{a} \cdot [\overline{b}]) \end{aligned}$$

Using lemma 3 we can derive the convertibility relations for the applied terms A_1, A_2, A_3 that will have no freshness obligations:

$$A_1 \cdot A \cdot B \approx F_1 \cdot A \tag{13}$$

$$A_2 \cdot A \cdot B \cdot C \approx F_2 \cdot A \cdot B \cdot [C] \tag{14}$$

$$A_3 \cdot A \cdot B \approx F_3 \cdot A \cdot [B] \tag{15}$$

Lemma 10. *A_1, A_2 and A_3 are closed terms:*

$$\mathit{supp}\ A_1 = \varnothing \qquad \mathit{supp}\ A_2 = \varnothing \qquad \mathit{supp}\ A_3 = \varnothing$$

Proof. Using lemma 9 and the properties of support.

Finally we define *Num*; the definition follows the idea from the beginning of this section:

Definition 11

$$\mathit{Num} = [[A_1,\ A_2,\ A_3]]$$

From the support of the terms A_1, A_2, A_3 and the fact that finite sequences preserve support, we also get that *Num* is a closed term:

$$\mathit{supp}\ \mathit{Num} = \varnothing$$

6 The Proof of the Second Fixed Point Theorem

In this Section we present to proof of the Second Fixed Point Theorem together with the most interesting lemma that combines all the definitions from the previous Section. Namely, all the constants F_{1-3} and A_{1-3} have been defined in such a way that the encoding of terms is λ-defined by *Num*. Here we will show that *Num* applied to an encoded a term is convertible to its encoding.

For the proofs presented in this Section we decide on using the Isabelle rendering of the formal proofs. We use the number of lemmas in the paper as the lemma names in Isabelle). The simplifier is set up with the lemmas about empty support and support preservation shown before in the paper. Given a λ-term *M* we prove the property by induction on the structure of *M*. For each of the three cases the proof proceeds by transforming the left-hand side using the convertibility rules and lemmas proven in the previous section obtaining the right-hand side:

lemma num_numeral:
  shows $\mathit{Num} \cdot \ulcorner M \urcorner \approx \ulcorner\ulcorner M \urcorner\urcorner$
proof (induct M)
  case $\overline{n}$
  have $\mathit{Num} \cdot \ulcorner(\overline{n})\urcorner = \mathit{Num} \cdot (\mathit{Var} \cdot \overline{n})$ **by** simp
  also have ... $= [[A_1,\ A_2,\ A_3]] \cdot (\mathit{Var} \cdot \overline{n})$ **by** simp
  also have ... $\approx \mathit{Var} \cdot \overline{n} \cdot [A_1,\ A_2,\ A_3]$ **using** 8 .

8 **also have** ... $\approx [A_1, A_2, A_3] \cdot U^2_2 \cdot \overline{n} \cdot [A_1, A_2, A_3]$ **using** 5 .
9 **also have** ... $\approx A_1 \cdot \overline{n} \cdot [A_1, A_2, A_3]$ **using** 9 **by** simp
10 **also have** ... $\approx F_1 \cdot \overline{n}$ **using** 13 .
11 **also have** ... $\approx \mathit{App} \cdot \ulcorner \mathit{Var} \urcorner \cdot (\mathit{Var} \cdot \overline{n})$ **using** 10 .
12 **also have** ... $= \ulcorner\ulcorner(\overline{n})\urcorner\urcorner$ **by** simp
13 **finally show** $\mathit{Num} \cdot \ulcorner(\overline{n})\urcorner \approx \ulcorner\ulcorner(\overline{n})\urcorner\urcorner$.
14 **next**
15 **case** $M \cdot N$
16 **assume** IH: $\mathit{Num} \cdot \ulcorner M \urcorner \approx \ulcorner\ulcorner M \urcorner\urcorner$ $\mathit{Num} \cdot \ulcorner N \urcorner \approx \ulcorner\ulcorner N \urcorner\urcorner$
17 **have** $\mathit{Num} \cdot \ulcorner(M \cdot N)\urcorner = \mathit{Num} \cdot (\mathit{App} \cdot \ulcorner M \urcorner \cdot \ulcorner N \urcorner)$ **by** simp
18 **also have** ... $= [[A_1, A_2, A_3]] \cdot (\mathit{App} \cdot \ulcorner M \urcorner \cdot \ulcorner N \urcorner)$ **by** simp
19 **also have** ... $\approx \mathit{App} \cdot \ulcorner M \urcorner \cdot \ulcorner N \urcorner \cdot [A_1, A_2, A_3]$ **using** 8 .
20 **also have** ... $\approx [A_1, A_2, A_3] \cdot U^2_1 \cdot \ulcorner M \urcorner \cdot \ulcorner N \urcorner \cdot [A_1, A_2, A_3]$ **using** 6 .
21 **also have** ... $\approx A_2 \cdot \ulcorner M \urcorner \cdot \ulcorner N \urcorner \cdot [A_1, A_2, A_3]$ **using** 9 **by** simp
22 **also have** ... $\approx F_2 \cdot \ulcorner M \urcorner \cdot \ulcorner N \urcorner \cdot \mathit{Num}$ **using** 14 **by** simp
23 **also have** ... $\approx \mathit{App} \cdot (\mathit{App} \cdot \ulcorner \mathit{App} \urcorner \cdot (\mathit{Num} \cdot \ulcorner M \urcorner)) \cdot (\mathit{Num} \cdot \ulcorner N \urcorner)$ **using** 11 .
24 **also have** ... $\approx \mathit{App} \cdot (\mathit{App} \cdot \ulcorner \mathit{App} \urcorner \cdot \ulcorner\ulcorner M \urcorner\urcorner) \cdot (\mathit{Num} \cdot \ulcorner N \urcorner)$ **using** IH **by** simp
25 **also have** ... $\approx \ulcorner\ulcorner(M \cdot N)\urcorner\urcorner$ **using** IH **by** simp
26 **finally show** $\mathit{Num} \cdot \ulcorner(M \cdot N)\urcorner \approx \ulcorner\ulcorner(M \cdot N)\urcorner\urcorner$.
27 **next**
28 **case** $\lambda x.\ P$
29 **assume** IH: $\mathit{Num} \cdot \ulcorner P \urcorner \approx \ulcorner\ulcorner P \urcorner\urcorner$
30 **have** $\mathit{Num} \cdot \ulcorner(\lambda x.\ P)\urcorner = \mathit{Num} \cdot (\mathit{Abs} \cdot (\lambda x.\ \ulcorner P \urcorner))$ **by** simp
31 **also have** ... $= [[A_1, A_2, A_3]] \cdot (\mathit{Abs} \cdot (\lambda x.\ \ulcorner P \urcorner))$ **by** simp
32 **also have** ... $\approx \mathit{Abs} \cdot (\lambda x.\ \ulcorner P \urcorner) \cdot [A_1, A_2, A_3]$ **using** 8 .
33 **also have** ... $\approx [A_1, A_2, A_3] \cdot U^2_0 \cdot (\lambda x.\ \ulcorner P \urcorner) \cdot [A_1, A_2, A_3]$ **using** 7 .
34 **also have** ... $\approx A_3 \cdot (\lambda x.\ \ulcorner P \urcorner) \cdot [A_1, A_2, A_3]$ **using** 9 **by** simp
35 **also have** ... $\approx F_3 \cdot (\lambda x.\ \ulcorner P \urcorner) \cdot [[A_1, A_2, A_3]]$ **using** 15 .
36 **also have** ... $= F_3 \cdot (\lambda x.\ \ulcorner P \urcorner) \cdot \mathit{Num}$ **by** simp
37 **also have** ... $\approx \mathit{App} \cdot \ulcorner \mathit{Abs} \urcorner \cdot (\mathit{Abs} \cdot (\lambda x.\ \mathit{Num} \cdot ((\lambda x.\ \ulcorner P \urcorner) \cdot \overline{x})))$
38 **by** (rule 12) simp_all
39 **also have** ... $\approx \mathit{App} \cdot \ulcorner \mathit{Abs} \urcorner \cdot (\mathit{Abs} \cdot (\lambda x.\ \mathit{Num} \cdot \ulcorner P \urcorner))$ **using** 4 **by** simp
40 **also have** ... $\approx \mathit{App} \cdot \ulcorner \mathit{Abs} \urcorner \cdot (\mathit{Abs} \cdot (\lambda x.\ \ulcorner\ulcorner P \urcorner\urcorner))$ **using** IH **by** simp
41 **also have** ... $= \ulcorner\ulcorner(\lambda x.\ P)\urcorner\urcorner$ **by** simp
42 **finally show** $\mathit{Num} \cdot \ulcorner(\lambda x.\ P)\urcorner \approx \ulcorner\ulcorner(\lambda x.\ P)\urcorner\urcorner$.
43 **qed**

The proof is similar to the text proof formalized, with one exception. In the λ-abstraction case, when transforming the application of F_3 (lines 36-38) a new variable name is necessary. This also comes from lemmas 2, 3, where we cannot remove freshness assumptions about variables introduced inside the term. In the above proof, it is also the only place where we need to use two methods to convince Isabelle that the transformation is correct. The application of F_3 (method rule 12) leaves proof obligations about freshness of the introduced variable. In this case the variable is supposed to be fresh for a closed term, which the simplifier can deduce automatically (method simp_all), however this may not always be the case (and it will not be the case in the final theorem).

The second fixed point theorems states, that for any λ-term F, there exists a term X, such that X is convertible to the application of F to the term representing the encoding of X. The proof defines the term X explicitly and shows that is has the desired property by transforming the term using convertibility rules and properties shown before:

44 **theorem** second_fixed_point_theorem:
45   **fixes** $F :: lam$
46   **shows** $\exists X.\ X \approx F \cdot \ulcorner X \urcorner$
47 **proof** -
48   **obtain** $x :: var$ **where** $x \# F$ **using** obtain_fresh **by** blast
49   **def** $W = \lambda x.\ F \cdot (App \cdot \overline{x} \cdot (Num \cdot \overline{x}))$
50   **def** $X = W \cdot \ulcorner W \urcorner$
51   **have** a: $X = W \cdot \ulcorner W \urcorner$ **unfolding** X_def ..
52   **also have** $\ldots = (\lambda x.\ F \cdot (App \cdot \overline{x} \cdot (Num \cdot \overline{x}))) \cdot \ulcorner W \urcorner$ **unfolding** W_def ..
53   **also have** $\ldots \approx F \cdot (App \cdot \ulcorner W \urcorner \cdot (Num \cdot \ulcorner W \urcorner))$ **by** simp
54   **also have** $\ldots \approx F \cdot (App \cdot \ulcorner W \urcorner \cdot \ulcorner\ulcorner W \urcorner\urcorner)$ **by** simp
55   **also have** $\ldots \approx F \cdot \ulcorner (W \cdot \ulcorner W \urcorner) \urcorner$ **by** simp
56   **also have** $\ldots = F \cdot \ulcorner X \urcorner$ **unfolding** X_def ..
57   **finally show** $X \approx F \cdot \ulcorner X \urcorner$ ..
58 **qed**

The proof is again similar to a textbook one, however as the first step of the proof we need to add line 48, even before the definitions of W and X. The term W introduces a new abstraction together with a new variable x. In the formal proof we assume that this variable is fresh for the original term F. This is indeed necessary, if x was bound in F then the first step of convertibility reasoning would leave the term $F[x := W]$ on the left side of the application until the end of the proof, and the property would not hold.

7 Conclusion

We compared a number of approaches for defining terms in the λ-calculus and showed a convenient way of defining constants. We start with a function that returns different atoms, define constants using those atoms and show their convertibility to the forms with freshness obligations. We use the simplifier only for convertibility equations, obtaining proofs where there are very few freshness obligations making them similar to paper proofs.

We formalized the second fixed point theorem of the λ-calculus, using Nominal Isabelle. We showed how to define constants and their properties, so that the proof resembles a paper proof. The main difference is obtaining a variable fresh for the given term. We show that if the variable is not fresh for the given term, the usual textbook proof does not hold.

In the proof we give the term that satisfies the second fixed point theorem explicitly; it remains to be seen how a meta-proof that talks about λ-definability could be formalized. The approach presented here could be compared with using standard combinator terms used by Norrish [11], this could perheapes simplify the equations for definitions with internal abstractions, in our case study this could be preferable in case of property 12.

7.1 Related Work

A number of proofs about the λ-calculus have already been performed with Nominal Isabelle. Examples include the Church-Rosser property, strong normalization. More examples are described in [13].

Norrish [10] derives an infrastructure for the λ-calculus manually in the HOL4 system, using properties from Nominal Logic. He proves a number of properties about the λ-calculus following [1] and the book by Hankin [6].

References

1. Barendregt, H.P.: The Lambda Calculus: Its Syntax and Semantics. Studies in Logic and the Foundations of Mathematics, vol. 103. Elsevier (2001)
2. Bengtson, J., Johansson, M., Parrow, J., Victor, B.: Psi-calculi: a framework for mobile processes with nominal data and logic. Logical Methods in Computer Science 7(1) (2011)
3. Bengtson, J., Parrow, J.: Formalising the pi-calculus using nominal logic. Logical Methods in Computer Science 5(2) (2008)
4. Blanchette, J.C., Böhme, S., Paulson, L.C.: Extending sledgehammer with smt solvers. Automated Deduction (2011) (accepted)
5. Böhm, C., Piperno, A., Guerrini, S.: Lambda-definition of function(al)s by normal forms. In: Sannella, D. (ed.) ESOP 1994. LNCS, vol. 788, pp. 135–149. Springer, Heidelberg (1994)
6. Hankin, C.: Lambda Calculi: A Guide for Computer Scientists. Graduate Texts in Computer Science, vol. 3. Clarendon Press (1993)
7. Huffman, B., Urban, C.: A New Foundation for Nominal Isabelle. In: Kaufmann, M., Paulson, L.C. (eds.) ITP 2010. LNCS, vol. 6172, pp. 35–50. Springer, Heidelberg (2010)
8. Kaliszyk, C., Urban, C.: Quotients revisited for Isabelle/HOL. In: Chu, W.C., Wong, W.E., Palakal, M.J., Hung, C.-C. (eds.) Proc. of the 26th ACM Symposium on Applied Computing (SAC 2011), pp. 1639–1644. ACM (2011)
9. Minamide, Y., Okuma, K.: Verifying CPS transformations in Isabelle/HOL. In: Proc. of the Workshop on Mechanized reasoning about languages with variable binding (MERLIN 2003). ACM (2003)
10. Norrish, M.: Mechanising λ-calculus using a classical first order theory of terms with permutations. Higher-Order and Symbolic Computation 19, 169–195 (2006)
11. Norrish, M.: Mechanised Computability Theory. In: van Eekelen, M., Geuvers, H., Schmaltz, J., Wiedijk, F. (eds.) ITP 2011. LNCS, vol. 6898, pp. 297–311. Springer, Heidelberg (2011)
12. Pitts, A.M.: Nominal Logic, A First Order Theory of Names and Binding. Information and Computation 183, 165–193 (2003)

13. Urban, C., Tasson, C.: Nominal Techniques in Isabelle/HOL. In: Nieuwenhuis, R. (ed.) CADE 2005. LNCS (LNAI), vol. 3632, pp. 38–53. Springer, Heidelberg (2005)
14. Urban, C., Berghofer, S.: A Recursion Combinator for Nominal Datatypes Implemented in Isabelle/HOL. In: Furbach, U., Shankar, N. (eds.) IJCAR 2006. LNCS (LNAI), vol. 4130, pp. 498–512. Springer, Heidelberg (2006)
15. Urban, C., Cheney, J., Berghofer, S.: Mechanizing the Metatheory of LF. In: Proc. of the 23rd LICS Symposium, pp. 45–56 (2008)
16. Urban, C., Kaliszyk, C.: General Bindings and Alpha-Equivalence in Nominal Isabelle. In: Barthe, G. (ed.) ESOP 2011. LNCS, vol. 6602, pp. 480–500. Springer, Heidelberg (2011)

Simple, Functional, Sound and Complete Parsing for All Context-Free Grammars

Tom Ridge

University of Leicester

Abstract. Parsers for context-free grammars can be implemented directly and naturally in a functional style known as "combinator parsing", using recursion following the structure of the grammar rules. However, naive implementations fail to terminate on left-recursive grammars, and despite extensive research the only complete parsers for general context-free grammars are constructed using other techniques such as Earley parsing. Our main contribution is to show how to construct simple, sound and complete parser implementations directly from grammar specifications, for all context-free grammars, based on combinator parsing. We then construct a generic parser generator and show that generated parsers are sound and complete. The formal proofs are mechanized using the HOL4 theorem prover. Memoized parsers based on our approach are polynomial-time in the size of the input. Preliminary real-world performance testing on highly ambiguous grammars indicates our parsers are faster than those generated by the popular Happy parser generator.

1 Introduction

Parsing is central to many areas of computer science, including databases (database query languages), programming languages (syntax), network protocols (packet formats), the internet (transfer protocols and markup languages), and natural language processing. Context-free grammars are typically specified using a set of rules in Backus-Naur Form (BNF). An example[1] of a simple grammar with a single rule for a nonterminal `E` (with two alternative expansions) is `E -> "(" E "+" E ")" | "1"`. A parse tree is a finite tree where each node is formed according to the grammar rules. We can concatenate the leaves of a parse tree *pt* to get a string (really, a substring option) substring_of *pt* accepted by the grammar, see Fig. 1. A parser for a grammar is a program that takes an input string and returns parse trees for that string.

A popular parser implementation strategy is combinator parsing, wherein sequencing and alternation are implemented using the infix combinators `**>` and `|||` (higher-order functions that take parsers as input and produce parsers as output). For example[2]

[1] Real BNF requires a nonterminal such as `E` to be written as `<E>`.

[2] The examples are based on real OCaml implementations, but are formally pseudocode because OCaml function names must start with a lowercase letter.

J.-P. Jouannaud and Z. Shao (Eds.): CPP 2011, LNCS 7086, pp. 103–118, 2011.

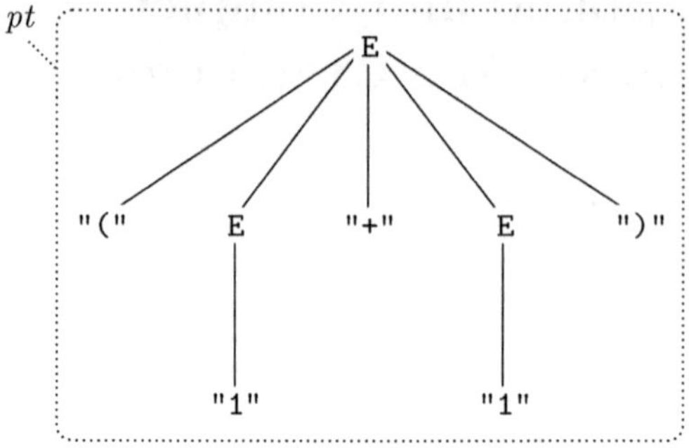

substring_of pt = SOME "(1+1)"

Fig. 1.

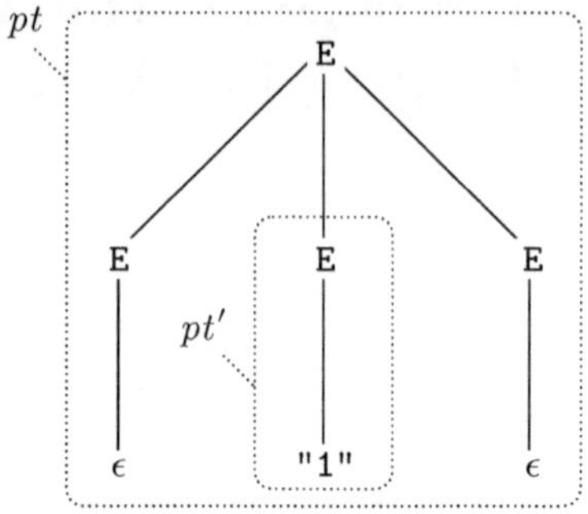

substring_of pt = substring_of pt'

Fig. 2.

```
let rec E = fun i ->
  ((a "(") **> E **> (a "+") **> E **> (a ")") ||| (a "1")) i
```

The code works by first consuming a "(" character from the input, then calling itself recursively to parse an E, then consuming a "+" character, and so on. Termination is clear because recursive calls to E are given strictly less input to parse.

Combinator parsing cannot be used directly if the grammar contains rules such as E -> E E E that are left-recursive. For example, a naive attempt to implement the grammar E -> E E E | "1" | ϵ[3] gives

```
let rec E = fun i ->
  ((E **> E **> E) ||| (a "1")  ||| (a "")) i
```

This code would attempt to parse an E by first expanding to E E E, and then recursively attempting to parse an E on the same input, leading to non-termination. One solution is to alter the original grammar specification to avoid left recursion, but this is undesirable for several reasons, and may not always be possible. Despite these drawbacks combinator parsing is conceptually simple, almost trivial to implement, and integrates smoothly with the host language (typically a simply-typed functional programming language). For these reasons *combinator parsing is extremely popular*, and many systems include hand-crafted parsers based on combinator parsing. In contrast, the complicated implementations of other parsing techniques have many associated drawbacks. For example, Yacc produces error messages, such as "shift/reduce conflict", that are *incomprehensible* without a good knowledge of the underlying implementation technology.

Contribution. The main contribution of our work is to show how to implement simple, terminating, sound and complete parsers for arbitrary context-free grammars using combinator parsing. The heart of our contribution is a parser wrapper (a function from parsers to parsers) `check_and_upd_lctxt` which wraps

[3] The terminal representing the empty string "" is usually written ϵ.

grammar_to_parser p_of_tm g sym i = case sym of
TM $tm \rightarrow ((p\_of\_tm\ tm) \gg (\lambda\ v.\ \mathsf{LF}(tm, v)))\ i$ || NT $nt \rightarrow$
let $rules$ = FILTER $(\lambda\ (nt', rhs).\ nt' = nt)\ g$ in
let $alts1$ = (FLAT ∘ (MAP SND)) $rules$ in
let $alts2$ = MAP (MAP (λ sym. grammar_to_parser p_of_tm g sym)) $alts1$ in
let p = or_list (MAP (then_list2 nt) $alts2$) in
check_and_upd_lctxt nt p i)

The parser generator grammar_to_parser is parameterized by: a function p_of_tm which gives a parser for each terminal; the grammar g (a list of BNF-type rules); and sym, the symbol corresponding to the parser that should be generated. If sym is a terminal tm then p_of_tm tm gives the appropriate parser. If sym is a nonterminal nt then the relevant $rules$ are filtered from the grammar, the right hand sides are combined into a list of alternatives $alts1$, grammar_to_parser is recursively mapped over $alts1$, and finally the results are combined using the parser combinators or_list and then_list2 to give a parser p. In order to prevent nontermination p is wrapped by check_and_upd_lctxt.

Fig. 3. A verified, sound and complete parser generator (HOL4)

the body of an underlying parser and eliminates some parse attempts whilst preserving completeness. For example, for the grammar `E -> E E E | "1" |` ϵ, a terminating, sound and complete parser can be written as follows:

```
let rec E = fun i -> check_and_upd_lctxt "E"
  ((E **> E **> E) ||| (a "1") ||| (a "")) i
```

The first argument `"E"` to `check_and_upd_lctxt` is necessary to indicate which nonterminal is being parsed in case the grammar contains more than one nonterminal. In Fig. 3 we define a parser generator for arbitrary context-free grammars based on this parser wrapper (the reader should not expect to understand the code at this point). We prove the parser generator correct using the HOL4 theorem prover. Our approach retains the simplicity of combinator parsing, including the ability to incorporate standard extensions such as "semantic actions". The worst-case time complexity of our algorithm when memoized is $O(n^5)$. In real-world performance comparisons on highly ambiguous grammars, our parsers are consistently faster than those generated by the Happy parser generator [1].

Key Ideas. Consider the highly ambiguous grammar `E -> E E E | "1" |` ϵ. This gives rise to an infinite number of parse trees. A parser cannot hope to return an infinite number of parse trees in a finite amount of time. However, many parse trees pt have proper subtrees pt' such that both pt and pt' are rooted at the same nonterminal, and substring_of pt = substring_of pt', see Fig. 2. *This is the both the cause of the infinite number of parse trees, and the underlying cause of non-termination in implementations of combinator parsing.*

We call a parse tree bad if it *contains* a subtree such as pt. If we rule out bad trees we can still find a good tree for any parse-able input. Moreover, given a context-free grammar g and input s, it turns out that there are at most a *finite* number of good parse trees pt such that substring_of pt = SOME s. Thus for a given grammar we have identified a class of parse trees (the good parse trees) that is complete (any input that can be parsed, can be parsed to give a good parse tree) and moreover is *finite*.

At the implementation level, we construct a function `check_and_upd_lctxt` which wraps the body of an underlying parser and eliminates parse attempts that would lead to nontermination by avoiding bad parse trees. This requires the parser input type to be slightly modified to include information about the parsing context (those parent parses that are currently in progress), but crucially this is invisible to the parser writer who simply makes use of standard parser combinators. Generalizing this approach gives the parser generator in Fig. 3.

Structure of the Paper. In Sect. 2 we define the types used in later sections, and give a brief description of the formalization of substrings. Sect. 3 discusses the relationship between grammars and parse trees, whilst Sect. 4 discusses the relationship between parse trees and the parsing context. The standard parsing combinators are defined in Sect. 5. The new functions relating to the parsing context, including check_and_upd_lctxt, are defined in Sect. 6. The remainder of the body of the paper is devoted to correctness. In Sect. 7 we discuss termination and soundness. In Sect. 8 we formalize informal notions of completeness, and in Sect. 9 we show that our parser generator produces parsers that are complete. In Sect. 10 we discuss implementation issues, such as memoization and performance. Finally we discuss related work and conclude. Our implementation language is OCaml and the complete OCaml code and HOL4 proof script are available online[4], together with example grammars and sample inputs. For reasons of space, in this paper we occasionally omit definitions of straightforward well-formedness predicates such as wf_grammar. We give outlines of the proofs, including the main inductions, but do not discuss the proofs in detail. The interested reader will find all definitions and proofs in the mechanized HOL4 proof scripts online.

Notation. BNF grammars are written using `courier`, as is OCaml code and pseudo-code. Mechanized HOL4 definitions are written using sans_serif for defined constants, and *italic* for variables. Common variable names are displayed in Fig. 4, but variations are also used. For example, if x is a variable of type α then xs is a variable of type α list. Similarly suffixing and priming are used to distinguish several variables of the same type. For example, s, s', s_pt, s_rem and s_tot are all common names for variables of type substring. For presentation purposes, we occasionally blur the distinction between strings and substrings. Records are written $\langle$ fld $= v$; ... $\rangle$. Update of record r is written r with $\langle$ fld $= v$ $\rangle$. Function application is written $f\ x$. List cons is written $x :: xs$. The empty list is []. List membership is written MEM x xs. Other HOL4 list functions should be comprehensible to readers with a passing knowledge of functional programming.

[4] `http://www.cs.le.ac.uk/~tr61`

$$
\begin{aligned}
s &: \mathsf{string} \\
l, h &: \mathsf{num} \\
s &: \mathsf{substring} \\
tm &: \mathsf{term} = \mathsf{ty\_term} \\
nt &: \mathsf{nonterm} = \mathsf{ty\_nonterm} \\
sym &: \mathsf{symbol} = \mathsf{TM\ of\ term} \mid \mathsf{NT\ of\ nonterm} \\
rhs, alts &: \mathsf{(symbol\ list)\ list} \\
r, rule &: \mathsf{parse\_rule} = \mathsf{nonterm} \times \mathsf{((symbol\ list)\ list)} \\
g &: \mathsf{grammar} = \mathsf{parse\_rule\ list} \\
pt &: \mathsf{parse\_tree} = \mathsf{NODE\ of\ nonterm} \times \mathsf{parse\_tree\ list} \mid \mathsf{LF\ of\ term} \times \mathsf{substring} \\
q &: \mathsf{simple\_parser} = \mathsf{substring} \rightarrow \mathsf{parse\_tree\ list} \\
lc &: \mathsf{context} = \mathsf{(nonterm} \times \mathsf{substring)\ list} \\
i &: \mathsf{ty\_input} = \langle\ \mathsf{lc}\ :\ \mathsf{context};\ \mathsf{sb}\ :\ \mathsf{substring}\ \rangle \\
p &: \alpha\ \mathsf{parser} = \mathsf{ty\_input} \rightarrow (\alpha \times \mathsf{substring})\ \mathsf{list} \\
ss\_of\_tm &: \mathsf{ty\_ss\_of\_tm} = \mathsf{term} \rightarrow \mathsf{substring\ set} \\
p\_of\_tm &: \mathsf{ty\_p\_of\_tm} = \mathsf{term} \rightarrow \mathsf{substring\ parser}
\end{aligned}
$$

Fig. 4. Common variable names for elements of basic types, with type definitions

$\mathsf{string}\ s = \mathsf{let}\ (s,l,h) = s\ \mathsf{in}\ s$
$\mathsf{low}\ s = \mathsf{let}\ (s,l,h) = s\ \mathsf{in}\ l$
$\mathsf{high}\ s = \mathsf{let}\ (s,l,h) = s\ \mathsf{in}\ h$
$\mathsf{len}\ s = \mathsf{let}\ (s,l,h) = s\ \mathsf{in}\ h\ -\ l$
$\mathsf{wf\_substring}\ (s,l,h) = l \leq h \wedge h \leq |s|$

$\mathsf{inc\_low}\ n\ s = \mathsf{let}\ (s,l,h) = s\ \mathsf{in}\ (s,l+n,h)$
$\mathsf{dec\_high}\ n\ s = \mathsf{let}\ (s,l,h) = s\ \mathsf{in}\ (s,l,h-n)$
$\mathsf{inc\_high}\ n\ s = \mathsf{let}\ (s,l,h) = s\ \mathsf{in}\ (s,l,h+n)$
$\mathsf{full}\ s = (s,0,|s|)$
$\mathsf{toinput}\ s = \langle\ \mathsf{lc} = [];\ \mathsf{sb} = s\ \rangle$

$\mathsf{concatenate\_two}\ s1\ s2 = \mathsf{if}\ (\mathsf{string}\ s1\ = \mathsf{string}\ s2) \wedge (\mathsf{high}\ s1\ = \mathsf{low}\ s2)\ \mathsf{then}$
$\quad \mathsf{SOME}\ ((\mathsf{string}\ s1, \mathsf{low}\ s1, \mathsf{high}\ s2))\ \mathsf{else\ NONE}$

$\mathsf{concatenate\_list}\ ss = \mathsf{case}\ ss\ \mathsf{of}\ [] \rightarrow \mathsf{NONE} \mid\mid [s1] \rightarrow (\mathsf{SOME}\ s1) \mid\mid s1 :: ss1 \rightarrow ($
$\quad \mathsf{case\ concatenate\_list}\ ss1\ \mathsf{of\ NONE} \rightarrow \mathsf{NONE} \mid\mid \mathsf{SOME}\ s2 \rightarrow \mathsf{concatenate\_two}\ s1\ s2)$

Fig. 5. Common functions on substrings

2 Types and Substrings

Figure 4 gives the basic types we require. In the following sections, it is formally easier to work with substrings rather than strings. A substring (s, l, h) represents the part of a string s between a low index l and a high index h. The type substring consists only of well-formed triples (s, l, h). Common substring functions, including the well-formedness predicate, are defined in Fig. 5. Returning to Fig. 4, the type of terminals is term; the type of nonterminals is nonterm. Formally terminals and nonterminals are kept abstract, but in the OCaml implementation they are strings. Symbols are the disjoint union of terminals and nonterminals. A parse rule such as `E -> E E E | "1" |` ϵ consists of a nonterminal l.h.s. and several alternatives on the r.h.s. (an alternative is simply a list of symbols). A grammar is a list of parse rules (really, a finite set) and a parse tree consists of nodes (each decorated with a nonterminal), or leaves (each decorated with a terminal and the substring that was parsed by that terminal). A simple parser takes an input substring and produces a list of parse trees.

Combinator parsers typically parse prefixes of a given input, and return (a list of) a result value paired with the substring that remains to be parsed: α preparser = substring $\longrightarrow$ ($\alpha \times$ substring) list. Rather than taking just a substring as input, our parsers need additional information about the context. The context, type context = (nonterm $\times$ substring) list, records information about which nonterminals are already in the process of being parsed, and the substring that each parent parse took as input. The input i for a parser is just a record with two fields: the usual substring i.sb, and the context i.lc. We emphasize that this slight increase in the complexity of the input type is invisible when using our parser combinators as a library: the only code that examines the context is `check_and_upd_lctxt`.

Whilst BNF grammars clearly specify how to expand nonterminals, in practice the specification of terminal parsers is more-or-less arbitrary. Formally, we should keep terminal parsers loosely specified. The set pts_of ss_of_tm g of parse trees for a grammar is therefore parameterized by a function ss_of_tm such that LF(tm, s) is a parse tree only if $s \in ss\_of\_tm\ tm$. A function of type ty_p_of_tm gives a parser for each terminal.

3 Grammars and Parse Trees

Parse trees pt and pt' match if they have the same root symbol, and substring_of pt = substring_of pt'. A parse tree has a bad root if it contains a proper subtree that matches it. For example, in Fig. 2, pt has a bad root because subtree pt' matches it. A good tree is a tree such that no *subtrees* have bad roots. The following theorem implies that any parse-able input can be parsed to give a good tree.

Theorem 1 (good_tree_exists_thm). *Given a grammar g, for any parse tree pt one can construct a good tree pt' that matches.*

$$\begin{array}{l} \text{good\_tree\_exists\_thm} = \forall\ ss\_of\_tm.\ \forall\ g.\ \forall\ pt.\ \exists\ pt'. \\ \quad pt \in (\text{pts\_of}\ ss\_of\_tm\ g) \\ \quad \longrightarrow pt' \in (\text{pts\_of}\ ss\_of\_tm\ g) \wedge (\text{matches}\ pt\ pt') \wedge (\text{good\_tree}\ pt') \end{array}$$

Proof. If pt_0 is not good, then it contains a subtree pt and a proper subtree pt' of pt that matches pt. If we replace pt by pt' we have reduced the number of subtrees which have bad roots. The transformed tree is well-formed according to the grammar and matches the original. If we repeat this step, we can eliminate all subtrees with bad roots.

4 Parse Trees and the Parsing Context

In this section we define an inductive relationship admits lc pt between parsing contexts lc and parse trees pt. We use a context during parsing to eliminate parse attempts that would lead to bad parse trees and potential nontermination. If lc is the empty context [], then the function $\lambda\ pt.$ admits [] pt

actually characterizes good trees ie admits [] $pt \leftrightarrow$ good_tree pt. The definition of good_tree is wholly in terms of parse trees, whilst the parse trees returned by our parsers depend not only on the parsing context, but on complicated implementation details of the parsers themselves. The definition of admits serves as a bridge between these two, incorporating the parsing context, but usefully omitting complicated parser implementation details.

```
admits lc pt =
  let s_pt = THE(substring_of pt) in
  case pt of
    NODE(nt, pts) → (¬(MEM (nt, s_pt) lc) ∧ EVERY (admits ((nt, s_pt) :: lc)) pts)
    || LF(_, _) → T
```

The function THE is the projection from the option type: THE(SOME x) $= x$. Let s_pt = THE(substring_of pt). This definition states that, for a parse tree pt with root nt to be admitted, the pair (nt, s_pt) must not be in the context lc, *and moreover* if we extend the context by the pair (nt, s_pt) to get a context lc', then every immediate subtree of pt must be admitted by lc'. Leaf nodes are always admitted. As an example, consider the bad parse tree pt in Fig. 2 which is not admitted by the empty context, ¬ (admits [] pt). In this case, $(nt, s\_pt) =$ (E, "1"), so that $lc' =$ [(E, "1")] , and clearly ¬ (admits lc' pt'). Our parsers return all parse trees admitted by the empty context, which is initially empty. The following theorem guarantees that this includes all good trees.

Theorem 2 (admits_thm). *Good parse trees are admitted by the empty context.*

```
admits_thm = ∀ pt. wf_parse_tree pt ⟶ good_tree pt ⟶ admits [] pt
```

Proof. Induction on the size of pt.

5 Terminal Parsers and Parser Combinators

The basic parser combinators are defined in Fig. 6. The standard definition of the alternative combinator ||| appends the output of one parser to the output of another. The sequential combinator **> uses the first parser to parse prefixes of the input, then applies the second parser to parse the remaining suffixes. This definition is almost standard, except that the parsing context i.lc is the same for *p1* as for *p2*. The "semantic action" combinator ≫ simply applies a function to the results returned by a parser. We generalize the basic combinators to handle lists (then_list, then_list2 and or_list).

The definition of grammar_to_parser in Fig. 3 is parametric over a function p_of_tm which gives a parser $p\_tm = p\_of\_tm\ tm$ for each terminal tm. At the implementation level, p_tm can be more-or-less arbitrary. However, for our results to hold, p_of_tm is required to satisfy a well-formedness requirement wf_p_of_tm. For soundness, parse trees pt returned by terminal parser p_tm must be such that substring_of pt is a prefix of the input. For completeness the parse trees produced by a terminal parser p_tm for a given prefix of the input should not change when the input is extended. These conditions are very natural, but the completeness condition is subtle, and has some interesting consequences, for example, it rules out lookahead which is common in lexer implementations.

```
p1 **> p2 = λ i.
  let f (e1, s1) =
    MAP (λ (e2, s2). ((e1, e2), s2)) (p2 ⟨ lc=i.lc; sb=s1 ⟩)
  in
  (FLAT ∘ (MAP f) ∘ p1) i

p1 ||| p2 = λ i. APPEND (p1 i) (p2 i)

p ≫ f = (MAP (λ (e, s). (f e, s))) ∘ p

always = (λ i. [([], substr i)]) : α list parser

never = (λ i. []) : α parser

then_list (ps : α parser list) = case ps of
  [] → always
  || p::ps → ((p **> (then_list ps))
     ≫ (λ (x, xs). x :: xs))

or_list (ps : α parser list) = case ps of
  [] → never
  || p::ps → (p ||| (or_list ps))

then_list2 nt = λ ps.
  then_list ps ≫ (λ xs. NODE(nt, xs))
```

Fig. 6. Parser combinators

```
update_lctxt nt (p : α parser) = λ i.
  p (i with ⟨ lc=(nt, i.sb)::i.lc ⟩)

ignr_last (p : α parser) = λ i.
  if len (substr i) = 0 then [] else
  let dec = dec_high 1 in
  let inc (e, s) = (e, inc_high 1 s) in
  ((MAP inc) ∘ p ∘ (lift dec)) i

check_and_upd_lctxt nt (p : α parser) = λ i.
  let should_trim =
    EXISTS ((=) (nt, i.sb)) i.lc in
  if should_trim ∧ (len i.sb = 0) then
    []
  else if should_trim then
    (ignr_last (update_lctxt nt p)) i
  else
    (update_lctxt nt p) i
```

Fig. 7. Updating the parsing context

6 Updating the Parsing Context

The parsing context is used to eliminate parse attempts that might lead to non-termination. In Fig. 7, update_lctxt nt is a parser wrapper parameterized by a nonterminal nt. During a parse attempt the nonterminal nt corresponds to the node that is currently being parsed, that is, all parse trees returned by the current parse will have root nt. The parser p corresponds to the parser that will be used to parse the *immediate subtrees* of the current tree. The wrapper update_lctxt nt ensures that the context i.lc is extended to $(nt, i.\mathsf{sb}) :: i.\mathsf{lc}$ before calling the underlying parser p on the given input i.

The parser wrapper ignr_last calls an underlying parser p on the input *minus the last character* (via dec_high); the unparsed suffix of the input then has the last character added back (via inc_high) before the results are returned. The purpose of ignr_last is to force termination when recursively parsing the same nonterminal, by successively restricting the length of the input that is available to parse.

The heart of our contribution is the parser wrapper check_and_update_lctxt *nt* which is also parameterized by a nonterminal *nt*. This combinator uses the context to eliminate parse attempts. As before, *nt* corresponds to the node that is currently being parsed. The boolean should_trim is true iff the context i.lc contains a pair (nt, i.sb). If this is the case, then we can safely restrict our parse attempts to *proper prefixes* of the input i.sb, by wrapping update_lctxt *nt p* in ignr_last. Theorem main_thm in Sect. 9 guarantees that this preserves completeness. At this point we have covered the definitions required for the parser generator in Fig. 3.

7 Termination, Soundness and Prefix-Soundness

In this section we show that the definition of grammar_to_parser in Fig. 3 is well-formed and terminating by giving a well-founded measure that decreases with each recursive call. We then define formally what it means for a parser to be sound. We also define the stronger property of prefix-soundness. The parser generator grammar_to_parser generates prefix-sound parsers.

The following well-founded measure function is parameterized by the grammar g and gives a natural number for every input i. In Fig. 3, recursive calls to grammar_to_parser are given inputs with strictly less measure, which ensures that the definition is well-formed and that all parses terminate. The function SUM computes the sum of a list of numbers.

```
measure g i =
  let nts = nonterms_of_grammar g in
  let f nt = len i.sb + (if MEM (nt, i.sb) i.lc then 0 else 1) in
  SUM(MAP f nts)
```

Theorem 3. *Recursive calls to* grammar_to_parser *are given inputs i' whose measure is strictly less than the measure of the input i provided to the parent.*

Proof. The proof proceeds in two steps. First, we show by analysis of cases that the invocation of p when evaluating check_and_upd_lctxt $nt\ p\ i$ is called with an input i' with strictly less measure than that of i.

```
((λ i. [(i, s0)]) = p)
⟶ MEM nt (nonterms_of_grammar g)
⟶ MEM (i', s) (check_and_upd_lctxt nt p i) ⟶ measure g i' < measure g i
```

Second, we observe that recursive calls to grammar_to_parser are nested under then_list2 *nt*, so that each recursive call receives an input i'' where either i''.sb = i'.sb or len i''.sb < len i'.sb. In the latter case we have

```
len s'' < len s' ⟶ measure g ⟨ lc = lc; sb = s'' ⟩ ≤ measure g ⟨ lc = lc; sb = s' ⟩
```

We now turn our attention to soundness. The simplest form of soundness requires that any parse tree pt that is returned by a parser q_sym for a symbol sym when called on input s should conform to the grammar g, have a root symbol sym, and be such that substring_of pt = SOME s.

```
sound ss_of_tm g sym q_sym = ∀ s. ∀ pt.
  wf_grammar g
  ∧ MEM pt (q_sym s)
  ⟶
  pt ∈ (pts_of ss_of_tm g)
  ∧ (root pt = sym)
  ∧ (substring_of pt = SOME s)
```

Standard implementations of the sequential combinator attempt to parse all prefixes *s_pt* of a given input substring *s_tot* and return a (list of pairs of) a parse tree *pt* and the remainder *s_rem* of the input that was not parsed. In this case, we should ensure that concatenating *s_pt* and *s_rem* gives the original input *s_tot*.

```
prefix_sound ss_of_tm g sym p_sym = ∀ s_tot. ∀ pt. ∀ s_rem. ∃ s_pt.
  wf_grammar g
  ∧ MEM (pt, s_rem) (p_sym (toinput s_tot))
  ⟶
  pt ∈ (pts_of ss_of_tm g)
  ∧ (root pt = sym)
  ∧ (substring_of pt = SOME s_pt)
  ∧ (concatenate_two s_pt s_rem = SOME s_tot)
```

Theorem 4 (prefix_sound_grammar_to_parser_thm). *Parsers generated by grammar_to_parser are prefix-sound.*

```
prefix_sound_grammar_to_parser_thm = ∀ p_of_tm. ∀ g. ∀ sym.
  let p = grammar_to_parser p_of_tm g sym in
  let ss_of_tm = ss_of_tm_of p_of_tm in
  wf_p_of_tm p_of_tm ∧ wf_grammar g ⟶ prefix_sound ss_of_tm g sym p
```

Proof. Unfolding the definition of prefix_sound, we need to show a property of parse trees *pt*. The formal proof proceeds by an outer induction on the size of *pt*, and an inner structural induction on the list of immediate subtrees of *pt*.

We now observe that a prefix-sound parser can be easily transformed into a sound parser: just ignore those parses that do not consume the whole input. For this we need simple_parser_of *p*, which returns those parse trees produced by *p* for which the entire input substring was consumed.

```
simple_parser_of (p : parse_tree parser) =
  λ s. (MAP FST ∘ FILTER (λ (pt, s'). substring_of pt = SOME s)) (p (toinput s))
```

Theorem 5 (sound_grammar_to_parser_thm). *Parsers generated by grammar_to_parser are sound when transformed into simple parsers.*

```
sound_grammar_to_parser_thm = ∀ p_of_tm. ∀ g. ∀ sym.
  let p = grammar_to_parser p_of_tm g sym in
  let ss_of_tm = ss_of_tm_of p_of_tm in
  wf_p_of_tm p_of_tm ∧ wf_grammar g ⟶ sound ss_of_tm g sym (simple_parser_of p)
```

8 Completeness and Prefix-Completeness

In previous sections we have talked informally about completeness. In this section we define what it means for a parser to be complete with respect to a grammar. We also define the stronger property of prefix-completeness.

The simplest form of completeness requires that any parse tree *pt* that conforms to a grammar *g* and has a root symbol *sym* should be returned by the parser *q_sym* for *sym* when called on a suitable input string.

```
unsatisfactory_complete ss_of_tm g sym q_sym = ∀ s. ∀ pt.
  pt ∈ (pts_of ss_of_tm g)
  ∧ (root pt = sym)
  ∧ (substring_of pt = SOME s)
  ⟶
  MEM pt (q_sym s)
```

A grammar *g* and an input *s* can give rise to a potentially infinite number of parse trees *pt*, but a parser can only return a finite list of parse trees in a finite amount of time. For such non-trivial grammars, no parser can be complete in the sense of the definition above. Thus, this definition of completeness is unsatisfactory. If we accept that some parse trees must be omitted, we can still require that any input that can be parsed is actually parsed, and *some* parse tree *pt′* is returned.

```
complete ss_of_tm g sym q_sym = ∀ s. ∀ pt. ∃ pt′.
  pt ∈ (pts_of ss_of_tm g)
  ∧ (root pt = sym)
  ∧ (substring_of pt = SOME s)
  ⟶
  matches pt pt′ ∧ MEM pt′ (q_sym s)
```

Of course, our strategy is to return parse trees *pt′* as witnessed by good_tree_exists_thm.A more precise definition of completeness would require a parser to return all good trees, and main_thm from Sect. 9 shows that our parsers are complete in this sense, given the characterization of good trees via admits. One advantage of the definition above is that one does not need to understand the definition of good tree to understand our statement of completeness. We now introduce the related notion of prefix-completeness. As in the previous section, prefix-complete parsers can be transformed into complete parsers.

```
prefix_complete ss_of_tm g sym p_sym = ∀ s_tot. ∀ s_pt. ∀ s_rem. ∀ pt. ∃ pt′.
  (concatenate_two s_pt s_rem = SOME s_tot)
  ∧ pt ∈ (pts_of ss_of_tm g)
  ∧ (root pt = sym)
  ∧ (substring_of pt = SOME s_pt)
  ⟶
  matches pt pt′ ∧ MEM (pt′, s_rem) (p_sym (toinput s_tot))
```

Theorem 6 (prefix_complete_complete_thm). *If p is prefix-complete, then simple_parser_of p is complete.*

```
prefix_complete_complete_thm = ∀ ss_of_tm. ∀ g. ∀ sym. ∀ p.
  prefix_complete ss_of_tm g sym p ⟶ complete ss_of_tm g sym (simple_parser_of p)
```

9 Parser Generator Completeness

Theorem 7 (main_thm). *A parser p for symbol sym generated by* grammar_to_parser *is complete for prefixes s_pt of the input, in the sense that p returns all parse trees pt that are admitted by the context lc.*

$$\begin{array}{l}
\mathsf{main_thm} = \forall\ p_of_tm.\ \forall\ g.\ \forall\ pt.\ \forall\ sym.\ \forall\ s_pt.\ \forall\ s_rem.\ \forall\ s_tot.\ \forall\ lc.\\
\quad \mathsf{let}\ p = \mathsf{grammar_to_parser}\ p_of_tm\ g\ sym\ \mathsf{in}\\
\quad \mathsf{let}\ ss_of_tm = \mathsf{ss_of_tm_of}\ p_of_tm\ \mathsf{in}\\
\quad \mathsf{wf_p_of_tm}\ p_of_tm\\
\quad \wedge\ \mathsf{wf_grammar}\ g\\
\quad \wedge\ \mathsf{wf_parse_tree}\ pt\\
\quad \wedge\ pt \in (\mathsf{pts_of}\ ss_of_tm\ g)\\
\quad \wedge\ (\mathsf{root}\ pt = sym)\\
\quad \wedge\ (\mathsf{substring_of}\ pt = \mathsf{SOME}\ s_pt)\\
\quad \wedge\ (\mathsf{concatenate_two}\ s_pt\ s_rem = \mathsf{SOME}\ s_tot)\\
\quad \wedge\ \mathsf{admits}\ lc\ pt\\
\quad \longrightarrow\\
\quad \mathsf{MEM}\ (pt, s_rem)\ (p\ \langle\ \mathsf{lc} = lc;\ \mathsf{sb} = s_tot\ \rangle)
\end{array}$$

Proof. The proof is by an outer induction on the size of pt, with an inner structural induction on the list of immediate subtrees of pt.

A parser is initially called with an empty parsing context ie input i is such that $i.\mathsf{lc} = []$. We can use admits_thm to change the assumption admits $lc\ pt$ in the statement of main_thm to the assumption good_tree pt. We can further use good_tree_exists_thm to give the following corollary to the main theorem:

Corollary 1. *Parsers generated by* grammar_to_parser *are prefix-complete.*

$$\begin{array}{l}
\mathsf{corollary} = \forall\ p_of_tm.\ \forall\ g.\ \forall\ sym.\\
\quad \mathsf{let}\ ss_of_tm = \mathsf{ss_of_tm_of}\ p_of_tm\ \mathsf{in}\\
\quad \mathsf{let}\ p = \mathsf{grammar_to_parser}\ p_of_tm\ g\ sym\ \mathsf{in}\\
\quad \mathsf{wf_p_of_tm}\ p_of_tm \wedge \mathsf{wf_grammar}\ g \longrightarrow \mathsf{prefix_complete}\ ss_of_tm\ g\ sym\ p
\end{array}$$

This combined with prefix_complete_complete_thm gives:

Theorem 8 (complete_grammar_to_parser_thm). *Parsers generated by* grammar_to_parser *are complete when transformed into simple parsers.*

$$\begin{array}{l}
\mathsf{complete_grammar_to_parser_thm} = \forall\ p_of_tm.\ \forall\ g.\ \forall\ sym.\\
\quad \mathsf{let}\ ss_of_tm = \mathsf{ss_of_tm_of}\ p_of_tm\ \mathsf{in}\\
\quad \mathsf{let}\ p = \mathsf{grammar_to_parser}\ p_of_tm\ g\ sym\ \mathsf{in}\\
\quad \mathsf{wf_p_of_tm}\ p_of_tm \wedge \mathsf{wf_grammar}\ g\\
\quad \longrightarrow \mathsf{complete}\ ss_of_tm\ g\ sym\ (\mathsf{simple_parser_of}\ p)
\end{array}$$

10 Implementation Issues

Code extraction. The HOL4 definitions required for grammar_to_parser are executable within HOL4 itself, using either basic term rewriting or the more efficient strategies embodied in `EVAL_CONV`. We should expect that evaluating code inside a theorem prover is relatively slow compared to interpreting similar code using one of the usual functional languages (OCaml, SML, Haskell). Our OCaml implementations are manually extracted from the HOL4 definitions. An alternative is to use the HOL4 code extraction facilities which involves pretty-printing the HOL4 definitions as OCaml, but it is not clear that this step preserves soundness (the problem is HOL4 type definitions and their relation to ML type definitions).

Terminal parsers. Terminal parsers are required to satisfy a well-formedness requirement, but are otherwise more-or-less arbitrary. The OCaml implementation includes several common terminal parsers that arise in practice. For example, the function `parse_AZS` is a terminal parser that parses a sequence of capital letters. Verification of these terminal parsers is left for future work.

Parsing a grammar specification. The parser generator is parameterized by a grammar g (a list of rules). However, grammars are typically written concretely using BNF syntax which must itself be parsed. We therefore define the following syntax of BNF. We have adopted two features from Extended BNF: nonterminals do not have to be written within angled brackets, and arbitrary terminals can be written within question marks. The terminal `?ws?` accepts non-empty strings of whitespace, `?notdquote?` (resp. `?notsquote?`) accepts strings of characters not containing a double (resp. single) quote character, `?AZS?` accepts non-empty strings of capital letters, and `?azAZs?` accepts non-empty strings of letters.

```
RULES -> RULE | RULE ?ws? RULES
RULE -> SYM ?ws? "->" ?ws? SYMSLIST
SYMSLIST -> SYMS | SYMS ?ws? "|" ?ws? SYMSLIST
SYMS -> SYM | SYM ?ws? SYMS
SYM -> '"' ?notdquote? '"' | "'" ?notsquote? "'" | ?AZS?
  | "?" ?azAZs? "?"
```

Implementing a parser for this grammar is straightforward. The top-level parser for `RULES` returns a grammar. To turn the grammar into a parser, we use the parser generator in Fig. 3.

Memoization. For efficient implementations it is necessary to use memoization on the function `grammar_to_parser`. Memoization takes account of two observations concerning the argument `i`. First, as mentioned previously, the context `i.lc` is implemented as a list but is used as a set. Therefore care must be taken to ensure that permutations of the context are treated as equivalent during memoization. The simplest approach is to impose an order on elements in `i.lc` and ensure that `i.lc` is always kept in sorted order. Second, the only elements `(nt,s)` in `i.lc` that affect execution are those where `s = i.sb`. Thus, before memoization, we discard all elements in `i.lc` where this is not the case. For future work it should be straightforward to add the memoization table as an extra argument to grammar_to_parser and then prove correctness.

Theoretical performance. Many grammars generate an exponential number of good parse trees in terms of the size of the input string. Any parser that returns all such parse trees must presumably take an exponential amount of time to do so. However, several parsing techniques claim to be able to parse arbitrary context-free grammars in sub-exponential time. In fact, these parsing techniques do not return parse trees, but instead return a "compact representation" of all parse trees in polynomial time, from which a possibly infinite number of actual parse trees can be further constructed. The compact representation records which symbols could be parsed for which parts of the input: it is, in effect, a list of pairs, where each pair consists of a symbol and a substring. If we modify our parsers so that they return a dummy value instead of parse trees, then the memoization table is itself a form of compact representation. If we further assume that terminal parsers execute in constant time, then the time complexity of our algorithm is $O(n^5)$ in the length of the input, since there are $O(n^2)$ substrings, each appearing as input in at most $O(n^2)$ calls to the parser, each of which takes time $O(n)$ to execute[5]. Absolute real-world performance is better than this would suggest, because most calls to the parser simply involve looking up pre-existing values in the memoization table, and so execute very quickly.

Real-world performance Roughly speaking, the larger the class of grammar that a parsing technique can handle, the worse the performance. For example, Packrat parsing [5] takes time linear in the size of the input, but cannot deal with even simple non-ambiguous grammars such as `S -> "x" S "x" | "x"`. Of the three existing verified parsers, only the Packrat-based TRX parser [11] has any performance data, a comparison with the purpose-built Aurochs XML parser and the similar xml-light: as expected TRX is significantly slower. Preliminary testing using a simple XML grammar indicates that our parsers are competitive: an unmemoized version of our algorithm can parse a 1.4MB XML file in 0.31 seconds (better than Aurochs, and slightly worse than xml-light). More importantly, our algorithm is linear time in the size of the input. Aurochs and xml-light are purpose built XML parsers, and TRX does not handle all context-free grammars, however, there are some techniques such as GLR parsing that can handle arbitrary context-free grammars. There are very few implementations, but the popular Happy parser generator [1] is one such. Executing a compiled version of our memoized parser generator (which interprets the grammar) and comparing the performance with a compiled version of a parser produced by Happy in GLR mode (where the parser code directly encodes the grammar) on the grammar `E -> E E E | "1" |` ϵ, with input a string consisting solely of `1`s, gives the following figures. Noticeably, the longer the input, the better our parsers perform relative to Happy parsers. In fact, parsers generated by Happy in GLR mode appear to be $O(n^6)$ although GLR is theoretically $O(n^3)$ in the worst case. We leave investigation of this discrepancy, and further real-world performance analysis and tuning, to future work.

Input size/# characters	Happy parse time/s	Our parse time/s	Factor
20	0.19	0.11	1.73
40	9.53	3.52	2.71
60	123.34	30.46	4.05

[5] The time complexity is not obvious, and was informed by careful examination of real-world execution traces. For comparison, the time complexity of Earley parsers, CYK parsers, and GLR parsers is $O(n^3)$.

11 Related Work

A large amount of valuable research has been done in the area of parsing. We cannot survey the entire field here, but instead aim to give references to work that is most closely related to our own. A more complete set of references is contained in our previous work [15].

The first parsing techniques that can handle arbitrary context-free grammars are based on dynamic programming. Examples include CYK parsing [10] and Earley parsing [4]. In these early works, the emphasis is on implementation concerns, and in particular completeness is often not clear. For example [16] notes that Earley parsing is not correct for rules involving ϵ. Later the approach in [16] was also found to be incorrect. However, it is *in principle* clear that *variants* of these approaches can be proved complete for arbitrary context-free grammars. Combinator parsing and related techniques are probably folklore. An early approach with some similarities is [14]. Versions that are clearly related to the approach taken in this paper were popularized in [9].

The first approach to use the length of the input to force termination is [12]. The work most closely related to ours is that of Frost et al. [8,6,7], who limit the depth of recursion to $m * (1 + |s|)$, where m is the number of nonterminals in the grammar and $|s|$ is the length of the input. They leave correctness of their approach as an open question. For example, they state: "Future work includes proof of correctness ..." [7]; and "We are constructing formal correctness proofs ..." [8]. A major contribution of this paper, and the key to correctness, is the introduction of parsing context and the definition of good_tree. Amazingly, the measure function from Sect. 7 gives the same worst-case limit on the depth of recursion as that used by Frost et al. (although typically our measure function decreases faster because it takes the context into account), and so *this work can be taken as proof that the basic approach of Frost et al. is correct.*

The mechanical verification of parsers, as here, is a relatively recent development. Current impressive examples such as [2,11,3] cannot handle all context-free grammars. Recent impressive work on verified compilation such as [13] is complementary to the work presented here: our verified parser can extend the guarantees of verified compilation to the front-end parsing phase.

12 Conclusion

We presented a parser generator for arbitrary context-free grammars, based on combinator parsing. The code for a minimal version of the parser generator is about 20 lines of OCaml. We proved that generated parsers are terminating, sound and complete using the HOL4 theorem prover. The time complexity of the memoized version of our algorithm is $O(n^5)$. Real-world performance comparisons on the grammar `E -> E E E | "1" |` ϵ indicate that we are faster than the popular Happy parser generator running in GLR mode across a wide range of inputs.

There is much scope for future work, some of which we have mentioned previously. One option is to attempt to reduce the worst case time complexity from

$O(n^5)$. In an ideal world this could be done whilst preserving the essential beauty and simplicity of combinator parsing; in reality, it may not be possible to reduce the time complexity further without significantly complicating the underlying implementation.

References

1. Happy, a parser generator for Haskell, `http://www.haskell.org/happy/`
2. Barthwal, A., Norrish, M.: Verified, Executable Parsing. In: Castagna, G. (ed.) ESOP 2009. LNCS, vol. 5502, pp. 160–174. Springer, Heidelberg (2009)
3. Danielsson, N.A.: Total parser combinators. In: Hudak, P., Weirich, S. (eds.) ICFP, pp. 285–296. ACM (2010)
4. Earley, J.: An efficient context-free parsing algorithm. Commun. ACM 13(2), 94–102 (1970)
5. Ford, B.: Packrat parsing: simple, powerful, lazy, linear time, functional pearl. In: ICFP 2002: Proceedings of the Seventh ACM SIGPLAN International Conference on Functional Programming, New York, NY, USA, vol. 37/9, pp. 36–47. ACM (2002)
6. Frost, R.A., Hafiz, R., Callaghan, P.: Parser Combinators for Ambiguous Left-Recursive Grammars. In: Hudak, P., Warren, D.S. (eds.) PADL 2008. LNCS, vol. 4902, pp. 167–181. Springer, Heidelberg (2008)
7. Frost, R.A., Hafiz, R., Callaghan, P.C.: Modular and efficient top-down parsing for ambiguous left-recursive grammars. In: IWPT 2007: Proceedings of the 10th International Conference on Parsing Technologies, Morristown, NJ, USA, pp. 109–120. Association for Computational Linguistics (2007)
8. Hafiz, R., Frost, R.A.: Lazy Combinators for Executable Specifications of General Attribute Grammars. In: Carro, M., Peña, R. (eds.) PADL 2010. LNCS, vol. 5937, pp. 167–182. Springer, Heidelberg (2010)
9. Hutton, G.: Higher-order functions for parsing. J. Funct. Program. 2(3), 323–343 (1992)
10. Kasami, T.: An efficient recognition and syntax analysis algorithm for context-free languages. Technical Report AFCRL-65-758, Air Force Cambridge Research Laboratory, Bedford, Massachusetts (1965)
11. Koprowski, A., Binsztok, H.: TRX: A Formally Verified Parser Interpreter. In: Gordon, A.D. (ed.) ESOP 2010. LNCS, vol. 6012, pp. 345–365. Springer, Heidelberg (2010)
12. Kuno, S.: The predictive analyzer and a path elimination technique. Commun. ACM 8(7), 453–462 (1965)
13. Leroy, X.: Formal verification of a realistic compiler. Communications of the ACM (April 2009)
14. Pratt, V.R.: Top down operator precedence. In: Proceedings ACM Symposium on Principles Prog. Languages (1973)
15. Ridge, T.: Simple, functional, sound and complete parsing for all context-free grammars (2010) unpublished draft, `http://www.cs.le.ac.uk/~tr61`
16. Tomita, M.: Efficient Parsing for Natural Language: A Fast Algorithm for Practical Systems. Kluwer, Boston (1986)

A Decision Procedure for Regular Expression Equivalence in Type Theory*

Thierry Coquand and Vincent Siles

University of Gothenburg
{coquand,siles}@chalmers.se

Abstract. We describe and formally verify a procedure to decide regular expressions equivalence: two regular expressions are equivalent if and only if they recognize the same language. Our approach to this problem is inspired by Brzozowski's algorithm using derivatives of regular expressions, with a new definition of finite sets. In this paper, we detail a complete formalization of Brzozowki's derivatives, a new definition of finite sets along with its basic meta-theory, and a decidable equivalence procedure correctly proved using Coq and Ssreflect.

Introduction

The use of regular expressions is common in programming languages to extract data from strings, like the `scanf` function of the C-language for example. As shown in recent works [4,11] the equational theory of regular expressions can also be important for interactive theorem provers as providing a convenient tool for reasoning about binary relations. The fundamental result which is used there is the *decidability* of the problem whether two regular expressions are *equivalent*, i.e. recognize the same language, or not.

The purpose of this paper is to represent in type theory the elegant algorithm of Brzozowski [5] to test this equivalence. In an intuitionistic framework such as type theory, this in particular amounts to show that the equivalence between two regular expressions is a decidable relation. For this, we define in type theory a boolean valued function corresponding to Brzozowski's algorithm, and we show that this function *reflects* [18] equivalence: this function returns true on two regular expressions if and only if they are equivalent.

Brzozowski's algorithm has already been formally investigated but it never has been *completely* proved correct: in [11], the authors did not proved formally the termination of their algorithm, and in [1], the authors did not finished the proof of correctness of the procedure. In this paper, we describe a *complete* formalization of Brzozowski decision procedure based on derivatives of regular expressions.

In order to achieve this formal representation, we introduce a new definition of finiteness in type theory, which may have an interest in itself. This definition

* The research leading to these results has received funding from the European Union's 7th Framework Programme under grant agreement nr. 243847 (ForMath).

J.-P. Jouannaud and Z. Shao (Eds.): CPP 2011, LNCS 7086, pp. 119–134, 2011.

is not equivalent constructively to the usual definition, which expresses that one can list all elements in this set. (Intuitively this new definition expresses that if we keep removing elements from a finite set, this will stop eventually.) We believe that this notion is useful to express in type theory algorithms relying on the computation of least fixed point of finite sets (see Coro. 1), such as the computation of minimal automata or the computation of a deterministic automaton associated to a non deterministic automaton.

In Sect. 1, we describe a new definition of finite sets called *inductively* finite sets, along with some basic properties required by the decision procedure correctness. In Sect. 2, we recall the definition of regular expressions and the notion of Brzozowski derivatives. The third section is dedicated to Brzozowski's proof that the set of derivatives is finite (*inductively* finite in our case). The decision algorithm for equivalence is then defined by recursion over this proof of finiteness and we discuss its representation in type theory. This uses an elegant idea of Barras for representing in type theory functions defined by well-founded recursion, which keeps logical soundness while having a satisfactory computational behavior. The last section presents some test cases. The whole development has been formalized[1] using Coq [7] and Ssreflect [18], and can be found at [8].

1 Finite Sets in Type Theory

1.1 Informal Motivations

The notion of *finiteness* is basic in mathematics and was one of the first notion to be formalized by Dedekind and Frege. A 1924 paper by Tarski [19] describes different possible axiomatizations in the framework of Zermelo set theory. One definition comes from Russell and Whitehead [17]: a subset of a set is finite if it can be inductively generated from the empty set by the operation of adding one singleton. A set A is then finite if A itself is a finite subset. Tarski shows that this is equivalent to another definition, in the framework of classical set theory: a set is finite if the relation of strict inclusion on subsets is well-founded. (By taking complement, this is equivalent to the fact that the relation $X \supsetneq Y$ is well-founded, i.e. there is no infinite sequences $X_0 \subsetneq X_1 \subsetneq X_2 \subsetneq \ldots$) On the other hand, these two definitions are only equivalent to Dedekind definition (a finite set is such that any endomap which is injective is a bijection) in presence of the axiom of choice.

In intuitionistic frameworks, the most commonly used definition seems to be the one of being Kuratowski finite [10], which is a small variation of Russell-Whitehead definition: a subset is finite if it can be inductively generated from the empty set, the singleton and the union operations. In type theory, this definition takes an equivalent more concrete form: a subset is finite if and only if it

[1] There are no axioms neither in this development nor in Ssreflect libraries. However, due to a flaw in the Print Assumptions command, one might think there are. This command considers definitions made opaque by signature ascription to be axioms, which is not the case.

can be enumerated by a list. The situation is complex if we consider sets with a non necessarily decidable equality. In this paper however, we limit essentially ourselves to *discrete* sets where the equality is decidable. With this extra assumption, to be Kuratowski finite is equivalent to be in bijection with a set N_k where N_k is defined recursively by $N_{k+1} = N_k + N_1$ where N_0 is the empty type and N_1 the unit type. So, given the information that A is Kuratowski finite, we can compute from it the cardinality of A, and, in particular, we can decide if A is empty of not.

In this paper, we explore another inductive definition of finiteness: a set A is *noetherian* (or *inductively finite*) if and only if for all a in A the set $A - \{a\}$ is noetherian. Intuitively, if we keep picking distinct elements in A, eventually we reach the empty set. It should be intuitive then that if A is Kuratowski finite then A is noetherian (by induction on the cardinality of A), but also that, only from the information that A is noetherian, one cannot decide if A is empty or not. So to be noetherian is intuitionistically weaker than being Kuratowski finite. (See the reference [9] which analyzes these notions in the framework of Bishop mathematics.) We have shown formally that to be noetherian is equivalent in type theory to a variation of Tarski definition: A is noetherian if and only if the relation $X \supsetneq Y$ is well-founded on subsets of A given by lists. Independently of our work, corresponding definitions of finiteness have been recently considered in the work [2].

1.2 Inductively Finite Sets in Type Theory

From now on, we use predicates over a given type A to represent subsets. Predicates over A are `Prop`-valued functions, the universe of propositions of Coq. Given a binary relation R and an element x of type A, $R\ x$ represents the set of the elements $\{y : A \mid R\ x\ y \text{ holds }\}$.

Given a type A, a set E and a binary relation R over A, we say that **Bar** $R\ E$ holds if and only if for any x in E, **Bar** $R\ (E \cap (R\ x))$ holds. This is an inductive definition[2] , which expresses intuitively that we can not find an infinite sequence $x_0,\ x_1,\ x_2, \ldots$ of elements satisfying E and such that we have $R\ x_i\ x_j$ if $i < j$.

This definition is closely connected to the notion of well-quasi-ordering [16]. Indeed R is a well-quasi-ordering on E if and only if it is transitive and decidable and its complement R' is such that **Bar** $R'\ E$ holds. If we start from a type A with a decidable equality $=_A$ then we define E to be (inductively) finite if **Bar** $(\lambda\ x.\lambda\ y.\neg\ (x =_A y))\ E$. Intuitively, it expresses that for any infinite sequence $x_0,\ x_1,\ x_2, \ldots$ there exists $i < j$ such that $x_i =_A x_j$.

As explained above, the usual definition of finite (or "Kuratowski finite") is that we can list all elements in E (see Fig. 2). It can be shown that E is inductively finite if it is finite, but the converse does not hold. Therefore, we capture more sets with this definition, but in general, it is not possible to describe an inductively finite set as the list of its elements.

[2] This is a particular case of *Bar* induction [16], and we kept the name.

```
Variable A:Type.
Definition gset := A → Prop.
Definition grel := A → A → Prop.

Inductive Bar (R: grel A ) (E: gset A ) : Prop :=
 | cBar : (∀ x:A, E x → Bar R (Intersection E (R x))) → Bar R E.

Definition lFinite (R : grel A) (E : gset A) := Bar (neq R) E.
```

Fig. 1. Definition of **Bar**

Lemma 1. Basic properties of inductively finite sets

- *If* **Bar** R F *and* $E \subseteq F$ *then* **Bar** R E.
- *If* **Bar** R E *and* **Bar** R F *then* **Bar** R $(E \cup F)$.
- *If* **Bar** R E *and* **Bar** S E *then* **Bar** $(R \cup S)$ E.
- *If* **Bar** R E *and* **Bar** S F *then* **Bar** $(R \times S)$ $(E \times F)$, *where* $(E \times F)$ (x,y) *means* $E\ x \wedge F\ y$, *and* $(R \times S)$ (x_0,y_0) (x_1,y_1) *means* $R\ x_0\ x_1 \vee S\ y_0\ y_1$.

The proof of the last point is essentially the same as the proof that a product of two well-quasi-ordering is a well-quasi-ordering [16].

For any set E over a type A, a list of elements of E can be interpreted as a finite subset of E. Any list defines a subset of E by defining a membership function InA:

- InA [ ] x is always false
- InA $(hd :: tl)$ x holds if and only if $x =_A hd$ or InA tl x.

Using this membership function, we can describe in type theory what it means to be Kuratowski finite. E is Kuratowski finite when there is a list X that enumarates the elements of E:

$$\exists X : [A], \forall x : A, E\ x \leftrightarrow \mathsf{InA}\ X\ x$$

If A is a type with decidable equality $=_A$, InA is a decidable predicate, and we can define a decidable equality $=_{[A]}$ on the type $[A]$ of lists (also written eql in the code) such that $X_0 =_{[A]} X_1$ holds if and only if X_0 and X_1 represent the same subset of E. If X is of type $[A]$, we define $[E]$ X (or gpred_list E X) to mean that all elements in X satisfy the predicate E.

Since we are working with an abstract equality over the type A, a natural condition on E is to require it to be *compatible* with the equality over A, that is for all x, y such that $x =_A y$ and $E\ x$ holds, then $E\ y$ also holds.

Proposition 1. *(Bar_gpred_list) If E is inductively finite on A and compatible then [E] is inductively finite on [A].*

This can be expressed as the result that the collection of all subsets (given by lists) of an inductively finite set is inductively finite. The proof is reminiscent of the constructive proof of Higman's Lemma about well-quasi-ordering in [16].

```
Definition KFinite (eqA : grel A) (E : gset A) : Prop :=
 ∃ X, (∀ x:A, E x ↔ INA eqA X x).

Definition gpred_list (E: gset A ) : gset (seq A) :=
 fix aux l : Prop :=
  match l with
  | nil ⇒ True
  | x :: xs ⇒ E x ∧ aux xs
end.
```

Fig. 2. Definition of Kuratowski finite and gpred_list

```
Definition E_compat (eqA : grel A) (E:gset A) := ∀ x y, eqA x y → E x → E y.

Lemma Bar_gpred_list : ∀ (eqA : grel A) (E : gset A),
  E_compat eqA E → IFinite eqA E → IFinite eql (gpred_list E).
```

Fig. 3. Finiteness of gpred_list

Proposition 2. *(Bar_fun) If f is a function preserving equality*

$$\forall x\ y, x =_A y \rightarrow fx =_B fy$$

and if E is inductively finite, then fE, the image of E by f, is also inductively finite.

Both Prop. 1 and 2 are important for the proof of the main Lemma 2.

```
Variable f : A → B.
Variable eqA : grel A.
Variable eqB : grel B.

Definition f_set (E: gset A) : gset B := fun (y:B) ⇒ exists2 x, E x & eqB (f x) y.
Variable f_compat : ∀ (a a': A), eqA a a' → eqB (f a) (f a').
Lemma Bar_fun : ∀ E , IFinite eqA E → IFinite eqB (f_set E).
```

Fig. 4. Definition of the property Bar_fun

A major result for proving Prop. 1 is the fact that if E is inductively finite on A then the relation of strict inclusion (sup) between subsets of E enumerated by a list is well-founded.

Theorem 1. *(lFinite_supwf) For all compatible set E, the relation* sup *is well founded if and only if E is inductively finite.*

Corollary 1. *If E is inductively finite, any monotone operator acting on subsets of E enumerated by a list has a least fixed-point.*

This is proved by building this list by well-founded recursion.

```
Theorem lFinite_supwf : ∀ (eqA : grel A) (E : gset A) , E_compat eqA E →
  (lFinite eqA E ↔ well_founded sup (gpred_list E)).
```

Fig. 5. Strict inclusion of lists

2 Regular Expressions

Now that we know how to encode inductively finite sets in type theory, we focus on the main purpose of this paper, deciding regular expressions equivalence. It is direct to represent the type of all regular expressions on a given alphabet Σ as an inductive type. Following Brzozowski's approach, we work with extended regular expressions, having conjunction and negation as constructors, and a "." constructor that matches any letter of the alphabet.

$$E, E_1, E_2 ::= \emptyset \mid \epsilon \mid a \mid . \mid E_1 + E_2 \mid E^* \mid E_1 E_2 \mid E_1 \& E_2 \mid \neg E$$

It is a remarkable feature of Brzozowski's algorithm that it extends directly the treatment of negation. If one uses finite automata instead, the treatment of negation is typically more difficult, since one would have to transform an automaton to a deterministic one in order to compute its complement.

To each regular expression E we associate a boolean predicate (using Ssreflect's pred) $L(E)$ on the set of words Σ^* such that a word u satisfies $L(E)$ if and only if u is recognized by E. So the boolean function $L(E)$ reflects the predicate of being recognized by the language E. We can then write "u \in E" (this is a notation for mem E u) to express that the word u is recognized by E. We consider that two languages are equal if they contain the same words:

$$\forall L_1\ L_2,\ L_1 = L_2 \quad \leftrightarrow \quad \forall u \in \Sigma^*, u \in L_1 = u \in L_2$$

Two regular expressions are equivalent if their associated languages are equal.

It is direct to define a boolean $\delta(E)$ (or has_eps E in our formalization) which tests whether the empty word ϵ is in $L(E)$ or not (see Fig. 6).

```
Variable symbol : eqType.
Definition word := seq symbol.
Definition language := pred word.
Inductive regular_expression :=
  | Void
  | Eps
  | Dot
  | Atom of symbol
  | Star of regular_expression
  | Plus of regular_expression & regular_expression
  | And of regular_expression & regular_expression
  | Conc of regular_expression & regular_expression
  | Not of regular_expression.

Definition EQUIV (E F:regexp) := ∀ s:word, (s \in E ) = (s \in F).
Notation "E ≡ F" := (EQUIV E F) ( at level 30).

Fixpoint has_eps (e: regular_expression) :=
  match e with
  | Void ⇒ false
  | Eps ⇒ true
  | Dot ⇒ false
  | Atom x ⇒ false
  | Star e1 ⇒ true
  | Plus e1 e2 ⇒ has_eps e1 || has_eps e2
  | And e1 e2 ⇒ has_eps e1 && has_eps e2
  | Conc e1 e2 ⇒ has_eps e1 && has_eps e2
  | Not e1 ⇒ negb (has_eps e1)
  end.
```

Fig. 6. Definition of regular expressions and the δ operator

2.1 Derivatives

Given a letter a in Σ and a regular expression E, we define E/a (or der a E), the derivative of E by a, by induction on E (see Fig. 7 for a direct encoding in type theory, or [5] for the original definition). A word u is in $L(E/a)$ if and only if the word au is in $L(E)$: $L(E/a)$ is called the *left-residual* of $L(E)$ by a. It is then possible to define E/u (or wder u E) for any word u by recursion on u

$$E/\epsilon = E \qquad E/(au) = (E/a)/u$$

The function δ and derivation operators give us a way to check whether a word is recognized by a regular expression. With the previous definitions, a word u is in $L(E)$ if and only if ϵ is in $L(E/u)$, which is equivalent to $\delta(E/u) = true$.

```
Fixpoint der (x: symbol) (e: regular_expression) :=
  match e with
  | Void ⇒ Void
  | Eps ⇒ Void
  | Dot ⇒ Eps
  | Atom y ⇒ if x == y then Eps else Void
  | Star e1 ⇒ Conc (der x e1) (Star e1)
  | Plus e1 e2 ⇒ Plus (der x e1) (der x e2)
  | And e1 e2 ⇒ And (der x e1) (der x e2)
  | Conc e1 e2 ⇒ if has_eps e1 then Plus (Conc (der x e1) e2) (der x e2)
    else (Conc (der x e1) e2)
  | Not e1 ⇒ Not (der x e1)
  end.

Fixpoint wder (u: word) (e: regular_expression) :=
  if u is x :: v then wder v (der x e) else e.

Fixpoint mem_der (e: regular_expression) (u: word) :=
  if u is x :: v then mem_der (der x e) v else has_eps e.

Lemma mem_derE : ∀ (u: word) (e: regular_expression), mem_der E u = (u \in E).

Lemma mem_wder : ∀ (u: word) (e: regular_expression),
  mem_der E u = has_eps (wder u E).
```

Fig. 7. Definition of the der operator and some of its properties

2.2 Similarity

Brzozowski proved in [5] that there is only a finite number of derivatives for a regular expression, up to the following rewriting rules:

$$E + E \sim E \qquad E + F \sim F + E \qquad E + (F + G) \sim (E + F) + G$$

This defines a decidable equality over regular expressions, called similarity, which also satisfies $L(E) = L(F)$ if $E \sim F$.

The exact implementation of these rewriting rules is not relevant to show that the set of derivatives is inductively finite. We provide two implementations in our formalization, one exactly matching these three rules, and a more efficient one which also includes the following rules:

$$\begin{array}{lll}
E + \emptyset \sim E & (EF)G \sim E(FG) & E \,\&\, E \sim E \\
E \,\&\, \emptyset \sim \emptyset & E \,\&\, (F \,\&\, G) \sim (E \,\&\, F) \,\&\, G & E^{**} \sim E^* \\
E \,\&\, \top \sim E & E \,\&\, F \sim F \,\&\, E & \neg\neg E \sim E \\
\epsilon E \sim E\epsilon \sim E & & \emptyset^* \sim \epsilon^* \sim \epsilon
\end{array}$$

The regular expression $\top$ stands for the regular expression that recognize any word, which we implemented as $\neg\emptyset$.

Our implementation is close to the one in [15]. To enforce these additional simplifications, we introduce a notion of canonical form (with a boolean predicate wf_re for being "a well-formed canonical expression") and a normalization function canonize in such a way that $E \sim F$ is defined as canonize E = canonize F (where = is the structural equality). This function relies on the use of smart-constructors which perform the previous rewriting rules. For example, the rewriting rules of Plus are enforced by keeping a strictly sorted lists of all the regular expressions linked by a "+". We then prove that these smart-constructors indeed satisfy the similarity requirements (see Fig. 8). In [11], the idea of normalizing regular expression to enforce the rules is also used, with the exact same idea of keeping sorted lists of regular expression. However, they do not create a different structure and just modify the existing regular expressions.

```
Fixpoint canonize c : canonical_regexp := match c with
  | Void ⇒ CVoid
  | Eps ⇒ CEps
  | Dot ⇒ CDot
  | Atom n ⇒ CAtom n
  | Star c' ⇒ mkStar (canonize c')
  | Plus c1 c2 ⇒ mkPlus (canonize c1) (canonize c2)
  | And c1 c2 ⇒ mkAnd (canonize c1) (canonize c2)
  | Conc c1 c2 ⇒ mkConc (canonize c1) (canonize c2)
  | Not c1 ⇒ mkNot (canonize c1)
end.

Lemma mkPlusC : ∀ r1 r2, mkPlus r1 r2 = mkPlus r2 r1.

Lemma mkPlus_id : ∀ r, wf_re r → mkPlus r r = r.

Lemma mkPlusA : ∀ r1 r2 r3, wf_re r1 → wf_re r2 → wf_re r3 →
  mkPlus (mkPlus r1 r2) r3 = mkPlus r1 (mkPlus r2 r3).
```

Fig. 8. canonize function and some properties of the Plus smart constructor

Brzozowski's result [5] that any regular expression has only a finite number of derivatives is purely existential. It is not so obvious how to extract from it a computation of a list of all derivatives up to similarity: even if we have an upper bound on the number of derivatives of a given regular expression E, it is not clear when to stop if we start to list all possible derivatives of E. In type theory, this will correspond to the fact that we can prove that the set of all derivatives of E is *inductively* finite up to similarity, but we can not prove without a further hypothesis on similarity (namely that similarity is closed under derivatives) that this set of *Kuratowski* finite up to similarity. On the other hand, we can always show that the set of derivatives is Kuratowski finite up to equivalence.

3 Brzozowski Main Result

The key property of Brzozowski we need to build the decision procedure is the fact that the set of derivatives of a regular expression is inductively finite (with respect to similarity). An interesting point is that we do not actually need Σ to be finite to prove this fact. However, we need Σ to be finite in order to effectively compute all the derivatives and compare two regular expressions.

The proof uses the following equalities about derivatives (see [5], Annexe II for a detailed proof):

$$(E+F)/u = E/u + F/u \qquad (E\,\&\,F)/u = E/u\,\&\,F/u \qquad \neg(E/u) = (\neg E)/u$$

If $u = a_1 \dots a_n$

$$(EF)/u = (E/u)F + \delta(E/a_1 \dots a_{n-1})F/a_n +$$

$$\delta(E/a_1 \dots a_{n-2})F/a_{n-1}a_n + \cdots + \delta(E)F/a_1 \dots a_n$$

and finally

$$E^*/u \sim (E/u)E^* + \Sigma\ \delta(E/u_1)\dots\delta(E/u_{p-1})(E/u_p)E^*$$

for any decomposition of u in non-empty words $u = u_1 \dots u_p$.

We represent the set of all derivatives of a given regular expression E by the predicate

$$\mathcal{D}er\ E = \{F \mid \exists u : \Sigma^*, E/u \sim F\}$$

We proved formally that this set was *inductively* finite with respect to similarity.

Lemma 2. The set of derivatives is inductively finite

For any regular expression E, the set $\mathcal{D}er\ E$ is inductively finite with respect to the similarity:

$$\forall (E : \mathsf{regexp}), \mathit{IFinite}\ \sim\ (\mathcal{D}er\ E)$$

Proof. The proof is done by induction on E, with a combination of the lemmas Bar_gpred_list and Bar_fun described in Sect. 1. We only describe here the case for Conc, which is the most difficult case. All the other ones are done in a similar way.

By induction, we know that $\mathcal{D}er\ E$ and $\mathcal{D}er\ F$ are inductively finite, and we want to prove that $\mathcal{D}er\ EF$ is too. Equality of regular expressions is performed using the $\sim$ operator with its extension $[\sim]$ to list of regular expressions [regexp]. Let us consider the following function:

$$\begin{array}{ll} fConc & : \mathsf{regexp} \times \mathsf{regexp} \times [\mathsf{regexp}] \rightarrow \mathsf{regexp} \\ fConc\ (e, f, L) = e\ f + L_1 + \cdots + L_n & \end{array}$$

- Using the equalities of derivatives we just stated, we first show that

$$\mathcal{D}er(EF) \subseteq fConc\ (\mathcal{D}er\ E, \{F\}, [\mathcal{D}er\ F]) \quad (1)$$

- The set $[\mathcal{D}er F]$ is inductively finite for $[\sim]$ thanks to lemma Bar_gpred_list, and the singleton set $\{F\}$ is obviously inductively finite for $\sim$.
- Using Brzozowski minimal set of rewriting rules, it is direct to show that $fConc$ preserves equality:

$$\forall e\, e'\, f\, f'\, l\, l', e \sim e' \wedge f \sim f' \wedge l\,[\sim]\, l' \rightarrow fConc\,(e, f, l) \sim fConc\,(e', f', l')$$

Then the image of the set $\mathcal{D}er\ E \times \{F\} \times [\mathcal{D}er\ F]$ by $fConc$ is inductively finite thank to lemma Bar_fun. Thanks to Lemma 1 and (1), we can conclude that $\mathcal{D}er(EF)$ is inductively finite.

To simplify we assume that Σ is now the type with two elements $\{0, 1\}$, but it would work with any finite set. The particular instance of **regular_expression** over this alphabet is named regexp.

As we said, it is not possible in the general case to enumerate any inductively finite set with a list, but in this particular case, it is possible to do so.

Lemma 3. Enumeration of the set of derivatives

For any regular expression E, it is possible to build a list L_E such that:

- $L_E \subseteq \mathcal{D}er\ E$
- $E\ \in\ L_E$
- $\forall(e : \textsf{regexp})(e \in L_E)\,\exists(e' : \textsf{regexp}), (e' \in L_E) \wedge (e' \sim e/0)$
- $\forall(e : \textsf{regexp})(e \in L_E)\,\exists(e' : \textsf{regexp}), (e' \in L_E) \wedge (e' \sim e/1)$

To build such a list, and prove Lemma 3, we apply Coro. 1 on the mononotone function:

$$\textsf{deriv}\ l = \textsf{map}(\textsf{der}\ 0)\ l \mathbin{++} \textsf{map}(\textsf{der}\ 1)\ l$$

The list L_E is the least fixpoint of deriv above the singleton list $[E]$.

As a consequence of these properties, we can show that any derivative of E is represented inside L_E:

Theorem 2. The list L_E is complete

For any regular expression E and any word u, there is a regular expression e in L_E such that $L(E/u) = L(e)$.

Another way to state it is to say is that the set of all all derivatives of E is KFinite *up to* equivalence.

Proof. The proof goes by induction on the length of the word u:

- If the length of u is 0, then $u = \epsilon$, and we have $E/\epsilon = E$. We can close this case since E is in L_E.
- If the length of u is $n + 1$, then $u = vi$ where its last letter i is either 0 or 1. By induction, there is e in L_E such that $L(e) = (E/v)$. Using the two last properties of L_E as described in the previous lemma, there is e' in L_E such that $e' \sim e/i$, which implies $L(e') = L(e/i)$. If we combine both conclusions, we get that $L(E/u) = L((E/v)/i) = L(e/i) = L(e')$, which ends this proof.

What we prove is that the set of all derivatives is Kuratowski finite up to equivalence. Contrary to what one may thought at first, it does not seem possible to show that this set is Kuratowski finite up to similarity. In order to be able to prove it, we need a priori a stronger condition on $\sim$, that we have $A/0 \sim B/0$ and $A/1 \sim B/1$ whenever $A \sim B$. This is the case for Brzozowski minimal set of rules, but it is not the case for our efficient implementation of the similarity. (As it turned out, to have a list up to equivalence is sufficient to get a decision for equivalence.) In particular, the rule $E^{**} \sim E^*$ is not stable by derivation, it would require to add $E^* E^{**} \sim E^*$ to our set of rules.

4 Description of the Decision Procedure

From the definition of regular expressions equivalence $\equiv$ and the basic properties of the δ operator, we can derive another specification for being equivalent:

$$\begin{aligned} \forall E\ F,\ E \equiv F &= L(E) = L(F) \\ &\leftrightarrow \forall u \in \Sigma^*, u \in L(E) = u \in L(F) \\ &\leftrightarrow \forall u \in \Sigma^*, \delta(E/u) = \delta(F/u) \end{aligned}$$

```
Definition delta2 (ef:regexp × regexp) := let (e,f) := ef in has_eps e == has_eps f.

Definition build_list_fun : regexp → regexp → seq (regexp× regexp).

Definition regexp_eq (r1 r2: regexp) : bool := (all delta2 (build_list_fun r1 r2)).

Lemma regexp_eqP : ∀ (r1 r2:regexp), reflect (r1 ≡ r2) (regexp_eq r1 r2).
```

Fig. 9. Decision procedure with its correctness proof

For any regular expressions E and F, we consider the set

$$\mathcal{D}er_2\ E\ F = \{(e, f) \mid \exists (u : word), e \sim E/u \wedge f \sim F/u\}$$

This set is included inside $\mathcal{D}er\ E \times \mathcal{D}er\ F$, so with Lemmas 1 and 2, we can conclude that $\mathcal{D}er_2\ E\ F$ is inductively finite for any $E\ F$. A similar approach to the proof Lemma 3 and Thm. 2 allows us to conclude that $\mathcal{D}er_2\ E\ F$ can be enumerated by a list $L_{E,F}$ and for all word u in Σ^*, there is (e, f) in $L_{E,F}$ such that $L(e) = L(E/u)$ and $L(f) = L(F/u)$.

This property of $L_{E,F}$ is enough to decide the equivalence: we know that $L(E) = L(F) \quad\leftrightarrow\quad \forall u \in \Sigma^*, \delta(E/u) = \delta(F/u)$ and we proved that, for any u in Σ^*, there is (e, f) in L_{EF} such that $L(e) = L(E/u)$ and $L(f) = L(F/u)$. Since $\delta(e) = \delta(E/u)$ and $\delta(f) = \delta(F/u)$, we can show that

$$L(E) = L(F) \quad\leftrightarrow\quad \forall (e, f) \in L_{EF}, \delta(e) = \delta(f)$$

which is a decidable predicate.

5 Representation in Type Theory

While we have been carrying out our formal development in the systems Coq [7] and Ssreflect [18], we never use in an essential way the impredicativity of the sort of propositions. So all our proofs could have been done as well in a predicative system with universes, such as the one presented in [12], extended with inductive definitions.

One key issue is for the representation of the function regexp_eq which is defined by well-founded recursion, itself defined (see Fig. 10) by saying that all elements are accessible [14]. Indeed, this function is defined by recursion on the fact that the set of derivatives of a regular expression is inductively finite, which can be expressed, as we have seen above, by the fact that a relation is well-founded. This representation, theoretically sound, is problematic operationally: the computation of this function requires a priori the computation of the proof that an element is accessible. This computation is heavy, and furthermore seems irrelevant, since the accessibility predicate has only one constructor. In order to solve this problem, we follow the solution of Barras, refined by Gonthier, which allows us to keep the logical soundness of the representation with a satisfactory computational behaviour. The main idea is to "guard" the accessibility proofs by adding constructors (see Fig. 10). If we replace a proof that a relation is well-founded wf by its guarded version guard 100 wf, we add in a lazy way 2^{100} constructors, and get in this way a new proof that the relation is well-founded which has a reasonable computational behavior.

```
Inductive Acc (E:gset A) (R:grel A) (x:A) : Prop :=
  Acc_intro : (∀ y, E y → R y x → Acc E R y) → Acc E R x.
Definition well_founded (E : gset A) (R : grel A) := ∀ a: A, Acc E R a.
Fixpoint guard (E: gset A) (R : grel A) n (wfR: well_founded E R):
  well_founded E R :=
  match n with
    | 0 ⇒ wfR
    | S n ⇒ fun x ⇒ Acc_intro (fun y _ _ ⇒ guard E R n (guard E R n wfR) y)
  end.
```

Fig. 10. Guarded version of accessibility proof

6 Some Examples

One important feature of our formalization is the fact that we obtain the decision procedure as a type theoretic boolean function, with which you can compute directly without extracting to ML code. We can then use this function to build in type theory other tactics to solve problems which can be encoded in the language of regular expressions.

The following tests have been performed on an Intel Core2 Duo 1.6 GHz, with 2 GB of memory, running `Archlinux` with kernel 2.6.39. We used Coq v8.3pl2 and Ssreflect v1.3pl1 to formalize the whole development. It is direct to reduce the problem of inclusion to the problem of equivalence by expressing $E \subseteq F$ as $E + F \equiv F$.

The first example we tested is due to Nipkow and Krauss in [11]. The theories of Thiemann and Sternagel [20] contains the lemma which reduces to the following inclusion of regular expressions

$$0(00^*1^* + 1^*) \subseteq 00^*1^*$$

Our implementation of similarity answers `true` in 0.007 seconds.

The second example is extracted from the following predicate

$$\forall n \geqslant 8, \exists x\ y, n = 3x + 5y$$

Is this predicate true or false ? This can be rewritten as the following regular expression problem:

$$000000000^* \subseteq (000 + 00000)^*$$

Our implementation answers `true` in 1.6 seconds.

Some more examples can be found in the file `ex.v` at [8]. Since we are only looking for building a tactic on top of this decision procedure, like in the reference [11], both results are within acceptable range for this purpose.

Conclusion and Future Work

The main contributions of this work are

- a complete formalization in type theory of Brzowoski's algorithm for testing the equivalence of regular expressions
- a new definition of finiteness and formal proofs of its basic closure properties, which may have an interest in itself
- the experimental verification that it is feasible to define in type theory functions by well-founded induction and to prove their properties, obtaining programs that have a reasonable operational behavior[3]

As a direct extension of Brzozowski's procedure, we also defined and proved correct a decision algorithm for the inclusion of regular expressions, that we have tested on some simple examples.

While doing this formalization, we discovered two facts about Brzozowski's algorithm that may not be obvious at first, and which are examples of what one

[3] As far as we know, this approach to representation of terminating general recursive function in type theory has not been tested before. For instance this approach is explicitly rejected in the reference [4], as being "inconvenient", since it "requires mixing some non-trivial proofs with the codes" while our work shows that it is reasonable in practice and theoretically satisfactory.

may learn from formalization (and new to us, though we have been teaching the notion of Brzozowski's derivatives for a few years). First, the number of derivatives is finite even if the alphabet is not. (However in practice, one has to restrict to finite alphabets if one wants to extract the list describing the derivatives.) Second, it is not so obvious how to extract from Brzozowski's purely existential result an actual computation of a list of all derivatives up to *similarity* (as one may have expected at first; without the further assumption that similarity is closed under derivatives we obtain only a list of derivatives up to equivalence).

There are other notions of derivatives that are worth investigating, like in [21] where they use partial derivatives known as *Antimorov*'s derivatives. A natural extension of this work would be, like in the references [4,11] to use it for a reflexive tactic for proving equalities in relation algebra. We don't expect any problem there, following [4,11]. A more ambitious project will be to use this work for writing a decision procedure for the theory WS1S [6], where formulae of this language are interpreted by regular expressions. Since we use extended regular expression, we have a direct interpretation of all boolean logical connectives, and what is missing is the interpretation of the existential quantification. For giving this interpretation, one possible Lemma would be to show that any extended regular expression is equivalent to a regular expression which uses only the operator of union, concatenation and Kleene star. This in turn should be a simple consequence of the fact that the set of derivatives of a given expression is Kuratowski finite up to equivalence. Using this result, we can then define given any map $f : \Sigma_1 \to \Sigma_2$ extended to words $f^* : \Sigma_1^* \to \Sigma_2^*$, and given a regular expression E over Σ_1, a new regular expression $f^*(E)$ over Σ_2 such that $L(f^*(E)) = f^*(L(E))$. It is then possible to interpret existential quantification using this operation.

References

1. Almeida, J.B., Moreira, N., Pereira, D., de Sousa, S.M.: Partial Derivative Automata Formalized in Coq. In: Domaratzki, M., Salomaa, K. (eds.) CIAA 2010. LNCS, vol. 6482, pp. 59–68. Springer, Heidelberg (2011)
2. Bezem, M., Nakata, K., Uustalu, T.: On streams that are finitely red (submitted, 2011)
3. Braibant, T., Pous, D.: A tactic for deciding Kleene algebras. In: First Coq Workshop (August 2009)
4. Braibant, T., Pous, D.: An Efficient Coq Tactic for Deciding Kleene Algebras. In: Kaufmann, M., Paulson, L.C. (eds.) ITP 2010. LNCS, vol. 6172, pp. 163–178. Springer, Heidelberg (2010)
5. Brzozowski, J.A.: Derivatives of regular expressions. JACM 11(4), 481–494 (1964)
6. Büchi, J.R.: Weak second order arithmetica and finite automata. Zeitscrift Fur Mathematische Logic und Grundlagen Der Mathematik 6, 66–92 (1960)
7. The Coq Development Team, `http://coq.inria.fr`
8. Coquand, T., Gonthier, G., Siles, V.: Source code of the formalization, `http://www.cse.chalmers.se/~siles/coq/regexp.tar.bzip2`
9. Coquand, T., Spiwack, A.: Constructively finite? In: Laureano Lambán, L., Romero, A., Rubio, J. (eds.) Scientific Contributions in Honor of Mirian Andrés Gómez Servicio de Publicaciones, Universidad de La rioja, Spain (2010)

10. Johnstone, P.: Topos theory. Academic Press (1977)
11. Krauss, A., Nipkow, T.: Proof Pearl: Regular Expression Equivalence and Relation Algebra. Journal of Automated Reasoning (March 2011) (published online)
12. Martin-Löf, P.: An intuitionistic type theory: predicative part. In: Logic Colloquium 1973, pp. 73–118. North-Holland, Amsterdam (1973)
13. Mirkin, B.G.: An algorithm for constructing a base in a language of regular expressions. Engineering Cybernetics 5, 51–57 (1996)
14. Nordström, B.: Terminating general recursion BIT, vol. 28, pp. 605–619 (1988)
15. Owens, S., Reppy, J., Turon, A.: Regular-expression Derivatives Re-examined. Journal of Functional Programming 19(2), 173–190
16. Richman, F., Stolzenberg, G.: Well-Quasi-Ordered sets. Advances in Mathematics 97, 145–153 (1993)
17. Russell, B., Whitehead, A.N.: Principia Mathematica. Cambridge University Press (1910)
18. Gonthier, G., Mahboubi, A.: An introduction to small scale reflection in Coq. Journal of Formalized Reasoning 3(2), 95–152 (2010)
19. Tarski, A.: Sur les ensembles finis. Fundamenta Mathematicae 6, 45–95 (1924)
20. Thiemann, R., Sternagel, C.: Certification of Termination Proofs Using CeTA. In: Berghofer, S., Nipkow, T., Urban, C., Wenzel, M. (eds.) TPHOLs 2009. LNCS, vol. 5674, pp. 452–468. Springer, Heidelberg (2009)
21. Wu, C., Zhang, X., Urban, C.: A Formalisation of the Myhill-Nerode Theorem Based on Regular Expressions (Proof Pearl). In: van Eekelen, M., Geuvers, H., Schmaltz, J., Wiedijk, F. (eds.) ITP 2011. LNCS, vol. 6898, pp. 341–356. Springer, Heidelberg (2011)

A Modular Integration of SAT/SMT Solvers to Coq through Proof Witnesses*

Michael Armand[1], Germain Faure[2], Benjamin Grégoire[1], Chantal Keller[2], Laurent Théry[1], and Benjamin Werner[2]

[1] INRIA Sophia-Antipolis
{Michael.Armand,Benjamin.Gregoire,Laurent.Thery}@inria.fr
[2] INRIA - Saclay–Île-de-France at LIX, École Polytechnique
{Germain.Faure,Chantal.Keller,Benjamin.Werner}@inria.fr

Abstract We present a way to enjoy the power of SAT and SMT provers in Coq without compromising soundness. This requires these provers to return not only a yes/no answer, but also a proof witness that can be independently rechecked. We present such a checker, written and fully certified in Coq. It is conceived in a modular way, in order to tame the proofs' complexity and to be extendable. It can currently check witnesses from the SAT solver ZChaff and from the SMT solver veriT. Experiments highlight the efficiency of this checker. On top of it, new reflexive Coq tactics have been built that can decide a subset of Coq's logic by calling external provers and carefully checking their answers.

1 Introduction

When integrating a technology like SAT/SMT solvers in type theoretical provers like Coq, one classically has the choice between two ways, which Barendregt and Barendsen [4] named the *autarkic* and the *skeptical* approach.

In the autarkic approach, all the computation is performed in the proof system. In this case, this means implementing a whole, sufficiently efficient, SAT/SMT solver as a Coq function, and then proving it correct. This approach is followed in the Ergo-Coq effort [10].

In the skeptical approach, the external tool, here the SAT/SMT solver, is instrumented in order to produce not only a yes/no answer but also a *proof witness*, or a *trace* of its computation. It is the approach we follow here. The main contribution of the paper is to propose a modular and effective checker for SAT and SMT proof witnesses, written in Coq and *fully certified*.

In general, the choice between the autarkic and the skeptical approach depends on the considered problem. Typically, when the problem is solved by a greedy algorithm or something similar requiring no backtracking, the autarkic approach is generally to be preferred. In the case of SAT/SMT solvers, where a lot of time is devoted to actually finding the proof path, the skeptical approach can have an edge in terms of efficiency. Another advantage of the skeptical

* This work was supported in part by the french ANR DECERT initiative.

J.-P. Jouannaud and Z. Shao (Eds.): CPP 2011, LNCS 7086, pp. 135–150, 2011.

approach may be that it requires much less effort to certify a checker only than a whole prover. A difficulty is that the external prover and the internal checker have to speak a common language; indeed, finding the best possible format for the proof witnesses is the crucial step which determines the whole architecture of this prover-checker interaction.

Let us note that we see two benefits of this work. The first one being, as mentioned above, the addition to Coq of powerful and yet sound automation tactics. A second one is that it gives a means to enhance the reliability of automatic provers, by offering the possibility to have their results checked, *a posteriori* in Coq.

One keypoint for success in computational proofs is making the best possible usage of the, somewhat limited, computational power available inside the prover (Coq). This concern underlies the whole work presented here. An important advantage is the new addition to Coq of updatable arrays [3] which we use extensively in this work.

A wholly different approach to the integration of SAT/SMT to provers is to transform the proof witnesses into deductions. This approach has been taken for the Isabelle and HOL Light provers. We provide some comparison in Section 7.

The paper is organized as follows. The next section recalls the basic principles of SAT and SMT solvers. Section 3 describes the modular structure of the checker written in Coq; Sections 4 and 5 detail its different components, dealing respectively with the SAT parts, and two specific theories. Part 6 describes how the different parts are linked together in order to provide a practical tool. Finally, Section 7 is devoted to benchmarks and comparison with other works.

The source code of the checker and information on its usage can be found online [1].

2 The SAT and SMT Problems

2.1 SAT Solvers

SAT solvers deal with propositional formulas given in Conjunctive Normal Form (CNF); they decide whether or not there exists an assignment of the variables satisfying the formula. We recall basic definitions. A *literal* is a variable or its negation, a *clause* is a disjunction of literals, noted $l_1 \vee \cdots \vee l_n$. Finally, a formula in CNF is given by a *finite set of clauses* $\mathcal{S}$, seen as their conjunction.

A valuation ρ associating a Boolean to each variable straightforwardly induces an interpretation of a set of clauses ($[\![S]\!]_\rho$) as a Boolean. A set of clauses $\mathcal{S}$ is satisfiable if and only if there exists a valuation ρ such that $[\![\mathcal{S}]\!]_\rho = \top$. Conversely, $\mathcal{S}$ is unsatisfiable if and only if for any valuation ρ, $[\![\mathcal{S}]\!]_\rho = \bot$.

Modern SAT solvers rely on variants of the DPLL algorithm which can be customized to generate a proof witness [12]. The witness is:

- either an assignment of the variables to $\top$ and $\bot$ in order to satisfy all the clauses, in the case where the set of clauses is satisfiable;
- or a proof by resolution of the empty clause, in the case where the formula is unsatisfiable.

We recall that the refutationally complete resolution rule is:

$$\frac{v \vee C \qquad \bar{v} \vee D}{C \vee D}$$

where v is called the *resolution variable*. A *comb tree* is a tree where at least one child of every inner node is a leaf. A comb tree of resolution rules is called a *resolution chain*.

From the point of view of result certification, the case of unsatisfiability is the more challenging one. The format used by most SAT solvers for proof witnesses of unsatisfiability is a list of resolution chains. This list should be understood as a shared representation of the resolution tree: each resolution chain derives a new clause (that can be used later by other resolution chains), and the last resolution chain should derive the empty clause. It corresponds exactly to a subset of the learned clauses that the algorithm encountered during its run.

2.2 SMT Solvers

SMT solvers decide an extension of the SAT problem in which positive literals are not only Boolean variables but also atomic propositions of some first-order theory (possibly multisorted). Given a signature Σ containing simple types, and function and predicate symbols with their types, a *theory* $\mathcal{T}$ is a set of formulas of type *bool* written using this signature, variables and logical connectives. Those formulas are called *theory lemmas.*

The standard architecture for SMT solvers is an interaction between a SAT solver and decision procedures for the theories [12]. The SAT solver generates models and the theory solvers try to refute them. When a SAT model is consistent with all the theories, the initial problem is found satisfiable. Otherwise, a new clause corresponding to a theory lemma is added to the SAT problem in order to rule out the model. The SAT solver can then be called again to generate another model. Since there are only a finite number of SAT models, this enumeration eventually terminates. If the empty clause is derived by the SAT solver, the initial problem is unsatisfiable.

In this setting, a proof witness for unsatisfiability is a resolution tree of the empty clause where leaves are not only initial clauses but also theory lemmas.

To our knowledge, at least three existing SMT solvers can deliver informative proof witnesses of this kind: Z3, CVC3 and veriT (some other solvers provide less informative witnesses). Although there are some output format differences, each of these three provers does return a resolution tree with theory lemmas in the case of unsatisfiability. They also give witnesses for satisfiability.

3 A Modular and Efficient Coq Checker for SAT and SMT Proof Witnesses

We have developed a general framework to verify SAT and SMT proof witnesses for unsatisfiability proofs. During the SAT/SMT process, new clauses are generated until reaching the empty clause. These new clauses are either propositional consequences of the initial clauses, theory lemmas, or related to the CNF conversion and to various simplifications. A *small certificate* is generated to explain how each of the new clauses that are useful to obtain unsatisfiability was produced.

Our checker is defined by bringing together small checkers, each dedicated to one aspect of the verification of the resolution tree (the propositional reasoning, the theory lemmas of a given theory, the CNF conversion etc.). This modularity is a key aspect of our work. Small checkers are then independent pieces of code that can be composed in a very flexible way. Section 4.1 is dedicated to checking resolution chains, Section 4.2 to checking CNF computation. For theory lemmas, Section 5.2 describes what has been done for congruence closure and Section 5.3 for linear integer arithmetic. In each section, we present exactly the certificate format, how the checker works and is proved correct. The actual connection between Coq and the SAT and SMT provers is presented only later in Section 6.

The common aim underlying these different parts is preserving efficiency, in time and space. The main difficulty is the very large number of clauses that may need to be handled. We therefore strongly rely on the new *persistent arrays* feature of Coq, described in [3]. Schematically, checkers can be understood as sharing a global *state*, holding the current set of clauses, and implemented by an array. One typical optimization will be to keep this array as small as possible, by re-using a cell as soon as the clause it holds is known to be not used anymore for further computations.

In order to achieve modularity, we restrict ourselves to a very lightweight interface for the small checkers. Our implementation is based on four main data types: $\mathcal{S}$, $\mathcal{C}$, clauseId, and scertif. The first one, $\mathcal{S}$, represents the state. Initially, the only clause in the state is the singleton clause that contains the formula to be proved unsatisfiable. The type for clauses is $\mathcal{C}$. An element of type clauseId is an identifier that refers to a clause. The get and set functions let us access and modify the state and the main checker only needs to be able to check whether a clause is empty, i.e it represents $\perp$:

$$\mathsf{get} : \mathcal{S} \rightarrow \mathsf{clauseId} \rightarrow \mathcal{C} \qquad \mathsf{set} : \mathcal{S} \rightarrow \mathsf{clauseId} \rightarrow \mathcal{C} \rightarrow \mathcal{S} \qquad \mathsf{isFalse} : \mathcal{C} \rightarrow \mathsf{bool}$$

A *small checker* in our setting is just a Coq program that, given a state and a small certificate c, returns a new clause C. It is *correct* if the new clause that is produced is a consequence of the state S:

$$\text{for any interpretation } \rho,\ [\![S]\!]_\rho \Rightarrow [\![C]\!]_\rho$$

The type scertif is an enumeration type that collects all the possible small certificates. Associated to this type, there is a dispatching function scheck that, depending on the small certificate it receives, calls the appropriate checker.

Since small checkers just generate clauses, the only information we have to provide when gluing together small certificates is where the new clauses have to be stored. Finally, at the end, the initial formula should be proved unsatisfiable, so the empty clause must have been derived. So, its actual location must be given by the main certificate. The type of such certificate is then

$$\mathsf{certif} := \mathsf{list}\ (\mathsf{clauseId} * \mathsf{scertif}) * \mathsf{clauseId}$$

The main checker is now trivially defined by

```
check S cert =
  let (l, k) := cert in
  let S' := List.fold_left (fun S (p, c) ⇒ set S p (scheck S c)) S l in
  isFalse (get S' k)
```

It takes an initial state S and a certificate *cert* and sequentially calls small checkers to compute new clauses extending the state with the generated clauses. Provided all small checkers are correct and the check returns true and since satisfiability is preserved, reaching an absurd state implies that the initial state was indeed unsatisfiable.

As hinted above, the get and set functions are built upon the persistent arrays of Coq and one such array is used in the state to store clauses. clauseIds are thus array indexes, i.e. 31-bit integers. So, access and update to the set of clauses are performed in constant time. Since we use arrays, in the following, (get S n) is written as $S.[n]$.

It is very unlikely that a given SMT solver will output exactly our kind of certificates. A pre-processing phase is required in order to translate proof-witnesses generated by the SMT solver into our format. In particular, the precise clause allocation that is required by our format is usually not provided by the SMT solver. Finding out such an allocation is a post-processing seen from the SMT solver, and a pre-processing seen from Coq. It involves techniques similar to register allocation in compilation. First, the maximal number of clauses that need to be alive at the same time is computed. Then, a cell is explicitly allocated to each clause, in such a way that two clauses that need to be alive at the same time do not share the same cell. In practice, this has a big impact on the memory required to check a certificate.

4 Small Checkers for SAT

4.1 A Small Checker for Resolution Chains

As explained in Section 2.1, the SAT contribution is represented in the proof witness by chains of resolutions. The constructor (res_certif $[| n_1; \ldots; n_i |]$) in scertif represents these chains. Let $R(C_1, C_2)$ be the resolution between the clause C_1 and the clause C_2. Given this certificate and a state S, the corresponding small checker iteratively applies resolution to eventually produce the new clause $R(\ldots(R(S.[n_1],\ S.[n_2]),\ldots),\ S.[n_i])$.

This efficient treatment of resolution chains requires a careful encoding of clauses and literals. First, we encode propositional variables as 31-bit integers. We follow the usual convention reserving location 0 for the constant true, which means that the interpretation of propositional variables ρ always comes with the side-condition that $\rho(0) = \text{true}$. Literals are also encoded as 31-bit integers, taking advantage of parity. The interpretation for literals is built such that:

$$[\![l]\!]_\rho = \text{if even}\, l \text{ then } \rho(l/2) \text{ else } \neg\rho(l/2).$$

The point being that parity check and division by two are very fast since they are directly performed by machine integer operations as explained in [3].

Clauses are represented by lists of literals. The interpretation $[\![c]\!]_\rho$ of a clause c is then the disjunction of the interpretation of its literals. The interpretation $[\![S]\!]_\rho$ of a state S is defined as the conjunction of the interpretation of its clauses.

To give a concrete example, consider the interpretation of proposition variables: $\rho(0) = \mathsf{true}$, $\rho(1) = x$, $\rho(2) = y$. If $S = [|\, [2;4];\ [5];\ [3;4]\, |]$ we have

$$\begin{aligned}[\![S]\!]_\rho &= [\![\, [2;4] \,]\!]_\rho \wedge [\![\, [5] \,]\!]_\rho \wedge [\![\, [3;4] \,]\!]_\rho = ([\![2]\!]_\rho \vee [\![4]\!]_\rho) \wedge [\![5]\!]_\rho \wedge ([\![3]\!]_\rho \vee [\![4]\!]_\rho) \\ &= (\rho(1) \vee \rho(2)) \wedge \neg\rho(2) \wedge (\neg\rho(1) \vee \rho(2)) = (x \vee y) \wedge \neg y \wedge (\neg x \vee y)\end{aligned}$$

In this setting, the interpretation of a set of clauses is always a CNF formula. A proof of unsatisfiability of this formula is the following chain of resolutions

$$\dfrac{\dfrac{\dfrac{x \vee y \qquad \neg y}{x} \qquad \neg x \vee y}{y} \qquad \neg y}{\bot}$$

This corresponds in our format to certificate $([0, \mathsf{res\_certif}\ [0;1;2;1]], 0)$.

4.2 Small Checkers for CNF Computation

With our previous small checker, proof witnesses for SAT problems in CNF can be checked in Coq. The next step is to be able to verify the transformation of a formula into an equisatisfiable formula in CNF. This is usually done using a technique proposed by Tseitin [14]. This involves generating a new variable for every subterm of the formula; with these new variables, the CNF transformation is linear. It is this idea that we are going to implement in our setting. Naming subterms corresponds to a form of hash-consing. A hashed formula is either an atom, true, false, or a logical connective. Sub-formulas of connectives are literals (i.e., a variable or its negation):

```
Type hform =
  | Fatom (a : atom) | Ftrue | Ffalse
  | Fand (ls : array lit) | For (ls : array lit) | Fxor (l1 l2 : lit)
  | Fimp(l1 l2 : lit) | Fite (l1 l2 l3 : lit) | Fiff (l1 l2 : lit) | Fdneg (l : lit).
```

Note that the connectives Fand and For are n-ary operators which allows a more efficient subsequent computation. Note also that we have no primitive constructor for negation, which has to be pushed to the literals (with little cost, using the odd/even coding described above). However, double negation is explicit and primitive, in order to represent the formula $\neg\neg x$ faithfully.

For computation, the state of the checker is extended with a new array `ftable` containing the table of the hashed formulas. For example, the formula $\neg((x \wedge y) \vee \neg(x \wedge y))$ can be encoded by the literal 9 using the formula table:

$$[|\,\mathsf{Ftrue};\ \mathsf{Fatom}\ 0;\ \mathsf{Fatom}\ 1;\ \mathsf{Fand}\ [|2;4|];\ \mathsf{For}\ [|6;7|]\,|]$$

with the interpretation for atoms defined by $\rho_A(0) = x, \rho_A(1) = y$. Three things are worth noticing. First, the sub-formula $x \wedge y$ appears twice in the formula but is shared in the table (at location 3); indeed, this representation allows maximal sharing. Second, we have to ensure that our table does not contain infinite terms, so our state can be interpreted. This is done by preserving some well-formedness on the table: if a literal m appears in the formula stored at location n, we always have $m/2 < n$. It is this condition that actually allows us to define the Boolean interpretation $[\![f]\!]_{\rho A}$ recursively over the formula f, where ρ is the interpretation of the variables.

Finally, the tables always have Ftrue at location 0. The interpretation of propositional variables of the previous section is simply defined as $\rho(n) = [\![\mathsf{ftable}.[n]]\!]_{\rho_A}$.

Tseitin identifies 40 generic tautology schemes used in the transformation to CNF. In the case of our example $\neg((x \wedge y) \vee \neg(x \wedge y))$, the transformation invokes the following tautology $\neg(A_0 \vee \cdots \vee A_i \vee \cdots \vee A_n) \Rightarrow \bar{A}_i$. For each of these tautologies, we have written a specific test function which verifies that the corresponding tautology can actually be applied. In this case, the certificate is written (nor_certif m i) and the corresponding checker verifies that $S.[m]$ is a singleton clause $[k]$ with k being odd and that $\mathsf{ftable}.[k/2]$ is a disjunction (For $[|l_1; \ldots;\ l_n|]$). If these conditions are met, it produces the clause $[\bar{l_i}]$ if $i < n$. If the verification fails, the true clause is returned. This trick of using the true clause as default will be used for all the other small checkers.

The full certificate of unsatisfiability for our example is then:

$$([(1, \mathsf{nor\_certif}\ 0\ 0);\ (0, \mathsf{nor\_certif}\ 0\ 1);\ (0, \mathsf{res\_certif}\ [|1;\ 0|]], 0).$$

The computation of the final set of clauses proceeds like this:

$$[|\,[9];\ [0]\,|] \overset{1,\mathsf{nor\_certif}\ 0\ 0}{\leadsto} [|\,[9];\ [7]\,|] \overset{0,\mathsf{nor\_certif}\ 0\ 1}{\leadsto} [|\,[6];\ [7]\,|] \overset{0,\mathsf{res\_certif}\ [|1;\ 0|]}{\leadsto} [|\,[\,];\ [7]\,|]$$

At the end, we find the empty clause at location 0, which ensures the initial formula is unsatisfiable.

Let us finally remark that our format is compatible with lazy CNF transformation and also that it is possible to delegate the CNF computation to the SMT solver.

5 Small Checkers for Congruence Closure and Linear Arithmetic

5.1 Refining the Term Representation

In order to handle theories, we need to provide a proper representation for atoms. Atoms can represent objects of different types, so we also need a proper representation for types. Theories like EUF manipulate uninterpreted functions, so we also need uninterpreted base types. Here is our representation:

Type btype = Tidx (n : int) | Tbool | TZ | ...
Type cst = Zcst (n : Z) | ...
Type op = Oidx (n : int) | Oeq (t : btype) | OZle | OZlt | OZplus | ...
Type hatom = Avar (v : avar) | Acst (c : cst) | Aapp (o : op) (a_s : list atom).

As for formulas, our encoding uses a table atable, so the type atom is an abbreviation for int. A hashed atom is either a variable (Avar), a constant of a theory (Acst), or an application of an operator to its arguments (list atom). Operators are uninterpreted functions or predicates (Oidx), or a function (OZplus) or a predicate (OZle) of a given theory. Base types are either uninterpreted (Tidx) or a type of a given theory (TZ).

To illustrate our representation, let us consider the formula $f\,x < 1 \vee g\,(y{+}1) < 1$ over the Coq integer Z where the Coq type of x is α and is left uninterpreted. We have the following tables:

$$\begin{aligned}\mathsf{ftable} &= [|\mathsf{Ftrue};\ \mathsf{Fatom}\ 5;\ \mathsf{Fatom}\ 7;\ \mathsf{For}\ [|2;4|]|]\\ \mathsf{atable} &= [|\mathsf{Avar}\ 0;\ \mathsf{Avar}\ 1;\ \mathsf{Acst}\ (\mathsf{Z\_cst}\ 1);\ \mathsf{Aapp}\ (\mathsf{Oidx}\ 0)\ [0];\ \mathsf{Aapp}\ \mathsf{OZplus}\ [1;2];\\ &\qquad \mathsf{Aapp}\ \mathsf{OZlt}\ [3;2];\ \mathsf{Aapp}\ (\mathsf{Oidx}\ 1)\ [4];\ \mathsf{Aapp}\ \mathsf{OZlt}\ [7;2]|]\end{aligned}$$

Interpreting types is easy. We just need a table ttable associating a Coq type to every type index. We denote by $[\![T]\!]_t$ the interpretation of a base type T with respect to this table. Interpreting atoms is more difficult, since we must build well-typed Coq terms. In particular, different elements of the table may be interpreted into different types. Therefore, our interpretation function returns a dependent pair (T, v) where T has type btype and v has type $[\![T]\!]_t$. The interpretation of atoms $[\![A]\!]$ uses two tables. The first one (vtable) is a valuation associating (T, v) to a variable index. The second one (otable) associates a pair $(([T_1, \ldots, T_n], T), f)$ to an operator index, where T_i, T have type btype and f has type $[\![T_1]\!]_t \to \ldots [\![T_n]\!]_t \to [\![T]\!]_t$. With these tables, defining the interpretation $[\![A]\!]$ is straightforward. We simply check that all applications are well-typed, if not we return (Tbool, true). This makes our interpretation a total function. Here are the three tables used by the interpretation for the previous example:

$$\begin{aligned}\mathsf{ttable} &= [|\alpha|]\\ \mathsf{vtable} &= [|(\mathsf{Tidx}\ 0, x); (\mathsf{TZ}, y)|]\\ \mathsf{otable} &= [|(([\mathsf{Tidx}\ 0], \mathsf{TZ}), f); (([\mathsf{TZ}], \mathsf{TZ}), g)|]\end{aligned}$$

The interpretation of atoms of the previous section is simply defined as $\rho_A(a) = [\![\mathsf{atable}.[a]]\!]$.

We need some side conditions on the different tables to be able to complete the proof of our small checkers. First, the hashed atom contained at position k of the atable should refer to atoms strictly smaller than k (this ensures that the interpretation terminates). Second, the atable should only contain well-typed hashed atoms with respect to vtable and otable. This last condition allows to reject formulas like $\neg(1 = \mathsf{true}) \vee \neg(\mathsf{true} = 2) \vee (1 = 2)$ which is correct from the transitivity point of view but is interpreted in Coq by $\mathsf{false} \vee \mathsf{false} \vee 1 = 2$

5.2 A Small Checker to Compute Congruence Closure

The theory of congruence closure is at the heart of all SMT solvers. In our term representation, equality is represented as a binary operator (Oeq) that is parameterised by the representation of the type on which equality operates. Consider the proof of unsatisfiability of the formula $\neg(f\,a = f\,b) \wedge b = c \wedge a = c$, which belongs to the congruence closure fragment. It creates the following tables:

$$\begin{aligned}
\mathsf{ftable} &= [|\ \mathsf{Ftrue};\ \mathsf{Fatom}\ 5;\ \mathsf{Fatom}\ 6;\ \mathsf{Fatom}\ 7;\ \mathsf{Fand}\ [|3;\ 4;\ 6|];\ \mathsf{Fatom}\ 8\ |];\\
\mathsf{atable} &= [|\ \mathsf{Avar}\ 0;\ \mathsf{Avar}\ 1;\ \mathsf{Avar}\ 2;\ \mathsf{Aapp}\ (\mathsf{Oidx}\ 0)\ [0];\ \mathsf{Aapp}\ (\mathsf{Oidx}\ 0)\ [1];\\
&\qquad \mathsf{Aapp}\ (\mathsf{Oeq}\ \mathsf{TZ})\ [3;4];\ \mathsf{Aapp}\ (\mathsf{Oeq}\ \mathsf{TZ})\ [1;2];\\
&\qquad \mathsf{Aapp}\ (\mathsf{Oeq}\ \mathsf{TZ})\ [0;2];\ \mathsf{Aapp}\ (\mathsf{Oeq}\ \mathsf{TZ})\ [0;1]\ |]\\
\mathsf{vtable} &= [|\ (\mathsf{TZ},\ a);(\mathsf{TZ},\ b);(\mathsf{TZ},\ c)\ |]\quad \mathsf{otable} = [|\ (([\mathsf{TZ}],\ \mathsf{TZ}),\ f)\ |]\quad \mathsf{ttable} = [|\ |]
\end{aligned}$$

where the formula is at location 4 of ftable. Note that location 5 is not necessary to encode the formula but for its proof (the same thing happens for the location 8 of atable). This is explained later.

Our checker is only capable of producing clauses obtained by instantiating one of these three theorems:

- transitivity: $x_1 \neq x_2 \vee \cdots \vee x_{n-1} \neq x_n \vee x_1 = x_n$
- function congruence: $x_1 \neq y_1 \vee \cdots \vee x_n \neq y_n \vee f\ x_1\ \ldots\ x_n = f\ y_1\ \ldots\ y_n$
- predicate congruence: $x_1 \neq y_1 \vee \cdots \vee x_n \neq y_n \vee \neg P\ x_1\ \ldots\ x_n \vee P\ y_1\ \ldots\ y_n$

The small certificates are (eq_trans c), (eq_congr c) and (eq_congr_pred c) where c represents the candidate clause to be produced. This explains why the tables of our previous example had more than the atoms of the initial formula. They also contain the atoms of the theory lemmas. The small checkers for these certificates only have to verify that c is indeed an instantiation of their corresponding theorem. For instance, the small checker for eq_trans $[l_1;\ldots;l_n;l]$ verifies that:

– l is even and each l_i is odd;
– $l/2$ refers to Aapp (Oeq t) $[a;b]$ and each $l_i/2$ refers to Aapp (Oeq t_i) $[a_i;b_i]$;
– a equals a_1, b equals b_n and for $1 \leq i < n$, a_i equals b_{i+1}.

Note that all the equality tests over atoms are just equalities over integers thanks to our maximal sharing of atoms. Furthermore, we do not need to check type equality between the t_i since the small checkers assume that the atom table is always well-typed.

Our modular checker can combine these three simple rules and the resolution checker to derive the empty clause from an unsatisfiable formula. In our example, the checker starts with the initial formula $\neg(f\,a = f\,b) \wedge b = c \wedge a = c$. After evaluating the part of the certificate dedicated to CNF computation, the state contains the clauses $[|[3];[4];[6];[0];[0]|]$ and what is left to be evaluated is

$$([(3, \mathsf{eq\_trans}\ [5;7;10]);(4, \mathsf{eq\_congr}\ [11;2]);(0, \mathsf{res\_certif}[|3;1;2;4;0|])], 0).$$

The computation then proceeds like this:

$$[|[3];[4];[6];[0];[0]|] \overset{3,\mathsf{eq\_trans}\ [5;7;10]}{\rightsquigarrow} [|[3];[4];[6];[5;7;10];[0]|] \overset{4,\mathsf{eq\_congr}\ [11;2]}{\rightsquigarrow}$$

$$[|[3];[4];[6];[5;7;10];[11;2]|] \overset{0,\mathsf{res\_certif}[|3;1;2;4;0|]}{\rightsquigarrow} [|[];[4];[6];[5;7;10];[11;2]|]$$

5.3 A Small Checker for Linear Arithmetic

The tactic lia [5] of Coq proves any valid formula of linear arithmetic. It already contains a checker based on Farkas' certificates:

$$\mathsf{lia\_check} \;:\; \mathsf{lia\_formula} \to \mathsf{lia\_certif} \to \mathsf{bool},$$

Note that lia uses a different representation (lia_formula). It also provides a proof of correction for this checker. We choose to use lia_check to build a small checker for our modular interface. Thus, the small certificate for linear arithmetic is simply (lia_certif c F) where c is the candidate clause and F of type lia_certif.

In order to validate the clause c, the small checker first calls the function lia_formula_of to translate c into an equisatisfiable formula f of type lia_formula, and then calls (lia_check f F). The correctness of this small checker relies on the correctness of the lia checker and on the correctness of our translation.

5.4 The Simplifier Small Checker

For more efficiency, most SMT solvers use on-the-fly term rewriting and simplification and do not give proof witnesses for these simplifications. Furthermore, sometimes the formula needs to be preprocessed before being sent to an external solver, again for efficiency reasons. In consequence, the formula f' proved by the proof witness can be slightly different from the initial formula f one wanted to prove. We have thus developed a dedicated small checker that verifies that a formula f is equivalent to f'. Our checker is able to prove equivalence through associativity of conjunction and disjunction, double negation, symmetry of equality, and simple rewriting of linear inequations (such as: $a \geq b \equiv b \leq a$). It is implemented by a simple simultaneous recursive descent of f and f'. Only the symmetry of equality requires some backtracking.

6 Building a Coq Tactic

To build an actual tactic out of our certified checker, we follow the usual steps for reflexive tactics. The first step is reification: given a formula f in Coq on a decidable domain, we have to build 5 tables and a literal l, such that the interpretation of l with respect to these tables is $\neg f$. The second step is to find a certificate that shows that $[l]$ is unsatisfiable. This is done by calling the SAT or SMT solver. We need to translate the problem into the solver input format. Then, the solver returns a proof witness, that we transform into a certificate.

During the first translation, we sometimes need to do some pre-processing. For example, ZChaff only accepts CNF formulas, so the CNF transformation is done before sending it. Also, the CNF transformation of veriT is more efficient if disjunctions and conjunctions are considered as n-ary operators (and not binary like in Coq), so we flatten the formula before sending it. The justification of this pre-processing is the prelude of the certificate.

The transformation of proof witnesses into certificate requires more work. We first need to update our tables so they contain all the formulas of the theory lemmas. Second, we have to transform each step of the proof witness into a sequence of small certificates. In the easiest cases, the solver gives *exactly* what we expect. This is the case, for instance, of the resolution chains produced by SAT solvers and the CNF transformation produced by veriT. In other cases, very lightweight modifications are necessary. For instance, the format for congruence closure of veriT automatically removes duplicates literals. It generates $\neg x = y \vee f\ x\ x = f\ y\ y$ while our certificate expects $\neg x = y \vee \neg x = y \vee f\ x\ x = f\ y\ y$. Finally, in the worst cases, we may have to rebuild completely the certificate. This is the case for veriT where theory lemmas for linear arithmetic come without justification. In this precise example, we use the external solver of `lia` to produce the Farkas' certificate.

Finally, the compactness of the certificate is very important. So, when it is not done by the solver, we prune it by removing all the justifications of unused clauses. It is during this phase that we compute the minimal required size of the array of clauses and perform clause allocation.

This work results into two new Coq tactics called `zchaff` and `verit` .

7 Results and Comparison with Other Works

7.1 Related Works

The work presented here extends what is described in [3] in several ways. First, we complemented the SAT checker with checkers for CNF computation, congruence closure, differential logic and linear arithmetic. To do so with a great modularity, the format of certificates has been rethought: the idea of small and main checkers makes it easier to add new theories. Second, we can formally check proof witnesses generated by the veriT solver, which combines all these theories. Finally, we use our checker to enhance the automation of Coq by defining new safe reflexive decision procedures.

Several SAT and SMT solvers have been integrated in LCF style interactive theorem provers including CVC Lite in HOL Light [11], haRVey in Isabelle/HOL [8], Z3 in HOL and Isabelle/HOL [6]. To our knowledge, our work is the first integration relying on proof witnesses in a proof assistant based on Type Theory. In the following, we will focus on the comparison with proof reconstruction in Isabelle/HOL of ZChaff [15] and Z3 [6] (this corresponds to the state-of-the-art). We point out that comparison for theories has to be considered with care since we do not use the same SMT solver.

Another approach is to write the SAT or SMT solver directly inside the proof assistant and formally prove its correctness. This is the approach followed in [10]. It has the advantage to validate the algorithms at work in the prover but is sensitive to any change, any optimization in the proof search. We compare the two approaches.

7.2 Experiments

All the experiments have been conducted on an Intel Quad Core processor with 2.66GHz and 4Gb RAM, running Linux. Our code which served for the experiments is available online [1]. It requires the native version of Coq [7]. It represents around 6,000 lines of Coq code and 8,000 lines of Ocaml code. The Coq code for our shared term representation is about 1,000 lines, the SAT part 1,200, the CNF part 1,500, the EUF 600, the LIA part 1,500 and the simplifier part 500. The complete checker corresponds to 1,000 lines of Coq code, the other 5,000 are for specifications and proofs.

SAT verification. We first compare our combination of the main checker with the small checker of resolution chains for ZChaff in Coq with proof reconstruction for ZChaff in Isabelle/HOL written by Alwen Tiu and Tjark Weber. We use Isabelle 2009-1 (running with Poly/ML 5.2) and ZChaff 2007.3.12.

We run ZChaff on a database of 151 unsatisfiable industrial benchmarks from SAT Race'06 and '08 with a timeout of 300 seconds. These benchmarks range from 300 to 2.3 million variables and from 1,800 to 8.9 million clauses. When ZChaff succeeds in the given time, it produces a proof witness whose size range from 41Kb to 205Mb. In that case, we run our checker and the Isabelle/HOL checker on it with a timeout of 300 seconds. Table 1 presents the number of benchmarks solved by ZChaff, and among them, the number of proof witnesses successfully checked by Isabelle/HOL and Coq. The times are the mean of the times for the 57 benchmarks on which ZChaff, Coq and Isabelle/HOL all succeeded, in seconds. Errors in Isabelle/HOL were due to timeouts.

It appears that Coq can check all the proof witnesses given by ZChaff in the given time. This is not surprising since our checker appears to be faster than ZChaff itself. However, the Isabelle/HOL checker is slower than ZChaff, which explains that only 72% of the proof witnesses can be checked without timeout.

The three curves on the left of Figure 1 present the number of benchmarks solved along the time by ZChaff, Isabelle and Coq. It clearly shows that the Coq checker is far faster verifying results than ZChaff is building them; the main time consumed by our combination is taken by ZChaff. However, the limiting factor of the ZChaff and Isabelle/HOL combination is Isabelle/HOL.

SMT verification. We now compare our combination of the main checker with the small checkers of resolution chains, CNF computation, congruence closure, differential logic and linear integer arithmetic for veriT in Coq with proof reconstruction for Z3 in Isabelle/HOL written by Sascha Böhme and Tjark Weber. We use Isabelle 2009-1 (running with Poly/ML 5.2), Z3 2.19 and the development version of veriT.

We took a database of unsatisfiable industrial benchmark from the SMT-LIB [2] for theories QF_UF (congruence closure), QF_IDL (differential logic) and QF_LIA (linear integer arithmetic). It is important to notice that veriT is not completely proof producing for QF_LIA, so we selected a subset of the benchmarks where veriT returns either *unknown* or *unsatisfiable* with a proof witness. On the one hand, we run veriT, followed by our Coq checker when veriT succeeds. On the other hand, we run Z3, followed by the Isabelle/HOL checker when Z3 succeeds. Each run has a timeout of 300 seconds. The mean of the sizes of Z3 proof witnesses is 12Mb, and the mean of the sizes of veriT proof

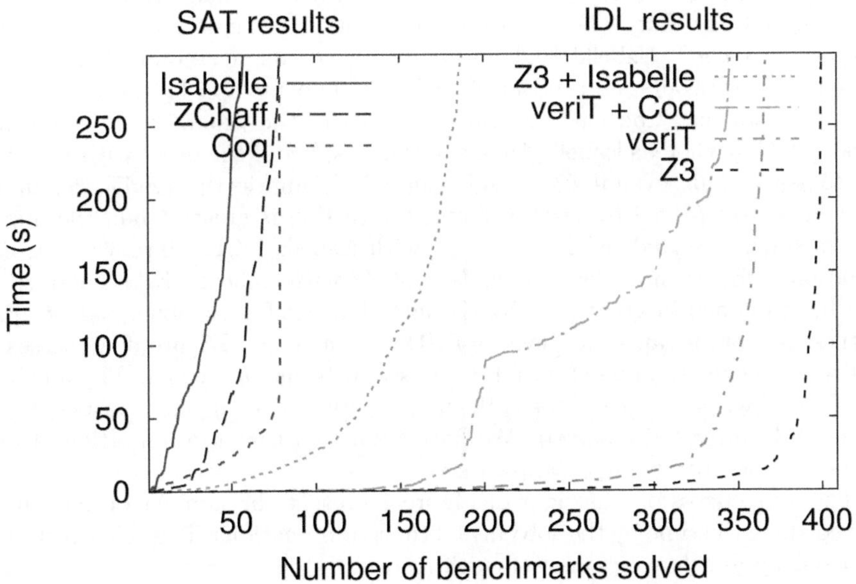

Fig. 1. Experiments on industrial benchmarks

Table 1. SAT benchmarks

Solved ZChaff			Isabelle/HOL checker			Coq checker		
#	%	Time	#	%	Time	#	%	Time
79	52	51.9	57	38	100.	79	52	17.5

Table 2. SMT benchmarks

Benchmarks		Solved Z3			Solved veriT			Isabelle/HOL checker			Coq checker		
Logic	#	#	%	Time	#	%	Time	#	%	Time	#	%	Time
QF_UF	1852	1834	99	2.5	1816	98	6.5	1775	96	25.8	1804	97	1.4
QF_IDL	409	402	98	0.6	368	90	6.3	190	46	55.2	349	85	37.8
QF_LIA	116	107	92	0.7	98	84	11.6	96	83	46.6	98	84	3.1

Table 3. Comparison with Ergo in Coq

	dplln	zchaff
H_7	28.0	0.2
H_8	262.7	1.2
H_9	-	1.6
H_{10}	-	6.7

	dplln	zchaff
deb_{700}	111.5	0.8
deb_{800}	147.9	1.0
deb_{900}	201.6	1.2
deb_{1000}	260.4	1.5

	cc	verit
$F(13, 5, 8)$	0.5	0.1
$F(25, 13, 1)$	1.3	0.1
$F(25, 15, 5)$	0.5	0.2
$F(25, 24, 24)$	16.9	0.1

	cc	verit
D_5	2.3	0.3
D_8	24.9	1.1
D_{10}	118.7	2.2
D_{15}	-	45.7

witnesses is 7.7Mb. Table 2 presents the number of benchmarks solved by Z3 and veriT, and among them, the number of proof witnesses successfully checked by Isabelle/HOL and Coq. The times are the mean of the times for the benchmarks on which Z3, veriT, Isabelle/HOL and Coq all succeeds, in seconds. Errors in Coq were due to timeouts, and in Isabelle/HOL to timeouts and failures.

It appears that Coq can check a large part of the proof witnesses given by veriT (98.6%) whereas Isabelle/HOL can check 88.0% of the proof witnesses given by Z3. As a result, even if Z3 can solve more benchmarks than veriT, the number of benchmarks solved by veriT combined with Coq is greater than the number of benchmarks solved by Z3 combined with Isabelle/HOL. Moreover, our combination is faster than the combination of Z3 with Isabelle/HOL. These results can be explained in great part by the fact that veriT gives much smaller proof witnesses. For instance, for logic QF_IDL, in average, Z3 proof witnesses are 7.9 times bigger than veriT proof witnesses in terms of storing. The quality of veriT proof witnesses strengthens the fact we use it, even if there exists currently more performing SMT solvers. We have been told that the limitation of proof witnesses for LIA should disappear soon.

The four curves on the right of Figure 1 present the number of benchmarks solved along the time by the solvers and their combinations. They clearly indicate that our approach compares well with respect to [6].

Tactics. We compare our `zchaff` and `verit` tactics with the reflexive tactics `dplln` and `cc` from Stephane Lescuyer's SMT solver Ergo written in Coq. To do so, we use the same formulas that are presented in Section 11.2 of [10]:

- for SAT:
  - the famous pigeon hole formulas which are unsatisfiable
  - the de Bruijn formulas: $deb_n = \forall x_0, \dots, x_{2n}, (x_{2n} \leftrightarrow x_0) \vee \bigvee_{i=0}^{2n-1} (x_i \leftrightarrow x_{i+1})$
- for EUF:
  - the formulas $FP(n,m,k) = \forall f x, f^n(x) = x \rightarrow f^m(x) = x \rightarrow f^k(x) = x$ which are true for any n, m, k such that k is a multiple of $\gcd(n,m)$
  - the formulas $D_n =$
    $$\forall f, \left(\bigwedge_{i=0}^{n-1} (x_i = y_i \wedge y_i = f(x_{i+1})) \vee (x_i = z_i \wedge z_i = f(x_{i+1})) \right) \rightarrow x_0 = f^n(x_n)$$

Results are presented in Table 3. Times are in seconds. We see that our `zchaff` and `verit` tactics here clearly outperform `dplln` and `cc`. This is not surprising since ZChaff and veriT have more efficient algorithms than Ergo. Note it may be difficult to change Ergo's algorithm since it would involve redoing many correctness proofs; the certificate approach is more flexible here. If we store proof witnesses, `zchaff` and `verit` get faster at rather small storage cost: in our examples, the largest proof witness is 41Mb large for D_{15}.

Regarding other existing Coq tactics, `zchaff` is far faster than `tauto`, and `verit` is similar to `congruence`. However, these latter ones do not solve the same goals, since `verit` can solve goals including congruence and propositional reasoning, and `congruence` can deal with inductive data-types.

8 Conclusion and Future Works

Compared to what the authors call "proof reconstruction" in [6], what we have presented here is much closer to program verification. We have developed inside Coq a program that checks traces generated by SAT and SMT solvers. A particular care has been given to the efficiency of the data representation for clauses and atoms. Even with the limited computing power available inside Coq, the checker is rather efficient: it is able to check in reasonable time huge proof witnesses coming from challenging benchmarks and it compares well with state of the art implementations in Isabelle/HOL.

From the methodology point of view, what we have done is very close to [13]. In this work, the authors have developed a checker for SMT proofs using LFSC. The main difference is that they delegate the verification of the SAT part to an external checker (written in C++). Here, we do everything within the logic.

We also took a special care in being generic: for example the same checker is used for ZChaff and veriT. This relies on the generic format we use after translating proof witnesses into certificates. So we expect the checker to be easily extensible. For the moment, veriT is the only SMT solver that is connected with Coq. We hope that our format of certificate could also be used successfully to connect to other proof producing solvers. Our next step is to integrate Z3.

The checker has also been proved correct. Using the technique of *proof by reflection*, it made it possible to derive a safe and automatic proof procedure within Coq. Formulas usually proved in Coq are rather small so a far less efficient checker could have been sufficient. Still, we believe that our work opens interesting new perspectives of using brute force methods for doing proof with Coq automatically. For example, one could encode the small problem she/he has to prove as a huge Boolean formula that our SAT tactic can solve instantaneously.

Surprisingly the difficult part of this work was more the actual design of the certificate and obtaining a good computational behaviour for the checker than performing the correctness proofs. This is largely due to the fact that we are not proving the full functional correctness of the checker. We are just proving that if the checker replies true, the theorem is valid. This makes a big difference for the proof effort. This reduces drastically the size of the invariants we had to prove and clearly makes the proof of such a large piece of code tractable in Coq.

For future works, our priority is clearly to increase the expressiveness of the formulas we can deal with. In particular, if we want our tool to be widely used by the Coq community, being able to deal with quantified formulas and user-defined functions is a must-have. For quantifiers, it has not been done yet mostly because the current version of veriT does not produce proof witnesses. Though, this should be available in the next version of the system. For definitions, more work has to be done since the type system of Coq is more powerful than the one proposed by the SMT-LIB standard. Other extensions we envision concern non-linear arithmetic, arrays and bit vectors.

Acknowledgments. Pascal Fontaine's responsiveness was crucial for this work. We wish to thank Sascha Böhme and Tjark Weber for the details and source code they gave us to reproduce the Isabelle experiments. We finally thank Christine Paulin, Guillaume Melquiond and Sylvain Conchon for their help concerning the comparison with `dplln` and `cc` in Coq. Anonymous referees provided helful constructive remarks.

References

1. Source code of the development, `http://www.lix.polytechnique.fr/~keller/Recherche/smtcoq.html`
2. SMT-LIB, `http://www.smtlib.org`
3. Armand, M., Grégoire, B., Spiwack, A., Théry, L.: Extending Coq with Imperative Features and Its Application to SAT Verification. In: Kaufmann and Paulson [9], pp. 83–98
4. Barendregt, H., Barendsen, E.: Autarkic Computations in Formal Proofs. J. Autom. Reasoning 28(3), 321–336 (2002)
5. Besson, F.: Fast Reflexive Arithmetic Tactics the Linear Case and Beyond. In: Altenkirch, T., McBride, C. (eds.) TYPES 2006. LNCS, vol. 4502, pp. 48–62. Springer, Heidelberg (2007)
6. Böhme, S., Weber, T.: Fast LCF-Style Proof Reconstruction for Z3. In: Kaufmann and Paulson [9], pp. 179–194
7. Dénès, M.: Coq with native compilation, `https://github.com/maximedenes/native-coq`
8. Fontaine, P., Marion, J.-Y., Merz, S., Nieto, L.P., Tiu, A.F.: Expressiveness + Automation + Soundness: Towards Combining SMT Solvers and Interactive Proof Assistants. In: Hermanns, H. (ed.) TACAS 2006. LNCS, vol. 3920, pp. 167–181. Springer, Heidelberg (2006)
9. Kaufmann, M., Paulson, L.C. (eds.): ITP 2010. LNCS, vol. 6172. Springer, Heidelberg (2010)
10. Lescuyer, S., Conchon, S.: Improving Coq Propositional Reasoning Using a Lazy CNF Conversion Scheme. In: Ghilardi, S., Sebastiani, R. (eds.) FroCoS 2009. LNCS, vol. 5749, pp. 287–303. Springer, Heidelberg (2009)
11. McLaughlin, S., Barrett, C., Ge, Y.: Cooperating Theorem Provers: A Case Study Combining HOL-Light and CVC Lite. ENTCS 144(2), 43–51 (2006)
12. Nieuwenhuis, R., Oliveras, A., Tinelli, C.: Solving SAT and SAT Modulo Theories: From an abstract Davis–Putnam–Logemann–Loveland procedure to DPLL. J. ACM 53(6), 937–977 (2006)
13. Oe, D., Stump, A.: Extended Abstract: Combining a Logical Framework with an RUP Checker for SMT Proofs. In: Lahiri, S., Seshia, S. (eds.) Proceedings of the 9th International Workshop on Satisfiability Modulo Theories, Snowbird, USA (2011)
14. Tseitin, G.S.: On the complexity of proofs in propositional logics. Automation of Reasoning: Classical Papers in Computational Logic (1967-1970) 2 (1983)
15. Weber, T.: SAT-based Finite Model Generation for Higher-Order Logic. Ph.D. thesis, Institut für Informatik, Technische Universität München, Germany (April 2008), `http://www.cl.cam.ac.uk/~tw333/publications/weber08satbased.html`

Modular SMT Proofs for Fast Reflexive Checking Inside Coq*

Frédéric Besson, Pierre-Emmanuel Cornilleau, and David Pichardie

INRIA Rennes – Bretagne Atlantique, France

Abstract. We present a new methodology for exchanging unsatisfiability proofs between an untrusted SMT solver and a sceptical proof assistant with computation capabilities like Coq. We advocate modular SMT proofs that separate boolean reasoning and theory reasoning; and structure the communication between theories using Nelson-Oppen combination scheme. We present the design and implementation of a Coq reflexive verifier that is modular and allows for fine-tuned theory-specific verifiers. The current verifier is able to verify proofs for quantifier-free formulae mixing linear arithmetic and uninterpreted functions. Our proof generation scheme benefits from the efficiency of state-of-the-art SMT solvers while being independent from a specific SMT solver proof format. Our only requirement for the SMT solver is the ability to extract unsat cores and generate boolean models. In practice, unsat cores are relatively small and their proof is obtained with a modest overhead by our proof-producing prover. We present experiments assessing the feasibility of the approach for benchmarks obtained from the SMT competition.

1 Introduction

During the past few years, interactive proof assistants have been very successful in the domain of software verification and formal mathematics. In these areas the amount of formal proofs is impressive. For Coq, one of the mainstream proof assistants, it is particularly impressive to see that so many proofs have been done with so little automation. In his POPL'06 paper on verified compilation [19, page 12], Leroy gives the following feedback on his use of Coq:

> Our proofs make good use of the limited proof automation facilities provided by Coq, mostly eauto (Prolog-style resolution), omega (Presburger arithmetic) and congruence (equational reasoning). However, these tactics do not combine automatically and significant manual massaging of the goals is necessary before they apply.

Yet, efficient algorithms exist to combine decision procedures for arithmetic and equational reasoning. During the late '70s, Nelson and Oppen have proposed a cooperation schema for decision procedures [23]. This seminal work, joint with the advances in SAT-solving techniques, has greatly influenced the design

* This work was partly funded by the ANR DeCert, FNRAE ASCERT and Région Bretagne CertLogS projects.

J.-P. Jouannaud and Z. Shao (Eds.): CPP 2011, LNCS 7086, pp. 151–166, 2011.

of modern SMT solvers [11,4,8]. Nowadays, these solvers are able to discharge enormous formulae in a few milliseconds. A proof assistant like Coq would gain a lot in usability with only a small fraction of this speed and automation.

Integrating such algorithms in a proof assistant like Coq is difficult. Coq is a *sceptical* proof assistant and therefore every decision procedure must justify its verdict to the proof kernel with an adequate typable proof term. We distinguish between two different methods for integrating a new decision procedure in a system like Coq. First, we can rely on an external tool, written in an other programming language than Coq, that builds a Coq proof term for each formula it can prove. The main limit of this approach is the size of the exchanged proof term, especially when many rewriting steps are required [17]. Second, we can *verify the prover* by directly programming it in Coq and mechanically proving its soundness. Each formula is then proved by running the prover inside Coq. Such a *reflexive* approach [17] leads to short proof terms but the prover has to be written in the constrained environment of Coq. Programming a state-of-the-art SMT solver in a purely functional language is by itself a challenging task; proving it correct is likely to be impractical — with a reasonable amount of time.

Our implementation is a trade-off between the two previous extreme approaches: we program a reflexive verifier that uses hints (or certificates) given by an untrusted prover (programmed in OCaml). Such an approach has the following advantages: 1) The verifier is simpler to program and prove correct in Coq than the prover itself; 2) Termination is obtained for free as the number of validation steps is known beforehand; 3) The hint conveys the minimum amount of information needed to validate the proof and is therefore smaller than a genuine proof term. This last point is especially useful when a reasoning takes more time to explain than the time to directly perform it in the Coq engine. Recall that the Coq reduction engine [16] allows the evaluation of Coq programs with the same efficiency as OCaml programs. This design allows us to find a good trade-off between proof time checking and proof size.

The mainstream approach for validating SMT proofs [15,20,6] requires a tight integration with an explanation-producing SMT solver. The drawbacks are that explanations may contain too much or too little details and are solver specific. Despite on-going efforts, there is no standard SMT proof format. In contrast, our methodology for generating unsatisfiability proofs is based on a coarser-grained interaction with the SMT solver. Our current implementation only requires an SMT solver producing unsat cores and boolean models. In practice, unsat cores are relatively small and their proofs are obtained with a modest overhead by our hand-crafted proof-producing prover. Our prover is co-designed with the Coq verifier and therefore has the advantage of generating the exact level of details needed to validate the proof. The contributions of this work can be summarised as follows:

– A new methodology for exchanging unsatisfiability proofs between an untrusted SMT solver and a sceptical proof assistant with computation capabilities like Coq. Our proof format is modular. It separates boolean reasoning from theory reasoning, and structures the communication between theories using the Nelson-Oppen combination scheme.

- A modular reflexive Coq verifier that allows for fine-tuned theory specific verifiers exploiting as much as possible the efficient Coq reduction engine. The current verifier is able to verify proofs for quantifier-free formula mixing linear arithmetic and uninterpreted functions.
- A proof-generation scheme that uses state-of-the-art SMT solvers in a black-box manner and only requires the SMT solvers to extract unsat-cores and boolean models. (These features are standardised by the SMT-LIB 2 format.)
- A proof-producing multi-theory prover that generate certificates to discharge theory lemmas, *i.e.*, unsat-cores. It is based on a standard Nelson-Oppen combination of a simplex prover for linear arithmetic and a congruence closure engine for uninterpreted functions.

To discharge SAT proofs, we use the reflexive boolean SAT verifier developed by Armand *et. al.* [2,1]. We only consider ground formula and therefore quantifier instantiation is not in the scope of this paper.

Our Coq development, our proof-producing prover and the benchmarks of Section 6 are available at `http://www.irisa.fr/celtique/ext/chk-no`.

2 Overview

This section gives an overview of some concepts used in state-of-the-art SMT solvers. It presents the SMT solver approach in three layers. Our proof format follows closely this layered presentation. We focus on formulae that must be proved unsatisfiable. We take as running example the following quantifier free *multi-theory* formula, that mixes specifically the theories of equality and Uninterpreted Functions (UF) and Linear Real Arithmetic (LRA).

$$f(f(x) - f(y)) \neq f(z) \wedge x \leq y \wedge ((y + z \leq x \wedge z \geq 0) \vee (y - z \leq x \wedge z < 0)) \quad (1)$$

For UF, a literal is an equality between multi-sorted ground terms and a formula is a conjunction of positive and negative literals. The axioms of this theory are reflexivity, symmetry and transitivity, and the congruence axiom $\forall a \forall b, a = b \Rightarrow f(a) = f(b)$ for functions. For LRA, a literal is a linear constraint $c_0 + c_1 \cdot x_1 + \cdots + c_n \cdot x_n \bowtie 0$ where $(c_i)_{i=0..n} \in \mathbb{Q}$ is a sequence of rational coefficients, $(x_i)_{i=1..n}$ is a sequence of real unknowns and $\bowtie \in \{=, >, \geq\}$. Following Simplify [14], disequality is managed on the UF side. Therefore, a formula is a conjunction of positive literals.

From input formula to unsat multi-theory conjunctions. The lazy SMT solver approach [13] abstracts each atom of the unsatisfiable input formula by a distinct propositional variable, uses a SAT solver to find a propositional model of the formula, and then checks that model against the theory. Models that are incompatible with the theories are discarded by adding a proper lemma to the original formula. This process is repeated until all possible propositional models have been explored. For the given running example, the initial boolean abstraction (2) is $A \wedge B \wedge ((C \wedge D) \vee (E \wedge \neg D))$ with the following mapping

A	B	C	D	E
$f(f(x) - f(y)) \neq f(z)$	$x \leq y$	$y + z \leq x$	$z \geq 0$	$y - z \leq x$

(3)

The first boolean model, $A{:}True, B{:}True, C{:}True, D{:}True, E{:}False$, corresponds to the conjunction

$$(f(f(x) - f(y)) \neq f(z)) \wedge (x \leq y) \wedge (y + z \leq x) \wedge (z \geq 0) \wedge \neg(y - z \leq x)$$

and can be proved unsatisfiable by a multi-theory solver. Hence the boolean model is discarded by adding the theory lemma $\neg(A \wedge B \wedge C \wedge D \wedge \neg E)$ to the original boolean formula. The process is repeated until no more boolean model can be found, showing that the current boolean formula is unsatisfiable.

This process can be speed up with several optimisations. First, theory lemmas can by obtained from *unsat cores*, *i.e.*, minimal subsets of a propositional model still unsatisfiable for the theories. Some SMT solvers also check partial models incrementally against the theory in order to detect conflicts earlier. Second, the multi-theory solver may discover propagation lemmas, *i.e.*, theory literals that are consequence of partial models. In a boolean form, such lemmas allow the SAT solver to reduce further its search tree. In all cases, a witness of unsatisfiability of the input formula is given by a proof of unsatisfiability of a boolean formula composed of the boolean abstraction of the input formula, plus boolean lemmas that correspond to negation of unsatisfiable multi-theory conjunctions. This leads to the first proof rule of our proof format:

$$\frac{\begin{array}{c} f^{\mathbb{B}}, \neg C_1^{\mathbb{B}}, \ldots, \neg C_n^{\mathbb{B}} \vdash_{Boolean} cert^{\mathbb{B}} : \text{False} \\ \forall i = 1, \ldots, n,\ \sigma(C_i^{\mathbb{B}}) \vdash_{NO} cert_i : \text{False} \end{array}}{\sigma(f^{\mathbb{B}}) \vdash_{SMT} (\sigma, (cert^{\mathbb{B}} : f^{\mathbb{B}}), [(cert_1 : C_1^{\mathbb{B}}), \ldots, (cert_n : C_n^{\mathbb{B}})]) : \text{False}}$$

In the following, a judgement of the form $\Gamma \vdash cert : F$ means that formula F can be deduced from hypotheses in Γ, using certificate *cert*. In the judgement $\sigma(f^{\mathbb{B}}) \vdash_{SMT} cert$: False, the certificate *cert* is composed of three elements: a mapping σ between propositional variables and theory literals, a boolean abstraction $f^{\mathbb{B}}$ of F and a list $C_1^{\mathbb{B}}, \ldots, C_n^{\mathbb{B}}$ of conjunctions of boolean variables. For this judgement to establish that the ground formula F is unsatisfiable, several premises have to be verified by the reflexive checker. First, $\sigma(f^{\mathbb{B}})$ must be reducible to F. It means that the boolean abstraction is just proposed by the untrusted prover and checked correct by the reflexive verifier. Second, the conjunction of $f^{\mathbb{B}}$ and all the negation $\neg C_1^{\mathbb{B}}, \ldots, \neg C_n^{\mathbb{B}}$ must be checked unsatisfiable with a boolean verifier. This verifier can be helped with a dedicated certificate $cert^{\mathbb{B}}$ (for example taking the form of a refutation tree). As explained before, the current paper does not focus on this specific part. We instead rely on the reflexive tactic proposed by Armand *et al.*, [2,1]. At last, every multi-theory conjunction $\sigma(C_i^{\mathbb{B}})$ must be proved unsatisfiable with a dedicated certificate $cert_i$. This is done with the judgement $\vdash_{NO}$ which is explained in the next subsection. For our example, the certificate would be composed of the mapping (3), the boolean abstraction (2), and the conjunctions $(A \wedge B \wedge C \wedge D)$ and $(B \wedge \neg D \wedge E)$.

Generation of SMT proofs. To generate our SMT proof format, we implement the simple SMT loop discussed earlier using SMT-LIB 2 scripts to interface with off-the-shelf SMT solvers. The SMT-LIB 2 [3] exposes a rich API for SMT solvers that makes this approach feasible. More precisely, SMT-LIB 2 defines scripts that are sequence of commands to be run by SMT solvers. The `asserts f` command adds the formula `f` to the current context and the `check-sat` command checks the satisfiability of the current context. If the context is satisfiable (`check-sat` returns `sat`), the `get-model` command returns a model. Otherwise, the `get-unsat-core` command returns a so-called unsat core that is a minimised unsatisfiable subset of the current context.

The SMT loop we described is implemented using SMT-LIB 2 compatible off-the-shelf SAT and SMT solvers (we chose Z3 for both). Given an initial unsatisfiable formula, the protocol is the following. To begin with, the boolean abstraction of the input formula is computed and sent to the SAT solver. For each boolean model returned by the SAT solver, the SMT solver is asked for an unsat core, whose negation is sent to the SAT solver. The loop terminates when the SAT solver returns an `unsat` status. Once all the unsat cores have been discovered, our OCaml prover generate certificates for them using the proof system described in Section 3 and Section 4. This untrusted certifying prover implements the Nelson-Oppen algorithm [23] described below. Overall, unsat cores tend to be very small (10 literals on average) and therefore our certifying prover is not the bottleneck. The boolean proof is obtained by running an independent certifying SAT solver. Unlike SMT solvers, DPLL-based SAT solvers have standardised proofs: *resolution proofs.*

Our prototype could be optimised in many ways. For instance, a boolean proof could be obtained directly without re-running a SAT solver. Our scheme would also benefit from a SMT-LIB 2 command returning all the theory lemmas (unsat cores are only a special kind of those) needed to reach a proof of unsatisfiability.

From unsat multi-theory conjunctions to unsat mono-theory conjunctions. In the previous steps, the theory solvers have been fed with conjunctions of multi-theory literals. We now explain the Nelson-Oppen (NO) algorithm that is a sound and complete decision procedure for combining infinitely stable theories with disjoint signatures [23]. Figure 1 presents the deduction steps of this procedure on the previous first theory conflict (corresponding to the boolean conjunction $(A \wedge B \wedge C \wedge D)$). We start from the formula at the top of Figure 1 and first apply a *purification* step that introduces sufficiently many intermediate variables to flatten each terms and dispatch *pure* formulae to each theory. Then each theory exchanges new equalities with the others until a contradiction is found.

Theory exchange is modelled by the Nelson-Oppen proof rule given below.

$$\frac{\Gamma_i \vdash_{T_i} cert_i : (\Gamma_i', eqs) \qquad \bigwedge_{x_k = y_k \in eqs} (\Gamma_1[j \mapsto x_k = y_k], \ldots, \Gamma_i', \ldots, \Gamma_n[j \mapsto x_k = y_k] \vdash_{NO} sons[k] : \text{False})}{\Gamma_1, \ldots, \Gamma_n \vdash_{NO} (cert_i, sons) : \text{False}}$$

We assume here a collection of n theories $T_1, \ldots, T_n$. In this judgement Γ_i represents an environment of pure literals of theory T_i. Each theory is equipped with its own deduction judgement $\Gamma_i \vdash_{T_i} cert_i : (\Gamma_i', eqs)$ where Γ_i and Γ_i' are environments of theory T_i, $cert_i$ is a certificate specific to theory T_i and *eqs* is a list of equalities between variables. Such a judgement reads as follows: assuming that all the literals in Γ_i hold, we can prove (using certificate $cert_i$) that all the literals in Γ_i' hold and that the disjunction of equalities between variables

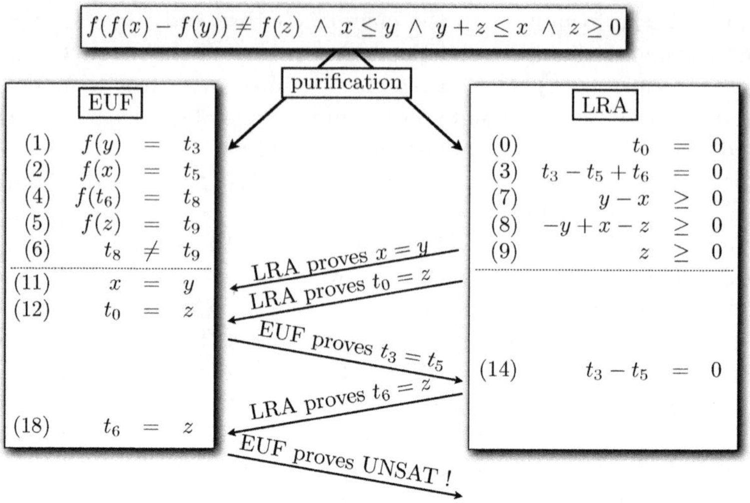

Fig. 1. Example of Nelson-Oppen equality exchange

in *eqs* can be proved. The judgement $\Gamma_1, \ldots, \Gamma_n \vdash_{NO} (cert_i, sons)$: False holds if given an environment $\Gamma_1, \ldots, \Gamma_n$ of the joint theory $T_1 + \ldots + T_n$, the certificate $(cert_i, sons)$ allows to exhibit a contradiction, *i.e.*, False. Suppose that certificate $cert_i$ establishes a judgement of the form $\Gamma_i \vdash_{T_i} cert_i : (\Gamma_i', eqs)$. If the list *eqs* is empty (*i.e.*, represents an empty disjunction), we have a proof that Γ_i is contradictory and therefore the joint environment $\Gamma_1, \ldots, \Gamma_n$ is contradictory and the judgement holds. An important situation is when the list *eqs* is always a singleton during a proof. This corresponds to the case of convex theories for which the Nelson-Oppen algorithm never needs to perform case-splits [23]. In the general case, we recursively exhibit a contradiction for each equality $(x_k = y_k)$ using the kth certificate of *sons*, *i.e.*, *sons*$[k]$ for a joint environment $(\Gamma_1[j \mapsto x_k = y_k], \ldots, \Gamma_i', \ldots, \Gamma_n[j \mapsto x_k = y_k])$ enriched with the equality $(x_k = y_k)$. For completeness, the index j used to store the equality $(x_k = y_k)$ should be fresh. The judgement holds if all the branches of the case-split over the equalities in *eqs* lead to a contradiction.

For the example given in Figure 1, we start with the sets Γ_{LRA} and Γ_{UF} of LRA hypotheses (resp. UF hypotheses). A first certificate $cert_1^{\text{LRA}}$ is required to prove the equality $x = y$, then a certificate $cert_1^{\text{UF}}$ to prove $t_3 = t_5$, then a certificate $cert_2^{\text{LRA}}$ to prove the equality $t_6 = z$, and at last a certificate $cert_2^{\text{UF}}$ to find a contradiction. The whole reasoning is hence justified by the following certificate: $(cert_1^{\text{LRA}}, \{(cert_1^{\text{UF}}, \{(cert_2^{\text{LRA}}, \{(cert_2^{\text{UF}}, \{\})\})\})\})$.

Discharging unsat mono-theory conjunctions. Each part of the NO proof is theory-specific: each theory must justify either the equalities exchanged or the contradiction found. A LRA proof of $a = b$ is made of two Farkas proofs [27] of $b - a \geq 0$ and $a - b \geq 0$. Each inequality is obtained by a linear combination of hypotheses that preserves signs. For example, the previous certificate $cert_1^{\text{LRA}}$

explains that hypothesis (7) gives $y - x \geq 0$ and (8) + (9) gives $x - y \geq 0$. A UF proof of $a = b$ is made of a sequence of rewrites that allows to reach b from a. For example, the certificate $cert_1^{\mathrm{UF}}$ explains the equality $t_3 = t_5$ with the following rewritings: $t_3 \xrightarrow{\texttt{trans. with (1)}} f(y) \xrightarrow{\texttt{congr. with (11)}} f(x) \xrightarrow{\texttt{trans. with (2)}} t_5$.

The rest of the paper is organised as follows. Section 3 presents the certificate format for the UF theory. Section 4 presents the certificate format for linear arithmetic. We present the modular Nelson-Oppen verifier in Section 5 and give some experiments in Section 6. We discuss related work in Section 7 and conclude in Section 8 with a discussion on further work.

3 Certificate Checking and Generation for UF

In this section we introduce the certificate language and checker for UF and present an overview of the certifying prover.

Certificate language. A certificate is a list of commands executed in sequence. Each command operates on the state of the checker which is a pair (Γ, eq). The assumption set Γ is a mapping from indices to assumptions, written $\Gamma(i) \mapsto a = b$, and eq is the *current equality*, *i.e.*, the last one proved. Each command corresponds to an axiom or a combination of axioms of the UF theory.

```
Inductive command :=
  | Refl (t : term) | Trans (i : index) (sym : bool)
  |Congr (i : index) (sym : bool) (pos : index) | Push (i : index).
```

The semantics is given by rules on judgements of the form $(\Gamma, eq) \xrightarrow{\texttt{cmd}} (\Gamma', eq')$ where (Γ', eq') is the state obtained after executing the command cmd from the state (Γ, eq). The boolean s in `Trans` and `Congr` commands makes symmetry explicit: if $\Gamma(i) \mapsto t = t'$ then we define $\Gamma(i)^{true} \mapsto t' = t$ and $\Gamma(i)^{false} \mapsto t = t'$.

$$\frac{}{\Gamma, . = . \xrightarrow{\texttt{Refl}(y)} \Gamma, y = y} \qquad \frac{\Gamma(i)^s \mapsto t = t'}{\Gamma, x = t \xrightarrow{\texttt{Trans}(i,s)} \Gamma, x = t'} \qquad \frac{\Gamma' = \Gamma[i \mapsto x = t]}{\Gamma, x = t \xrightarrow{\texttt{Push(i)}} \Gamma', x = t}$$

$$\frac{\Gamma(i)^s \mapsto a_p = a'_p}{\Gamma, x = f(a_0..a_p..a_n) \xrightarrow{\texttt{Congr}(i,p,s)} \Gamma, x = f(a_0..a'_p..a_n)}$$

The command $\texttt{Refl}(y)$ corresponds to the reflexivity axiom and initialises the current equality with the tautology $y = y$, whatever the previous equality. Subsequent commands will rewrite the right hand side of this equality. The command $\texttt{Trans}(i, s)$ updates the right hand side of the current equality: if we can prove that $x = t$ (current equality) and we know that $t = t'$ (equality indexed by i) then we can deduce $x = t'$. The command $\texttt{Congr}(i, p, s)$ rewrites a sub-term of the right hand side: in any given context if we can prove $x = f(y)$ (current equality) and we know that $y = z$ (equality indexed by i) then we can deduce $x = f(z)$ and make it the new current equality. The parameter p is used to determine where to rewrite. The command $\texttt{Push}(i)$ is used to update the assumption set Γ

with the current equality $x = t$, creating a new context $\Gamma' = \Gamma[i \mapsto x = t]$ to be used to evaluate the next commands. It allows us to factorise sub-proofs and is mandatory to keep invariant the depth of terms.

The relation $\Gamma \vdash_{UF} cert_{UF} : (\Gamma', eqs)$ implements the theory specific judgement seen in Section 2.

$$\frac{\Gamma, z = z \xrightarrow{cert}_* \Gamma', x = y}{\Gamma \vdash_{UF} \texttt{UF\_Eq}(cert) : (\Gamma', [x = y])} \qquad \frac{\Gamma, z = z \xrightarrow{cert}_* \Gamma', x = y \quad \Gamma(i) \mapsto x \neq y}{\Gamma \vdash_{UF} \texttt{UF\_False}(i, cert) : (\Gamma', nil)}$$

Suppose that we obtain a state $(\Gamma, x = y)$ after processing a list *cert* of commands. The certificate $\texttt{UF\_False}(i, cert)$ deduces a contradiction if $\Gamma(i) \mapsto x \neq y$ and the certificate $\texttt{UF\_Eq}(cert)$ deduces the equality $x = y$.

Certificate generation. follows closely [24] where the certifying prover maintains a *proof forest* that keeps track of the reasons why two nodes are merged. Besides the usual *merge* and *find* operations, the data structure has a new operator $\texttt{explain}(a, b, \texttt{forest})$ which outputs a *proof* that $a = b$ based on `forest`. In our case the proofs are certificates, while in the original approach they were non-redundant unsatisfiable unordered sets of assumptions.

We show below the proof forest corresponding to the UF part of the example of Figure 1. Trees represent equivalence classes and each edge is labelled by assumptions. The prover updates the forest with each *merge*. Two distinct classes can be merged for two reasons: an equality between variables is added or two terms are equal by congruence.

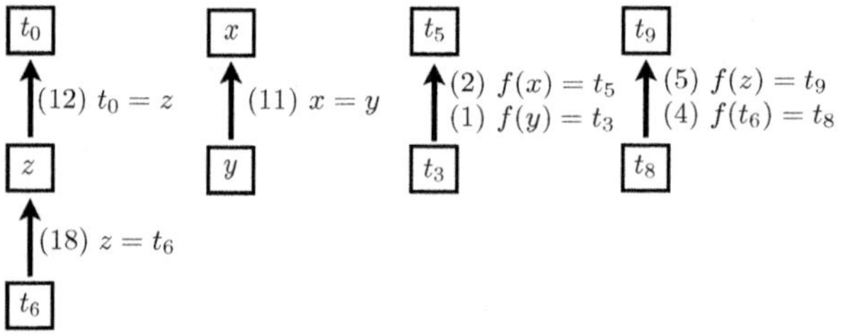

Suppose for example that the problem contains (2) $f(x) = t_5$ and (1) $f(y) = t_3$ and we add the equality (11) $x = y$. First we have to add an edge between x and y, labelled by the reason of this merge, *i.e.*, assumption (11). Then we have to add an edge between t_3 and t_5, and label it with the two assumptions that triggered that merge by congruence, *i.e.*, (1) and (2).

To output a certificate that two variables are equal, we travel the path between the two corresponding nodes, and each edge yields a list of commands. An edge labelled by an equality corresponds to a transitivity: $t_6 \xrightarrow{(18)} z$ yields $[\texttt{Trans}(18, true)]$. An edge labelled by two equalities uses congruence: $t_3 \xrightarrow{(1)(2)} t_5$ yields $[\texttt{Trans}(1, false); \texttt{Congr}(11, 1, true); \texttt{Trans}(2, true)]$. If the equality that triggered the congruence was *discovered* by UF and not an assumption, we have to explain it first, then update the environment accordingly using the `Push` command, and finally use the stored equality with the `Congr` command.

4 Certificate Checking and Generation for LRA and LIA

In this section we introduce the certificate language and proof system for linear arithmetic and describe its certifying prover. Literals are of the form $e \bowtie 0$ with e a linear expression manipulated in (Horner) normal form and $\bowtie \in \{\geq, >, =\}$.

Certificate language. Since our initial work [5], we are maintaining and enhancing reflexive tactics for real arithmetic (`psatz`) and linear integer arithmetic (`lia`). Those tactics, which are now part of the Coq code-base, are based on the *Positivstellensatz* [28], a rich proof system which is complete for non-linear (real) polynomial arithmetic. Those reflexive verifiers are at the core of our current theory verifiers for linear real arithmetic (LRA) and linear integer arithmetic (LIA). We present here simplified proof systems specialised for linear arithmetic.

For linear real arithmetic Farkas' lemma provides a sound and complete notion of certificate for proving that a conjunction of linear constraints is unsatisfiable [27, Corollary 7.1e]. It consists in exhibiting a positive linear combination of the hypotheses that is obviously unsatisfiable, *i.e.*, deriving $c \bowtie 0$ for $\bowtie \in \{>, \geq, =\}$ and c a constant such that $c \bowtie 0$ does not hold. To construct such a contradiction, we start with a sub-proof system that allows to derive an inequality with a list of commands (a Farkas certificate). Each command is a pair $\texttt{Mul}(c, i)$ where c is a coefficient (in type $\mathbb{Z}$) and i the index of an assumption in the current assumption set. Such a command is used below in a judgement $\Gamma, e \bowtie 0 \xrightarrow{\texttt{Mul}(c,i)} \Gamma', e' \bowtie' 0$ with $\bowtie$ and $\bowtie'$ in $\{\geq, >\}$. $\Gamma \cup \{e \bowtie 0\}$ is the current set of assumptions, $e' \bowtie' 0$ is a new deduced inequality and Γ' is an enriched set of assumptions. For LIA, the proof system is augmented with a `Cut` command to generate *cutting planes* [27, chapter 23] and a rule for case-splitting `Enum`. We also need a `Push` and a `Get` command in order to update the environment and retrieve an already derived formula. The semantics of the commands is given in Figure 2. The operators $[*], [+], [-]$ model the standard arithmetic operations but maintain the normalised form of the linear expressions. The rules for the `Mul` command follow the standard sign rules in arithmetic: for example, if e' is positive we can add it c times to the right part of the inequality $e \bowtie 0$, assuming c is strictly positive. To implement the `Cut` rule, the constant g is obtained by computing the greatest common divisor of the coefficient of the linear expression. For inequalities, the rule allows to *cut* the constant. For equalities, it allows to detect a contradiction if g does not divide d ($\neg(g \mid d)$).

A LRA certificate is then either a proof of $0 > 0$ given by a list of commands or a proof of $x = y$ given by two lists of commands (one for $x - y \geq 0$ and one other for $y - x \geq 0$.

```
Inductive LRA_certificate :=
|LRA_False (l : list command) |LRA_Eq (l1 l2 : list command)
```

$$\frac{\Gamma \vdash l : 0 > 0}{\Gamma \vdash_{LRA} (\texttt{LRA\_False}(l)) : (\Gamma, nil)} \qquad \frac{\Gamma \vdash l_1 : e \geq 0 \quad e = x[-]y \quad \Gamma \vdash l_2 : [-]e \geq 0}{\Gamma \vdash_{LRA} (\texttt{LRA\_Eq}(l_1, l_2)) : (\Gamma, [x = y])}$$

$$\frac{c > 0 \quad \Gamma(i) \mapsto e' \geq 0}{\Gamma, e \bowtie 0 \xrightarrow{\texttt{Mul}(c,i)} \Gamma, (c[*]e'[+]e) \bowtie 0} \qquad \frac{\Gamma(i) \mapsto e' = 0}{\Gamma, e \bowtie 0 \xrightarrow{\texttt{Mul}(c,i)} \Gamma, (c[*]e'[+]e) \bowtie 0}$$

$$\frac{c > 0 \quad \Gamma(i) \mapsto e' > 0}{\Gamma, e \bowtie 0 \xrightarrow{\texttt{Mul}(c,i)} \Gamma, (c[*]e'[+]e) > 0} \qquad \frac{\Gamma(i) = e' \bowtie 0}{\Gamma, e \bowtie' 0 \xrightarrow{Get(i)} \Gamma, e' \bowtie 0}$$

$$\frac{\Gamma' = \Gamma[i \mapsto e \bowtie 0]}{\Gamma, e \bowtie 0 \xrightarrow{Push(i)} \Gamma', e \bowtie 0} \qquad \frac{g > 0}{\Gamma, (g[*]e[-]d) \geq 0 \xrightarrow{\texttt{Cut}} \Gamma, (e[-]\lceil d/g \rceil) \geq 0}$$

$$\frac{g \mid d}{\Gamma, (g[*]e[-]d) = 0 \xrightarrow{\texttt{Cut}} \Gamma, (e[-](d/g)) = 0} \qquad \frac{\neg(g \mid d)}{\Gamma, (g[*]e[-]d) = 0 \xrightarrow{\texttt{Cut}} \Gamma, 0 > 0}$$

$$\frac{\begin{array}{c}\Gamma(i_1) \mapsto e[-]l \geq 0 \quad \Gamma(i_2) \mapsto h[-]e \geq 0 \\ \forall v \in [l,h], \Gamma, e = v \xrightarrow{c_{v-l}}_{*} \Gamma'_v, e' \bowtie' 0\end{array}}{\Gamma, \cdot \bowtie 0 \xrightarrow{\texttt{Enum}(i_1,i_2,[c_0;\ldots;c_{h-l}])} \Gamma, e' \bowtie' 0}$$

Fig. 2. LRA and LIA proof rules

Because the theory LIA is non-convex, it is necessary to deduce contradictions but also disjunction of equalities.

```
Inductive LIA_certificate :=
| LIA_False (l : list command)
| LIA_Eq (eqs : list (var * var)) (l : list (list command))
```

Proving equalities is done by performing a case-split and each list of commands $l \in l$ is used to prove that a case is unsatisfiable.

Certificate generation. In order to produce Farkas certificates efficiently, we have implemented the Simplex algorithm used in Simplify [14]. This variant of the standard linear programming algorithm does not require all the variable to be non-negative, and directly handles (strict and large) inequalities and equalities. Each time a contradiction is found, one line of the Simplex tableau gives us the expected Farkas coefficients. The algorithm is also able to discover new equalities between variables. In this case again, the two expected Farkas certificates are read from the current tableau, up to trivial manipulations.

For LIA, we use a variant of the Omega test [26]. The Omega test lacks a way to derive equalities but the number of shared variables is sufficiently small to allow an exhaustive search. Moreover, an effective heuristics is to pick as potential equalities the dis-equalities present in the unsat core.

5 Design of a Modular Nelson-Oppen Proof-Verifier

This section presents the design of a reflexive Coq verifier for a Nelson-Oppen style combination of theories. Section 5 presents the main features of the theory interface. Section 5 explains the data-structures manipulated by the Nelson-Oppen proof-checker, *i.e.*, its dependently typed environment and its certificates.

Theory interface. A theory `T` defines a type for sorts `sort`, terms `term` and formulae `form`. Sorts, terms and formulae are equipped with interpretation functions `isort`, `iterm` and `iform`. The function `isort:sort→Type` maps a sort to a Coq type. Terms and formulae are interpreted with respect to a typed environment `env∈Env` defined by `Env:=var→∀(s:sort),isort s`. Each theory uses an environment Γ `∈Gamma` to store formulae. Environments expose the following API:

```
Record GammaAPI : Type := {|
 empty : Gamma ; add : form → Gamma → Gamma;
 ienv : Env → Gamma → Prop;
 ienv_empty : ∀ env, ienv env empty;
 ienv_add : ∀ (f : form) (s : Gamma) (env : Env), ienv env s →
                        iform env f → ienv env (add f s) |}.
```

Environments are equipped with an interpretation function `ienv`. The `empty` environment represents an empty conjunction of formulae, *i.e.*, the assertion *true* and is such that `ienv env empty` holds for any environment. The operation `add` models the addition of a formula and is compatible with the interpretation `iform` of formulae. Our instantiations exploit the fact that environments are kept abstract: for UF, environments are radix trees allowing a fast look-up of formulae; for LRA, they are simple lists but arithmetic expressions are normalised (put in Horner normal form) by the `add` operation.

The key feature provided by a theory T is a proof-checker `Checker`. It takes as argument an environment Γ and a certificate *cert*. Upon success, the checker returns an updated environment Γ' and a list $eqs = (x_1 =_{s_1} x'_1, \ldots, x_n =_{s_n} x'_n)$ of equalities between sorted variables. In such cases, `Checker_sound` establishes that $\Gamma \vdash_T cert : (\Gamma', eqs)$ is a judgement of the Nelson-Oppen proof system (see Section 2). A representative theory record is given below.

```
Record Thy := {|
 sort : Type; term : Type;  form : Type;
 sort_of_term : term → sort;  isort : sort → Type;
 Env := var → ∀ (s:sort), isort s;
 iterm : Env → ∀ (t : term), isort (sort_of_term t);
 iform : Env → form → Prop
 ...
 Checker : Gamma → Cert → option(Gamma * (list (Eq.t sort)));
 Checker_sound : ∀ cert Γ Γ' eqs, Checker Γ cert = Some(Γ', eqs)
  → ∀ (env : Env), ienv env Γ → ienv env Γ'
         /\∃s, ∃x, ∃y, (x =s y) ∈ eqs /\ env x s = env y s |}.
```

Nelson-Oppen proof-checker. Given a list of theories $\mathtt{T}_1,\ldots,\mathtt{T}_n$ the environment of the Nelson-Oppen proof-checker is a dependently typed list such that the ith element of the list is an environment of type $\mathtt{T}_i$`.(Gamma)`. Dependently typed lists are defined as follows:

```
Inductive dlist (A : Type) (typ : A → Type) : list A  →  Type :=
| dnil : dlist A typ nil
| dcons : ∀ (x : A) (e : typ x) (lx : list A) (le : dlist lx),
                  dlist A typ (x::lx).
```

A term `dcons x e lx le` constructs a list with head `e` and tail `le`. The type of `e` is `typ x` and the type of the elements of `le` is given by `(List.map typ lx)`.

It follows that the environment of the Nelson-Oppen proof-checker has type:

$$\texttt{dlist Thy Gamma } (\texttt{T}_1\texttt{::...::T}_n)$$

A single proof-step consists in checking a certificate `JCert` of the joint theory defined by $\texttt{JCert := T}_1\texttt{.(Cert) + ... + T}_n\texttt{.(Cert)}$.

Each certificate triggers the relevant theory proof-checker and derives an eventually empty list of equalities, *i.e.*, a proof of non-satisfiability. Each equality $\texttt{x} =_s \texttt{y}$ is cloned for each sort `s'` such that `isort s = isort s'` and propagated to the relevant theory. Each equality of the list is responsible for a case-split that may be recursively closed by a certificate (see Section 2). A certificate for the Nelson-Oppen proof-checker is therefore a tree of certificates defined by:

```
Inductive Cert := Mk (cert : JCert) (lcert : list Cert).
```

The Nelson-Oppen verifier consumes the certificate and returns `true` if the last deduced list of equalities is empty. In all other cases, the verification aborts and the verifier returns `false`.

6 Experiments

The purpose of our experiments is twofold. They first show that our SMT format is viable and can be generated for a substantial number of benchmarks. The experiments also assess the efficiency of our Coq reflexive verifier. We have evaluated our approach on quantifier-free first-order unsatisfiable formulae over the combinations of the theory of equality and uninterpreted functions (UF), linear real arithmetic (LRA), linear integer arithmetic (LIA) and real difference logic (RDL). All problems are unsatisfiable SMT-LIB 2 benchmarks selected from the SMT-LIB repository that are solved by Z3 in less than 30 seconds.

Table 1 shows our results sorted by logic. For each category, we measure the average running time of Z3 (Solved), the average running time of our certificate generation (Generation). The *Solved* time can be seen as a best-case scenario: the certifying prover uses Z3 and provide proofs that can be checked in Coq, so we do not expect faster results than the standalone state-of-the-art solver. We also measure the time it takes Coq to type-check our proof term (Qed) and have isolated the time spent by our Coq reflexive verifier validating theory lemmas (Thy). The generation phase (Generation) and the checking phases (Checking) have an individual timeout of 150 seconds. These timeouts account for most of the failures, the remaining errors come from shortcomings of the prototype.

Overall, the theory specific checkers account for less then 7% of checking time. However, this average masks big differences. For UFLRA, the checker spends less

Table 1. Experimental results for selected SMT-LIB logics

Logic	Solved (Z3)		Generation		Checking		
	#	Time (s)	Success	Time (s)	Success	Thy (s)	Qed (s)
UF	613	0.96	31.3%	42.55	100%	0.29	16.81
LRA	248	0.65	79.4%	6.79	69.5%	0.28	4.02
UFLRA	407	0.11	100%	0.72	98.8%	0.02	3.56
LIA	401	1.86	74.3%	9.05	46.0%	2.26	7.02
UFLIA	159	0.05	97.5%	8.15	96.1%	0.33	2.91
RDL	79	4.01	38.0%	11.24	53.3%	0.14	3.64
Total	1907	0.87	67.1%	11.02	80.8%	0.45	6.45

than 1% of its time in the theories, but for the integer arithmetic fragments it represents 11% of checking time for UFLIA and 32% for LIA. For UFLRA it can be explained by the simplicity of the problems : 80% of these formulae are unsatisfiable even if we only consider their boolean abstractions. For integer arithmetic the success ratio is rather low. It is hard to know whether this is due to the inherent difficulty of the problems or whether it pinpoints an inefficiency of the checker. The fault might also lie on the certifying prover side. In certain circumstances, it performs case-splits that are responsible for long proofs.

Sometimes, our simple SMT loop fails to produce certificates before timeout. For UF and RDL we only generate certificates for a third of the formulae. The generation of certificates could be optimised further. A more clever proof search strategy could improve both certificate generation and checking times: smaller certificates could be generated faster and checked more easily. Yet, the bottleneck is the reflexive verifier, which achieves 100% success ratio for UF only. Currently, we observe that our main limiting factor is not time but the memory consumption of the Coq process. A substantial amount of our timeouts are actually due to memory exhaustion. We are investigating the issue, but the objects we manipulate (formulae, certificates) are orders of magnitude larger than those manipulated on a day-to-day basis by a proof-assistant. We know we are reaching the limits of the system. Actually, to perform our experiments we already overcome certain inefficiencies of Coq. For instance, to construct formulae and certificates we by-pass Coq front-end, which is not efficient enough for this application, and use homemade optimised versions of a few Coq tactics.

7 Related Work

The area of proof-generating decision procedure has been pioneered by Boulton for the HOL system [7] and Necula for Proof Carrying Code [21]. In the context of the latter, the Touchstone theorem prover [22] generates LF proof terms. In our approach, each decision procedure comes with its own certificate language, and a reflexive checker. It allows us to choose the level of details of the certificates without compromising correctness. Several authors have examined UF proofs [12,24]. They extend a pre-existing decision procedure with proof-producing mechanism without degrading its complexity and achieving a certain level of irredundancy. However, their notion of proof is reduced to unsatisfiable cores of literals rather than proof trees. Our certificate generation builds on such works to produce detailed explanations. SMT solvers such as CVC3 [4], veriT [8] and Z3 [10] all generate proofs in their own proof language. Many rules reflect the internal reasoning with various levels of precision: certain rules detail each computation step, some others account for complex reasoning with no further details. Such solvers aim at discharging large and/or hard problems, at the price of simplicity. Our approach here differs because our proof rules are specific to the decision procedure we have implemented in our prover. We do not sacrifice soundness since our proof verifier is proved correct (and executable) in Coq.

Several approaches have been proposed to integrate new decision procedures in sceptical proof assistants for various theories. First-order provers have been integrated in Isabelle [25], HOL [18] or Coq [9]. These works rely generally on resolution proof trees. Similar proof formats have been considered to integrate Boolean satisfiability checking in a proof assistant. Armand et al. [2] have extended the Coq programming language with machine integers and persistent array and have used these new features to directly program in Coq a reflexive SAT checker. On a similar topic, Weber and Amjad [29] have integrated a state-of-the-art SAT solver in Isabelle/HOL, HOL4 and HOL Light using translation from SAT resolution proofs to LCF-style proof objects.

Previous work has been devoted to reconstruct SMT solvers proofs in proof assistants. McLaughlin et al. [20] have combined CVC Lite and HOL light for quantifier-free first-order logic with equality, arrays and linear real arithmetic. Ge and Barrett have continued that work with CVC3 and have extended it to quantified formulae and linear integer arithmetic. This approach highlighted the difficulty for proof reconstruction. Independently Fontaine et al. [15] have combined haRVey with Isabelle/HOL for quantifier free first-order formulae with equality and uninterpreted functions. In their scheme, Isabelle solves UF sub-proofs with hints provided by haRVey. Our UF certificate language is more detailed and does not require any decision on the checker side. Böhme and Weber [6] are reconstructing Z3 proofs in the theorem provers Isabelle/HOL and HOL4. Their implementation is particularly efficient but their fine profiling shows that a lot of time is spent re-proving sub-goals for which the Z3 proof does not give sufficient details. Armand et al. [2] have recently extended their previous work [2] to check proofs generated by the SMT solver veriT [8]. Our approaches are similar and rely on proof by reflexion. A difference lies in the certificate generation scheme. Their implementation is tied to a specific SMT solver and its ability to generate proofs. In our approach, we do not require SMT solvers to generate proofs but instead designed our own proof-producing prover to discharge theory lemmas.

8 Conclusion and Perspectives

We have presented a reflexive approach for integrating a SMT solver in a sceptical proof assistant like Coq. It is based on a SMT proof format that is independent from a specific SMT solver. We believe our approach is robust to changes in the SMT solvers but allows nonetheless to benefit from their improvements. For most usages, the overhead incurred by our SMT loop is acceptable. It could even be reduced if SMT solvers gave access to the theory lemmas they use during their proof search. We are confident that such information could be generated by any SMT solver with little overhead. Implementing our approach necessitates proof-producing decision procedures. However, the hard job is left to the SMT solver that extracts unsat cores. A fine-grained control over the produced proof has the advantage of allowing to optimise a reflexive verifier and of ensuring the completeness of the verifier with respect to the prover. Our Nelson-Oppen Coq verifier is both reflexive and parametrised by a list of theories. This design is modular and easy to extend with new theories. Our prototype implementation is perfectible but already validates SMT formulae of industrial size. Such extreme experiments test the limits of the proof-assistant and will eventually help at improving its scalability. In the future, we plan to integrate new theories such as the theory of arrays and bit-vectors. Another theory of interest is the theory of constructors that would be useful to reason about inductive types.

References

1. Armand, M., Faure, G., Gregoire, B., Keller, C., Théry, L., Werner, B.: A Modular Integration of SAT/SMT Solvers to Coq Through Proof Witnesses. In: Jouannaud, J.-P., Shao, Z. (eds.) CPP 2011. LNCS, vol. 7086, pp. 135–150. Springer, Heidelberg (2011)
2. Armand, M., Grégoire, B., Spiwack, A., Théry, L.: Extending Coq with Imperative Features and its Application to SAT Verification. In: Kaufmann, M., Paulson, L.C. (eds.) ITP 2010. LNCS, vol. 6172, pp. 83–98. Springer, Heidelberg (2010)
3. Barret, C., Stump, A., Tinelli, C.: The SMT-LIB standard: Version 2.0 (2010)
4. Barrett, C.W., Tinelli, C.: CVC3. In: Damm, W., Hermanns, H. (eds.) CAV 2007. LNCS, vol. 4590, pp. 298–302. Springer, Heidelberg (2007)
5. Besson, F.: Fast Reflexive Arithmetic Tactics the Linear Case and Beyond. In: Altenkirch, T., McBride, C. (eds.) TYPES 2006. LNCS, vol. 4502, pp. 48–62. Springer, Heidelberg (2007)
6. Böhme, S., Weber, T.: Fast LCF-Style Proof Reconstruction for Z3. In: Kaufmann, M., Paulson, L.C. (eds.) ITP 2010. LNCS, vol. 6172, pp. 179–194. Springer, Heidelberg (2010)
7. Boulton, R.J.: Efficiency in a Fully-Expansive Theorem Prover. PhD thesis, University of Cambridge Computer Laboratory, Technical Report 337 (1994)
8. Bouton, T., de Oliveira, D.C.B., Déharbe, D., Fontaine, P.: veriT: An Open, Trustable and Efficient SMT-Solver. In: Schmidt, R.A. (ed.) CADE-22. LNCS, vol. 5663, pp. 151–156. Springer, Heidelberg (2009)
9. Contejean, E., Corbineau, P.: Reflecting Proofs in First-Order Logic with Equality. In: Nieuwenhuis, R. (ed.) CADE 2005. LNCS (LNAI), vol. 3632, pp. 7–22. Springer, Heidelberg (2005)
10. de Moura, L.M., Bjørner, N.: Proofs and Refutations, and Z3. In: LPAR 2008 Workshops: KEAPPA. CEUR-WS.org, vol. 418 (2008)
11. de Moura, L., Bjørner, N.S.: Z3: An Efficient SMT Solver. In: Ramakrishnan, C.R., Rehof, J. (eds.) TACAS 2008. LNCS, vol. 4963, pp. 337–340. Springer, Heidelberg (2008)

12. de Moura, L.M., Rueß, H., Shankar, N.: Justifying equality. ENTCS 125(3), 69–85 (2005)
13. de Moura, L., Rueß, H., Sorea, M.: Lazy Theorem Proving for Bounded Model Checking Over Infinite Domains. In: Voronkov, A. (ed.) CADE 2002. LNCS (LNAI), vol. 2392, pp. 438–455. Springer, Heidelberg (2002)
14. Detlefs, D., Nelson, G., Saxe, J.B.: Simplify: a theorem prover for program checking. J. ACM 52(3), 365–473 (2005)
15. Fontaine, P., Marion, J.-Y., Merz, S., Nieto, L.P., Tiu, A.F.: Expressiveness + Automation + Soundness: Towards Combining SMT Solvers and Interactive Proof Assistants. In: Hermanns, H. (ed.) TACAS 2006. LNCS, vol. 3920, pp. 167–181. Springer, Heidelberg (2006)
16. Grégoire, B., Leroy, X.: A compiled implementation of strong reduction. In: ICFP 2002, pp. 235–246. ACM (2002)
17. Grégoire, B., Mahboubi, A.: Proving Equalities in a Commutative Ring Done Right in Coq. In: Hurd, J., Melham, T. (eds.) TPHOLs 2005. LNCS, vol. 3603, pp. 98–113. Springer, Heidelberg (2005)
18. Hurd, J.: Integrating Gandalf and HOL. In: Bertot, Y., Dowek, G., Hirschowitz, A., Paulin, C., Théry, L. (eds.) TPHOLs 1999. LNCS, vol. 1690, pp. 311–322. Springer, Heidelberg (1999)
19. Leroy, X.: Formal certification of a compiler back-end or: programming a compiler with a proof assistant. In: POPL 2006, pp. 42–54. ACM (2006)
20. McLaughlin, S., Barrett, C., Ge, Y.: Cooperating theorem provers: A case study combining HOL-Light and CVC Lite. ENTCS 144(2), 43–51 (2006)
21. Necula, G.C.: Compiling with Proofs. PhD thesis, CMU (1998)
22. Necula, G.C., Lee, P.: Proof Generation in the Touchstone Theorem Prover. In: McAllester, D. (ed.) CADE 2000. LNCS, vol. 1831, pp. 25–44. Springer, Heidelberg (2000)
23. Nelson, G., Oppen, D.C.: Simplification by cooperating decision procedures. ACM Trans. Program. Lang. Syst. 1, 245–257 (1979)
24. Nieuwenhuis, R., Oliveras, A.: Proof-Producing Congruence Closure. In: Giesl, J. (ed.) RTA 2005. LNCS, vol. 3467, pp. 453–468. Springer, Heidelberg (2005)
25. Paulson, L.C., Susanto, K.W.: Source-Level Proof Reconstruction for Interactive Theorem Proving. In: Schneider, K., Brandt, J. (eds.) TPHOLs 2007. LNCS, vol. 4732, pp. 232–245. Springer, Heidelberg (2007)
26. Pugh, W.: The omega test: a fast and practical integer programming algorithm for dependence analysis. In: SC, pp. 4–13 (1991)
27. Schrijver, A.: Theory of Linear and Integer Programming. Wiley (1998)
28. Stengle, G.: A nullstellensatz and a positivstellensatz in semialgebraic geometry. Mathematische Annalen 207(2), 87–97 (1973)
29. Weber, T., Amjad, H.: Efficiently checking propositional refutations in HOL theorem provers. J. Applied Logic 7(1), 26–40 (2009)

Tactics for Reasoning Modulo AC in Coq

Thomas Braibant and Damien Pous*

LIG, UMR 5217, CNRS, INRIA, Grenoble

Abstract. We present a set of tools for rewriting modulo associativity and commutativity (AC) in Coq, solving a long-standing practical problem. We use two building blocks: first, an extensible reflexive decision procedure for equality modulo AC; second, an OCaml plug-in for pattern matching modulo AC. We handle associative only operations, neutral elements, uninterpreted function symbols, and user-defined equivalence relations. By relying on type-classes for the reification phase, we can infer these properties automatically, so that end-users do not need to specify which operation is A or AC, or which constant is a neutral element.

1 Introduction

Motivations. Typical hand-written mathematical proofs deal with commutativity and associativity of operations in a liberal way. Unfortunately, a proof assistant requires a formal justification of all reasoning steps, so that the user often needs to make boring term re-orderings before applying a theorem or using a hypothesis. Suppose for example that one wants to rewrite using a simple hypothesis like `H`: $\forall$`x`, `x+-x` $= 0$ in a term like `a+b+c+-(c+a)`. Since Coq standard **`rewrite`** tactic matches terms syntactically, this is not possible directly. Instead, one has to reshape the goal using the commutativity and associativity lemmas:

```
                                                (* ⊢ ((a+b)+c)+-(c+a) = ... *)
rewrite (add_comm a b), ← (add_assoc b a c).   (* ⊢ (b+(a+c))+-(c+a) = ... *)
rewrite (add_comm c a), ← add_assoc.           (* ⊢ b+((a+c)+-(a+c)) = ... *)
rewrite H.                                     (* ⊢ b+0              = ... *)
```

This is not satisfactory for several reasons. First, the proof script is too verbose for such a simple reasoning step. Second, while reading such a proof script is easy, writing it can be painful: there are several sequences of rewrites yielding to the desired term, and finding a reasonably short one is difficult. Third, we need to copy-paste parts of the goal to select which occurrence to rewrite using the associativity or commutativity lemmas; this is not a good practice since the resulting script breaks when the goal is subject to small modifications. (Note that one could also select occurrences by their positions, but this is at least as difficult for the user which then has to count the number of occurrences to skip, and even more fragile since these numbers cannot be used to understand the proof when the script breaks after some modification of the goal.)

In this paper, we propose a solution to this short-coming for the Coq proof-assistant: we extend the usual rewriting tactic to automatically exploit associativity and commutativity (AC), or just associativity (A) of some operations.

* Supported by "Choco", ANR-07-BLAN-0324 and "PiCoq", ANR-10-BLAN-0305.

J.-P. Jouannaud and Z. Shao (Eds.): CPP 2011, LNCS 7086, pp. 167–182, 2011.

Trusted unification vs untrusted matching. There are two main approaches to implementing rewriting modulo AC in a proof-assistant. First, one can extend the unification mechanism of the system to work modulo AC [20]. This option is quite powerful, since most existing tactics would then work modulo AC. It however requires non-trivial modifications of the kernel of the proof assistant (e.g., unification modulo AC does not always yield finite complete sets of unifiers). As a consequence, this obfuscates the meta-theory: we need a new proof of strong normalisation and we increase the trusted code base. On the contrary, we can restrict ourselves to pattern matching modulo AC and use the core-system itself to validate all rewriting steps [8]. We chose this option.

Contributions, scope of the library. Besides the facts that such tools did not exist in Coq before and that they apparently no longer exist in Isabelle/HOL (see §6.1 for a more thorough discussion), the main contributions of this work lie in the way standard algorithms and ideas are combined together to get tactics that are efficient, easy to use, and covering a large range of situations:

- We can have any number of associative and possibly commutative operations, each possibly having a neutral element. For instance, we can have the operations `min`, `max`, $+$, and $*$ on natural numbers, where `max` and $+$ share the neutral element 0, $*$ has neutral element 1, and `min` has no neutral element.
- We deal with arbitrary user-defined equivalence relations. This is important for rational numbers or propositions, for example, where addition and subtraction (resp. conjunction and disjunction) are not AC for Leibniz equality, but for rational equality, `Qeq` (resp. propositional equivalence, `iff`).
- We handle "uninterpreted" function symbols: n-ary functions for which the only assumption is that they preserve the appropriate equivalence relation—they are sometimes called "proper morphisms". For example, subtraction on rational numbers is a proper morphism for `Qeq`, while pointwise addition of numerators and denominators is not. (Note that any function is a proper morphism for Leibniz equality.)
- The interface we provide is straightforward to use: it suffices to declare instances of the appropriate type-classes [22] for the operations of interest, and our tactics will exploit this information automatically. Since the type-class implementation is first-class, this gives the ability to work with polymorphic operations in a transparent way (e.g., concatenation of lists is declared as associative once and for all.)

Methodology. Recalling the example from the beginning, an alternative to explicit sequences of rewrites consists in making a transitivity step through a term that matches the hypothesis' left-hand side syntactically:

```
                                   (* ⊢ ((a+b)+c)+-(c+a) = ... *)
transitivity (b+((a+c)+−(a+c))).   (*   ⊢ ((a+b)+c)+-(c+a) = b+((a+c)+-(a+c)) *)
  ring. (* aac_reflexivity *)      (* ⊢ b+((a+c)+-(a+c)) = ... *)
rewrite H.                         (* ⊢ b+0 = ... *)
```

Although the `ring` tactic [14] solves the first sub-goal here, this is not always the case (e.g., there are AC operations that are not part of a ring structure). Therefore, we have to build a new tactic for equality modulo A/AC: `aac_reflexivity`.

Another drawback is that we also have to copy-paste and modify the term manually, so that the script can break if the goal evolves. This can be a good practice in some cases: the transitivity step can be considered as a robust and readable documentation point; in other situations we want this step to be inferred by the system, by pattern matching modulo A/AC [15].

All in all, we proceed as follows to define our `aac_rewrite` rewriting tactic. Let $\equiv_{AC}$ denote equality modulo A/AC; to rewrite using a universally quantified hypothesis of the form $H\colon \forall \tilde{x}, p\tilde{x} = q\tilde{x}$ in a goal G, we take the following steps, which correspond to building the proof-tree on the right-hand side:

1. choose a context C and a substitution σ such that $G \equiv_{AC} C[p\sigma]$ (pattern matching modulo AC);
2. make a transitivity step through $C[p\sigma]$;
3. close this step using a dedicated decision procedure (`aac_reflexivity`);
4. use the standard `rewrite`;
5. let the user continue the proof.

$$\dfrac{\dfrac{}{G \equiv_{AC} C[p\sigma]}\,3 \qquad \dfrac{H \qquad \dfrac{\vdots}{C[q\sigma]}\,5}{C[p\sigma]}\,4}{G}\,2$$

For the sake of efficiency, we implement the first step as an OCaml oracle, and we check the results of this (untrusted) matching function in the third step, using the certified decision procedure `aac_reflexivity`. To implement this tactic, we use the standard methodology of *reflection* [8,1,14]. Concretely, this means that we implement the decision procedure as a Coq function over "reified" terms, which we prove correct inside the proof assistant. This step was actually quite challenging: to our knowledge, `aac_reflexivity` is the first reflexive Coq tactic that handles uninterpreted function symbols. In addition to the non-trivial reification process, a particular difficulty comes from the (arbitrary) arity of these symbols. To overcome this problem in an elegant way, our solution relies on a dependently typed syntax for reified terms.

Outline. We sketch the user interface (§2) before describing the decision procedure (§3) and the algorithm for pattern matching modulo AC (§4). We detail our handling of neutral elements and subterms separately (§5). We conclude with related works and directions for future work (§6).

2 User Interface, Notation

Declaring A/AC operations. We rely on type-classes [22] to declare the properties of functions and A/AC binary operations. This allows the user to extend both the decision procedure and the matching algorithm with new A/AC operations or units in a very natural way. Moreover, this is the basis of our reification mechanism (see §3.2).

The classes corresponding to the various properties that can be declared are given in Fig. 1: being associative, commutative, and having a neutral element. Basically, a user only needs to provide instances of these classes in order to

```
Variables (X: Type) (R: relation X) (op: X → X → X).
Class Associative := law_assoc: ∀x y z, R (op x (op y z)) (op (op x y) z).
Class Commutative := law_comm: ∀x y, R (op x y) (op y x).
Class Unit (e: X) := { law_id_left: ∀x, R (op e x) x; law_id_right: ∀x, R (op x e) x }.
```

```
Instance plus_A: Associative eq plus.
Instance plus_C: Commutative eq plus.
Instance plus_U: Unit eq plus 0.

Instance app_A X: Associative eq (app X).
Instance app_U X: Unit eq (app X) (nil X).
```

```
Instance and_A: Associative iff and.
Instance and_C: Commutative iff and.
Instance and_U: Unit iff and True.
Instance and_P: Proper (iff ⇒iff ⇒iff) and.
Instance not_P: Proper (iff ⇒iff) not.
```

Fig. 1. Classes for declaring properties of operations, example instances

use our tactics in a setting with new A or AC operations. These classes are parameterised by a relation (R): one can use an arbitrary equivalence relation.

Fig. 1 also contains examples of instances. Polymorphic values (app, nil) are declared in a straightforward way. For propositional connectives (and, not), we also need to show that they preserve equivalence of propositions (iff), since this is not Leibniz equality; we use for that the standard Proper type-class—when the relation R is Leibniz equality, these instances are inferred automatically. Of course, while we provide these instances, more can be defined by the user.

Example usage. The main tactics we provide are aac_rewrite, to rewrite modulo A/AC, and aac_reflexivity to decide an equality modulo A/AC. Here is a simple example where we use both of them:

```
H1: ∀x y z, x∩y ∪ x∩z = x∩(y∪z)
H2: ∀x y, x∩x = x
a, b, c, d: set
=====================
(a∩c ∪ b∩c∩d) ∩ c = (a ∪ d∩b) ∩ c
```

```
Proof.
  aac_rewrite H1; (* c ∩ (a ∪ b∩d) ∩ c = ... *)
  aac_rewrite H2; (* c ∩ (a ∪ b∩d)     = ... *)
  aac_reflexivity.
Qed.
```

As expected, we provide variations to rewrite using the hypothesis from right to left, or in the right-hand side of the goal.

Listing instances. There might be several ways of rewriting a given equation: several subterms may match, so that the user might need to select which occurrences to rewrite. The situation can be even worse when rewriting modulo AC: unlike with syntactical matching, there might be several ways of instantiating the pattern so that it matches a given occurrence. (E.g., matching the pattern $x + y + y$ at the root of the term $a + a + b + b$ yields two substitutions: $\{x \mapsto a + a; y \mapsto b\}$ and the symmetrical one—assuming there is no neutral element.) To help the user, we provide an additional tactic, aac_instances, to display the possible occurrences together with the corresponding instantiations. The user can then use the tactic aac_rewrite with the appropriate options.

Notation and terminology. We assume a signature Σ and we let $f, g, h, \ldots$ range over function symbols, reserving letters $a, b, c, \ldots$ for constants (function symbols of arity 0). We denote the set of *terms* by $T(\Sigma)$. Given a set V of variables, we let $x, y, z, \ldots$ range over (universally quantified) variables; a *pattern* is a term with variables, i.e., an element of $T(\Sigma + V)$. A *substitution* (σ) is a partial function that maps variables to terms, which we extend into a partial function from patterns to terms, as expected. Binary function symbols (written with an infix symbol, $\diamond$) can be associative (axiom A) and optionally commutative (axiom C); these symbols may be equipped with a left and right unit u (axiom $U_{u,\diamond}$):

$$A_\diamond : x \diamond (y \diamond z) \equiv (x \diamond y) \diamond z \quad C_\diamond : x \diamond y \equiv y \diamond x \quad U_{u,\diamond} : x \diamond u \equiv x \wedge u \diamond x \equiv x$$

We use $+_i$ (or $+$) for associative-commutative symbols (AC), and $*_i$ (or $*$) for associative only symbols (A). We denote by $\equiv_{AC}$ the equational theory generated by these axioms on $T(\Sigma)$. For instance, in a non-commutative semi-ring $(+, *, 0, 1)$, $\equiv_{AC}$ is generated by A_+, C_+, A_* and $U_{1,*}, U_{0,+}$.

3 Deciding Equality Modulo AC

In this section, we describe the `aac_reflexivity` tactic, which decides equality modulo AC, is extensible through the definition of new type-class instances, and deals with uninterpreted function symbols of arbitrary arity. For the sake of clarity, we defer the case where binary operations have units to §5.1.

3.1 The Algorithm and Its Proof

A two-level approach. We use the so called 2-level approach [4]: we define an inductive type `T` for terms and a function `eval: T → X` that maps reified terms to user-level terms living in some type `X` equipped with an equivalence relation `R`, which we sometimes denote by $\equiv$. This allows us to reason and compute on the syntactic representation of terms, whatever the user-level model.

We follow the usual practice which consists in reducing equational reasoning to the computation and comparison of normal forms: it then suffices to prove that the normalisation function is correct to get a sound decision procedure.

```
Definition norm: T → T := ...
Lemma eval_norm: ∀u, eval (norm u) ≡ eval u.
Theorem decide: ∀u v, compare (norm u) (norm v) = Eq → eval u ≡ eval v.
```

This is what is called the *autarkic way*: the verification is performed inside the proof-assistant, using the conversion rule. To prove `eval u ≡ eval v`, it suffices to apply the theorem `decide` and to let the proof-assistant check by computation that the premise holds. The algorithm needs to meet two objectives. First, the normalisation function (`norm`) must be efficient, and this dictates some choices for the representation of terms. Second, the evaluation function (`eval`) must be simple (in order to keep the proofs tractable) and total: ill-formed terms shall be rejected syntactically.

Packaging the reification environment. We need Coq types to package information about binary operations and uninterpreted function symbols. They are given in Fig. 2, where `respectful` is the definition from Coq standard library for declaring proper morphisms. We first define functions to express the fact that

```
(* type of n-ary homogeneous functions *)
Fixpoint type_of (X: Type) (n: nat): Type :=
  match n with 0 ⇒ X | S n ⇒ X → type_of X n end.

(* relation to be preserved by n-ary functions *)
Fixpoint rel_of (X: Type) (R: relation X) (n: nat): relation (type_of X n) :=
  match n with 0 ⇒ R | S n ⇒ respectful R (rel_of n) end.
```

```
Module Bin.
 Record pack X R := {
  value: X → X → X;
  compat: Proper (R ⇒R ⇒R) value;
  assoc: Associative R value;
  comm: option (Commutative R value) }.
End Bin.
```

```
Module Sym.
 Record pack X R := {
  ar: nat;
  value: type_of X ar;
  compat: Proper (rel_of X R ar) value }.
End Sym.
```

Fig. 2. Types for symbols

n-ary functions are proper morphisms. A "binary package" contains a binary operation together with the proofs that it is a proper morphism, associative, and possibly commutative (we use the type-classes from Fig. 1). An "uninterpreted symbol package" contains the arity of the symbol, the corresponding function, and the proof that this is a proper morphism.

The fact that symbols arity is stored in the package is crucial: by doing so, we can use standard finite maps to store all function symbols, irrespective of their arity. More precisely, we use two environments, one for uninterpreted symbols and one for binary operations; both of them are represented as non-dependent functions from a set of indices to the corresponding package types:

```
Variables (X: Type) (R: relation X).
Variable e_sym: idx → Sym.pack X R.
Variable e_bin: idx → Bin.pack X R.
```

(The type `idx` is an alias for `positive`, the set of binary positive numbers; this allows us to define the above functions efficiently, using binary trees).

Syntax of reified terms. We now turn to the concrete representation of terms. The first difficulty is to choose an appropriate representation for AC and A nodes, to avoid manipulating binary trees. As it is usually done, we flatten these binary nodes using variadic nodes. Since binary operations do not necessarily come with a neutral element, we use non-empty lists (resp. non-empty multi-sets) to reflect the fact that A operations (resp. AC operations) must have at least one argument. (We could even require A/AC operations to have at least two arguments but this would slightly obfuscate the code and prevent some sharing for multi-sets.) The second difficulty is to prevent ill-formed terms, like "`log` 1 2 3", where a unary function is applied to more than one argument. One could define a predicate stating that terms are well-formed [11], and use this extra hypothesis in later reasonings. We found it nicer to use dependent types to enforce the constraint that symbols are applied to the right number of arguments: it suffices to use vectors of arguments rather than lists. The resulting data-type for reified terms is given in Fig. 3; it depends on the environment for

```
(* non-empty lists/multisets *)
Inductive nelist A :=
| nil: A → nelist A
| cons: A → nelist A → nelist A.

Definition nemset A :=
  nelist (A*positive).

(* reified terms *)
Inductive T: Type :=
| bin_ac: idx → nemset T → T
| bin_a : idx → nelist T → T
| sym: ∀i, vect T (Sym.ar (e_sym i)) → T.
```

```
Fixpoint eval (u: T): X :=
match u with
| bin_ac i l ⇒ let o:=Bin.value (e_bin i) in
    nefold_map o (fun(u,n)⇒copy o n (eval u)) l
| bin_a i l ⇒ let o:=Bin.value (e_bin i) in
    nefold_map o eval l
| sym i v ⇒ xeval v (Sym.value (e_sym i))
end
with xeval i (v: vect T i): Sym.type_of i→X :=
match v with
| vnil ⇒(fun f ⇒ f)
| vcons u v ⇒(fun f ⇒ xeval v (f (eval u)))
end.
```

Fig. 3. Data-type for terms, and related evaluation function

uninterpreted symbols (`e_bin`). This definition allows for a simple implementation of `eval`, given on the right-hand side. For uninterpreted symbols, the trick consists in using an accumulator to store the successive partial applications.

As expected, this syntax allows us to reify arbitrary user-level terms. For instance, take `(a*S(b+b))−b`. We first construct the following environments where we store information about all atoms:

e_sym	e_bin
1 ⇒ ⦇ ar := 1; value := S; compat := ... ⦈	1 ⇒ ⦇ value := plus; compat := ... ;
2 ⇒ ⦇ ar := 0; value := a; compat := ... ⦈	assoc := _ ; comm := Some ... ⦈
3 ⇒ ⦇ ar := 0; value := b; compat := ... ⦈	_ ⇒ ⦇ value := mult; compat := ... ;
_ ⇒ ⦇ ar := 2; value := minus; compat := ... ⦈	assoc := _ ; comm := None ⦈

These environment functions are total: they associate a semantic value to indices that might be considered as "out-of-the-bounds". This requires environments to contain at least one value, but this makes the evaluation function total and easier to reason about: there is no need to return an option or a default value in case undefined symbols are encountered. We can then build a reified term whose evaluation in the above environments reduces to the starting user-level terms:

```
Let t := sym 4 [bin_a 2 [(sym 2 [])]; (sym 1 [bin_ac 1 [(sym 3 [],1);(sym 3 [],1)]])]; sym 3 []].
Goal eval e_sym e_bin t = (a*S(b+b))−b. reflexivity. Qed.
```

Note that we cannot split the environment `e_bin` into two environments `e_bin_a` and `e_bin_ac`: since they would contain at least one binary operation with the proof that it is A or AC, it would not be possible to reify terms in a setting with only A or only AC operations. Moreover, having a single environment for all binary operations makes it easier to handle neutral elements (see §5.1).

Normalisation of reified terms in Coq. Normal forms are computed as follows: terms are recursively flattened under A/AC nodes and arguments of AC nodes are sorted. We give excerpts of this Coq function below, focusing on AC nodes: `bin_ac'` is a smart constructor that prevents building unary AC nodes, and `norm_msets norm i` normalises and sorts a multi-set, ensuring that none of its children starts with an AC node with index `i`.

```
Definition bin_ac' i (u: nemset T): T := match u with nil (u,1) ⇒ u | _ ⇒ bin_ac i u end.
Definition extract_ac i (s: T): nemset T :=
  match s with bin_ac j m when i = j ⇒ m | _ ⇒ [s,1] end.
Definition norm_msets norm i (u: nemset T): nemset T :=
  nefold_map merge_sort (fun (x,n) ⇒ copy_mset n (extract_ac i (norm x))) u
...
Fixpoint norm (u: T): T := match u with
| bin_ac i l ⇒ if is_commutative e_bin i then bin_ac' i (norm_msets norm i l) else u
| bin_a i l ⇒ bin_a' i (norm_lists norm i l)
| sym i l ⇒ sym i (vector_map norm l)
end.
```

Note that `norm` depends on the information contained in the environments: the look-up `is_commutative s_bin i` in the definition of `norm` is required to make sure that the binary operation `i` is actually commutative (remember that we need to store A and AC symbols in the same environment, so that we might have AC nodes whose corresponding operation is not commutative). Similarly, to handle neutral elements (§5.1), we will rely on the environment to detect whether some value is a unit for a given binary operation.

Correctness and completeness. We prove that the normalisation function is sound. This proof relies on the above defensive test against ill-formed terms: since invalid AC nodes are left intact, we do not need the missing commutativity hypothesis when proving the correctness of `norm`. We did not prove completeness. First, this is not needed to get a sound tactic. Second, this proof would be quite verbose (in particular, it requires a formal definition of equality modulo AC on reified terms). Third, we would not be able to completely prove the completeness of the resulting tactic anyway, since one cannot reason about the OCaml reification and normalisation functions in the proof-assistant [14,7].

3.2 Reification

Following the reflexive approach to solve an equality modulo AC, it suffices to apply the above theorem `decide` (§3.1) and to let Coq compute. To do so, we still need to provide two environments `e_bin` and `e_sym` and two terms `u` and `v`, whose evaluation is convertible to the starting user-level terms.

Type-class based reification. We do not want to rely on annotations (like projections of type-classes fields or canonical structures) to guess how to reify the terms: this would force the users to use our definitions and notations from the beginning. Instead, we let the users work with their own definitions, and we exploit type-classes to perform reification. The idea is to query the type-class resolution mechanism to decide whether a given subterm should be reified as an AC operation, an A operation, or an uninterpreted function symbol.

The inference of binary A or AC operations takes place first, by querying for instances of the classes `Associative` and `Commutative` on all binary applications. The remaining difficulty is to discriminate whether other applications should be considered as a function symbol applied to several arguments, or as a constant. For instance, considering the application `f (a+b) (b+c) c`, it suffices to query for `Proper` instances in the following order:

1.	`Proper (R ⇒ R ⇒ R ⇒ R)`	`(f)`	?
2.	`Proper (R ⇒ R ⇒ R)`	`(f (a+b))`	?
3.	`Proper (R ⇒ R)`	`(f (a+b)(b+c))`	?
4.	`Proper (R)`	`(f (a+b)(b+c) c)`	?

The first query that succeeds tells which partial application is a proper morphism, and with which arity. Since the relation `R` is reflexive, and any element is proper for a reflexive relation, the inference of constants—symbols of arity 0—is the catch-all case of reification. We then proceed recursively on the remaining arguments; in the example, if the second call is the first to succeed, we do not try to reify the first argument (`a+b`): the partial application `f(a+b)` is considered as an atom.

Reification language. We use OCaml to perform this reification step. Using the meta-language OCaml rather than the meta-language of tactics LTAC is a matter of convenience: it allows us to use more efficient data-structures. For instance, we use hash-tables to memoise queries to type-class resolution, which would have been difficult to mimic in LTAC or using canonical structures. The resulting code is non-trivial, but too technical to be presented here. Most of the difficulties come from the fact that we reify uninterpreted functions symbols using a dependently typed syntax, and that our reification environments contain dependent records: producing such Coq values from OCaml can be tricky. Finally, using Coq's plug-in mechanism, we wrap up the previous ideas in a tactic, `aac_reflexivity`, which automates this process, and solves equations modulo AC.

Efficiency. The dependently typed representation of terms we chose in order to simplify proofs does not preclude efficient computations. The complexity of the procedure is dominated by the merging of sorted multi-sets, which relies on a linear comparison function. We did not put this decision procedure through an extensive testing; however, we claim that it returns instantaneously in practice. Moreover, the size of the generated proof is linear with respect to the size of the starting terms. By contrast, using the tactic language to build a proof out of associativity and commutativity lemmas would usually yield a quadratic proof.

4 Matching Modulo AC

Solving a matching problem modulo AC consists in, given a pattern p and a term t, finding a substitution σ such that $p\sigma \equiv_{AC} t$. There are many known algorithms [11,12,15,18]; we present here a simple one.

Naive algorithm. Matching modulo AC can easily be implemented non-deterministically. For example, to match a sum $p_1 + p_2$ against a term t, it suffices to consider all possible decompositions of t into a sum $t_1 + t_2$. If matching p_1 against t_1 yields a solution (a substitution), it can be used as an initial state to match p_2 against t_2, yielding a more precise solution, if any. To match a variable x against a term t, there are two cases depending on whether or not the

```
val (≫=): α m → (α → β m) → β m
val (≫|): α m → α m → α m
val return: α → α m
val fail: unit → α m
```

Fig. 4. Search monad primitives

```
val split_ac: idx → term → (term * term) m
val split_a : idx → term → (term * term) m
```

Fig. 5. Search monad derived functions

```
mtch (p1 +i p2) t σ = split_ac i t ≫= (fun (t1,t2) → mtch p1 t1 σ ≫= mtch p2 t2)
mtch (p1 *i p2) t σ = split_a  i t ≫= (fun (t1,t2) → mtch p1 t1 σ ≫= mtch p2 t2)
mtch (f(p̄)) (f(ū)) σ = fold_2 (fun acc p t → acc ≫= mtch p t) (return σ) p̄ ū
mtch (var x) t σ when Subst.find σ x = None = return (Subst.add σ x t)
mtch (var x) t σ when Subst.find σ x = Some v = if v ≡AC t then return σ else fail()
```

Fig. 6. Backtracking pattern matching, using monads

variable has already been assigned in the current substitution. If the variable has already been assigned to a value v, we check that $v \equiv_{AC} t$. If this is not the case, the substitution must be discarded since x must take incompatible values. Otherwise, i.e., if the variable is fresh, we add a mapping from x to v to the substitution. To match an uninterpreted node $f(\overline{q})$ against a term t, it must be the case that t is headed by the same symbol f, with arguments $\overline{u}$; we just match $\overline{q}$ and $\overline{u}$ pointwise.

Monadic implementation. We use a monad for non-deterministic and backtracking computations. Fig. 4 presents the primitive functions offered by this monad: $\gg=$ is a backtracking bind operation, while $\gg|$ is non-deterministic choice. We have an OCaml type for terms similar to the inductive type we defined for Coq reified terms: applications of A/AC symbols are represented using their flattened normal forms. From the primitives of the monad, we derive functions operating on terms (Fig. 5): the function `split_ac i` implements the non-deterministic split of a term t into pairs (t_1, t_2) such that $t \equiv_{AC} t_1 +_i t_2$. If the head-symbol of t is $+_i$, then it suffices to split syntactically the multi-set of arguments; otherwise, we return an empty collection. The function `split_a i` implements the corresponding operation on associative only symbols. The matching algorithm proceeds by structural recursion on the pattern, which yields the code presented in Fig. 6 (using an informal ML-like syntax). A nice property of this algorithm is that it does not produce redundant solutions, so that we do not need to reduce the set of solutions before proposing them to the user.

Correctness. Following [11], we could have attempted to prove the correctness of this matching algorithm. While this could be an interesting formalisation work *per se*, it is not necessary for our purpose, and could even be considered an impediment. Indeed, we implement the matching algorithm as an oracle, in an arbitrary language. Thus, we are given the choice to use a free range of optimisations, and the ability to exploit all features of the implementation language. In any case, the prophecies of this oracle, a set of solutions to the matching problem, are verified by the reflexive decision procedure we implemented in §3.

```
Variable e_bin: idx → Bin.pack X R

Record binary_for (u: X) := {
 bf_idx: idx;
 bf_desc: Unit R (Bin.value (e_bin bf_idx)) u }.
```

```
Record unit_pack := {
 u_value: X;
 u_desc: list (binary_for u_value) }.

Variable e_unit: idx → unit_pack.
```

Fig. 7. Additional environment for terms with units

5 Bridging the Gaps

Combining the decision procedure for equality modulo AC and the algorithm for matching modulo AC, we get the tactic for rewriting modulo AC. We now turn to lifting the simplifying assumptions we made in the previous sections.

5.1 Neutral Elements

Adding support for neutral elements (or "units") is of practical importance:

- to let `aac_reflexivity` decide more equations (e.g., `max` 0 (`b`∗1)+`a` = `a`+`b`);
- to avoid requiring the user to normalise terms manually before performing rewriting steps (e.g., to rewrite using $\forall x, x \cup x = x$ in the term $a \cap b \cup \emptyset \cup b \cap a$);
- to propose more solutions to pattern matching problems (consider rewriting $\forall xy, x \cdot y \cdot x^{\perp} = y$ in $a \cdot (b \cdot (a \cdot b)^{\perp})$, where $\cdot$ is associative only with a neutral element: the variable y should be instantiated with the neutral element).

Extending the pattern matching algorithm. Matching modulo AC with units does not boil down to pattern matching modulo AC against a normalised term: $a \cdot b \cdot (a \cdot b)^{\perp}$ is a normal form and the algorithm of Fig. 6 would not give solutions with the pattern $x \cdot y \cdot x^{\perp}$. The patch is however straightforward: it suffices to let the non-deterministic splitting functions (Fig. 5) use the neutral element possibly associated with the given binary symbol. For instance, calling `split_a` on the previous term would return the four pairs $\langle 1, a \cdot b \cdot (a \cdot b)^{\perp} \rangle$, $\langle a, b \cdot (a \cdot b)^{\perp} \rangle$, $\langle a \cdot b, (a \cdot b)^{\perp} \rangle$, and $\langle a \cdot b \cdot (a \cdot b)^{\perp}, 1 \rangle$, where 1 is the neutral element.

Extending the syntax of reified terms. An obvious idea is to replace non-empty lists (resp. multi-sets) by lists (resp. multi-sets) in the definition of terms—Fig. 3. This has two drawbacks. First, unless the evaluation function (Fig. 3) becomes a partial function, every A/AC symbol must then be associated with a unit (which precludes, e.g., `min` and `max` to be defined as AC operations on relative numbers). Second, two symbols cannot share a common unit, like 0 being the unit of both `max` and `plus` on natural numbers: we would have to know at reification time how to reify 0, is it an empty AC node for `max` or for `plus`? Instead, we add an extra constructor for units to the data-type of terms, and a third environment to store all units together with their relationship to binary operations. The actual definition of this third environment requires a more clever crafting than the other ones. The starting point is that a unit is nothing by itself, it is a unit for some binary operations. Thus, the type of the environment for units has to depend on the `e_bin` environment. This type is given in Fig. 7. The record `binary_for` stores a binary operation (pointed to by its index `bf_idx`) and a proof that the parameter `u` is a neutral element for this operation. Then, each unit is bundled with a list of operations it is a unit for (`unit_pack`): like for the environment `e_sym` , these dependent records allow us to use plain, non-dependent maps. In the end, the syntax of reified terms depends only on the environment for uninterpreted symbols (`e_sym`), to ensure that arities are respected, while the environment for units (`e_unit`) depends on that for binary operations (`e_bin`).

Extending the decision tactic. Updating the Coq normalisation function to deal with units is fairly simple but slightly verbose. Like we used the `e_bin` environment to check that `bin_ac` nodes actually correspond to commutative operations, we exploit the information contained in `e_unit` to detect whether a unit is a neutral element for a given binary operation. On the contrary, the OCaml reification code, which is quite technical, becomes even more complicated. Calling type-class resolution on all constants of the goal to get the list of binary operations they are a unit for would be too costly. Instead, we perform a first pass on the goal, where we infer all A/AC operations and for each of these, whether it has a neutral element. We construct the reified terms in a second pass, using the previous information to distinguish units from regular constants.

5.2 Subterms

Another point of high practical importance is the ability to rewrite in subterms rather than at the root. Indeed, the algorithm of Fig. 6 does not allow to match the pattern $x+x$ against the terms $f(a+a)$ or $a+b+a$, where the occurrence appears under some context. Technically, it suffices to extend the (OCaml) pattern matching function and to write some boilerplate to accommodate contexts; the (Coq) decision procedure is not affected by this modification. Formally, subterm-matching a pattern p in a term t results in a set of solutions which are pairs $\langle C, \sigma\rangle$, where C is a context and σ is a substitution such that $C[p\sigma] \equiv_{AC} t$.

Variable extensions. It is not sufficient to call the (root) matching function on all syntactic subterms: the instance $a+a$ of the pattern $x+x$ is not a syntactic subterm of $a+b+a$. The standard trick consists in enriching the pattern using a *variable extension* [19,21], a variable used to collect the trailing terms. In the previous case, we can extend the pattern into $y+x+x$, where y will be instantiated with b. It then suffices to explore syntactic subterms: when we try to subterm-match $x+x$ against $(a+c)*(a+b+a)$, we extend the pattern into $y+x+x$ and we call the matching algorithm (Fig. 6) on the whole term and the subterms a, b, c, $a+c$ and $a+b+a$. In this example, only the last call succeeds.

The problem with subterms and units. However, this approach is not complete in the presence of units. Suppose for instance that we try to match the pattern $x+x$ against $a*b$, where $*$ is associative only. If the variable x can be instantiated with a neutral element 0 for $+$, then the variable extension trick gives four solutions:

$$a*b+[] \qquad (a+[])*b \qquad a*(b+[])$$

(These are the returned contexts, in which [] denotes the hole; the substitution is always $\{x \mapsto 0\}$.) Unfortunately, if $*$ also has a neutral element 1, there are infinitely many other solutions:

$$a * b * (1 + []) \qquad a * b + 0 * (1 + []) \qquad a * b + 0 * (1 + 0 * (1 + [])) \qquad \ldots$$

(Note that these solutions are distinct modulo AC, they collapse to the same term only when we replace the hole with 0.) The latter solutions only appear when the pattern can be instantiated to be equal to a neutral element (modulo A/AC). We opted for a pragmatic solution in this case: we reject these peculiar solutions, displaying a warning message. The user can still instantiate the rewriting lemma explicitly, or make the appropriate transitivity step using `aac_reflexivity`.

6 Conclusions

The Coq library corresponding to the tools we presented is available from [9]. We do not use any axiom; the code consists of about 1400 lines of Coq and 3600 lines of OCaml. We conclude with related works and directions for future work.

6.1 Related Works

Boyer and Moore [8] are precursors to our work in two ways. First, their paper is the earliest reference to reflection we are aware of, under the name "Meta-functions". Second, they use this methodology to prove correct a simplification function for cancellation modulo A. By contrast, we proved correct a decision procedure for equality modulo A/AC with units which can deal with arbitrary function symbols, and we used it to devise a tactic for rewriting modulo A/AC.

Ring. While there is some similarity in their goals, our decision procedure is incomparable with the Coq `ring` tactic [14]. On the one hand, `ring` can make use of distributivity and opposite laws to prove goals like $\mathtt{x}^2 - \mathtt{y}^2 = (\mathtt{x}-\mathtt{y}) * (\mathtt{x}+\mathtt{y})$, holding in any ring structure. On the other hand, `aac_reflexivity` can deal with an arbitrary number of AC or A operations with their units, and more importantly, with uninterpreted function symbols. For instance, it proves equations like $\mathtt{f(x \cap y)} \cup \mathtt{g(\emptyset \cup z)} = \mathtt{g\ z} \cup \mathtt{f(y \cap x)}$, where `f`, `g` are arbitrary functions on sets. Like with `ring`, we also have a tactic to normalise terms modulo AC.

Rewriting modulo AC in HOL and Isabelle. Nipkow [17] used the Isabelle system to implement matching, unification and rewriting for various theories including AC. He presents algorithms as proof rules, relying on the Isabelle machinery and tactic language to build actual tools for equational reasoning. While this approach leads to elegant and short implementations, what is gained in conciseness and genericity is lost in efficiency, and the algorithms need not terminate. The rewriting modulo AC tools he defines are geared toward automatic term normalisation; by contrast, our approach focuses on providing the user with tools to select and make one rewriting step efficiently.

Slind [21] implemented an AC-unification algorithm and incorporated it in the `hol90` system, as an external and efficient oracle. It is then used to build tactics for AC rewriting, cancellation, and modus-ponens. While these tools exploit pattern matching only, an application of unification is in solving existential goals. Apart from some refinements like dealing with neutral elements and A symbols, the most salient differences with our work are that we use a reflexive decision procedure to check equality modulo A/AC rather than a tactic implemented in the meta-language, and that we use type-classes to infer and reify automatically the A/AC symbols and their units.

Support for the former tool [17] has been discontinued, and it seems to be also the case for the latter [21]. To our knowledge, even though HOL-light and HOL provide some tactics to prove that two terms are equal using associativity and commutativity of a single given operation, tactics comparable to the ones we describe here no longer exist in the Isabelle/HOL family of proof assistants.

Rewriting modulo AC in Coq. Contejean [11] implemented in Coq an algorithm for matching modulo AC, which she proved sound and complete. The emphasis is put on the proof of the matching algorithm, which corresponds to a concrete implementation in the CiME system. Although decidability of equality modulo AC is also derived, this development was not designed to obtain the kind of tactics we propose here (in particular, we could not reuse it to this end). Finally, symbols can be uninterpreted, commutative, or associative and commutative, but neither associative only symbols nor units are handled.

Gonthier et al. [13] have recently shown how to exploit a feature of Coq's unification algorithm to provide "less ad hoc automation". In particular, they automate reasoning modulo AC in a particular scenario, by diverting the unification algorithm in a complex but really neat way. Using their trick to provide the generic tactics we discuss here might be possible, but it would be difficult. Our reification process is much more complex: we have uninterpreted function symbols, we do not know in advance which operations are AC, and the handling of units requires a dependent environment. Moreover, we would have to implement matching modulo AC (which is not required in their example) using the same methodology; doing it in a sufficiently efficient way seems really challenging.

Nguyen et al. [16] used the external rewriting tool ELAN to add support for rewriting modulo AC in Coq. They perform term rewriting in the efficient ELAN environment, and check the resulting traces in Coq. This allows one to obtain a powerful normalisation tactic out of any set of rewriting rules which is confluent and terminating modulo AC. Our objectives are slightly different: we want to easily perform small rewriting steps in an arbitrarily complex proof, rather than to decide a proposition by computing and comparing normal forms.

The ELAN trace is replayed using elementary Coq tactics, and equalities modulo AC are proved by applying the associativity and commutativity lemmas in a clever way. On the contrary, we use the high-level (but slightly inefficient) `rewrite` tactic to perform the rewriting step, and we rely on an efficient reflexive decision procedure for proving equalities modulo AC. (Alvarado and Nguyen first proposed a version where the rewriting trace was replayed using reflection, but without support for modulo AC [2].)

From the user interface point of view, leaving out the fact that the support for this tool has been discontinued, our work improves on several points: thanks to the recent plug-in and type-class mechanisms of Coq, it suffices for a user to declare instances of the appropriate classes to get the ability to rewrite modulo AC. Even more importantly, there is no need to declare explicitly all uninterpreted function symbols, and we transparently support polymorphic operations (like `List.app`) and arbitrary equivalence relations (like `Qeq` on rational numbers, or `iff` on propositions). It would therefore be interesting to revive this tool using the new mechanisms available in Coq, to get a nicer and more powerful interface.

Although this is not a general purpose interactive proof assistant, the Maude system [10], which is based on equational and rewriting logic, also provides an efficient algorithm for rewriting modulo AC [12]. Like ELAN, Maude could be used as an oracle to replace our OCaml matching algorithm. This would require some non-trivial interfacing work, however. Moreover, it is unclear to us how to use these tools to get all matching occurrences of a pattern in a given term.

6.2 Directions for Future Works

Heterogeneous terms and operations. Our decision procedure cannot deal with functions whose range and domain are distinct sets. We could extend the tactic to deal with such symbols, to make it possible to rewrite using equations like $\forall uv, \|u+v\| \leq \|u\| + \|v\|$, where $\|\cdot\|$ is a norm in a vector space. This requires a more involved definition of reified terms and environments to keep track of type information; the corresponding reification process seems quite challenging.

We could also handle heterogeneous associative operations, like multiplication of non-square matrices, or composition of morphisms in a category. For example, matrix multiplication has type $\forall$ `n m p, X n m` $\rightarrow$ `X m p` $\rightarrow$ `X n p` (`X n m` being the type of matrices with size n, m). This would be helpful for proofs in category theory. Again, the first difficulty is to adapt the definition of reified terms, which would certainly require dependently typed non-empty lists.

Other decidable theories. While we focused on rewriting modulo AC, we could consider other theories whose matching problem is decidable. Such theories include, for example, the Abelian groups and the Boolean rings [6] (the latter naturally appears in proofs of hardware circuits).

Integration with other tools. Recently, tactics have been designed to exploit external SAT/SMT solvers inside Coq [3]. These tactics rely on a reflexive proof checker, used to certify the traces generated by the external solver. However, in the SMT case, these traces do not contain proofs for the steps related to the considered theories, so that one needs dedicated Coq decision procedures to validate these steps. Currently, mostly linear integer arithmetic is supported [3], using the `lia` tactic [5]; our tactic `aac_reflexivity` could be plugged into this framework to add support for theories including arbitrary A or AC symbols.

Acknowledgements. We would like to thank Matthieu Sozeau for his precious help in understanding Coq's internal API.

References

1. Allen, S.F., Constable, R.L., Howe, D.J., Aitken, W.E.: The Semantics of Reflected Proof. In: Proc. LICS, pp. 95–105. IEEE Computer Society (1990)
2. Alvarado, C., Nguyen, Q.-H.: ELAN for Equational Reasoning in Coq. In: Proc. LFM 2000. INRIA (2000) ISBN 2-7261-1166-1
3. Armand, M., Faure, G., Grégoire, B., Keller, C., Théry, L., Werner, B.: A Modular Integration of SAT/SMT Solvers to Coq Through Proof Witnesses. In: Jouannaud, J.-P., Shao, Z. (eds.) CPP 2011. LNCS, vol. 7086, pp. 135–150. Springer, Heidelberg (2011)
4. Barthe, G., Ruys, M., Barendregt, H.: A Two-Level Approach Towards Lean Proof-Checking. In: Berardi, S., Coppo, M. (eds.) TYPES 1995. LNCS, vol. 1158, pp. 16–35. Springer, Heidelberg (1996)
5. Besson, F.: Fast Reflexive Arithmetic Tactics the Linear Case and Beyond. In: Altenkirch, T., McBride, C. (eds.) TYPES 2006. LNCS, vol. 4502, pp. 48–62. Springer, Heidelberg (2007)
6. Boudet, A., Jouannaud, J.-P., Schmidt-Schauß, M.: Unification in Boolean Rings and Abelian groups. J. Symb. Comput. 8(5), 449–477 (1989)
7. Boutin, S.: Using Reflection to Build Efficient and Certified Decision Procedures. In: Ito, T., Abadi, M. (eds.) TACS 1997. LNCS, vol. 1281, pp. 515–529. Springer, Heidelberg (1997)
8. Boyer, R.S., Moore, J.S. (eds.): The Correctness Problem in Computer Science. Academic Press (1981)
9. Braibant, T., Pous, D.: Tactics for working modulo AC in Coq. Coq library (June 2010), `http://sardes.inrialpes.fr/~braibant/aac_tactics/`
10. Clavel, M., Durán, F., Eker, S., Lincoln, P., Martí-Oliet, N., Meseguer, J., Talcott, C.: The Maude 2.0 system. In: Nieuwenhuis, R. (ed.) RTA 2003. LNCS, vol. 2706, pp. 76–87. Springer, Heidelberg (2003)
11. Contejean, E.: A Certified AC Matching Algorithm. In: van Oostrom, V. (ed.) RTA 2004. LNCS, vol. 3091, pp. 70–84. Springer, Heidelberg (2004)
12. Eker, S.: Single Elementary Associative-Commutative Matching. J. Autom. Reasoning 28(1), 35–51 (2002)
13. Gonthier, G., Ziliani, B., Nanevski, A., Dreyer, D.: How to make ad hoc proof automation less ad hoc. In: Proc. ICFP, ACM (to appear, 2011)
14. Grégoire, B., Mahboubi, A.: Proving Equalities in a Commutative Ring Done Right in Coq. In: Hurd, J., Melham, T. (eds.) TPHOLs 2005. LNCS, vol. 3603, pp. 98–113. Springer, Heidelberg (2005)
15. Hullot, J.M.: Associative Commutative pattern matching. In: Proc. IJCAI, pp. 406–412. Morgan Kaufmann Publishers Inc. (1979)
16. Nguyen, Q.H., Kirchner, C., Kirchner, H.: External Rewriting for Skeptical Proof Assistants. J. Autom. Reasoning 29(3-4), 309–336 (2002)
17. Nipkow, T.: Equational reasoning in Isabelle. Sci. Comp. Prg. 12(2), 123–149 (1989)
18. Nipkow, T.: Proof transformations for equational theories. In: Proc. LICS, pp. 278–288. IEEE Computer Society (1990)
19. Peterson, G., Stickel, M.: Complete sets of reductions for some equational theories. J. ACM 28(2), 233–264 (1981)
20. Plotkin, G.: Building in equational theories. Machine Intelligence 7 (1972)
21. Slind, K.: AC Unification in HOL90. In: Joyce, J.J., Seger, C.-J.H. (eds.) HUG 1993. LNCS, vol. 780, pp. 436–449. Springer, Heidelberg (1994)
22. Sozeau, M., Oury, N.: First-Class Type Classes. In: Mohamed, O.A., Muñoz, C., Tahar, S. (eds.) TPHOLs 2008. LNCS, vol. 5170, pp. 278–293. Springer, Heidelberg (2008)

Reconstruction of Z3's Bit-Vector Proofs in HOL4 and Isabelle/HOL

Sascha Böhme[1], Anthony C.J. Fox[2], Thomas Sewell[3], and Tjark Weber[2]

[1] Fakultät für Informatik, TU München
boehmes@in.tum.de
[2] Computer Laboratory, University of Cambridge
{acjf3,tw333}@cam.ac.uk
[3] National ICT Australia
thomas.sewell@nicta.com.au

Abstract. The Satisfiability Modulo Theories (SMT) solver Z3 can generate proofs of unsatisfiability. We present independent reconstruction of unsatisfiability proofs for bit-vector theories in the theorem provers HOL4 and Isabelle/HOL. Our work shows that LCF-style proof reconstruction for the theory of fixed-size bit-vectors, although difficult because Z3's proofs provide limited detail, is often possible. We thereby obtain high correctness assurances for Z3's results, and increase the degree of proof automation for bit-vector problems in HOL4 and Isabelle/HOL.

1 Introduction

Interactive theorem provers, such as Isabelle/HOL [30] and HOL4 [21], have become powerful and trusted tools in formal verification. They typically provide rich specification logics that are suited to modelling the behaviour of complex systems. Deep theorems can be proved through user guidance. However, without the appropriate tool support, proving even simple theorems can be a tedious task when using interactive provers. Despite the merits of user guidance in proving theorems, there is a clear need for increased proof automation in interactive theorem provers.

In recent years, automated theorem provers have emerged for combinations of first-order logic with various background theories, e.g., linear arithmetic, arrays and bit-vectors. An overview of decision procedures for these domains can be found in [25]. These automated provers, called Satisfiability Modulo Theories (SMT) solvers, are of particular value in formal verification, where specifications and verification conditions can often be expressed as SMT formulas [11,7]. Interactive theorem provers can greatly benefit from the reasoning power of SMT solvers: proof obligations that are SMT formulas can be passed to the automated prover, which will solve them without further human guidance [5].

This paper focuses on the theory of bit-vectors. This is an important theory, since bit-vector problems often occur during hardware and software verification, e.g., arising from loop invariants, ranking functions, and from code/circuits that

J.-P. Jouannaud and Z. Shao (Eds.): CPP 2011, LNCS 7086, pp. 183–198, 2011.

involve machine arithmetic. Isabelle/HOL and HOL4 have internal decision procedures for solving bit-vector problems, however, their capabilities are exceeded by those of SMT solvers such as Z3 (see [34]), which is a state-of-the-art SMT solver developed by Microsoft Research, see [29]. However, there is a critical distinction in the design philosophies of these provers: interactive provers are highly conservative, placing proof soundness above efficiency/coverage, whereas SMT solvers are generally more liberal and place high emphasis upon performance. Almost every SMT solver is known to contain bugs [10]. When integrated naively, the SMT solver (and the integration) becomes part of the trusted code base: bugs could lead to inconsistent theorems in the interactive prover. For formal verification, where correctness is often paramount, this is undesirable.

This soundness problem can be solved by requiring the SMT solver to produce proofs (of unsatisfiability), and reconstructing these proofs in the interactive prover. In this paper, we present independent reconstruction of unsatisfiability proofs for bit-vector theories generated by Z3 in Isabelle/HOL and HOL4. LCF-style [20] theorem provers implement a relatively small trusted kernel (see Sect. 3), which provides a fixed set of simple inference rules. In contrast, Z3 uses a number of powerful inference rules in its proofs (see Sect. 4) and this makes proof reconstruction challenging. In this paper, we extend a previous implementation of proof reconstruction for Z3 [8] to the theory of fixed-size bit-vectors (as defined in the FIXED_SIZE_BITVECTORS theory of SMT-LIB [2]).

The motivation for our work is twofold. First, we increase proof automation in HOL4 and Isabelle/HOL by using Z3 as an automated prover back-end. Second, we obtain a high degree of confidence in Z3's results. Due to the LCF-style architecture of HOL4 and Isabelle/HOL, the trusted code base consists only of their relatively small inference kernels. In particular, there is no need to trust our (much more complex) proof checker. Any error in a proof will be uncovered during reconstruction. Thus our checker can be used to identify bugs in Z3, and to certify the status of unsatisfiable SMT-LIB benchmarks.

We describe our implementation in detail in Sect. 6. Evaluation is performed on a large number of SMT-LIB benchmarks from the QF_AUFBV, QF_BV, and QF_UFBV logics (see Sect. 7). Section 8 concludes.

2 Related Work

SMT solvers have been an active research topic for the past few years, and an integration with interactive theorem provers has been pursued by a number of researchers.

In oracle style integrations, see [14,31], the client interactive theorem prover simply trusts the SMT solver's results. While this allows for a fast and relatively simple integration, a bug in the SMT solver (or in the integration) could lead to inconsistent theorems in the interactive prover. Closer to our work are integrations that perform proof reconstruction.

McLaughlin et al. [26] describe a combination of HOL Light and CVC Lite for quantifier-free first-order logic with equality, arrays and linear real arithmetic.

Ge and Barrett [19] present the continuation of that work for CVC3 [3], the successor of CVC Lite, supporting also quantified formulas and linear integer arithmetic. CVC Lite's and CVC3's proof rules are much more detailed than the ones used by Z3. For instance, CVC3 employs more than 50 rules for the theory of real linear arithmetic alone.

Conchon et al. [12] integrated their prover Ergo with the Coq [4] interactive theorem prover. Unlike most SMT solvers, Ergo supports polymorphic first-order logic. Proof reconstruction, however, is restricted to congruence closure and linear arithmetic.

Fontaine et al. [15] describe an integration of the SMT solver haRVey with Isabelle/HOL [30]. Their work is restricted to quantifier-free first-order logic with equality and uninterpreted functions. Hurlin et al. [24] extend this approach to quantified formulas. Background theories (e.g., linear arithmetic, arrays) are not supported.

At SMT'09, Böhme [6] presented proof reconstruction for Z3 in Isabelle/HOL. Böhme and Weber [8] recently extended this to HOL4, improving both reconstruction speed and completeness (i.e., correct coverage of Z3's inference rules). Their highly optimized implementation supports uninterpreted functions with equality, quantifiers, arrays, linear integer and real arithmetic.

Common to the above approaches is their lack of support for bit-vector operations. To our knowledge, this paper is the first to address LCF-style proof reconstruction for the background theory of fixed-size bit-vectors.

3 LCF-Style Theorem Proving

The term LCF-style [20] describes theorem provers that are based on a small inference kernel. Theorems are implemented as an abstract data type, and the only way to construct new theorems is through a fixed set of functions (corresponding to the underlying logic's axiom schemata and inference rules) provided by this data type. This design greatly reduces the trusted code base. Proof procedures based on an LCF-style kernel cannot produce unsound theorems, as long as the implementation of the theorem data type is correct.

Traditionally, most LCF-style systems implement a natural deduction calculus. Theorems represent *sequents* $\Gamma \vdash \varphi$, where Γ is a finite set of *hypotheses*, and φ is the sequent's *conclusion*. Instead of $\emptyset \vdash \varphi$, we simply write $\vdash \varphi$.

The LCF-style systems that we consider here, HOL4 and Isabelle/HOL, are popular theorem provers for polymorphic higher-order logic (HOL) [21], based on the simply-typed λ-calculus. Isabelle's type system is more sophisticated than HOL4's [22], but we do not require any of the advanced features for this work.

On top of their LCF-style inference kernels, HOL4 and Isabelle/HOL offer various automated proof procedures: notably a simplifier, which performs term rewriting, a decision procedure for propositional logic, tableau- and resolution-based first-order provers, and decision procedures for Presburger arithmetic and real algebra. We particularly use a recent decision procedure for bit-vectors based on bit-blasting (see Sect. 5).

The implementation language of HOL4 and Isabelle/HOL is Standard ML [27]. To benefit from the LCF-style design of these provers and the reasoning tools built on top of their inference kernels, we must use this language to implement proof reconstruction.

Both HOL4 and Isabelle provide a primitive inference rule that performs substitution of type and term variables. Substitution is typically much faster than (re-)proving a theorem's specific instance. General theorems (which we will call *schematic*) can, therefore, play the role of efficient additional inference rules.

4 Z3: Language and Proof Terms

Succinct descriptions of Z3's language and proof terms have been given in [28,6]. We briefly review the key features, expanding on previous descriptions where necessary.

Z3's language is many-sorted first-order logic, based on the SMT-LIB language [2]. Basic sorts include Bool, Int and Real. Interpreted functions include arithmetic operators ($+$, $-$, $\cdot$), Boolean connectives ($\vee$, $\wedge$, $\neg$), constants $\top$ and $\bot$, first-order quantifiers ($\forall$, $\exists$), array operations select and store, the distinct predicate and equality. Proof reconstruction for these has been described before [8].

The present paper focuses on the theory of fixed-width bit-vectors. This adds basic sorts BitVec m for every $m > 0$, bit-vector constants like #b0, and various operations on bit-vectors: concatenation (concat), sub-vector extraction (extract), bit-wise logical operations (bvnot, bvand, bvor), arithmetic operations (bvneg, bvadd, bvmul, bvudiv, bvurem), shift operations (bvshl, bvlshr), unsigned comparison (bvult), and several derived operations. The theory is described in full detail in the `Fixed_Size_BitVectors` and `QF_BV` files[1] of SMT-LIB.

Z3's proof terms encode natural deduction proofs. The deductive system used by Z3 contains 16 axioms and inference rules.[2] These range from simple rules like **mp** (modus ponens) to rules that abbreviate complex reasoning steps. To adapt our previous implementations of proof reconstruction [8] to the theory of bit-vectors, we need to look at two rules in particular: **rewrite** for equality reasoning involving interpreted functions, and **th-lemma-bv** for arbitrary lemmas specific to the theory of bit-vectors. We discuss these in more detail in Sect. 6.

Z3's proofs are directed acyclic graphs (DAGs). Each node represents application of a single axiom or inference rule. It is labelled with the name of that axiom or inference rule and its conclusion. The edges of a proof graph connect conclusions with their premises. The hypotheses of sequents are not given explicitly. A designated root node concludes $\bot$.

[1] Available at `http://combination.cs.uiowa.edu/smtlib/logics/QF_BV.smt2`.

[2] Another 18 rules are described in the Z3 documentation, but were not exercised in any of the benchmarks used for evaluation (see Sect. 7). Although we omit these rules from our presentation, our implementations can handle them as well [8].

In 2010, version 2 of the SMT-LIB language was introduced [2]. It is worth noting that Z3's concrete syntax for proofs of SMT-LIB 2 benchmarks is vastly different from its syntax for proofs of SMT-LIB 1.2 benchmarks. While the SMT-LIB 1.2 proof syntax was line-based (with one inference or term definition per line), the SMT-LIB 2 proof syntax of Z3 resembles the SMT-LIB 2 benchmark syntax and represents proofs as S-expressions.

We have written a recursive-descent parser for (a large subset of) the SMT-LIB 2 language in Standard ML. The parser translates formulas in SMT-LIB syntax into formulas in higher-order logic.[3] We have also written a Standard ML parser for the new Z3 proof format that utilizes our SMT-LIB 2 benchmark parser internally.

At present, Z3's new proof syntax still contains a few quirks and incompatibilities with SMT-LIB 2 (e.g., different names for certain constants, missing parentheses around argument lists). We hope that these syntax issues, which currently complicate proof parsing, will be addressed in a future version of Z3.

5 Bit-Vectors in Higher-Order Logic

Isabelle/HOL's latest theory of machine words (bit-vectors) was developed by Dawson [13], and is based on constructing an isomorphic type for the finite set $\{0, \ldots, 2^n - 1\}$.

As of 2005 HOL4's theory of bit-vectors utilises Harrison's technique for modelling the n-dimensional Euclidean space in HOL Light, see [23].[4] Harrison's approach is based on considering the function space $N \rightarrow A$, where N is constrained to be finite. For bit-vectors we consider a Boolean co-domain, i.e., $A = \mathbb{B}$. Isabelle/HOL has a more advanced type system than HOL4 and HOL Light; however, they all support parametric polymorphism and this is sufficient to provide a workable theory of bit-vectors.

To aid clarity, this section will simply focus on HOL4's bit-vector library. We give an overview of this library from the end-user perspective.

Bit-vectors in HOL4 are represented by the type $\alpha\,\mathsf{word}$. For example, 8-bit words have type $8\,\mathsf{word}$, which can also be written as $\mathsf{word8}$. The numeric type 8 has exactly eight members, which gives us the required word length. The function $\mathsf{dimindex}$ returns the word length, and $\mathsf{dimword}$ returns the number of elements, e.g., $\mathsf{dimindex}(:8) = 8$ and $\mathsf{dimword}(:8) = 256$. The bit-vector library supports a broad collection of standard operations, including, but not limited to:

1. *Signed and unsigned arithmetic operations.* Examples include bit-vector negation, addition, subtraction, multiplication and less-than.
2. *Bitwise/logical operations.* Examples include complement, bitwise-and (&&), bitwise-or (!!), as well as various shifts and rotations.

---

[3] Our parser identified numerous conformance issues in SMT-LIB 2 benchmarks. We have reported these to the benchmark maintainers.

[4] Prior to this a quotient type construction was used in HOL4.

3. *Signed and unsigned casting maps.* Examples include an embedding from naturals to words ($\mathsf{n2w}$), zero-extension ($\mathsf{w2w}$), sign-extension, word extraction ($><$), and word concatenation.

Bit-vector literals are denoted with a 'w' suffix. Standard number bases are supported, for example 0xAw, 0b1010w and 10w all denote a word literal with value ten.

Importantly, all SMT-LIB bit-vector operations have corresponding definitions in the bit-vector libraries of HOL4 and Isabelle/HOL.

The word library provides a number of simplification sets (*simp-sets*) (which control the behaviour of the simplifier), *conversions* (which construct an equivalence theorem for an input term) and semi-decision procedures. These provide the building blocks for carrying out interactive proofs and for developing further (more powerful) tools. Most simp-sets primarily consist of collections of conditional rewrite rules, but they may also apply conversions, decision procedures and implement other functionality. The main bit-vector simp-set carries out basic algebraic simplification, such as associative-commutative (AC) rewriting, covering the arithmetic and bitwise operations.

Two main bit-vector semi-decision procedures are available:

- `WORD_DECIDE`. This procedure has somewhat limited coverage but it does provide a fairly quick means to discharge basic bit-vector problems. There are three stages: algebraic simplification, bit expansion for non-arithmetic operations, and finally propositional and bounds-based reasoning. By default the final stage makes use of HOL4's standard natural number decision procedure, which enables it to prove, for instance,

$$\vdash \ \forall a : \mathsf{word8}.\ a >_{+} 253\mathrm{w} \implies (a = 254\mathrm{w}) \vee (a = 255\mathrm{w})$$

  by utilizing the constraint $0 \leq n < 256$, where n is the numeric value of the bit-vector a.
- `BBLAST`. This is a semi-decision procedure that offers better coverage than `WORD_DECIDE`. However, it is still essentially propositional in nature, covering pure bit-vector problems of the form: $\forall w_1 \ldots w_n.\ P(w_1, \ldots, w_n)$ or $\exists w_1 \ldots w_n.\ P(w_1, \ldots, w_n)$. As before, the procedure starts by applying algebraic simplifications, but this time the second stage also carries out bit-expansion for addition (which in turn subsumes subtraction and the word orderings). The final stage involves calling a SAT solver. One advantage of this approach is that counterexamples can be provided when goals are invalid. The main limitations are that the procedure does not handle nested quantification (or, more generally, first-order reasoning), and goals that require non-trivial reasoning about multiplication/division. `BBLAST` is described in more detail in [17].

When carrying out interactive proofs, human guidance and additional tools (such as first-order provers) provide the means to tackle goals that are more complex than these individual semi-decision procedures can handle on their own.

6 Proof Reconstruction

Proof reconstruction for the theory of bit-vectors extends our previous work on LCF-style proof reconstruction for Z3 [8]. The general setup remains the same. We translate (negated) proof goals from HOL4 or Isabelle/HOL into SMT-LIB 2 syntax and apply Z3. If Z3 determines the negated goal to be unsatisfiable, we then parse its proof (using our parser for the SMT-LIB 2 language and its extension to the Z3 proof language, see Sect. 4) to obtain the encoded information as a value in Standard ML.[5]

Proofs are represented by a balanced tree (with lookup in $O(\log n)$ time) that maps node identifiers to proof nodes. Proof nodes are given by a disjoint union. Initially, each node contains the information that is recorded explicitly in the Z3 proof: the axiom or inference rule used at the node, the node identifiers of premises, and the rule's conclusion. Once the inference step has been checked in HOL4 or Isabelle/HOL, this information is replaced by a corresponding theorem.

To reconstruct a proof, we start at its designated root node and perform a (depth-first) post-order traversal of the proof DAG. Each node's premises are derived as theorems in HOL4 or Isabelle/HOL. Then these theorems are used to derive the node's conclusion. Ultimately, the root node's inference step, which derives $\bot$ from the root's premises, is reconstructed. We obtain a theorem that proves $\bot$ from the given assumptions, i.e., that shows unsatisfiability of the negated HOL4 or Isabelle/HOL proof goal.

Out of the 16 axioms and inference rules used by Z3, 14 perform propositional and first-order reasoning. These rules are independent of any background theory. Proof reconstruction for them has been described in [8]. It is intricate, but does not require adaptation for the theory of bit-vectors.

Only two rules—incidentally, the most complicated ones in Z3's deductive system—involve bit-vector reasoning: **rewrite** and **th-lemma-bv**. The former is used for equality reasoning about interpreted functions (including not just bit-vector operations, but also logical operators and other interpreted functions). The latter is used to derive arbitrary theory-specific lemmas. It is this rather vague specification of their semantics—and the fact that neither rule provides additional justifications, e.g., trace information—that makes proof reconstruction challenging. We now discuss our implementations of proof reconstruction for **rewrite** and **th-lemma-bv** in detail.

Schematic theorems. Matching a theorem's conclusion against a given term and, if successful, instantiating the theorem accordingly is typically much faster than deriving the specific instance from first principles. By studying the actual usage of **rewrite** in Z3's proofs, we identified about 20 useful schematic theorems for bit-vector reasoning.[6] Examples include associativity and commutativity of

[5] It is important that this round trip from higher-order logic to SMT-LIB 2 and back constitutes an identity transformation. Otherwise, proof reconstruction would derive the wrong formula.

[6] This is in addition to over 230 schematic theorems for propositional and arithmetic reasoning identified earlier [8].

bit-wise operations, e.g., $(x \mathbin{\&\&} y) \mathbin{\&\&} z = x \mathbin{\&\&} (y \mathbin{\&\&} z)$, $x \mathbin{\&\&} y = y \mathbin{\&\&} x$, neutrality of 0w for bit-wise disjunction, $0\text{w} \mathbin{!!} x = x$, and simplification rules for bit extraction, e.g., $(7 >\!\!< 0)(x : \mathsf{word8}) = x$. We store all schematic theorems in a term net to allow faster search for a match.

Schematic theorems are, in fact, our main workhorse for **rewrite**. On the benchmarks used for evaluation (see Sect. 7), **rewrite** is invoked more than 1 million times. 92.5 % of the proof obligations presented to **rewrite** are solved by instantiation of a schematic theorem.

The theory of fixed-size bit-vectors, unlike other background theories considered in earlier work [8], requires conditional schematic theorems. For instance, converting a bit-vector literal x from type $\alpha\,\mathsf{word}$ to $\beta\,\mathsf{word}$ yields essentially the same literal, provided the literal could be represented in type $\alpha\,\mathsf{word}$ in the first place:

$$\vdash x < \mathsf{dimword}(:\alpha) \implies \mathsf{w2w}\,(\mathsf{n2w}\,x : \alpha\,\mathsf{word}) = (\mathsf{n2w}\,x : \beta\,\mathsf{word}).$$

We prove these conditions by recursive instantiation of (unconditional) schematic theorems, e.g., $\vdash 1 < \mathsf{dimword}(:\alpha)$, and in many cases by simplification: terms such as $\mathsf{dimindex}(:\alpha)$ and $\mathsf{dimword}(:\alpha)$ can be evaluated for numeric types α.

We also use schematic theorems in the implementation of **th-lemma-bv**, but there the impact is much smaller. **th-lemma-bv** is called over 50 million times on the benchmarks used for evaluation, but less than 0.1% of its proof obligations are solved by instantiation. We could increase this percentage by adding more schematic theorems (at the cost of increased memory usage and start-up time), but essentially the lemmas proved by **th-lemma-bv** are more diverse and benchmark dependent than those proved by **rewrite**. For **th-lemma-bv**, schematic theorems are mostly useful to prove corner cases not covered by one of the automated decision procedures discussed below.

Theorem memoization. Isabelle/HOL and HOL4 allow instantiating free variables in a theorem, while Z3 has to re-derive theorems that differ in their uninterpreted functions. Hence, there is more potential for theorem re-use in Isabelle/HOL and HOL4 than in Z3. We exploit this by storing theorems that **rewrite** or **th-lemma-bv** prove via computationally expensive bit-vector decision procedures (see below) in a term net. Since every theorem is also stored in a proof node anyway, this increases memory requirements only slightly: namely by the memory required for the net's indexing structure.

Before invoking a decision procedure on a proof obligation, we attempt to retrieve a matching theorem from the net. This succeeds for 4.5 % of all proof obligations presented to **rewrite**, and for an impressive 99.3 % of proof obligations presented to **th-lemma-bv**. Here we see that schematic theorems and theorem memoization largely complement each other. For **rewrite**, proof obligations that occur frequently are often available as schematic theorems already. For **th-lemma-bv**, however, few proof obligations seemed sufficiently generic to be included as schematic theorems, but individual benchmarks still prove instances of the same proof obligation many times. Therefore, theorem memoization shines.

Strategy selection. Schematic theorems can only prove formulas that have a specific (anticipated) structure. Theorem memoization is successful only when a matching lemma was added to the term net earlier. Initially, bit-vector proof obligations must be proved by other means.

We rely on HOL4's and Isabelle/HOL's existing automation for bit-vector logic (see Sect. 5). Both provers provide a toolbox of semi-decision procedures for bit-vector proof obligations. Further procedures may be programmed in Standard ML. This leads to an unbounded number of proof procedures, which will typically succeed on different (not necessarily disjoint) sets of bit-vector formulas, and exhibit vastly different timing behaviours both in success and failure cases. For instance, proving $x + y = y + x$ by rewriting is trivial if commutativity of bit-vector addition is available as a rewrite rule. Proving the same theorem strictly by bit-blasting alone is possible, but may take significantly longer if the number of bits in x and y is large.

Our current implementations use only four different proof procedures for **rewrite** and **th-lemma-bv**. For **rewrite**, we first try a simplification-based approach, expressing many word operations in terms of !! (disjunction), << (left shift) and $\gg\!\!<$ (word extract), and then unfolding the definition of bit-wise operators, i.e., considering each bit position separately. This is followed by the application of arithmetic rewrites, an evaluation mechanism for ground arithmetic terms, and a decision procedure for linear arithmetic. This powerful approach solves 98 % of bit-vector goals presented to **rewrite** that are not handled by schematic theorems or memoization. The remaining 2 % are solved by a decision procedure that converts word arithmetic expressions into a canonical form. In particular, we need to fix the sign of word equalities: for instance, $-x = y \iff x + y = 0\mathrm{w}$.

For **th-lemma-bv**, we first use simplification with a relatively large set of standard rewrite rules for (arithmetic and logical) word expressions, including unfolding of bit-wise operators. Over 99 % of goals presented to **th-lemma-bv** are thereby reduced to propositional tautologies. The remaining goals are solved by bit-blasting.

This choice of proof procedures is the result of careful optimization. Starting from a set of about 10 different proof procedures, applied in a hand-chosen order, we independently optimized our implementations of **rewrite** and **th-lemma-bv** using a greedy approach: based on detailed profiling data (see Sect. 7), we modified the order in which these proof procedures were applied to try those that had the shortest average runtime (per solved goal) first. We iterated this process until the number of timeouts was not reduced any further, and tried several different initial orders to avoid local optima. Each iteration required several days of CPU time.

Clearly, more sophisticated approaches than this variant of random-restart hill climbing could be employed. If wall time is considered to be more important than CPU time, we could simply apply a number of proof procedures in parallel, taking advantage of modern multi-core architectures. We could also devise a heuristic hardness model that analyses each proof goal to predict the proof procedure that is most likely to find a proof quickly. The SATzilla solver successfully uses a similar approach to decide propositional satisfiability [35].

However, one should keep in mind that this optimization problem is ultimately caused by a lack of detail in Z3's proofs for bit-vector theorems. Rather than devoting large amounts of resources to tuning the HOL4 and Isabelle/HOL implementations of bit-blasting and other bit-vector decision procedures, it would seem more worthwhile to modify Z3 itself to print more detailed certificates for the theory of bit-vectors.

7 Experimental Results

Evaluation was performed on SMT-LIB [2] problems comprising quantifier-free (QF) first-order formulas over (combinations of) the theories of arrays (A), equality and uninterpreted functions (UF), and bit-vectors (BV). SMT-LIB logic names are formed by concatenation of the theory abbreviations given in parentheses. We evaluated our implementations on all unsatisfiable bit-vector benchmarks in SMT-LIB.[7] At the time of writing, this comprises 4974 benchmarks from three logics: QF_AUFBV, QF_BV, and QF_UFBV. These benchmarks originate from a variety of sources. They constitute a diverse and well-balanced problem set for evaluation.

We obtained all figures[8] on a 64-bit Linux system with an Intel Core i7 X920 processor, running at 2 GHz. Measurements were conducted with Z3 2.19, the latest version of Z3 at the time of writing. As underlying ML environment, we used Poly/ML 5.4.1 for both Isabelle/HOL and HOL4. For comparability with earlier work [6,8], we restricted proof search to two minutes and proof reconstruction to five minutes, and limited memory usage for both steps to 4 GB. All measured times are CPU times (with garbage collection in Poly/ML excluded).

Beyond measuring success rates and runtimes of proof reconstruction, we also measured the performance of HOL4 bit-blasting for comparison, and we provide profiling data to give a deeper insight into our results. For space reasons, we do not show Isabelle/HOL results in detail, but they are roughly similar to the HOL4 results discussed below.

7.1 Proof Generation with Z3

Table 1 shows the results obtained from applying Z3 2.19 to all unsatisfiable bit-vector benchmarks in SMT-LIB. For every SMT-LIB logic, we show the number of benchmarks and the average benchmark size. We then measured the number of errors (e.g., segmentation faults), timeouts, and proofs generated by Z3. We also show the average solving time (excluding errors and timeouts), and the average size of generated proofs. We invoked Z3 with option `PROOF_MODE=2`, which enables proof generation.

---
[7] These benchmarks were obtained from `http://smtexec.org/exec/smtlib2_benchmarks.php` on June 13, 2011, using the query `logic~BV & status=unsat`.

[8] Our data is available at `http://www.cl.cam.ac.uk/~tw333/bit-vectors/`.

Proofs are typically much larger than the original SMT-LIB benchmark—almost 53 times as large on average. The total size of generated proofs is 34.9 GB, and the total CPU time for Z3 on all benchmarks (including errors and timeouts) is around 29.5 hours.

Table 1. Experimental results (Z3 2.19) for selected SMT-LIB logics

Logic	Benchmarks		Errors		Timeouts		Proofs			
	#	Size (avg)	#	Ratio	#	Ratio	#	Ratio	Time (avg)	Size (avg)
QF_AUFBV	3566	93 kB	0	0.0 %	118	3.3 %	3448	96.7 %	0.6 s	1.2 MB
QF_BV	1377	322 kB	12	0.9 %	630	45.8 %	735	53.4 %	17.3 s	41.1 MB
QF_UFBV	31	343 kB	0	0.0 %	15	48.4 %	16	51.6 %	37.1 s	29.2 MB
Total	4974	158 kB	12	0.2 %	763	15.3 %	4202	84.5 %	4.2 s	8.3 MB

7.2 Bit-Blasting in HOL4

Next, we show the results of bit-blasting in HOL4 for comparison. We used our SMT-LIB 2 parser (see Sect. 4) to translate benchmarks into higher-order logic. We then applied HOL4's `BBLAST` tactic to the same set of SMT-LIB benchmarks previously presented to Z3 (i.e., the number and size of benchmarks in Tab. 2 is the same as before). We used a timeout of five minutes per benchmark. Similar to before, we show the number of errors, timeouts, and proofs found by `BBLAST`, as well as the average solving time (excluding errors and timeouts). Every inference performed by `BBLAST` is checked by HOL4's inference kernel, but no persistent proof objects are generated. Therefore, there is no column for proof size in Tab. 2.

Table 2. Experimental results (`BBLAST`) for selected SMT-LIB logics

Logic	Benchmarks		Errors		Timeouts		Proofs		
	#	Size (avg)	#	Ratio	#	Ratio	#	Ratio	Time (avg)
QF_AUFBV	3566	93 kB	1089	30.5 %	474	13.3 %	2003	56.2 %	26.2 s
QF_BV	1377	322 kB	745	54.1 %	504	36.6 %	128	9.3 %	56.2 s
QF_UFBV	31	343 kB	31	100.0 %	0	0.0 %	0	0.0 %	—
Total	4974	158 kB	1865	37.5 %	978	19.7 %	2131	42.8 %	28.0 s

Errors mostly indicate that `BBLAST` gave up on the benchmark. To prove unsatisfiability, many benchmarks require combinations of bit-blasting and equality reasoning (e.g., congruence closure), which `BBLAST` is not capable of, or reasoning about specific bit-vector operations in ways not supported by `BBLAST`. Our results, therefore, show that Z3 is not only much faster than `BBLAST`, but also that it can solve a wider range of problems.

7.3 Proof Reconstruction in HOL4

We checked all proofs generated by Z3 in the HOL4 theorem prover, using a timeout of five minutes per benchmark. Table 3 shows our results. We present the number of errors, timeouts (including out-of-memory results), and successfully checked proofs, along with average HOL4 runtime for the latter. We also show total HOL4 runtime (including errors and timeouts) for each logic.

Errors are caused by unsound inferences in proofs, and in many cases by bugs in Z3's proof pretty-printer,[9] but also by shortcomings in our implementation of proof reconstruction, which fails on some corner cases.

Table 3. Experimental results (HOL4 proof reconstruction) for Z3's proofs

Logic	Proofs	Errors		Timeouts		Success			Overall time (approx)
	#	#	Ratio	#	Ratio	#	Ratio	Time (avg)	
QF_AUFBV	3448	587	17.0 %	54	1.6 %	2807	81.4 %	1.4 s	5.4 hrs
QF_BV	735	96	13.1 %	356	48.4 %	283	38.5 %	18.8 s	31.0 hrs
QF_UFBV	16	0	0.0 %	16	100.0 %	0	0.0 %	—	1.2 hrs
Total	4202	683	16.3 %	426	10.1 %	3090	73.5 %	2.6 s	37.6 hrs

Although HOL4 achieves an overall success rate of 73.5 %, we see that this rate varies significantly with the SMT-LIB logic. QF_AUFBV contains a large number of relatively easy benchmarks, which can be solved quickly by Z3, have small proofs, and consequently can (in most cases) be checked successfully in HOL4. Table 1 indicates that QF_BV and QF_UFBV contain significantly harder problems. This is reflected by the performance of HOL4 on these logics, which can check a mere 38.5 % of benchmarks in QF_BV within the given time limit, and times out for all 16 proofs in QF_UFBV. However, proof reconstruction is more than an order of magnitude faster than BBLAST, and can solve 1.5 times as many SMT-LIB problems. Proof generation with Z3 is typically one to two orders of magnitude faster than proof reconstruction in HOL4.

7.4 Profiling

To further understand these results and to identify potential for future optimization, we present relevant profiling data for our HOL4 implementation. (Isabelle/HOL profiling data is roughly similar.) Figures 1 to 3 show bar graphs that indicate the percentage shares of total runtime (dark bars) for **rewrite**, **th-lemma-bv**, and Z3's other proof rules. Additionally, time spent on parsing proof files is shown as well (see Tab. 1 for average proof sizes). We contrast each proof rule's relative runtime with the mean frequency of that rule (light bars).

[9] We have notified the Z3 authors of the problems that we found. Additionally, we corrected some obvious syntax errors in proofs, e.g., unbalanced parentheses.

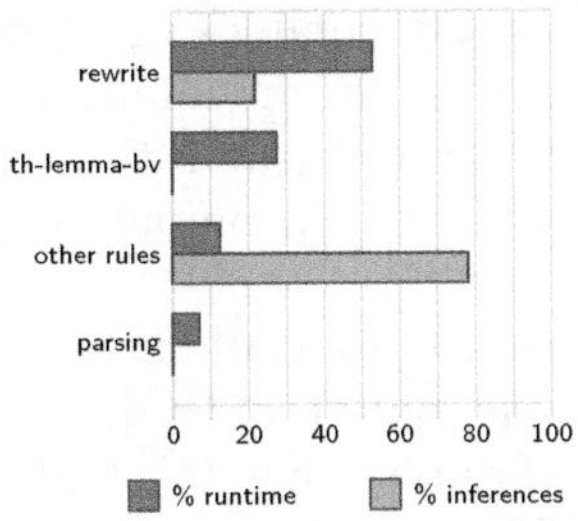

Fig. 1. QF_AUFBV

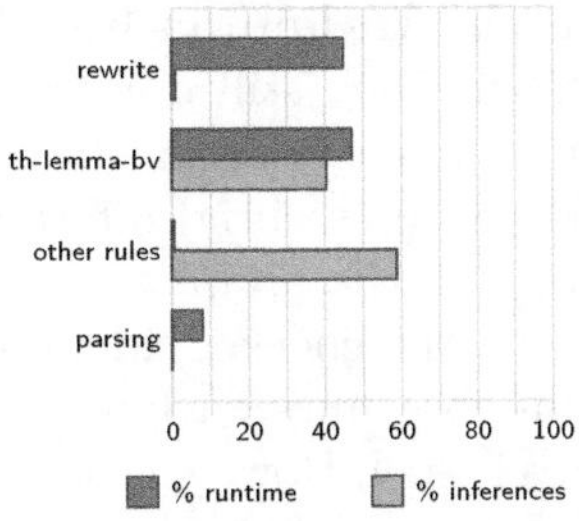

Fig. 2. QF_BV

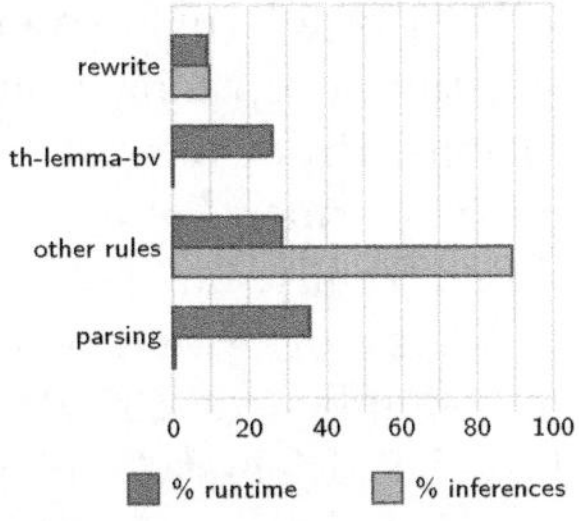

Fig. 3. QF_UFBV

We see that despite extensive optimization, proof reconstruction times are still dominated by **rewrite** and **th-lemma-bv**. Although less than 1% of all inferences in QF_AUFBV and QF_UFBV are applications of **th-lemma-bv**, checking these consumes over 26% of runtime. Even more extremely, **rewrite** in QF_BV accounts for less than 1% of inferences, but almost 45% of proof reconstruction time. In contrast, all other rules combined constitute the majority of proof inferences (between 59% for QF_BV and 89% for QF_UFBV), but they can be checked much more quickly: in 29% (for QF_UFBV) or less of total runtime.

Proof parsing takes less than 8% of total runtime for QF_AUFBV and QF_BV, but 36% for QF_UFBV. It times out on the largest proofs. Proofs for QF_BV are larger than proofs for QF_UFBV on average (see Tab. 1), but QF_BV contains many small proofs that can be parsed relatively quickly. The variation in proof size is much smaller for QF_UFBV. Median proof sizes are 3.7 MB for QF_BV and 22.5 MB for QF_UFBV, respectively.

8 Conclusions

Bit-vectors play an important role in hardware and software verification. They naturally show up in the verification of, e.g., 32- and 64-bit architectures and machine data types [18]. In this paper, we have extended a previous implementation of LCF-style proof reconstruction for Z3 [8] to the theory of fixed-size bit-vectors. To our knowledge, this paper is the first to consider independent checking of SMT solver proofs for bit-vector theories.

Even though Z3's proofs provide little detail about theory-specific reasoning, our experimental results (Sect. 7) show that LCF-style proof reconstruction for the theory of fixed-size bit-vectors is often possible. We have achieved an overall success rate of 73.5% on SMT-LIB benchmarks. We thereby obtain high correctness assurances for Z3's results. Checking Z3's proofs also significantly increases the degree of proof automation for bit-vector problems in HOL4 and Isabelle/HOL. Proof reconstruction is more powerful in scope and performance than built-in decision procedures, such as `BBLAST`, previously offered by these provers. Our implementations are freely available[10] and already in use [5].

[10] See `http://hol.sourceforge.net` and `http://isabelle.in.tum.de`.

Z3's proof rules **rewrite** and **th-lemma-bv** seem overly complex. Despite substantial optimization efforts, they still dominate runtime in our implementations. Proof reconstruction currently needs to duplicate proof search that Z3 has performed before, to re-obtain essential information that was computed by Z3 internally, but not included in the proof.

More work could be done on the checker side: for instance, we could attempt to re-implement Z3's decision procedure for bit-vectors [34] on top of HOL4's or Isabelle's LCF-style inference kernel. However, instead of duplicating Z3's highly tuned decision procedures in our proof checker, it would seem more sensible to modify Z3's proof format to include all relevant information [9].

Unfortunately, we could not do this ourselves because Z3 is closed source. We again [8] encourage the Z3 authors to (1) replace **rewrite** by a collection of simpler rules with clear semantics and less reconstruction effort, ideally covering specific rewriting steps of at most one theory, and (2) enrich **th-lemma-bv** with additional easily-checkable certificates or trace information guiding refutations to avoid invocations of expensive decision procedures (e.g., bit-blasting) in the checker.

Based on previous experience [32] we are confident that the techniques presented in this paper can be used to achieve similar performance for bit-vector reasoning in other LCF-style theorem provers for higher-order logic.

Future work should aim for improved reconstruction coverage (i.e., fewer errors) and improved performance, possibly after Z3's proof format has been modified as suggested above. We also intend to evaluate proof reconstruction for typical goals of Isabelle/HOL or HOL4; to implement parallel proof reconstruction [33], by checking independent paths in the proof DAG concurrently; and to investigate proof compression [1,16] for SMT proofs.

Acknowledgments. This research was partially funded by EPSRC grant EP/F067909/1. NICTA is funded by the Australian Government as represented by the Department of Broadband, Communications and the Digital Economy and the Australian Research Council through the ICT Centre of Excellence program. The authors are grateful to Nikolaj Bjørner and Leonardo de Moura for their help with Z3.

References

1. Amjad, H.: Data compression for proof replay. Journal of Automated Reasoning 41(3–4), 193–218 (2008)
2. Barrett, C., Stump, A., Tinelli, C.: The SMT-LIB Standard: Version 2.0. In: Gupta, A., Kroening, D. (eds.) Proceedings of the 8th International Workshop on Satisfiability Modulo Theories, Edinburgh, England (2010)
3. Barrett, C.W., Tinelli, C.: CVC3. In: Damm, W., Hermanns, H. (eds.) CAV 2007. LNCS, vol. 4590, pp. 298–302. Springer, Heidelberg (2007)
4. Bertot, Y.: A Short Presentation of Coq. In: Mohamed, O.A., Muñoz, C., Tahar, S. (eds.) TPHOLs 2008. LNCS, vol. 5170, pp. 12–16. Springer, Heidelberg (2008)

5. Blanchette, J.C., Böhme, S., Paulson, L.C.: Extending Sledgehammer with SMT Solvers. In: Bjørner, N., Sofronie-Stokkermans, V. (eds.) CADE 2011. LNCS, vol. 6803, pp. 116–130. Springer, Heidelberg (2011)
6. Böhme, S.: Proof reconstruction for Z3 in Isabelle/HOL. In: 7th International Workshop on Satisfiability Modulo Theories, SMT 2009 (2009)
7. Böhme, S., Moskal, M., Schulte, W., Wolff, B.: HOL-Boogie — An Interactive Prover-Backend for the Verifying C Compiler. Journal of Automated Reasoning 44(1–2), 111–114 (2010)
8. Böhme, S., Weber, T.: Fast LCF-Style Proof Reconstruction for Z3. In: Kaufmann, M., Paulson, L.C. (eds.) ITP 2010. LNCS, vol. 6172, pp. 179–194. Springer, Heidelberg (2010)
9. Böhme, S., Weber, T.: Designing proof formats: A user's perspective. In: First Workshop on Proof Exchange for Theorem Proving (to appear, 2011)
10. Brummayer, R., Biere, A.: Fuzzing and delta-debugging SMT solvers. In: 7th International Workshop on Satisfiability Modulo Theories, SMT 2009 (2009)
11. Collavizza, H., Gordon, M.: Integration of theorem-proving and constraint programming for software verification. Tech. rep., Laboratoire d'Informatique, Signaux et Systèmes de Sophia-Antipolis (2008)
12. Conchon, S., Contejean, E., Kanig, J., Lescuyer, S.: Lightweight integration of the Ergo theorem prover inside a proof assistant. In: AFM 2007: Proceedings of the Second Workshop on Automated Formal Methods, pp. 55–59. ACM Press (2007)
13. Dawson, J.: Isabelle theories for machine words. Electronic Notes in Theoretical Computer Science 250(1), 55–70 (2009); Proceedings of the Seventh International Workshop on Automated Verification of Critical Systems (AVoCS 2007)
14. Erkök, L., Matthews, J.: Using Yices as an automated solver in Isabelle/HOL. In: AFM 2008: Proceedings of the Third Workshop on Automated Formal Methods, pp. 3–13. ACM Press (2008)
15. Fontaine, P., Marion, J.-Y., Merz, S., Nieto, L.P., Tiu, A.F.: Expressiveness + Automation + Soundness: Towards Combining SMT Solvers and Interactive Proof Assistants. In: Hermanns, H. (ed.) TACAS 2006. LNCS, vol. 3920, pp. 167–181. Springer, Heidelberg (2006)
16. Fontaine, P., Merz, S., Woltzenlogel Paleo, B.: Compression of Propositional Resolution Proofs Via Partial Regularization. In: Bjørner, N., Sofronie-Stokkermans, V. (eds.) CADE 2011. LNCS, vol. 6803, pp. 237–251. Springer, Heidelberg (2011)
17. Fox, A.C.J.: LCF-Style Bit-Blasting in HOL4. In: van Eekelen, M., Geuvers, H., Schmaltz, J., Wiedijk, F. (eds.) ITP 2011. LNCS, vol. 6898, pp. 357–362. Springer, Heidelberg (2011)
18. Fox, A.C.J., Gordon, M.J.C., Myreen, M.O.: Specification and verification of ARM hardware and software. In: Hardin, D.S. (ed.) Design and Verification of Microprocessor Systems for High-Assurance Applications, pp. 221–248. Springer, Heidelberg (2010)
19. Ge, Y., Barrett, C.: Proof translation and SMT-LIB benchmark certification: A preliminary report. In: 6th International Workshop on Satisfiability Modulo Theories, SMT 2008 (2008)
20. Gordon, M., Wadsworth, C.P., Milner, R.: Edinburgh LCF. LNCS, vol. 78. Springer, Heidelberg (1979)
21. Gordon, M.J.C., Pitts, A.M.: The HOL logic and system. In: Towards Verified Systems. Real-Time Safety Critical Systems Series, vol. 2, ch. 3, pp. 49–70. Elsevier (1994)

22. Haftmann, F., Wenzel, M.: Constructive Type Classes in Isabelle. In: Altenkirch, T., McBride, C. (eds.) TYPES 2006. LNCS, vol. 4502, pp. 160–174. Springer, Heidelberg (2007)
23. Harrison, J.: A HOL Theory of Euclidean Space. In: Hurd, J., Melham, T.F. (eds.) TPHOLs 2005. LNCS, vol. 3603, pp. 114–129. Springer, Heidelberg (2005)
24. Hurlin, C., Chaieb, A., Fontaine, P., Merz, S., Weber, T.: Practical proof reconstruction for first-order logic and set-theoretical constructions. In: Proceedings of the Isabelle Workshop 2007, Bremen, Germany, pp. 2–13 (July 2007)
25. Kroening, D., Strichman, O.: Decision Procedures – An Algorithmic Point of View. Springer, Heidelberg (2008)
26. McLaughlin, S., Barrett, C., Ge, Y.: Cooperating theorem provers: A case study combining HOL-Light and CVC Lite. Electronic Notes in Theoretical Computer Science 144(2), 43–51 (2006)
27. Milner, R., Tofte, M., Harper, R., MacQueen, D.: The Definition of Standard ML–Revised. MIT Press (1997)
28. de Moura, L.M., Bjørner, N.: Proofs and refutations, and Z3. In: Proceedings of the LPAR 2008 Workshops, Knowledge Exchange: Automated Provers and Proof Assistants, and the 7th International Workshop on the Implementation of Logics. CEUR Workshop Proceedings, vol. 418, CEUR-WS.org (2008)
29. de Moura, L., Bjørner, N.S.: Z3: An Efficient SMT Solver. In: Ramakrishnan, C.R., Rehof, J. (eds.) TACAS 2008. LNCS, vol. 4963, pp. 337–340. Springer, Heidelberg (2008)
30. Nipkow, T., Paulson, L.C., Wenzel, M.T.: Isabelle/HOL. LNCS, vol. 2283. Springer, Heidelberg (2002)
31. Weber, T.: SMT solvers: New oracles for the HOL theorem prover. International Journal on Software Tools for Technology Transfer (to appear, 2011)
32. Weber, T., Amjad, H.: Efficiently checking propositional refutations in HOL theorem provers. Journal of Applied Logic 7(1), 26–40 (2009)
33. Wenzel, M.: Parallel proof checking in Isabelle/Isar. In: ACM SIGSAM 2009 International Workshop on Programming Languages for Mechanized Mathematics Systems (2009)
34. Wintersteiger, C.M., Hamadi, Y., de Moura, L.M.: Efficiently solving quantified bit-vector formulas. In: Bloem, R., Sharygina, N. (eds.) Proceedings of the 10th International Conference on Formal Methods in Computer-Aided Design, FMCAD 2010, Lugano, Switzerland, October 20-23, pp. 239–246. IEEE (2010)
35. Xu, L., Hutter, F., Hoos, H.H., Leyton-Brown, K.: SATzilla: Portfolio-based algorithm selection for SAT. J. Artif. Intell. Res (JAIR) 32, 565–606 (2008)

Teaching Experience: Logic and Formal Methods with Coq

Martin Henz and Aquinas Hobor*

National University of Singapore

Abstract. During the past three years we have been integrating mechanized theorem proving into a traditional introductory course on formal methods. We explain our goals for adding mechanized provers to the course, and illustrate how we have integrated the provers into our syllabus to meet those goals. We also document some of the teaching materials we have developed for the course to date, and what our experiences have been like.

1 Introduction

National University of Singapore's School of Computing teaches introductory formal methods as CS3234 (undergraduate) and CS5209 (graduate). In 2007 and 2008 the first author taught CS3234 using a traditional approach with the standard undergraduate textbooks *Mathematical Logic for Computer Science* [BA01] and *Logic in Computer Science* [HR00]. Sad to say, the results were equally "traditional":

1. The module was "hard" in the eyes the students due to the necessity of understanding of an unusual number concepts on several abstraction levels.
2. Students viewed formal systems as a subject far removed from useful applications.
3. Weaker students often found exercises and tutorials unusually "dry" and "boring".

The first point made for a steep learning curve, the second decreased the motivation of students to climb the curve, and the third posed further obstacles for those students who have enough motivation to even try. In short, there was clear room for improvement.

When the second author joined the team we proceeded to address these problems (after acknowledging the first one as only partially solvable). The goal was to shorten the gap between theory and practice by providing relevant and appealing applications and to implement a "hands-on" approach by introducing adequate didactic tools.

Several tools are popularly used to teach formal systems in computer science, including logic programming systems, model checkers, and SAT solvers. We found it difficult to justify the learning overhead that these tools require given that they are often only used for one or two sections of a module. Ideally, the same tool would be used *throughout* the module, reducing overhead to a minimum and allowing for more sophisticated use as the course progressed into more complex territory.

We determined to use the proof assistant Coq. While not having been developed specifically for didactic use, Coq's basic concepts have proved to be sufficiently easy for third year undergraduates (and even, sometimes, for graduate students). Initial results have been encouraging: the interactive discovery of proofs using Coq provided a useful

* Supported by a Lee Kuan Yew Postdoctoral Fellowship.

J.-P. Jouannaud and Z. Shao (Eds.): CPP 2011, LNCS 7086, pp. 199–215, 2011.

reinforcement of the conceptual material, and we have been successful in integrating the theorem prover into almost every part of the course. We feel that the consistent use of Coq has improved the students' comprehension of formal logic, and the quality of their proofs, as long as they are willing to invest the time for learning Coq.

Remainder of paper. We next go through our course syllabus, focusing for each topic on how we have added mechanized proving to a more traditional curriculum. We then describe the course format (*e.g.*, the number of assignments, weighting of various components in the final grade) and explain its rationale. We conclude with a discussion of the feedback we have received from our students and our own experiences.

Associated material. We developed a substantial amount of material (hundreds of pages) as part of modifying this course, including slides, lecture notes, homework assignments (both paper and Coq), laboratory exercises, Coq quizzes, and exams [HH10].[1] For much of the course this material was the primary reference for the students. When appropriate in what follows we shall provide pointers into specific parts of this material; readers are kindly reminded that this supplementary material is drawn from several iterations of the same course and is very much a work in progress. We eventually hope to package this material into some kind of book.

2 Syllabus

Orientation. The National University of Singapore (NUS) follows a relatively short 13-week semester. After week 13, students have a reading period before exams. In recent years, CS3234 has had between 30 and 37 students, with an unusually large number drawn from the strongest undergraduate students in the School of Computing. In contrast, CS5209 often has more than 50 students, largely because one of the qualifying exams (required to proceed with the PhD program) covers formal methods.

2.1 Traditional Logic: Weeks 1 and 2

Motivation. Usually, courses in formal methods in computer science start with propositional logic because it is the simplest modern formal logical system. The challenge is that students are presented very early with substantially new concepts on two levels.

The conceptual level: the distinction of syntax and semantics, what constitutes a proof, proof theory (natural deduction), and semantic arguments (models).

The logic-specific level: Propositional formulas as elements of an inductively defined set (or of a context-free grammar), introduction and elimination rules for propositional logic, and a valuation-based semantics (truth tables).

We found it desirable to pursue a gentler approach at the beginning of the course, aiming for a shallower learning curve. The idea is to start with a logic framework that enjoys very simple syntax, semantics and formal reasoning techniques, allowing the students to focus on and properly digest the conceptual components. This approach will also give us the opportunity to introduce the nuts and bolts of Coq gently.

[1] Note: Readers interested in seeing the solutions to the assignments and exams should contact us directly.

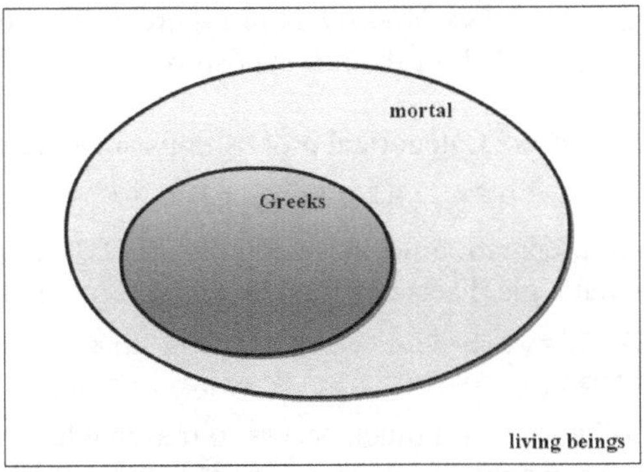

Fig. 1. Venn diagram for "All Greeks are mortal"

We believe that Aristotle's term logic [PH91] is appropriate for this purpose. Among several possible encodings of term logic in Coq, we chose a version that combines simplicity with consistency of presentation, compared to the next logic, propositional logic. Students get familiar with basic Coq concepts like definitions, proofs, and tactics, without getting bogged down in complex inductively defined syntax and semantics.

Basic components of term logic. The atomic unit of meaning in term logic are *categorical terms*, *e.g.*, `humans`, `Greeks`, and `mortal`. We encode this in Coq as follows:

```
Parameter Term : Type.
Parameters Greeks humans mortal : Term.
```

A *categorical proposition* then puts two such terms together as in the famous universal proposition "all Greeks are humans." Besides the quantity "universal" (all), we provide for "particular" (some) propositions, and beside the quality "affirmative", we provide for "negative" propositions, leading the the following definitions in Coq:

```
Structure Quantity : Type := universal | particular.
Structure Quality : Type := affirmative | negative.
```

Data structures of type `CategoricalProposition` are then constructed from a `Quantity`, a `Quality`, a subject `Term` and an object `Term`.

```
Record CategoricalProposition : Type := cp {
  quantity : Quantity;
  quality : Quality;
  subject : Term;
  object : Term }.
```

An appropriate Coq `Notation` enables the students to (most of the time) write propositions as natural English sentences such as `All Greeks are humans`.

Semantics from naïve set theory. A model $\mathcal{M}$ for a term logic can be given by providing a universe of objects $U^{\mathcal{M}}$, and a subset (or unary predicate) $t^{\mathcal{M}} \subseteq U^{\mathcal{M}}$, for each term t. The semantics of a universal proposition is then given by

$$(\texttt{All}\ \mathit{subject}\ \texttt{are}\ \mathit{object})^{\mathcal{M}} = \begin{cases} T & \text{if } \mathit{subject}^{\mathcal{M}} \subseteq \mathit{object}^{\mathcal{M}}, \\ F & \text{otherwise} \end{cases}$$

and can be visualized by a *Venn diagram* as in Figure 1. The reader can see [HH10, `notes/Traditional.pdf`] for the full exposition.

Introducing logical concepts. Categorical propositions are lifted into `Prop` using

```
Parameter holds : CategoricalProposition -> Prop.
```

consistent with and in preparation for more complex `holds` predicates introduced in propositional and modal logic. Facts can then be introduced interactively, as in:

```
Axiom HumansMortality: holds (All humans are mortal).
Axiom GreeksHumanity: holds (All Greeks are humans).
```

Such factual "axioms" allow for a quick access to reasonable examples. However, the rigidity of this style of reasoning does not escape the attentive students.

A graphical notation for axioms prepares the ground for natural deduction:

$$\frac{}{\texttt{All humans are mortal}}\,[\text{HumansMortality}]$$

A more interesting axiom—traditionally called B<u>a</u>rb<u>a</u>r<u>a</u> as a mnemonic device—expresses transitivity of the subset relation:

$$\frac{\texttt{All}\ \mathit{middle}\ \texttt{are}\ \mathit{major} \qquad \texttt{All}\ \mathit{minor}\ \texttt{are}\ \mathit{middle}}{\texttt{All}\ \mathit{minor}\ \texttt{are}\ \mathit{major}}\,[\text{Barbara}]$$

Its representation in Coq introduces conjunction and implication at the meta-level.

```
Axiom Barbara : forall major minor middle,
    holds (All middle are major) /\ holds (All minor are middle)
 -> holds (All minor are major).
```

Basic tactics such as `split` can be observed in action in this proof of Greek mortality:

```
Lemma GreeksMortality : holds (All Greeks are mortal).
Proof.
  apply Barbara with (middle := humans).
  split.
  apply HumansMortality.
  apply GreeksHumanity.
Qed.
```

Interactive proof sessions. Equipped with the basic reasoning techniques of traditional logic, students can now proceed to more complex proofs. An attractive realm of "applications" are Lewis Carroll's logical puzzles. For example, from the following premises

- No ducks waltz.
- No officers ever decline to waltz.
- All my poultry are ducks.

we should be able to conclude, naturally enough, that no officers are my poultry. After defining appropriate terms such as `things_that_waltz` and a complement constructor for negative terms (`non`), we can define the corresponding lemma in Coq:

```
Lemma No_Officers_Are_My_Poulty :
   holds (No ducks are things_that_waltz) /\
   holds (No officers are non things_that_waltz) /\
   holds (All my_poultry are ducks)
   ->
   holds (No officers are my_poultry).
```

The proof uses tactics that implement traditional Aristotelian reasoning techniques such as obversion and contraposition [Bor06]; the interested reader is referred to [HH10, `notes/Traditional.pdf`] for details on their implementation in Coq. Attentive students now realize that in the proof, assumptions play the role of earlier factual axioms, but have the advantage of being localized to the proof.

We are able to cover the basics of Aristotelian term logic in a week and a half (the first half week being reserved for standard course introductory material such as statements on course policy). Afterwards, the students have achieved a basic understanding of the syntax/semantics distinction, models, axioms, lemmas, proofs, and tactics (all of which of course to be repeatedly reinforced throughout the course), and are thus ready to focus on the logic-specific aspects of propositional logic.

Later in the course—equipped with predicate logic—students can go back to their first logic, represent categorical terms by unary predicates, and prove Aristotle's axioms such as Barbara as lemmas in an easy exercise:

```
Lemma BarbaraInPred: forall (major minor middle: term -> Prop),
  forall x, ((middle(x) -> major(x)) /\ (minor(x) -> middle(x)))
            ->
            (minor(x) -> major(x)).
```

2.2 Propositional Logic: Weeks 3 and 4

Prelude: rule-defined sets as data structures. First, we have to give some kind of intuition for what an inductive set is, so that we can define the syntax of the logic. However, we would prefer to defer formal discussion of such sets and their associated proof rules (*i.e.*, induction) until after we cover predicate logic in week 8 (§2.5).

We have discovered that the simplest way to give this intuition is to take advantage of the fact that we are teaching computer science majors, and make the connection between a (simple) inductive type and a (simple) data structure, such as a linked list. We provide some simple Java code that uses provides poor man's versions of the natural numbers (with `Zero` and `Succ(...)`) and binary trees; we then demonstrate the corresponding Coq code for these types as well [HH10, `notes/Induction.pdf`]. This would be simpler, of course, if our students knew ML, but we do not have that luxury. In practice demonstrating the idea in Java provides some intuition and reinforces the idea that logical formulas have a well-defined structure; in addition we can use the Java code to start to address interesting questions, *e.g.*, about cyclic data structures.

Encoding as an object logic. Our presentation of paper-based propositional logic is entirely standard: we give the syntax of formulas, define the semantics (*e.g.*, valuations, truth tables), give the natural deduction rules, and cover soundness/completeness. One small departure from the norm is that we bring up the idea of intuitionistic logic quite early, and informally explain its connection to computability.

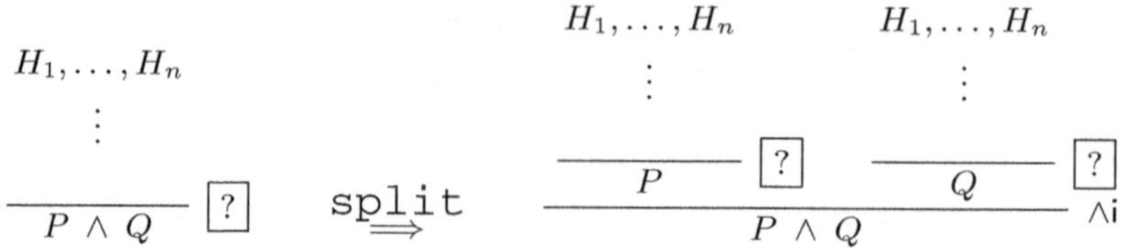

Fig. 2. A diagram that explains the `split` tactic

More interesting is how we cover the topic in Coq. Because we already introduced some basic Coq in week 2 (§2.1), we have the advantage that some basic Coq concepts and tactics (e.g., implication and `intros`) are already sinking in. To reinforce that idea, and to keep the concepts of object- and metalogic strictly separate, we first cover propositional logic as an object logic and hew closely to how we defined it on paper. That is, we inductively define the syntax of formulas, introduce the idea of a valuation, and then define an evaluator fixpoint that takes a formula and a valuation and produces a boolean value according to the standard truth table rules [HH10, `notes/Propositional_Logic.pdf`].

We then provide a module type that gives the various natural deduction rules, and assign two kinds of homework: first, we require that they use those rules to prove various standard problems in propositional logic using machine-checked natural deduction. One big advantage of the object-logic encoding is that they must use axioms explicitly (*e.g.* `apply Conj_I.`) instead of the typical Coq tactics (`split.`). We have found that the built-in tactics do a bit too much (*e.g.*, many overloadings for the `destruct` tactic), and by explicitly requiring named axioms we can match in Coq the exact shape of paper-based natural deduction proofs. For the second part of the homework, we require that they implement a module matching that module type, thereby proving the soundness of the rules [HH10, `coq/homework_02.v`]. For natural deduction, we encourage cross-pollination by assigning some of the same problems in both the paper portion of the homework and in the Coq portion.

Switching between the object logic and the meta logic. Once students have a handle on propositional logic, and have gotten a bit more of a feel for Coq, it is time to ask an obvious question: why are we defining our own version of conjunction, when Coq already provides it? In fact, it is extremely convenient to use Coq's built-in operators, since it greatly enhances automation possibilities[2]. We give some problems to ensure that students are familiar with the basic tactics; by this point, for most these are quite simple [HH10, `coq/Propositional_Logic_Lab2.v`].

Explaining what Coq is (approximately) doing. Students tend to be mystified by what Coq is exactly doing. This leads to undue "hacking", wherein students try tactics at random until for some unknown reason they hit on the right combination. After we have introduced propositional logic and the students have done some Coq homework, we try to explain what is going on by a series of diagrams like the one in Figure 2.

[2] And when we get to predicate logic the ability to offload binders onto Coq is a godsend.

This kind of diagram shows a proof state transformation with the pre-state to the left and the post-state to the right. Here we show the transformation for the `split` tactic; the goal of the pre-state, appropriately enough, is a conjunction $P \wedge Q$. We have a series of hypotheses $H_1, \ldots, H_n$, but Coq is not sure how to proceed from them to the goal; we symbolize this by labeling the rule with a question mark (boxed for emphasis). The `split` tactic tells Coq that the two parts will be proven independently from our hypotheses; accordingly, afterwards, Coq presents us with two fresh goals: P and Q, and again asks us how to proceed. Coq has inserted the conjunction introduction axiom ($\wedge$i) to connect those goals into a proof of the conjunction.

We have found that students understand the tactics much more clearly after we demonstrate the transformations they perform by using these kinds of diagrams. As an aside, one time we ran the course, we provided a series of tactics that followed the axioms of propositional logic a bit more clearly (*e.g.*, we defined a tactic `disj_e` that was identical to `destruct`). This turned out to be a bad idea: not only did it limit students' ability to look up documentation online, but it meant that they were not really able to use Coq in a standard style after completing the course.

Case study: tournament scheduling. Too often, formal systems appear to have little practical application. To combat this perception, we like to conclude our discussion of propositional logic with an example of using propositional logic to solve a computational problem, via its encoding as a propositional satisfiability problem: Hantao Zhang's encoding [Zha02] of the Atlantic Coast Conference 1997/98 benchmark [NT98], as a propositional formula. The fully automated proof of its satisfiability using the satisfiability checker SATO [Zha93] yields the solutions to the benchmark problem orders of magnitudes faster than techniques from operations research.

2.3 Predicate Logic: Weeks 5 and 6

Just as in the case for propositional logic, our presentation of pen-and-paper predicate logic is largely standard: syntax, semantics, proof rules, metatheory. We have found that one place where our pen-and-paper explanation is aided by the use of the theorem prover is in substitution. It is quite simple to create some formulas in Coq and then use the `rewrite` tactic to substitute equalities, observing how Coq manages the binders.

Since we have already made the distinction between object- and metalogics, we take full advantage of Coq's binder management. That is, while we carefully define substitution for paper methods, and demonstrate how Coq handles the situation as explained above, we entirely avoid defining any mechanized substitution methods ourselves[3]. Among other advantages, this allows us to completely sidestep the quicksand of computable equality testing, which would be needed to define substitution in Coq.

Most of the tactics for predicate logic in Coq (`exists`, `destruct`, and `intros`) are fairly simple; one exception is the tactic for universal elimination (`generalize`), which is a little bit weird (why has my goal just changed?). Although usually we prefer to just teach the Coq tactics as-is, in this case we define a custom tactic that does a universal elimination by combining a `generalize` with an `intro` and a `clear`.

[3] If students are interested we may briefly mention De Bruijn indices.

2.4 Midterm Exam and Case Study: Week 7

We find it convenient to give a midterm examination after predicate logic. This exam covers traditional, propositional, and predicate logic and is done entirely on paper. By this point, the students have already had several Coq quizzes, and so we are able to track their progress in the theorem prover that way. In addition, the logistics of running an exam in the laboratory are fairly complicated and so we only do so for the final (§3).

Network security analysis. After the midterm, the students are too jumpy to listen to anything formal, and so we do not want to start a fresh topic. Instead, just as with propositional logic, we like to present an example of applying predicate logic to a real-world problem, this time of network security analysis [OGA05].

2.5 Formal Induction: Week 8

After predicate logic, we return to the subject of induction. Whereas our treatment of induction in week 3 (§2.2) was informal and by analogy to data structures in computer science, by week 8 we are ready to be quite formal.

In previous years, we discovered that students had an extremely hard time understanding the nuances of formal structural induction; common errors include: not covering all cases, not proving the induction hypothesis in a case, assuming the wrong induction hypothesis in a case, failing to generalize the induction hypothesis, etc. The advantage of deferring the formal treatment of induction until after predicate logic is that students have enough familiarity with Coq to be able to use it to explore the topic. The payoff is substantial; indeed, the single biggest improvement in comprehension for an individual topic occurred after we introduced mechanized induction.

We were not able to find a textbook that covered structural induction in a way we were happy about; accordingly, we wrote some fairly extensive lecture notes on the subject [HH10, `notes/Induction.pdf`]. By doing so, we were able to develop the paper and mechanized versions of induction with similar notation and in parallel, which allows students to follow along with their own Coq session and experiment.

Another advantage of developing our own materials was that we are able to introduce several related topics that are "off the beaten track". For example, although we do not cover it in full detail, we find it useful to explain coinduction as a contrast to induction. We point out that for both inductive and coinductive types, case analysis forms the basic elimination rules and constructors form the basic introduction rules. Inductive types get a more powerful kind of elimination rule (fixpoints) whereas coinductive types get a more powerful kind of introduction rule (cofixpoints). We also point out the connection to nonterminating vs. terminating computation, a concept which connects back to earlier discussions about intuitionistic logic.

The end result was that most students were able to write extremely clear inductive proofs, even when the induction in question was not straightforward, *e.g.*, when it required a generalization of the induction hypothesis (including the often-confusing situations wherein quantifiers need rearrangement before induction).

Teaching with Coq becomes a bit entwined with teaching Coq. One of the challenges of using Coq as a didactic tool is that Coq is extremely complicated. It is amazing how easily one runs into all kinds of didactically-inconvenient topics at awkward moments. We try to sprinkle in some of these ideas ahead of time, so that when they come up later students already have some context. Moreover, covering the nitty-gritty details further a minor goal, which is to provide the students with a better understanding of Coq, in case they want to use it going forward for another class or a research project—and indeed, several did so. While discussing induction we also cover the ideas of pattern-matching[4], exhaustive/redundant matching, polymorphic types, and implicit arguments.

2.6 Modal Logic: Weeks 9 and 10

Introducing modal logic with Coq was a bit challenging. There are two main problems:

1. The semantics of modal logic is usually introduced on paper by defining a finite set of worlds, each of which is a finite set of propositional atoms. The relation between worlds is then a finite set of arrows linking the worlds. Immediately this runs into trouble in Coq—an example of the already mentioned propensity for Coq to force unpleasant didactic issues to the fore, *e.g.*, Coq does not have a simple way to encode finite sets without using library code and explaining the importance of constructive tests for equality (both of which we have avoided in the past).
2. Coq does not have a clean way to carry out natural deduction proofs in modal logic. The best method we have found, a clever encoding by deWind, is still clunky when compared to simple paper proofs [dW01]. Current research in Coq using modal logic tends to prefer semantic methods over natural deduction—that is, modal logic is used to *state properties and goals* rather than *prove theorems*.

In the end, although our initial explanation of modal logic on paper was given in the standard propositional style, on the Coq side we decided to plunge headlong into a higher-order encoding of modal logic. Modal formulae are functions from the parameterized type of worlds into `Prop`, and we lift the usual logical operators (conjunction, etc.) from the metalogic. With judicious use of `Notation`, the formulas in Coq can look pretty close to how we write them on paper. Here is a small sample of our setup:

```
Definition Proposition : Type := world -> Prop.

Definition holds_in (w : world) (phi : Proposition) :=
  phi w.
Notation "w ||- phi" := (holds_in w phi) (at level 50).

Definition And (phi psi : Proposition) : Proposition :=
  fun w => (w ||- phi) /\ (w ||- psi).
Notation "phi && psi" := (And phi psi).
```

[4] One detail we have largely avoided discussing is the distinction between computable and incomputable tests for equality—*i.e.*, those that live in `Type` vs. `Prop`. This might be a mistake; one of the advantages of using a mechanical theorem prover is that it is easy to demonstrate the importance of maintaining the computable/incomputable distinction by simply observing that Coq can do much less automation when computability is not maintained.

We also lift the universal and existential quantifiers from the metalogic, giving the students a first-order (at least) version of modal logic to play with[5]. Even better, if we are careful in how we lift the logical operators then the usual Coq tactics (`split`, etc.) work on modal logic formulas "as one might expect":

```
Goal forall w P Q,
  w ||- P && Q ->
  w ||- Q && P.
Proof.
  intros w P Q PandQholds.
  destruct PandQholds as [Pholds Qholds].
  split; [apply Qholds | apply Pholds].
Qed.
```

This is extremely useful since the cost of learning a new tactic is quite high to a student.

Since our students already have a grasp of quantification, they can understand when we define the modal box and diamond operators in the standard way (parameterized over some global binary relation between worlds `R`).

```
Definition Box (phi : Proposition) : Proposition :=
  fun w => forall w', R w w' -> (w' ||- phi).
Notation "[] phi" := (Box phi) (at level 15).

Definition Diamond (phi : Proposition) : Proposition :=
  fun w => exists w', R w w' /\ (w' ||- phi).
Notation "<> phi" := (Diamond phi) (at level 15).
```

To reason about these operators they must be `unfolded` and then dealt with in the metalogic, but in practice we find that easier than trying to duplicate paper natural deduction proofs. In any event, encoding modal logic in this way allows the students to prove standard modal facts without undue stress, and in addition gives a feel for modal logics with quantifiers. We also introduce multimodal logics—logics with multiple relations between worlds, by parameterizing Box and Diamond:

```
Definition BoxR (R' : world -> world -> Prop)
                (phi : Proposition) : Proposition :=
  fun w => forall w', R' w w' -> (w' ||- phi).
```

We return to this idea when we study the semantics of Hoare logic in week 12 (§2.7).

Multimodal logics also lead into our investigation of correspondence theory—*i.e.*, the connection between the worlds relation `R` and the modal axioms. Here we are able use our Coq encoding of modal logic to demonstrate some very elegant proofs of some of the standard equivalences (*e.g.*, reflexive with T, transitive with 4) in a way that demonstrates the power of higher-order quantification, giving students a taste of richer logics. For more details see [HH10, `notes/Modal_Logic.pdf`].

---

[5] In fact, we have given them something much more powerful: the quantification is fully impredicative, although we do not go into such details.

2.7 Hoare Logic: Weeks 11 and 12

We turn towards Hoare logic as we near the end of the semester. Our Coq integration was not very successful in helping students understand concrete program verifications. The problem seems to be that mechanically verifying even fairly simple programs leads to huge Coq scripts, and often into tedious algebraic manipulations (*e.g.*, $(n+1) \times m = n \times m + m$, where n and m are integers, not naturals). These kinds of goals tend to be obvious on paper, but were either boring or very frustrating for the students to prove in Coq. Accordingly, we did almost all of the program verifications on paper only.

There were two exceptions: first, we required the students to do a handful of extremely short (*e.g.*, two-command) program verifications in Coq, just to get a little taste of what they were like. Secondly, we showed them a verification of the 5-line factorial program given as the standard example of Hoare verification in Huth and Ryan [HR00]. Although the Coq verification was more than 100 lines, it was worth demonstrating, since it found a bug (or at least a woeful underspecification) in the standard textbook proof[6]. This got the key point across: one goes through the incredible hassle of mechanically checking programs because it is the most thorough way to find mistakes; see [HH10, `slides/slides_11_b.color.pdf`, 46–56] for more detail.

Success on the semantic side. We had much better luck integrating Coq into our explanation of the semantics of Hoare logic. This is a topic that several introductory textbooks skip or only cover informally, but we found that Coq allowed us to cover it in considerable detail. In the end, our students were able to mechanically prove the soundness of Hoare logics of both partial and total correctness for a simple language[7]. The difficulty of these tasks were such that we think they demonstrate that our students had reached both a certain familiarity with Coq and a deeper understanding of Hoare logic.

Part of the challenge with providing a formal semantics for Hoare logic is the amount of theoretical machinery we need to introduce (*e.g.*, operational semantics). A second challenge is producing definitions that are simple enough to make sense to the students, while still allowing reasonably succinct proofs of the Hoare axioms.

Finding the right balance was not so easy, but after several attempts we think we have developed a good approach. We use a big-step operational semantics for our language; for most commands this is quite simple. However, the `While` command is a bit trickier; here our step relation recurses inductively, which means that programs that loop forever cannot be evaluated. Our language is simple enough (*e.g.*, no input/output) that this style of operational semantics is defensible, even if it is not completely standard.

Hoare logic as a species of modal logic. We use modal logic to give semantics to the Hoare tuple in the style of dynamic logic [HKT00]. One obvious advantage of such a choice is that *Hoare logic becomes an application for modal logic*—that is, it increases students' appreciation of the utility of the previous topic. This style allows the definitions to work out very beautifully, as follows. Suppose our (big-)step relation, written

[6] The underspecification comes from not defining how the factorial function (in math, not in code) behaves on negative input, and the bug from not adjusting the verification accordingly.

[7] The proof of the `While` rule was extra credit. Several students solved this rule for the logic of partial correctness; to date we have not had any students solve the total correctness variant.

$c \vdash \rho \rightsquigarrow \rho'$, relates some starting context ρ to some terminal context ρ' after executing the command c. Define the family of context-relations indexed by commands S_c by

$$\rho\, S_c\, \rho' \quad \equiv \quad c \vdash \rho \rightsquigarrow \rho'$$

and the multimodal universal $\Box_c^S$ and existential $\Diamond_c^S$ operators as usual over S_c:

$$\begin{array}{lcl} \rho \models \Box_c^S P & \equiv & \forall \rho'.\ (\rho\, S_c\, \rho') \rightarrow (\rho' \models P) \\ \rho \models \Diamond_c^S P & \equiv & \exists \rho'.\ (\rho\, S_c\, \rho') \rightarrow (\rho' \models P) \end{array}$$

That is, if $\Box_c^S P$ holds on some state ρ, then P will hold on any state reachable after running the command c (recall that only terminating commands can be run); similarly, if $\Diamond_c^S P$ holds on some state ρ, then it is possible to execute the command c, and afterwards P will hold. Now we can give semantics to Hoare tuples as follows[8]:

$$\begin{array}{lcl} \{P\}\, c\, \{Q\} & \equiv & \forall \rho.\ \rho \models (P \Rightarrow \Box_c^S Q) \\ [P]\, c\, [Q] & \equiv & \forall \rho.\ \rho \models (P \Rightarrow \Diamond_c^S Q) \end{array}$$

Although this style of definition is not suitable for more complicated languages, they work very well here and we find them to be aesthetically pleasing. Moreover, they lead to extremely elegant proofs of the standard Hoare rules. In fact, with the exception of the `While` rule for total correctness, none of the Hoare axioms took us more than about 10 lines of Coq script to prove, which put them within reach of our students' efforts[9]. This allowed us to give the entire soundness proof of the Hoare logic as a (fairly long) homework assignment. For more details, see [HH10, `notes/Hoare.pdf`].

2.8 Other Topics: Week 13

The final week of the course is less formal and covers topics of interest to the instructors (*e.g.*, separation logic). Since there is no time to assign homework on topics covered, we do not want to get into huge amounts of detail, and any final exam questions on those topics are by convention fairly simple. In addition, we schedule part of the lecture for students' questions on material covered in the earlier part of the course.

2.9 What We Cut

We added several new topics to the standard topics covered in an introductory logic course (*e.g.*, by Huth and Ryan [HR00]): traditional (Aristotelian) logic, intuitionistic logic, more complex problems in predicate logic and induction, multimodal logic, the semantics of Hoare logic, and most of all the mechanical theorem prover Coq. Although we worked hard to preserve as much of the standard curriculum as we could, there were a few topics that we had to cut for reasons of time. While we covered what a SAT solver is, and explained that the problem was NP-complete, we did not explain any details of the kinds of heuristics involved to get good performance in practice. We also cut most of the material on model checking and temporal logic.

[8] Writing $\Rightarrow$ to mean (lifted) implication, *i.e.*, $\rho \models P \Rightarrow Q \;\equiv\; (\rho \models P) \rightarrow (\rho \models Q)$.

[9] A useful rule of thumb when setting assignments: if the instructors can solve something in n lines, most of the students can solve the same thing in fewer than $5n$ lines.

3 Course Format

We found it crucial for the students to acquire familiarity with Coq early in the course. Accordingly, we gave Coq assignments and quizzes. This resulted in a student workload that was significantly above average for comparable courses, since we did not compromise on the number of traditional paper-based assignments. As a result, the assessment components in the latest incarnation of CS3234 (Sem 1 2010/2011) included: 7 paper assignments (at 2% each), 5 Coq assignments (at 2% each), 6 twenty minute Coq quizzes (at 2% each), a one hour paper midterm (10%), and a two hour final with both Coq and paper problems (22% in Coq, 32% on paper). As one might imagine, preparing and grading this many assignments requires a serious commitment on the part of the instructors as well—and in addition, we were preparing course slides, lecture notes, and laboratory exercises. Fortunately, our department was able to allocate two teaching assistants to help giving the tutorials/laboratories and doing some of the grading; we ended up having one of the highest support/student ratios in the department. In the previous year (Sem 1 2009/2010) we did it all ourselves, and we had very little time to do other work. Of course, as we continue to develop and can begin to reuse the course materials, a good part of the labor is reduced.

When we last taught the graduate version CS5209 (Sem 2 2009/2010), we tried to assign less homework, hoping that graduate students would be able to learn the material without as much supervision. We were mistaken; quite a few of our graduate students had a very hard time with Coq, which was related to the lesser amount of homework. In the future we will assign more work in CS5209. We also tried to give some of the material as a group project; this also turned out to be a bad idea as some of the team members did not put in nearly enough work to do well on the Coq part of the final exam.

Academic honesty. Since the Coq scripts are usually quite short and appear to contain little idiosyncratic information, the temptation to copy solutions from other students seemed to be unusually high. We countered this temptation by conducting systematic cross-checking of scripts, introducing Coq quizzes, which are conducted in computer labs with internet access disabled and submitted at the end of the session, and adding a significant Coq component to the final exam, along with a traditional paper component.

4 Results of Course

It is extremely difficult to be properly scientific when analyzing didactic techniques. We can only run one small "experiment" per year, with numerous changes in curriculum, student quality, topics covered, and instructor experience. The numerous variables make us very cautious in drawing conclusions from quantitative values such as test scores. We are left with subjective opinions of students and instructors on the learning experience.

For our part, we believe that students *that were willing to put in the time required to become familiar with Coq* significantly increased their comprehension of the material. For example, we noticed a definite improvement in pen-and-paper solutions after the students had covered similar problems in Coq. We were also able to give more complex homework problems (*e.g.*, trickier induction and the semantics of Hoare logic) that we would not have been able to cover with pen-and-paper without leaving most of the class behind. We emphasize: both stronger and weaker students benefited from using Coq; the students that seemed to do the worst with the new approach were those that were unwilling to spend the (substantial) time required to become familiar with Coq.

For the students' part, we can do a fair before-and-after comparison of the student feedback for CS3234, because the two incarnations of the module before introduction of Coq were given by the first author in Semester 1 2007/2008 and Semester 1 2008/2009, and the two incarnations after the introduction of Coq were given by both authors in Semester 1 2009/2010 and Semester 1 2010/2011. At the National University of Singapore, students provide their general opinion of the module using scores ranging from 1 (worst) to 5 (best). The students also provide subjective feedback on the difficulty of the module, ranging from 1 (very easy) to 5 (very difficult). The following table includes the average feedback scores in these two categories, as well as the student enrollment and survey respondents in the listed four incarnations:

Semester	Coq Inside	Enrollment	Respondents	Opinion	Difficulty
Sem 1 2007/2008	No	37	24	3.58	3.87
Sem 1 2008/2009	No	33	20	3.55	3.95
Sem 1 2009/2010	Yes	32	17	4.17	4.00
Sem 1 2010/2011	Yes	30	19	3.84	4.05

Students can also give qualitative feedback; here is some of this feedback before Coq:

- "I would like to see more materials from a (real life) application"
- "dry module to me, cant see the link in what is taught and that i'd ever going to apply it. maybe can make it more real life applicable, and talk about how in real programming life would we use such logics. i mean we just learn the logics but dun really know where we will really be making use of it."
- "Quite good.. But everything is too theoretical .."
- "There are very complex ideas which are very difficult to explain."

Here is some of the feedback after the introduction of Coq:

- "Fantastic module. The workload is slightly heavy but that is fine. Learnt a lot."
- "Strengths: help students understand various aspects of logic and how it can be applied in computer science. Weakness: Only the surfaces of some topics. cannot appreciate their usefulness. Homeworks (paper + coq) consume a lot of time"
- "The strength of this module covers various topic on formal proving, giving me a deeper understand on the application of discrete structure that i had taken before. The lecture slides and some of the additional notes are clear and helpful. I like the idea of having Coq lab session, whereby we apply what we learn. However, some of the quiz are very challenging and i think we do need more extra practices (not included in CA marks) on the Coq besides just the homework. The workload is rather heavy and each assignment and homework is just 2%."
- "good module with many labs that can give me a good understanding of COQ"

We received an email from a student of CS5209 that nicely summarizes the benefits and challenges from Coq from the students' perspective: "I would like to thank you for the Automated Theorem Prover (Coq) you taught in CS5209 course. It makes life easy while trying to prove theorem as compared to paper part. In addition to this it saves life of student in Final exam. In the beginning for the course I hated Coq a lot, but slowly I start liking it as I understood the way tactic works and how to use them. Now it has become most favorite and interesting part of mine in this course."

5 Related Work

There has been extensive previous work in using proof assistants to teach formal methods. For example, certain logic courses at Carnegie Mellon have been using the ETPS system (a variant of the TPS proof system developed with a focus on education) since 1983 [ABB+03]. Some of the conclusions from using the ETPS system mirror our own: students have been able to prove more difficult theorems with the aid of a proof assistant, and "students show remarkable creativity in developing surprisingly awkward ways of doing things". However, while there is a considerable amount of material in the literature about the ETPS system as a piece of software, we have not found much in the way of experience reports in terms of how to integrate the system into a curriculum.

More recent work has largely focused on using a proof assistant to teach programming languages (including type theory) as opposed to introductory logic. SASyLF is an LF-based proof assistant designed specifically to enable mechanizing proofs about programming languages simple enough for the classroom [ASS08]. One primary advantage of a more specialized tool such as SASyLF is that the surface syntax can be much closer to paper proofs. Although we found that SASyLF allows for quite elegant statement of grammars and judgments, we found the actual proof scripts to be a bit verbose and cumbersome. The disadvantage, generally speaking, of specialized educational tools is that they tend to be "broad but shallow"—that is, they trade off expressive power for ease of use; in the case of SASyLF, for example, users are restricted to second-order logic. We wanted our students to have exposure to a software system that would allow them to explore further if they wished to (as, indeed, a number did). A related thread is the development of alternative general-purpose theorem provers with, hopefully, a simpler user experience, such as Matita [ACTZ07] or ACL2 Sedan [ACL].

Pierce and others at the University of Pennsylvania use the proof assistant Coq for teaching programming language semantics [Pie09], and observe several of the same general points that we do, *e.g.*, that teaching with Coq becomes entwined with teaching Coq. We suspect that teaching multiple courses (e.g., both logic and programming languages) with the same proof assistant would yield considerable advantages since the costs of teaching Coq would be spread over two semesters instead of one. We are not aware of any attempt to teach such a sequence.

6 Conclusion

We have outlined a migration of a traditional course on logic for computer science to a format that makes extensive use of the theorem prover Coq. Our approach resulted from teaching the material three times (twice in an undergraduate and once in a graduate setting). Along the way, we have found a number of didactic techniques to be useful:

- Introduction of Aristotelian term logic prior to propositional logic so that we can introduce the basic concepts of logic and Coq more gently.
- Keeping the object- and metalogics separate at the beginning; only transitioning to direct use of Coq's `Prop` once the distinction is clear.
- Delaying formal discussion of induction until after predicate logic, and then covering it in detail once students' familiarity with Coq can provide assistance.
- Presenting a full-powered modal logic in Coq instead of attempting to precisely duplicate the experience on paper; a significant exploration of correspondence theory.
- Giving a semantics for Hoare logic so that students can prove the Hoare axioms.
- Presenting several direct applications of formal systems to computational problems: resource scheduling for propositional logic; network security analysis for predicate logic; and Hoare logic's semantics for modal logic.

Comparing the student feedback from CS3234 before and after the migration, it is clear that the introduction of Coq was well received by the students, as shown by a significant improvement of the overall student opinion of the module, at the cost of a modest increase in module difficulty. Anecdotal evidence suggests that the students appreciated the additional learning opportunities afforded by the use of Coq throughout the courses. Overall, considering the available evidence, we believe that the use of Coq in these courses has improved the students' learning of formal logic considerably. The price to pay was additional time spent on learning Coq, which we consider a worth-while investment in its own right.

The material resulting from the migration (including an extensive collection of Coq assignments, quizzes and exam questions) is available online [HH10] for the benefit of the community of academics involved in teaching logic to computer science students.

References

[ABB+03] Andrews, P.B., Bishop, P., Brown, C.E., Issar, S., Pfenning, F., Xi, H.: ETPS: A system to help students write formal proofs (2003)

[ACL] The ACL2 Sedan, `http://acl2s.ccs.neu.edu/acl2s/doc`

[ACTZ07] Asperti, A., Coen, C.S., Tassi, E., Zacchiroli, S.: User interaction with the matita proof assistant. Journal of Automated Reasoning 39(2), 109–139 (2007)

[ASS08] Aldrich, J., Simmons, R.J., Shin, K.: SASyLF: An educational proof assistant for language theory. In: 2008 ACM SIGPLAN Workshop on Functional and Declarative Programming Education (FDPE 2008), Victoria, BC, Canada (2008)

[BA01] Ben-Ari, M.: Mathematical Logic for Computer Science. Springer, Heidelberg (2001)

[Bor06] Borchert, D.M. (ed.): Glossary of Logical Terms, 2nd edn. Encyclopedia of Philosophy. Macmillan (2006)

[dW01] de Wind, P.: Modal logic in Coq. VU University Amsterdam, IR-488 (2001), `http://www.cs.vu.nl/~tcs/mt/dewind.ps.gz`

[HH10] Henz, M., Hobor, A.: Course materials for cs3234/cs5209 (2010), `http://www.comp.nus.edu.sg/~henz/cs3234`

[HKT00] Harel, D., Kozen, D., Tiuryn, J.: Dynamic Logic. MIT Press (2000)

[HR00] Huth, M.R.A., Ryan, M.D.: Logic in Computer Science: Modelling and reasoning about systems. Cambridge University Press, Cambridge (2000)

[NT98] Nemhauser, G.L., Trick, M.A.: Scheduling a major college basketball conference. Operations Research 46(1), 1–8 (1998)

[OGA05] Ou, X., Govindavajhala, S., Appel, A.W.: MulVAL: A logic-based network security analyzer. In: 14th USENIX Security Symposium (2005)

[PH91] Parry, W.T., Hacker, E.A.: Aristotelian Logic. State University of New York Press (1991)

[Pie09] Pierce, B.C.: Lambda, the ultimate TA: Using a proof assistant to teach programming language foundations (2009)

[Zha93] Zhang, H.: SATO: A decision procedure for propositional logic. Association of Automated Resasoning Newsletters 22 (1993, updated version of November 29) (1997)

[Zha02] Zhang, H.: Generating college conference basketball schedules by a SAT solver. In: Proceedings of the Fifth International Symposium on Theory and Applications of Satisfiability Testing, Cincinnati, Ohio, pp. 281–291 (2002)

The Teaching Tool CalcCheck: A Proof-Checker for Gries and Schneider's "Logical Approach to Discrete Math"

Wolfram Kahl

McMaster University, Hamilton, Ontario, Canada
kahl@cas.mcmaster.ca

Abstract. Students following a first-year course based on Gries and Schneider's LADM textbook had frequently been asking: "How can I know whether my solution is good?"

We now report on the development of a proof-checker designed to answer exactly that question, while intentionally not helping to find the solutions in the first place. CalcCheck provides detailed feedback to LaTeX-formatted calculational proofs, and thus helps students to develop confidence in their own skills in "rigorous mathematical writing".

Gries and Schneider's book emphasises rigorous development of mathematical results, while striking one particular compromise between full formality and customary, more informal, mathematical practises, and thus teaches aspects of both. This is one source of several unusual requirements for a mechanised proof-checker; other interesting aspects arise from details of their notational conventions.

1 Introduction

When teaching a first-year course on Logic and Discrete Mathematics for Computer Science following Gries and Schneider's textbook "A Logical Approach to Discrete Math" ("LADM" for short) [GS93] for the first time, I obtained feedback from students feeling that the book did not contain sufficiently many worked examples, that insufficient solutions for exercises were available[1], and, especially, that they felt at a loss since they did not see any way of knowing how good their answers were before the marked assignment was returned to them.

The following year (2011), I therefore started to implement "CalcCheck", a tool intended mainly as a proof-checker for the calculational proof style taught by LADM. For the time being, the usage paradigm of CalcCheck is the same as that of Spivey's Z type-checker *f*uzz: CalcCheck also operates on LaTeX source by parsing and analysing the contents of specific formal environments, and providing feedback on those. Using LaTeX as input syntax has the advantage that students learn a general-purpose skill, with only very little formalism-specific overhead.

[1] An "Instructor's Manual" containing solutions exists, but is made available explicitly only to instructors, with the proviso "that answers to selected exercises may be used in lectures or distributed to students as answers to homeworks or tests".

J.-P. Jouannaud and Z. Shao (Eds.): CPP 2011, LNCS 7086, pp. 216–230, 2011.

For example, the following proof can be found on p. 46 of LADM (without the "**Proving**" line):

Proving (3.16) $(p \not\equiv q) \equiv (q \not\equiv p)$:

$$
\begin{array}{ll}
 & p \not\equiv q \\
= & \langle \text{Def. of } \not\equiv \text{ (3.10)} \rangle \\
 & \neg(p \equiv q) \\
= & \langle \text{Symmetry of } \equiv \text{ (3.2)} \rangle \\
 & \neg(q \equiv p) \\
= & \langle \text{Def. of } \not\equiv \text{ (3.10), with } p, q := q, p \rangle \\
 & q \not\equiv p
\end{array}
$$

Using the LaTeX macro package accompanying CALCCHECK, this proof rendering has been generated from the following LaTeX source:

```
\begin{calc}[(3.16) $(p \nequiv q) \equiv (q \nequiv p)$]
    p \nequiv q
  \CalcStep{=}{Def.~of $\nequiv$ (3.10)}
    \lnot(p \equiv q)
  \CalcStep{=}{Symmetry of $\equiv$ (3.2)}
    \lnot(q \equiv p)
  \CalcStep{=}{Def.~of $\nequiv$ (3.10), with $p, q \becomes q, p$}
    q \nequiv p
\end{calc}
```

The LaTeX macros have been kept as unobtrusive as possible, with the aim of letting the skill of producing CALCCHECK-checked proofs directly improve the skill of producing hand-written proofs in the exams.

Running CALCCHECK on an input file containing the above LaTeX fragment produces the following output to an HTML file, and also in Unicode to the terminal:[2]

Proving (3.16) (p ≢ q) ≡ (q ≢ p):

p ≢ q
= ⟨ Def.~of $\nequiv$ (3.10) ⟩
— CalcCheck-0.2.12: (3.10) Definition of ≢ — OK
¬(p ≡ q)
= ⟨ Symmetry of $\equiv$ (3.2) ⟩
— CalcCheck-0.2.12: (3.2) Symmetry of ≡ — OK (no change)
¬(q ≡ p)
= ⟨ Def.~of $\nequiv$ (3.10), with $p, q \becomes q, p$ ⟩
— CalcCheck-0.2.12: (3.10) Definition of ≢ — OK
q ≢ p

— CalcCheck-0.2.12: Proof matches goal — OK

[2] CALCCHECK output included in this paper has been rendered by a WWW browser from the CALCCHECK-generated HTML files.

This output is only produced if there are no syntax errors, and contains the relevant parts of the input together with additional annotations:

- The optional argument of the `{calc}` environment is the proof goal; in this case, the goal is recognised as one of the numbered LADM theorems.
- CALCCHECK attempts to verify that the whole proof, $(p \not\equiv q) = \ldots = \ldots = (q \not\equiv p)$ is actually a proof of the goal, assuming all steps are correct. LADM includes a number of different patterns how such calculational proofs can satisfy their goals (similar to the optional *method* argument of **proof** in Isabelle/Isar [Nip03], but rarely made explicit in LADM).
- In LADM, each proof step requires a "hint" stating the theorem(s) applied in this step; CALCCHECK attempts to verify for each proof step that it can be obtained from the theorems mentioned in the hint. Currently, CALCCHECK relies on the theorem numbers, e.g., "(3.10)", but it is planned to make it recognise also theorem names, e.g. also "Def. $\not\equiv$", that are perfectly acceptable in the context of hand-written mathematics.[3] Therefore, CALCCHECK first of all reports which theorems it recognises as mentioned in the hint, or "Could not extract information" if it recognised none. Following that, it adds "— OK" if it can derive the proof step from these theorems, and "could not justify this step" otherwise.

For an example of the latter, here is the CALCCHECK output for one student proof—the first "could not justify" should really have alerted the student to the simple typo here (v for r in the second expression), and looking closely at the second "could not justify" would have revealed that the referenced theorem number belongs to a different theorem:

  p ⇒ q ∧ r
= ⟨ (3.59) Alternative Definition of Implication ⟩
  — CalcCheck-0.2.12: (3.59) Definition of ⇒; could not justify this step!
  ¬p ∨ (q ∧ v)
= ⟨ (3.46) Distributivity of $\lor$ over $\land$ ⟩
  — CalcCheck-0.2.12: (3.46) Distributivity of ∧ over ∨; could not justify this step!
  (¬p ∨ q) ∧ (¬p ∨ r)
= ⟨ (3.59) Alternative Definition of Implication ⟩
  — CalcCheck-0.2.12: (3.59) Definition of ⇒ — OK
  (p ⇒ q) ∧ (p ⇒ r)

CALCCHECK is not complete, that is, it cannot justify all acceptable correct proof steps, and, due to the not-fully-formal nature of LADM proofs, also never will be

[3] The course website continues to list the same rules as in the previous year:
- Theorem numbers are never *necessary* for marks in this course
- Theorem numbers are nice for disambiguation — [...]
- Typically, a hint with just one of [name of the theorem], [theorem number], and [the theorem [...], that is, the Boolean expression] is acceptable, although not necessarily nice. [...]

complete. However, for the central LADM Chapter 3 "Propositional Calculus", CALCCHECK can certify all correct proofs that are given in sufficient detail, which is rarely more that given in typical LADM proofs.

For predicate logic (chapters 8–9) and the theories of sets, functions, and relations (chapters 11 and 14), occasionally more detail is required; for example in the following proof about the domain of relations, one would normally probably hope to be able to contract another two or three of the eight steps into larger steps:

$$
\begin{array}{ll}
& x \in \{p \mid p \in R \bullet \mathsf{fst}.p\} \\
= & \langle (11.3) \text{ Membership in set comprehension, with } \neg \mathit{occurs}(\text{`}p\text{'}, \text{`}b\text{'}) \rangle \\
& (\exists\, p \mid p \in R \bullet x = \mathsf{fst}.p) \\
= & \langle (8.21\text{p}) \text{ Pair dummy expansion} \rangle \\
& (\exists\, b, c \mid (p \in R)[p := \langle b, c \rangle] \bullet (x = \mathsf{fst}.p)[p := \langle b, c \rangle]) \\
= & \langle (14.4\text{p}) \text{ Pair projection} \rangle \\
& (\exists\, b, c \mid \langle b, c \rangle \in R \bullet x = b) \\
= & \langle (9.19) \text{ Trading for } \exists, (1.3) \text{ Symmetry of } = \rangle \\
& (\exists\, b, c \mid b = x \bullet \langle b, c \rangle \in R) \\
= & \langle (8.20) \text{ Nesting, } (8.14) \text{ One-point rule} \rangle \\
& (\exists c \mid\bullet \langle x, c \rangle \in R) \\
= & \langle \text{Changing relational notation} \rangle \\
& (\exists\, c \mid\bullet x\, R\, c) \\
= & \langle (11.7) \rangle \\
& x \in \{x \mid (\exists\, c \mid\bullet x\, R\, c)\} \\
= & \langle (14.16) \text{ Domain of relations} \rangle \\
& x \in \mathsf{Dom}.R
\end{array}
$$

The resulting CALCCHECK output below demonstrates some additional features:

- Provisos concerning variable binding are derived automatically from the theorem statement, and always documented in the CALCCHECK output.
- Proviso handling is still incomplete — $\neg \mathit{occurs}(\text{`}b, c\text{'}, \text{`}R\text{'})$ fails to interpret R as a meta-variable. This proviso should be a global assumption, but handling of such assumptions is also still missing. Nevertheless, the listing of the used $\neg \mathit{occurs}$ assumptions is helpful especially for students who are new to the intricacies of variable binding.
- In cases where CALCCHECK does not understand the hint ("Could not extract information"), it still accepts certain trivial steps, in the case here a change of input notation that is not reflected in the abstract syntax, and therefore also does not influence the CALCCHECK output. (Merging this "change of notation" step with one of the previous steps would of course be accepted, too, but has been left separate here for demonstration.)

- CalcCheck can evaluate substitutions — this happens here at the occurrence of the one-point rule (8.14). However, second-order matching is not yet implemented. Therefore, certain applications of rules involving substitution require the user to make this matching explicit; here, this is the case for the result of the second step, which uses the following rule not found in LADM:

  (8.21p) **Pair Dummy Expansion**: Provided $\neg occurs(`x, y', `R, P')$,

$$(\star\ p : t_1 \times t_2 \mid R \bullet P)$$
$$= \quad (\star\ x : t_1; y : t_2 \mid R[p := \langle x, y\rangle] \bullet P[p := \langle x, y\rangle])$$

(The CalcCheck output below also demonstrates some deviations from LADM notation: Quantification and set comprehension $\{\ldots \mid \ldots \bullet \ldots\}$ use a bullet instead of a colon, since the colon is used also for typing, and is "less visually separating". Also, pairs are displayed (x, y) instead of $\langle x, y\rangle$, but both notations are accepted in input.)

```
 x ∈ { p | p ∈ R • fst.p}
= ⟨ (11.3) Membership in set comprehension, with $\lnot\occurs{p}{b}$ ⟩
   — CalcCheck-0.2.12: (11.3) Set membership
    — OK: ¬occurs('p', 'x')
 (∃ p | p ∈ R • x = fst.p )
= ⟨ (8.21p) Pair dummy expansion ⟩
   — CalcCheck-0.2.12: (8.21p) Pair dummy expansion
    — OK: ¬occurs('b, c', 'x = fst.p, p ∈ R')
 (∃ b, c | (p ∈ R)[p := (b, c)] • (x = fst.p)[p := (b, c)] )
= ⟨ (14.4p) Pair projection ⟩
   — CalcCheck-0.2.12: (14.4p) Pair projection — OK
 (∃ b, c | (b, c) ∈ R • x = b )
= ⟨ (9.19) Trading for $\exists$, (1.3) Symmetry of $=$ ⟩
   — CalcCheck-0.2.12: (9.19) Trading for ∃, (1.3) Symmetry of = — OK
 (∃ b, c | b = x • (b, c) ∈ R )
= ⟨ (8.20) Nesting, (8.14) One-point rule ⟩
   — CalcCheck-0.2.12: (8.20) Quantification nesting, (8.14) One-point rule
    — OK: ¬occurs('b', 'x')
 (∃ c |• (x, c) ∈ R )
= ⟨ Changing relational notation ⟩
   — CalcCheck-0.2.12: Could not extract information — OK (no change)
 (∃ c |• (x, c) ∈ R )
= ⟨ (11.7) ⟩
   — CalcCheck-0.2.12: (11.7) x ∈ { x | R } ≡ R — OK
 x ∈ { x | (∃ c |• (x, c) ∈ R ) }
= ⟨ (14.16) Domain of relations ⟩
   — CalcCheck-0.2.12: (14.16) Domain
    — OK: ¬occurs('c, x', 'R')
 x ∈ Dom.R
```

In addition to this support for checking calculational proofs, CalcCheck also has initial support for checking declarations produced as part of formalisation exercises, or "English to Logic" translation (LADM chapters 2, 5, and sections 8.1 and 9.3).

Using CalcCheck during the work on their assignments does give students a useful first taste of proof certification, and increases their ability to produce and appreciate rigorous proofs.

Section 3 presents additional features of CalcCheck, and in Sect. 4 we further explain the use of CalcCheck in the course setting. Section 5 explains the main challenges encountered producing formal support for the particular kind of semi-formal mathematics practised in LADM, and in Sect. 6 we quickly describe the current implementation.

During the term of initial development, CalcCheck was made available to the students both as source code and as compiled executables for their most common computing platform; it is now available via `http://CalcCheck.McMaster.CA/`.

2 Related Work

The only related system I am currently aware of that uses a LaTeX-based input syntax is Spivey's Z type-checker *fuzz* [Spi08], which analyses declarations and expressions of the Z specification notation [Spi89], and performs syntax- and type-checking. An `argue` environment is provided for typesetting calculational proofs, but *fuzz* does no proof-checking, and also does not type-check the contents of `argue` environments. It is possible to turn `argue` proofs into legal `zed` expressions by commenting out the proof hints at the TeX-level; although these `zed` expressions can then be type-checked by *fuzz*, this is still an unsatisfactory kludge.

All other systems use their own specific input syntax.

A general-purpose proof assistant that has been used for teaching, including first-year courses [HR96,BZ07] is Mizar, which pioneered formalisation of the structure of conventional mathematical proofs. The resulting large proof structure language also appears to be a central topic of the Mizar-based courses, which makes that approach quite different in flavour than the emphasis of LADM on calculational proofs.

SASyLF [ASS08] is a proof checker designed specifically for teaching programming language theory and type theory (to graduate students); it has special syntax to present definitions of syntax, semantics, and typing rules of object languages, and checks structured proofs of language theoretical properties. Aldrich *et al.* [ASS08] report extensively on their efforts to evaluate the pedagogic effects of using their proof checker, and emphasise in particular the early feedback aspect.

Several systems are available that provide support for Hilbert-style proofs, including Tutch [ACP01] (which concentrates on intuitionistic logics), EPTS [ABP+04], and the Logic Daemon interactive website accompanying Allen and Hand's "Logic Primer" [AH01]. While ETPS seems to be used mainly via an interactive user interface, and the Logic Daemon is available only as a web service, Abel *et al.* argue [ACP01] that the batch-mode operation of Tutch, where editing is separate from proof checking, and the proof checker is used similarly to a programming language compiler, is advantageous for acquiring tool-independent proof skills. (The proof programming facilities of Tutch also allow more structured proofs.)

Yet another approach to tool support for teaching logic concentrates on model construction and exploration; several of the systems described in [GRB93] fall into this category.

3 CalcCheck Overview

The current usage paradigm of CalcCheck follows that of Spivey's Z type-checker *f*uzz [Spi08]: The user writes a LaTeX source file using a dedicated LaTeX package defining the rendering of special-purpose TeX macros, and while this file can directly be processed using LaTeX for typesetting, it can also be passed to CalcCheck for analysis of the formal content. Not all TeX mathematics is analysed, but only that contained in the following special environments:

- {`calc`} environments contain calculational proofs, and also displayed mathematical expressions (which could be understood as zero-step calculational proofs).
- {`decls`} environments contain declarations and definitions.

For declarations, inside the `decls` environment the following special macros are available:

- `\declType` for type declarations (type annotations in other contexts just use ":").
- `\declEquiv` for definition of propositions and predicates — "declared as equivalent"
- `\declEqu` for definition of other constants and functions — "declared as equal"
- `\remark` for remarks at the end of a line
- `\also` to separate multiple declarations

Furthermore, natural-language fragments are permitted in `\mbox{...}`, making it possible to assign, in a formal manner, informal meaning to formal identifiers, following the practise of LADM. (To avoid confusion with the use of the colon in type declarations and annotations, we render `\declEquiv` as "$:\equiv$" and `\declEqu` as "$:=$", whereas LADM tends to use just ":" there, too.)

With this, the formalisation of the LADM example sentence "Henry VIII had one son and Cleopatra had two" proceeds as follows:

`We declare:`
`\begin{decls}`
`  h \declEquiv \mbox{Henry VIII had one son}`
`\also`
`  c \declEquiv \mbox{Cleopatra had two sons}`
`\end{decls}`
`Then the original sentence is formalised as:`
`\begin{calc}`
`  h \land c`
`\end{calc}`
	We declare:
	$h :\equiv$ Henry VIII had one son
	$c :\equiv$ Cleopatra had two sons
	Then the original sentence is formalised as:
	$h \wedge c$

Relating formal identifiers to their informal meaning can be achieved via embedding informal material inside formal definitions in `\mbox{...}`, or by adding a `\remark{...}` to a formal declaration — both are ignored by CalcCheck.

```
\begin{decls}
   P \declEqu \mbox{set of persons}
\also
   A \declType P \remark{Alex}
\also
   J \declType P
\also
   J \declEqu \mbox{Jane}
\end{decls}
```

$P :=$ set of persons

$A \ : \ P$ — Alex

$J \ : \ P$

$J :=$ Jane

Functions and predicates can be introduced with conventional definitions, again either informal, that is, in `\mbox{...}`, of formal. For hard line breaks inside formal material, there is a `\BREAK` macro (that can also be used in `{calc}` environments). CalcCheck ignores most common LaTeX spacing commands.

```
\begin{decls}
   called \declType P \times P \tfun \BB
\also
   called(p,q)
  \declEquiv
   \mbox{$p$ called $q$}
\also
   lonely \declType P \tfun \BB
\also
   lonely . p
  \declEquiv
   \lnot (\exists \ q : P
          \BREAK \strut\;
          \withspot called(q,p) )
\end{decls}
```

$\mathit{called} \ : \ P \times P \rightarrow \mathbb{B}$

$\mathit{called}(p, q) :\equiv p$ called q

$\mathit{lonely} \ : \ P \rightarrow \mathbb{B}$

$\mathit{lonely}.p :\equiv \neg(\exists\ q : P$
$\quad\bullet\ \mathit{called}(q, p))$

Most features of the `{calc}` environment have already been introduced in Sect. 1. If the optional goal argument is provided, the goal may be shown also by proving it equal to an already-known theorem; the special macro `\ThisIs{...}` is used to refer to that theorem in what is typeset as a comment (following LADM practise), but checked by CalcCheck. Such a `\ThisIs{...}` annotation can follow either the first or the last line of a proof.

```
\begin{calc}[(3.5)
  Reflexivity of $\equiv$, $p \equiv p$
]
   p \equiv p
\CalcStep{=}{(3.3) Identity of $\equiv$}
   \true
  \ThisIs{(3.4)}
\end{calc}
```

Proving (3.5) Reflexivity of $\equiv$, $p \equiv p$:

$p \equiv p$

$=$ ⟨(3.3) Identity of $\equiv$⟩

true — This is (3.4)

Throughout these example, it should be obvious that the effort involved in producing CalcCheck input is almost completely contained in the effort necessary for producing LaTeX source for the desired output. CalcCheck occasionally prescribes the use of particular LaTeX macros, but rarely requires truly additional effort. Even with respect to the choice of LaTeX macros, CalcCheck is more lenient than *f*uzz, by allowing also "standard" LaTeX macros like `\wedge` and `\vee` instead of the more mnemonic `\land` and `\lor` proposed for use with CalcCheck. This decision was made to lower the friction for students who are not only new to CalcCheck, but at the same time also new to LaTeX, and, at least in some instances, tended to use the first macro they found in any LaTeX-related material for the symbol they had to produce.

4 Teaching with CalcCheck

This first time that CalcCheck was used, it was developed while the course was delivered. Once CalcCheck had been fully introduced into the course, the following rule was added to the weekly assignments:

> You **must** submit a LaTeX file with correct CalcCheck syntax — with CalcCheck syntax errors or LaTeX errors, your submission earns 0 points.

To emphasise the difference between the phases of syntax analysis and proof checking, CalcCheck produces, after successful parsing, the following message:

```
CalcCheck-0.2.11: No syntax errors.
CalcCheck-0.2.11: Now checking...
```

At the same time it was emphasised that the students retained full responsibility for the correctness of their submitted proofs: If CalcCheck were to "OK" an incorrect step it would still count as a mistake — this rule was stated only for pedagogical reasons, to alert students to the fact that even mechanised proving systems are not necessarily to be trusted; although CalcCheck is not formally verified, I still have high confidence that it is sound.

On the other hand, where CalcCheck "could not justify" a step that the markers found to be correct, it still earned full marks. In the context of propositional logic, such cases were limited to single proof steps involving more than two rewrite steps, since for certain rules, even two steps could involve a lengthy search. Therefore, for propositional logic, students never *had to* submit a proof with steps that CalcCheck "could not justify"; they always had the choice of making the intermediate steps explicit to obtain a fully checked proof (with the CalcCheck run finishing much faster). Some students nevertheless had the confidence to submit correct but uncertified larger steps.

Since rules with provisos were not implemented during the course, proofs in predicate logic and set theory were expected to contain steps that CalcCheck "could not justify". (And some had the confidence to submit incorrect steps, in files without syntax errors, so that one would expect that they had seen that CalcCheck "could not justify" their work.)

5 Formalising LADM

Even though LADM is certainly one of the most rigorous textbooks for discrete mathematics currently available, it makes no claim to present a formal system in full detail, and implementing mechanical support for LADM does in fact show up a number of issues that are not covered conclusively by the book.

For example, the conjunction and disjunction operators $\wedge$ and $\vee$ are assigned the same precedence, but no rule is given for aggregation among occurrences of *different* operators of the same precedence. The rule that "All nonassociative operators associate to the left, except ..." does not cover the expression $5-3+2$, although terms like that do occur in LADM, and are, as usual, interpreted as unambiguously denoting $(5-3)+2$. The current version of CalcCheck generalises this to arbitrary (non-right associating) operators of the same precedence, so that $p \vee q \wedge r$ (which does not occur in LADM, but is also not explicitly forbidden) denotes $(p \vee q) \wedge r$ by virtue of association to the left. However, since many students routinely omit any parentheses between $\wedge$ and $\vee$, no matter what the intended structure is, it is probably more useful to just forbid unparenthesised occurrences of these operators together. (In such cases, CalcCheck does show the inserted parentheses in its output, but at least some students do not use this as help.)

Another precedence-related issue concerns infix notation for membership in relations, where LADM says (p. 269):

> In general, for any relation ρ:
> $\langle b, c\rangle \in \rho$ and $b\ \rho\ c$ are interchangeable notations.
> [...] By convention, the precedence of a name ρ of a relation that is used as a binary infix operator is the same as the precedence of $=$; furthermore, ρ is considered to be conjunctional.

Now the name of a relation that is used as a binary infix operator can be a complex expression; LADM p. 272 contains a proof where not only "$(\sigma \circ \theta)$" is used as infix operator, but also "$(\rho \circ \sigma) \circ \theta$" (without enclosing parentheses), producing the expression "$a\ (\rho \circ \sigma) \circ \theta\ d$". Extending this, LADM appears to allow us to write

$$a\ (\rho \circ \sigma) \circ \theta\ (b + c) \in S\ ,$$

and, due to conjunctionality, this has to parse as

$$\langle a, b + c\rangle \in ((\rho \circ \sigma) \circ \theta) \quad \wedge \quad (b + c) \in S\ ,$$

although locally, both parenthesised expressions could also be arguments of function applications, which means that the following parse would be legal, too:

$$((a\ (\rho \circ \sigma)) \circ (\theta\ (b + c))) \in S$$

Therefore, the grammar resulting from a strict reading of the LADM rules for infix relations is ambiguous. Although this ambiguity probably can always be resolved via type checking, where sufficient typing information has been supplied, this is still not only non-trivial to implement, but also potentially quite confusing for students. Currently, CALCCHECK does not accept unparenthesised binary operator applications as infix relation names.

Another area full of pitfalls for any not-fully-formal system is that of variable binding. The approach of LADM towards variable binding is probably best characterised as first-order abstract syntax with implicit metavariable binding, and with a slight tendency to use object-level language also on the meta-level, and to treat substitutions as explicit, as demonstrated most clearly by the extension of the definition of "textual substitution" to cover quantification:

(8.11) Provided $\neg occurs(\text{`}y\text{'}, \text{`}x, F\text{'})$,

$$(\star\, y \mid R \bullet P)[x := F] \quad = \quad (\star\, y \mid R[x := F] \bullet P[x := F])$$

LADM introduces a general quantification syntax for arbitrary abelian monoids; if $\star$ is a symmetric and associative operator and has a unit, then "Expression $(\star x : X \mid R : P)$ denotes the application of operator $\star$ to the values P for all x in X for which range R is true."[4] They do point out that, as a result, not all quantifications are defined, and some theorems include definedness of certain quantifications as provisos, which will, in general, not be decidable for an automatic proof checker.

It appears to me that the provided axioms for quantification are insufficient to prove $(\star\, x, y \mid R \bullet P) \quad = \quad (\star\, y, x \mid R \bullet P)$ without side conditions restricting variable occurrences, but this is silently used in the proof of (8.22) on p. 151, so I assume this as an additional quantification axiom.

In the chapters introducing quantification (8 and 9), potential capture or rebinding of variables is dealt with carefully via explicit provisos formulated using the formal meta-level predicate $occurs(\text{`\_'}, \text{`\_'})$ taking a list of variables and a list of expressions as arguments. However, in the chapter on set theory, many necessary provisos are omitted from theorem statements without warning; it even happens that a proviso is checked in a proof hint where the invoked theorem was stated without provisos. As mentioned in the introduction, CALCCHECK calculates these binding-related provisos from the theorem statement by checking for each metavariable whether it occurs under different sets of binders. This calculation needs to take implicitly bound metavariables into account, too — for example, the following theorem needs no proviso:

(11.7) $x \in \{x \mid R \bullet x\} \quad \equiv \quad R$

This is because both occurrences of R are in the scope of a binder for x, where the binder for the RHS occurrence is the implicit meta-level binder induced by the free occurrence of x in the LHS.

An area where LADM is more explicitly informal is that of higher-level proof structures. Although proof techniques like assuming the antecedent and case analysis are introduced in chapter 4 with a formal-looking syntax, this syntax

[4] After putting this to a vote in the course, we replaced the second ":" in the quantification notation with the " $\bullet$ " also used in the Z notation in that place.

is not used in later applications of these techniques. It therefore appears to be sensible to refer to existing systems that do offer high-level proof structuring like Mizar [GKN10,NK09] or Isabelle/Isar [Nip03] for the purpose of designing an appropriate variant for future versions of CalcCheck.

6 Implementation Aspects

CalcCheck is currently implemented in less than 5000 lines of Haskell [HHPJW07]. The front-end uses the monadic parser combinator library Parsec [LM01].

Since LADM pushes its readers very early and very hard to think about theorems up to permutations of the arguments of associative and commutative operators, AC matching is required, and Contejean's AC matching algorithm [Con04] was adapted for the implementation. Currently, proof checking is almost purely based on breadth-first search in the rewriting relation generated by the non-AC laws. (AC laws do not need to be applied since they are identities on the abstract syntax for AC expressions.) This can fan out very quickly for certain rules in larger terms, but in most cases, performance is not an issue. The depth of the search is currently limited to two applications of rewrite rules induced by any of the theorems that CalcCheck could identify as referenced by the `\CalcStep` hint (currently only by their theorem number). As a matter of proof presentation, two steps are almost always adequate: One would occasionally wish for three or, in very large, repetitive expressions, even four rule application in a single proof step, but rarely for more.

Once all requirements are settled, we envisage a reimplementation that itself has a mechanised correctness proof, and might therefore move from Haskell to a dependently-typed programming language, for example Agda [Nor07].

7 Conclusion and Future Work

Although LADM essentially concentrates on teaching *rigorous informal* mathematics, at least large parts are accessible to formal treatment. Since there appears to be no previous mechanised support for LADM, we contributed a mechanised proof checker, CalcCheck, intended to be used for teaching with LADM, and therefore required to be useful without demanding significant extra effort of formality beyond the use of LaTeX.

The LADM logic is not intended for mechanisation, but rather for training students in successful communication of rigorous mathematical arguments. Forcing proofs to be integrated into LaTeX documents is, in our opinion, more conducive to this goal than using a stand-alone syntax, and is, in fact, very similar to the spirit of *literate programming* [Knu84,Knu92].

In addition, acquiring the "CV-listable skill" of using LaTeX for document formatting appears to be more attractive for many students than learning the special-purpose syntax of an academic tool. Since a CalcCheck-checked assignment submission is first of all a LaTeX document, and the CalcCheck-specific syntax has intentionally been designed as a set of minor constraints on the use of a particular set of LaTeX macros, the use of CalcCheck appears to be perceived by the students to come with the cost of having to learn (some) LaTeX, but otherwise "just do its job", namely to aid them with producing correct proofs, but without requiring non-reusable special-purpose skills.

Future continued development of CalcCheck will strive to at least preserve this accessibility.

For improving the user experience, we plan to more fully support Unicode source files; CalcCheck already parses Unicode representations of most LADM symbols where this is appropriate; work is now needed mostly on the LaTeX side. With that, not only the LaTeX and CalcCheck outputs will be similar in appearance to handwritten variant that students will continue to be expected to produce, but also the source file they are editing, hopefully further increasing the overall accessibility of the CalcCheck experience.

Another significant usability improvement would come from flexible "understanding" of theorem names in the proof step hints, so that students do not need to memorise or look up the theorem numbers all the time, but can instead concentrate of learning the theorem names, which will be much more useful in the long term.

As mentioned previously, support for higher-level proof structure is still missing, and this includes explicit assumption management also for the purpose of properly treating global $\neg occurs$ assumptions.

Proper dependency management needs to be added, and this goes beyond comparing theorem numbers, where usually, the proof of a theorem with number n may only use theorems with smaller numbers. However, if a theorem with a larger number $n + k$ can be shown using only theorems with numbers smaller than n, then theorem $n+k$ may be used in the proof of theorem n, and such detours can be important for didactic purposes. CalcCheck will therefore need to keep track of the precise dependencies between the proofs contained in the checked file in addition to the theorem number ordering of the reference theorem list.

Dependency management also affects matching, and even display of expressions: Only operators for which associativity and commutativity laws are in scope can be treated accordingly by the AC matching mechanism, and have parentheses omitted in output. It might be useful to add, for example, self-inverse operations like Boolean negation and relation converse, to special treatment by future extensions of the current AC matching mechanism.

Since substitution theorems like

$$\text{(3.84a)} \quad e = f \wedge E[z := e] \equiv e = f \wedge E[z := f]$$

are normally applied without making the substitution involved explicit, second-order matching is necessary. However, being able to switch it off may still be useful for didactic purposes.

Particularly pressing is the addition of type-checking, with understandable type error messages. For this, it should be possible to build on previous research concerning type error messages in programming languages, e.g. [Hee05]. Note that the LADM notations used for "universe" sets and set complements depend normally on implicit type arguments (and possibly on explicit "fixing the universe of discourse" to an arbitrary set, which does not need to be a type), so without type-checking, it is impossible to check most proofs involving properties of complement.

Sometimes, students appear to give up in their attempts of producing a fully "OK"-ed proof, and assume that their "could not justify" messages are due only to limitations of CALCCHECK, even in cases where the step in question is invalid, so that no possible hint can justify it. Not only in propositional logic, but also in purely propositional steps of predicate logic proofs, validity of proof steps is decidable, and reporting invalid steps will be a useful aid for students.

Although individual students reported that they found CALCCHECK taught them to know more precisely what they were doing when doing mathematics, the main measurable didactic effect of using CALCCHECK in the past year appears to have been that students now routinely produced syntactically correct formulae even on the hand-written exam and outside calculational proofs, that is, in particular in formalisation exercises — this was not the case in the previous year. Once the students are using an accessible system that is similarly strict in pointing out type errors and invalid proof steps, this should make a further noticeable difference in the resulting active language skills in the language of discrete mathematics.

References

ABP+04. Andrews, P.B., Brown, C.E., Pfenning, F., Bishop, M., Issar, S., Xi, H.: ETPS: A system to help students write formal proofs. Journal of Automated Reasoning 32, 75–92 (2004), doi:10.1023/B:JARS.0000021871.18776.94

ACP01. Abel, A., Chang, B.-Y.E., Pfenning, F.: Human-readable machine-verifiable proofs for teaching constructive logic. In: Proceedings of Workshop on Proof Transformation, Proof Presentation and Complexity of Proofs (PTP 2001). Università degli Studi Siena, Dipartimento di Ingegneria dell'Informazione, Tech. Report 13/0 (2001), http://www2.tcs.ifi.lmu.de/~abel/tutch/

AH01. Allen, C., Hand, M.: Logic Primer, 2nd edn. MIT Press (2001), http://logic.tamu.edu/

ASS08. Aldrich, J., Simmons, R.J., Shin, K.: SASyLF: An educational proof assistant for language theory. In: Huch, F., Parkin, A. (eds.) Proceedings of the 2008 International Workshop on Functional and Declarative Programming in Education, FDPE 2008, pp. 31–40. ACM (2008)

BZ07. Borak, E., Zalewska, A.: Mizar Course in Logic and Set Theory. In: Kauers, M., Kerber, M., Miner, R., Windsteiger, W. (eds.) MKM/CALCULEMUS 2007. LNCS (LNAI), vol. 4573, pp. 191–204. Springer, Heidelberg (2007)

Con04. Contejean, E.: A Certified AC Matching Algorithm. In: van Oostrom, V. (ed.) RTA 2004. LNCS, vol. 3091, pp. 70–84. Springer, Heidelberg (2004)

GKN10. Grabowski, A., Korniłowicz, A., Naumowicz, A.: Mizar in a nutshell. J. Formalized Reasoning 3(2), 153–245 (2010)

GRB93. Goldson, D., Reeves, S., Bornat, R.: A review of several programs for the teaching of logic. The Computer Journal 36, 373–386 (1993)

GS93. Gries, D., Schneider, F.B.: A Logical Approach to Discrete Math. Monographs in Computer Science. Springer, Heidelberg (1993)

Hee05. Heeren, B.: Top Quality Type Error Messages. PhD thesis, Universiteit Utrecht, The Netherlands (September 2005)

HHPJW07. Hudak, P., Hughes, J., Jones, S.P., Wadler, P.: A history of Haskell: Being lazy with class. In: Third ACM SIGPLAN History of Programming Languages Conference (HOPL-III), pp. 12–1–12–55. ACM (2007)

HR96. James Hoover, H., Rudnicki, P.: Teaching freshman logic with MIZAR-MSE. Mathesis Universalis, 3 (1996), `http://www.calculemus.org/MathUniversalis/3/`; ISSN 1426-3513

Knu84. Knuth, D.E.: Literate programming. The Computer Journal 27(2), 97–111 (1984)

Knu92. Knuth, D.E.: Literate Programming. CSLI Lecture Notes, vol. 27. Center for the Study of Language and Information (1992)

LM01. Leijen, D., Meijer, E.: Parsec: Direct style monadic parser combinators for the real world. Technical Report UU-CS-2001-27, Department of Computer Science, Universiteit Utrecht (2001), `http://www.cs.uu.nl/~daan/parsec.html`

Nip03. Nipkow, T.: Structured Proofs in Isar/HOL. In: Geuvers, H., Wiedijk, F. (eds.) TYPES 2002. LNCS, vol. 2646, pp. 259–278. Springer, Heidelberg (2003)

NK09. Naumowicz, A., Korniłowicz, A.: A Brief Overview of Mizar. In: Berghofer, S., Nipkow, T., Urban, C., Wenzel, M. (eds.) TPHOLs 2009. LNCS, vol. 5674, pp. 67–72. Springer, Heidelberg (2009)

Nor07. Norell, U.: Towards a Practical Programming Language Based on Dependent Type Theory. PhD thesis, Department of Computer Science and Engineering, Chalmers University of Technology (September 2007)

Spi89. Spivey, J.M.: The Z Notation: A Reference Manual. Prentice Hall International Series in Computer Science. Prentice Hall (1989), Out of print; available via `http://spivey.oriel.ox.ac.uk/mike/zrm/`

Spi08. Spivey, M.: The fuzz type-checker for Z, Version 3.4.1, and The fuzz Manual, 2 edn. (2008), `http://spivey.oriel.ox.ac.uk/mike/fuzz/` (last accessed June 17, 2011)

VeriSmall: Verified Smallfoot Shape Analysis

Andrew W. Appel

Princeton University

Abstract. We have implemented a version of the Smallfoot shape analyzer, calling upon a paramodulation-based heap theorem prover. Our implementation is done in Coq and is extractable to an efficient ML program. The program is verified correct in Coq with respect to our Separation Logic for C minor; this in turn is proved correct in Coq w.r.t. Leroy's operational semantics for C minor. Thus when our VeriSmall static analyzer claims some shape property of a program, an end-to-end machine-checked proof guarantees that the assembly language of the compiled program will actually have that property.

A static analysis algorithm or type checker takes as input a program, and checks that the program satisfies a certain assertion—or in some cases calculates an appropriate assertion. A static analysis algorithm is *sound* if, whenever it calculates syntactically that the program satisfies a certain assertion, then the corresponding property really does hold on executions of the program. One way to prove soundness is to demonstrate that whenever the static analysis makes a claim, then there is a derivation tree in a given program logic that the assertion is valid for the program. Some implementations of static analyses can produce proof witnesses; this is an example of *proof-carrying code (PCC),* i.e. the pairing of a program + the witness of some static analysis applied to it.

What is the form of a "proof" for PCC? One might think it must be a derivation in logic that can be checked by a proof checker. But such derivations are unacceptably large in practice. It is more practical to factor the static analysis into an untrusted "inference" part and a proved-correct "checker". The first infers invariants and annotates the input program with assertions, as often as once per extended basic block. The checker recomputes the static analysis applied to the program, but (because of the annotations) does not need to infer any invariants, so the checker is a much simpler program. The annotations—assertions—constitute the proof witness.

The checker simply *must* be correct, or else this scheme could not reasonably be called proof-carrying code. But such checkers are generally too complex to be trusted without proof. Therefore, *Foundational PCC* requires a machine-checked proof that the checker program is sound.

In 2003 we demonstrated this approach for safety checking of compiled ML programs [15]. The "inference" part was a type-preserving compiler for Standard ML, which output a program in Typed Assembly Language. The "checker" was a nonbacktracking Prolog program written in Twelf, with a soundness proof written in Higher-Order Logic embedded in Twelf. To absolutely minimize the "trusted computing base," we implemented a tiny proof-checker for LF with a tiny interpreter for deterministic Prolog; this "checker for the checker" was 1100 lines of C, and needed to be trusted, in the sense that bugs in that component could cause unsoundness of the system.

J.-P. Jouannaud and Z. Shao (Eds.): CPP 2011, LNCS 7086, pp. 231–246, 2011.

In this paper we turn our attention beyond type systems to shape analysis based on separation logic. The state of the art in logics and proof assistants now makes it fairly straightforward to take *algorithms* from the scientific literature and implement them as *programs* with machine-checked proofs of correctness. We show that static analysis *programs* (not just *algorithms*), and decision-procedure *programs* (e.g., for entailments in separation logic) can be proved correct, and thus need not produce proof derivations. Our verified algorithm is a functional program with a proof of correctness, much like a "proof by reflection" in Coq. Because it is not just a witness-generating "program" specified as a collection of inference rules, we can more easily focus on efficiency, asymptotic complexity, and constant factors. It appears, from Cachera and Pichardie's survey of certified static analysis [6], that no one has done this before.

On the other hand, complex static analyses (that might be impractical to prove correct) can produce witnesses in the form of annotations that can be checked by our simple static analysis.

Our implementations are done in Gallina, the pure functional programming language embedded in the Coq theorem prover. Our proofs of correctness are done in the standard Coq tactic system. From the Gallina programs we use Coq's extraction to obtain ML programs, which we compile with the Ocaml system.

2 Smallfoot

Smallfoot [2,3] is a shape analyzer based on a decidable fragment of separation logic. It takes as input a pointer-manipulating program in an imperative language with structured control flow, annotated with assertions in separation logic. The assertions specify the relation of program variables to *list segments* and *tree segments*, as well as equalities and inequalities on variables. Smallfoot does not infer loop invariants: the input to Smallfoot must be explicitly annotated with loop invariants and with assertions at the beginning and end of function bodies.

Deciding entailments. Smallfoot repeatedly calls upon a decision procedure for entailments in (a decidable fragment of) separation logic. We use our Gallina implementation of such a decision procedure, and its Coq soundness proof [13].

Isolating conjuncts. When Smallfoot encounters a load, store, or deallocate command that operates at some address e (where e is an expression of the source language), it must rearrange the current precondition to isolate a (separating) conjunct of the form $e \mapsto e'$. This may require unfolding a list segment or introducing a disjunction. We will describe our Gallina program to isolate conjuncts—the algorithms that Berdine et al. [2] call *rearrangement* and *exorcism*—and its soundness proof.

Symbolic execution. Static analysis proceeds by forward symbolic execution from each assertion, through straight-line commands and through if-statements until another assertion is reached. We will describe our Gallina implementation of symbolic execution, and its soundness proof.

Frame inference. Smallfoot infers separation-logic frames for function calls, but our current protype does not implement this.

Tuerk's *Holfoot* [14] is a Smallfoot-like tool implemented in the HOL4 proof assistant. It is *proof-generating* rather than *verified*. Holfoot moves smoothly from fully automatic "shape" proofs to semiautomatic functional correctness proofs, generating lemmas that a human being or an SMT solver must prove. Holfoot is not connected to the operational semantics of any particular programming language, but to an abstract local-action semantics. Holfoot is not so much a specific algorithm as the carefully ordered application of inference rules, along with a *consequence conversion* system.

Here, in contrast, we focus on an efficient and verifiable static analysis algorithm for a real programming language connected to a real compiler, but unlike Tuerk we do not (yet!) go beyond shape analysis into the realm of functional correctness.

3 Syntax of Separation Logic

Definition var := positive.
Inductive expr := Nil: expr | Var: var → expr.
Inductive pure_atom := Eqv : expr→ expr→ pure_atom | Neqv : expr→ expr→ pure_atom.
Inductive space_atom :=
  Next: expr → expr → space_atom | Lseg: expr → expr → space_atom.
Inductive assertion := Assertion: ∀ (Π: list pure_atom) (Σ: list space_atom), assertion.
Inductive entailment : Type := Entailment : assertion → assertion → entailment.

Above is our syntactic separation logic fragment. Variable-names are represented by positive numbers. An expression is either the literal Nil or a variable. A pure (nonspatial) atom is of the form $e_1 = e_2$ or $e_1 \neq e_2$; an assertion contains (the conjunction of) a list Π of pure atoms, and the *separating* conjunction of a list of *space atoms*. Each space atom describes either a list cell or a list segment (Smallfoot's space atoms also describe trees, which our current prototype does not handle). The list cell Next e1 e2 represents a cons cell at address e_1 whose tail-pointer contains the value e_2, or in primitive separation logic, $(e_1 \mapsto \_) * (e_1 + 1 \mapsto e_2)$. The list segment Lseg e1 e2 represents either $e_1 = e_2$ (meaning an empty segment) or a chain of one or more list cells, starting at address e_1, whose last tail-pointer is e_2, and where $e_1 \neq e_2$.

Smallfoot is a forward symbolic execution algorithm that takes a known precondition P in this fragment, along with a command c, and derives a postcondition Q such that $\{P\}c\{Q\}$. In cases where Q is a disjunction, the disjunction is always at top-level.

4 Semantics of Separation Logics

One application of our shape analysis is in our Verified Software Toolchain [1], where we have a separation logic for C minor, which is a source language for the CompCert verified C compiler [10]. Our higher-order impredicative Concurrent Separation Logic is proved sound with respect to the operational semantics of C minor; Leroy et al. have proved CompCert correct w.r.t. the same operational semantics.

We can also imagine many other uses for efficient, proved-correct decision procedures and shape analyses, so we do not want to tie our soundness result too closely to a particular model of separation logic. Figure 1 shows our general interface, specified as a Module Type, to practically any reasonable model of separation logic that could

```
Require Import msl.sepalg.
Parameter loc : Type.
Parameter val: Type.
Declare Instance val_sa : sepalg val.
Parameter val2loc: val → option loc.
Parameter nil_val : val.
Axiom nil_not_loc: val2loc nil_val = None.
Parameter empty_val: val.
Axiom emp_empty_val: ∀v,   identity v  ↔  v=empty_val.
Definition full (v: val) := ∀v2, joins v v2 → identity v2.
Axiom val2loc_full: ∀v l, val2loc v = Some l → full v.
Axiom nil_full: full nil_val.
Axiom empty_not_full: ~full empty_val.
Axiom val2loc_inj: ∀v1 v2 l, val2loc v1 = Some l → val2loc v2 = Some l → v1=v2.
Axiom loc_eq_dec: ∀l1 l2: loc, Decidable.decidable (l1=l2).
Axiom nil_dec: ∀v, Decidable.decidable (v=nil_val).

Definition var : Type := positive.
Parameter env : Type.
Parameter env_get: env → var → val.
Parameter env_set: var → val → env → env.
Axiom gss_env : ∀(x : var) (v : val) (s : env), env_get (env_set x v s) x = v.
Axiom gso_env : ∀(x y : var) (v : val) (s : env), x<>y →
                    env_get (env_set x v s) y = env_get s y.
Parameter empty_env : env.
Axiom env_gempty: ∀x, env_get empty_env x = empty_val.

Parameter heap : Type.
Declare Instance heap_sa : sepalg heap.
Parameter rawnext: ∀(x: loc) (y : val) (s : heap), Prop.
Parameter emp_at: ∀(l: loc) (h: heap), Prop.
Axiom heap_gempty: ∀h l, identity h → emp_at l h.
Definition nil_or_loc (v: val) := v=nil_val ∨ ∃l, val2loc v = Some l.
Axiom mk_heap_rawnext: ∀h x0 x y,
   val2loc x0 = Some x → nil_or_loc y → ∃h', rawnext x y h' ∧ comparable h h'.

Axiom rawnext_out: ∀x x0 x' y h,
   rawnext x y h → val2loc x0 = Some x' → x'<>x → emp_at x' h.

Definition rawnext' x y h := ∃h0, join_sub h0 h ∧ rawnext x y h0.
Axiom rawnext_at1: ∀x y h1 h2 h, rawnext' x y h1 → join h1 h2 h →
      emp_at x h2 ∧ rawnext' x y h.
Axiom rawnext_at2: ∀x y h1 h2 h, join h1 h2 h → rawnext' x y h →
      emp_at x h2 → rawnext' x y h1.
Axiom rawnext_not_emp: ∀x y h, rawnext' x y h → ~emp_at x h.
Axiom emp_at_join: ∀h1 h2 h, join h1 h2 h→∀l, (emp_at l h1∧emp_at l h2) ↔emp_at l h.
```

Fig. 1. Specification of models for separation logic

```
Inductive state := State: ∀(s: env) (h: heap), state.

Definition expr_denote (e : expr) (σ : state) : val :=
    match e , σ with Nil , _ ⇒ nil_val | Var x , State s _ ⇒ env_get s (Some x) end.
Definition expr_eq (x y : expr) (s : state) := expr_denote x s = expr_denote y s.

Definition spred := state → Prop.
Definition neg (P : spred) : spred := fun σ : state ⇒ ~P σ.

Definition pure_atom_denote (a : pure_atom) : spred :=
  match a with Eqv e1 e2 ⇒ expr_eq e1 e2 | Neqv e1 e2 ⇒ neg (expr_eq e1 e2) end.

Inductive lseg : val → val → heap → Prop :=
| lseg_nil : ∀x h, identity h → nil_or_loc x → lseg x x h
| lseg_cons : ∀x x' h h0 h1 z, x<>y → val2loc x = Some x' →
     rawnext x' z h0 → lseg z y h1 → join h0 h1 h → lseg x y h.

Definition space_atom_denote (a : space_atom) : spred := fun σ ⇒
  match a, σ with
  | Next x y , State _ h ⇒
      fun σ ⇒ match val2loc (expr_denote x σ) with
      | Some l' ⇒ rawnext l' (expr_denote y σ) h ∧ nil_or_loc (expr_denote y σ)
      | None ⇒ False
      end
  | Lseg x y, State _ h ⇒ lseg (expr_denote x σ) (expr_denote y σ) h
  end.

Fixpoint list_denote {A T : Type} (f : A → T) (g : T → T → T) (b : T) l : T :=
  match l with nil ⇒ b | x :: l' ⇒ g (f x) (list_denote l') end.

Definition assertion_denote (f : assertion) : spred :=
match f with Assertion Π Σ ⇒ list_denote pure_atom_denote (@intersection state) T Π
                              ∧ list_denote space_atom_denote sep_con emp Σ
end.
```

Fig. 2. Denotations

support list segments. **Import** msl.sepalg refers to the notion of Separation Algebras [9] from our Mechanized Semantic Library (`msl.cs.princeton.edu`).

We prove that our C minor separation logic satisfies this specification. Separating out the interface in this way causes some local pain, compared to a direct nonabstract model, but the improvement in modularity is well worth it.

Based on this semantic specification of the operators, we can define the denotations of syntactic expressions and assertions, as shown in Figure 2.

Remark. Berdine et al. assume an abstract addressing model such that if $p \neq q$ then the fields of p cannot possibly overlap with the fields of q; other presentations of Separation Logic assume an address-arithmetic model, in which records might overlap; e.g., $100 \mapsto x * 101 \mapsto y * 102 \mapsto z$ might contain the pair (x,y) overlapping with

```
Definition fresh {A} (f: A → positive) (a: A) (x: positive) : Prop := Zpos (f a) ≤ Zpos x.

Definition agree_except (x: var) (σ σ': state) : Prop :=
 match σ, σ' with State s h , State s' h' ⇒
   (∀ x', x' <> x → env_get s (Some x') = env_get s' (Some x')) ∧ h=h' end.

Definition existsv(x: var) (P: spred) : spred := fun σ ⇒ ∃σ', agree_except x σ σ' ∧ P σ'.

Definition |−− (P Q: spred) := ∀s, P s → Q s.        Infix "|−−".

Lemma pure_atom_denote_agree: ∀a σ σ' x, fresh freshmax_pure_atom a x →
  agree_except x σ σ' → pure_atom_denote a σ → pure_atom_denote a σ'.

Lemma space_atom_denote_agree: ∀a σ σ' x, fresh freshmax_space_atom a x →
  agree_except x σ σ' → space_atom_denote a σ → space_atom_denote a σ'.
```

Fig. 3. Freshness

the pair (y,z). We model Next so that it may be instantiated in either the abstract or the address-arithmetic style. But the Smallfoot inference rules assumed by Berdine et al. are sound only if such overlap cannot occur. The only way we know how to assure this, in an address-arithmetic setting, is to make the rather strong assumption that list cells are aligned on a multiple-of-size boundary.

5 Fresh Variables

When symbolic execution rewrites separation-logic assertions, it sometimes uses fresh variables, i.e. new variables that are not free in the current program or precondition.

We have functions freshmax_expr, freshmax_pure_atom, freshmax_space_atom, freshmax_assertion, freshmax_stmt, that traverse assertions and commands to find the highest-numbered variable in use (highest-numbered nonfresh variable).

Figure 3 gives some definitions and lemmas regarding freshness of variables. Let a be an expression (or pure atom, space atom, assertion, statement) and let f be the freshmax_expr function (or respectively, freshmax_pure_atom, freshmax_space_atom, etc.). Then we say that some variable x is fresh for a by writing fresh f a x.

Zpos injects from positive to the integers; for efficiency our program *computes* on positives, but for convenience in *proofs* we use tactics and lemmas on the integers.

We can say that two states σ and σ' agree_except at x, and we define existsv x P to mean that P almost holds (on a given state)—that is, there exists a value v such that P would hold on the state if only we would set $x := v$. Finally, if x is fresh for a, and two states σ and σ' agree_except at x, then a at σ is equivalent to a at σ'.

6 Paramodulation

Smallfoot makes repeated calls to decide entailments in separation logic. Berdine et al. [2] sketch an algorithm for deciding entailments in their fragment of separation logic.

Navarro and Rybalchenko [11] apply *paramodulation*, a resolution theorem-proving algorithm, to this decision problem, and get a program that is significantly faster than the original Smallfoot implementation. Paramodulation [12] permits modular introduction of theories; a standard such theory to add is the *superposition calculus*, a theory of equalities and inequalities. Navarro and Rybalchenko extend paramodulation with superposition and with the *spatial terms* of Berdine et al.'s decidable fragment of separation logic, yielding a "heap theorem prover."

Gordon Stewart, Lennart Beringer, and I have built an implementation in Gallina of this paramodulation-based heap theorem prover. Our proof of soundness is nearly finished, and we intend to prove termination and perhaps completeness. Preliminary measurements of the extracted ML code are competitive with Smallfoot's entailment decider (also implemented in ML). This is not nearly as good as it might seem, because in fact Navarro and Rybalchenko's implementation (in Prolog) is about 100x faster than Smallfoot in solving large entailments. We expect that we can improve our program with more efficient data structures for term indexing and priority queues (with attended proofs of soundness). We will report on paramodulation in a separate paper [13].

7 Isolation

Consider the command `a:=b.next`, which loads a field of record b. (Similar issues pertain to storing a field or deallocating a record.) Suppose precondition P has the form $\mathsf{Next}\, b\, c * F$ for some frame F. Assuming that the variable a is not free in expressions b, c or formula F, it's easy to derive a postcondition $(a = c) * \mathsf{Next}\, b\, c * F$.

Suppose instead that P is $F_1 * \mathsf{Next}\, b\, c * F_2$. and the separation-logic Hoare rule for assignment has a syntactic form that requires $\mathsf{Next}\, b\, c * F$. Clearly by the associative law, $P \vdash \mathsf{Next}\, b\, c * (F_1 * F_2)$. We can use the rule of consequence to strengthen the precondition to match the desired form.

A harder case is one where the precondition P is $b \neq d * \mathsf{Lseg}\, b\, d * F$. Because the list segment is not empty ($b \neq d$), we can unfold it once; we insert a fresh variable x (not free in a, b, d, F) as follows: $P \vdash \mathsf{Next}\, b\, x * b \neq d * \mathsf{Lseg}\, x\, d * F$.

In each of the cases above, we rearrange the precondition to *isolate* one field as required by a load (or store); in the case of a deallocation command we would have to isolate all the fields of a particular record together, but the issues would be the same. An important component of the Smallfoot algorithm is this rearrangement.

In the case where P is $\mathsf{Lseg}\, b\, d * F$, such that $\mathsf{Lseg}\, b\, d * F \not\vdash b \neq d$, then the list segment might possibly be empty, so we cannot unfold it; symbolic execution will be stuck here, unable to prove by shape analysis that the program is safe.

The hardest case (as explained by Berdine et al.) is the "spooky disjunction." Suppose P is $d \neq e * \mathsf{Lseg}\, b\, d * \mathsf{Lseg}\, b\, e * F$. We know that exactly one of the two segments is nonempty; if both are empty, then $d = e$, and if both are nonempty, then the segments (b, d) and (b, e) would overlap (would not separate). Whichever segment is nonempty, we should be able to unfold it, but we do not know which. Therefore we can derive

$$P \vdash (\mathsf{Next}\, b\, x * d \neq e * \mathsf{Lseg}\, x\, d * b = e * F) \vee (\mathsf{Next}\, b\, x * d \neq e * b = d * \mathsf{Lseg}\, x\, e * F).$$

```
Fixpoint exorcize (e: expr) (Π: list pure_atom) (Σ0 Σ: list space_atom) (x: var)
                                                    : option(list assertion) :=
 match Σ with
 | nil ⇒ if incon (Assertion Π (rev Σ0)) then Some nil else None
 | Lseg f f' :: Σ1 ⇒
     if oracle (Entailment (Assertion Π (rev Σ0 ++ (Lseg f f') :: Σ1))
                          (Assertion (Eqv e f :: nil) (rev Σ0 ++ Lseg f f' :: Σ1)))
     then match exorcize e (Eqv f f' :: Π) (Lseg f f' :: Σ0) Σ1 x with
           | Some l ⇒ Some (Assertion Π (Next e (Var x) :: Lseg (Var x) f' :: rev Σ0 ++ Σ1) ::l)
           | None ⇒ None
           end
     else exorcize e Π (Lseg f f' :: Σ0) Σ1 x
 | a :: Σ1 ⇒ exorcize e Π (a :: Σ0) Σ1 x
 end.

Fixpoint isolate' (e: expr) (Π : list pure_atom) (Σ0 Σ: list space_atom) (x: var) (count: nat)
                                                    : option(list assertion) :=
match Σ with
| nil ⇒ if count < 2 then None
        else if incon (Assertion (Eqv e Nil :: Π) (rev Σ0))
             then exorcize e Π nil (rev Σ0) x
        else None
| Next e1 e2 :: Σ1 ⇒
     if eq_expr e e1
     then Some [Assertion Π (Next e e2 :: rev Σ0 ++ Σ1)]
     else if oracle (Entailment (Assertion Π (rev Σ0 ++ (Next e1 e2) :: Σ1))
                               (Assertion (Eqv e e1 :: nil) (rev Σ0 ++ (Next e1 e2) :: Σ1)))
        then Some [Assertion Π (Next e e2 :: rev Σ0 ++ Σ1)
else isolate' e Π (Next e1 e2 :: Σ0) Σ1 x count
| Lseg f f' :: Σ1 ⇒
     if oracle (Entailment (Assertion Π (rev Σ0 ++ (Lseg f f') :: Σ1))
                          (Assertion (Eqv e f :: Neqv f f' :: nil ) (rev Σ0 ++ (Lseg f f') :: Σ1)))
     then Some [Assertion Π (Next e (Var x) :: Lseg (Var x) f' :: rev Σ0 ++ Σ1)]
     else if oracle (Entailment (Assertion Π (rev Σ0 ++ (Lseg f f') :: Σ1))
                               (Assertion (Eqv e f :: nil) nil (rev Σ0 ++ (Lseg f f') :: Σ1)))
          then isolate' e Π (Lseg f f' :: Σ0) Σ1 x (S count)
     else isolate' e Π (Lseg f f' :: Σ0) Σ1 x count
end.

Definition isolate (e: expr) (P: assertion) (x: var) : option (list assertion) :=
 match P with Assertion Π Σ ⇒ isolate' e Π nil Σ x 0 end.
```

Fig. 4. Exorcize and isolate

The algorithm for eliminating the "spooky disjunctions" is called *exorcism* by Berdine et al., and their *entire* description of it is thus:

> To deal with this in the rearrangement phase we rely on a procedure for exorcising these spooky disjunctions. In essence, exor$(\Pi|\Sigma, e)$ is a collection of assertions obtained by doing enough case analysis (adding equalities and inequalities to Π) so that the location of e within a $*$-conjunct is determined. This makes the rearrangement rules complete. We omit a formal definition of exor for space reasons.

This function for isolating a field (preparatory to a load or store) we will name *isolate*. It calls upon an auxiliary function *exorcize*. Our assertion syntax has no disjunction operator, so we formulate the output of these functions as an option(list(assertion)). The result None indicates that it was not possible to isolate the given field; the result Some(l) gives a list l of assertions, the disjunction of which is implied by the input assertion P.

The isolate' function walks down a list Σ of space atoms, tossing them into Σ_0 as it passes by, as follows: • In the last **else** clauses of the Next and Lseg clauses, where e1 or f can't be proved equivalent to e, this Next or Lseg is an irrelevant conjunct—the recursive call to isolate' simply moves it from Σ to Σ_0 and continues. • If e is syntactically identical to e1, or we can prove $\Pi, \Sigma \vdash$ e = e1, then the conjunct Next e1 e2 matches, and isolate succeeds. • If we can prove from Π, Σ that e=f and f≠f', then isolate succeeds by unfolding this list segment. • Finally, if $\Pi, \Sigma \vdash$ e = f but we cannot also prove f≠f', then the conjunct is a candidate for a spooky disjunction, so we toss it into Σ_0 and increment the count variable, which counts the number of spooky disjuncts.

If isolate' reaches the end of Σ with count>1, then there is a spooky disjunction. exorcize handles it (Figure 4) by performing case-splitting (empty or nonempty) on each relevant Lseg. The two cases appear as Eqv f f' and Next e (Var x) :: Lseg (Var x) f' :: ..., respectively. In the Eqv case, we must also case-split on all the remaining relevant Lsegs, but in the non-Eqv case, all the others must be empty.

Discussion. This is just straightforward functional programming: nothing remarkable about it, except that we can now use Gallina's proof theory (i.e., CiC) to prove the soundness. Termination is already proved, because **Fixpoint** must terminate.

8 Soundness of Isolate

```
Lemma exorcize_sound: ∀e Π Σ x,
      fresh freshmax_expr e x → fresh freshmax_assertion (Assertion Π Σ) x →
      ∀cl, (exorcize e Π nil Σ x) = Some cl →
  (assertion_denote (Assertion Π Σ) |−−
    fold_right (fun P ⇒ union (existsv x (assertion_denote P))) FF cl) ∧
        (∀ Q, In Q cl →
                match Q with
                |Assertion _ (Next e0 _ :: _) ⇒ e=e0
                | _ ⇒ False
                end ∧ fresh freshmax_assertion Q (Psucc x)).
```

```
Lemma isolate_sound: ∀e P x results,
      isolate e P x = Some results →
       fresh freshmax_expr e x → fresh freshmax_assertion P x →
      assertion_denote P |--
        fold_right (fun Q ⇒ union (existsv x (assertion_denote Q))) FF results ∧
      ∀Q, In Q results →
            match Q with
           |Assertion _ (Next e0 _ :: _) ⇒ e=e0
           | _ ⇒ False
           end ∧ fresh freshmax_assertion Q (Psucc x).
```

Given an assertion P and the desire to isolate a conjunct of the form Next e _ , and given x fresh for e and P, suppose isolate e P x returns Some results. Then we know:

- The denotation of P entails the union of all the disjuncts Q in results, provided that we set the variable x to some appropriate value.
- Every disjunct Q is of the form Assertion _ (Next e _ :: _).
- Every free variable in Q has name $\leq$ x, i.e., the *next* variable after x is fresh for Q.

9 Symbolic Execution

Symbolic execution proceeds on a C minor syntax annotated with assertions. The shape analyses will not interpret many of the C minor expressions it sees, but simple expressions such as variables and and the constant 0 (interpreted as Nil) will be relevant to symbolic execution. Thus we define the function Cexpr2expr that translates simple expressions from C minor to the language of our syntactic assertions, and ignores others:

```
Definition Cexpr2expr (e: Cminor.expr) : option expr :=
  match e with Evar i ⇒ Some (Var i)
                   | Eval (Vint z) ⇒ if Int.eq_dec z Int.zero then Some Nil else None
                   | _ ⇒ None
  end.

Definition getSome {A} (x: option A) (f: A → bool):=
  match x with Some y ⇒ f y | None ⇒ false end.

Definition Cexpr2assertions(e:Cminor.expr)(a:assertion)(f:assertion→ assertion→ bool):=
 match a with Assertion Π Σ ⇒
  match e with
  | Ebinop (Cminor.Ocmp Ceq) a b ⇒
       getSome (Cexpr2expr a) (fun a' ⇒ getSome (Cexpr2expr b) (fun b' ⇒
         f (Assertion (Eqv a' b' ::Π) Σ) (Assertion (Neqv a' b' ::p) Σ)))
  | Ebinop (Cminor.Ocmp Cne) a b ⇒
       getSome (Cexpr2expr a) (fun a' ⇒ getSome (Cexpr2expr b) (fun b' ⇒
         f (Assertion (Neqv a' b' ::Π) Σ) (Assertion (Eqv a' b'::Π) Σ)))
  | _ ⇒ getSome (Cexpr2expr e) (fun a' ⇒
             f (Assertion (Neqv a' Nil ::Π) Σ) (Assertion (Eqv a' Nil ::Π) Σ))
  end
 end.
```

```
Fixpoint check (P: assertion) (BR: list assertion) (c: stmt) (x': positive)
                              (cont: assertion → positive → bool) : bool :=
 if incon P then true
 else match c with
  | Sskip ⇒ cont P x'
  | Sassert Q ⇒ oracle (Entailment P Q) && cont Q x'
  | Sassign x (Evar i) ⇒
     match P with Assertion Π Σ ⇒
     let P':=Assertion(Eqv (Var x) (subst_expr x (Var x') (Var i)) :: subst_pures x (Var x') Π)
                              (subst_spaces x (Var x') Σ)
      in cont P' (Psucc x')
     end
  | Sassign x (Eload Mint32 (Ebinop Cminor.Oadd (Evar i) (Eval (Vint ofs)))) ⇒
      Int.eq ofs (Int.repr 4) &&
      getSome (isolate (Var i) P x') (fun l ⇒
         forallb(fun P' ⇒
               match P' with
               | Assertion Π' (Next _ f :: Σ') ⇒
                 cont (Assertion (Eqv (Var x) (subst_expr x (Var (Psucc x')) f)
                                 :: subst_pures x (Var (Psucc x')) Π')
                       (subst_spaces x (Var (Psucc x')) (Next (Var i) f ::Σ')))
                            (Psucc (Psucc x'))
               | _ ⇒ false
               end)
              l)
  | Sstore Mint32 (Ebinop Cminor.Oadd e1 (Eval (Vint ofs))) e2 ⇒
       Int.eq ofs (Int.repr 4) &&
       getSome (Cexpr2expr e1) (fun e1' ⇒ getSome (Cexpr2expr e2) (fun e2' ⇒
       getSome (isolate e1' P x') (fun l ⇒
        forallb(fun P' ⇒
               match P' with
               | Assertion Π' (Next _ f :: Σ') ⇒
                    cont (Assertion Π' (Next e1' e2' :: Σ')) (Psucc x')
               | _ ⇒ false
               end)
             l)))
  | Sexit n ⇒ oracle (Entailment P (nth n BR false_assertion))
  | Sblock (Sloop (Sblock (Sifthenelse e c1 c2))) ⇒ (* while loop! *)
      Cexpr2assertions e P (fun P1 P2 ⇒
          check P1 (P::P2::BR) c1 x' (fun R y' ⇒ false) &&
          check P2 (P::P2::BR) c2 x' (fun R y' ⇒ false) &&
          cont P2 x')
  | Sifthenelse e c1 c2 ⇒
      Cexpr2assertions e P (fun P1 P2 ⇒
           check P1 BR c1 x' cont && check P2 BR c2 x' cont)
  | Sseq c1 c2 ⇒ check P BR c1 x' (fun P' y' ⇒ check P' BR c2 y' cont)
  | _ ⇒ false
 end.
```

Fig. 5. Symbolic execution

Symbolic execution is flow-sensitive, and when interpreting an **if** statement, "knows" in the **then** clause that the condition was true, and in the **else** clause that the condition was false. For this purpose we define a function Cexpr2assertions e a f that takes C-minor expression e and assertion a, and generates two new assertions equivalent (more or less) to e ∧ a and ∼e ∧ a, and applies the continuation f to both of these assertions. We write "more or less" because e ∧ a is actually an illegitimate mixture of two different syntaxes; e must be properly translated into the assertion syntax, which is the purpose of Cexpr2assertions.

Symbolic execution relies on functions subst_expr x e e', subst_pures x e Π, and subst_spaces x e Σ that substitute expression e for the variable x in (respectively) an expression e', a pure term Π, or a space term Σ.

Smallfoot symbolic execution uses a restricted form of assertion without disjunction. Therefore when a disjunction would normally be needed, Smallfoot does multiple symbolic executions over the same commands. For example, for

(**if** e **then** c1 **else** c2); c3; c4; assert Q

with precondition P, Smallfoot executes the commands c1;c3;c4 with precondition e ∧ P and then executes c2;c3;c4 with precondition ∼e ∧ P. Because Berdine et al.'s original Smallfoot used only simple "if and while" control flow, this re-execution was easy to express.

C minor has a notion of nonlocal loop exit; that is, one can exit from any number of nested blocks (such as loops, loop bodies, or switch statements). One branch of an **if** statement might exit, while the other might continue normally. To handle this notion, the parameters of the check function include not only a precondition P but a break-condition list BR that gives exit-postconditions for all possible exit labels.

In order to handle re-execution mixed with multiple-level exit, we write the symbolic execution function in continuation-passing style. The argument cont is the check function's continuation. Once check has computed the postcondition Q for a given statement, it calls cont with Q. If it needs to call cont more than once, it may do so. For example, in the clause for Sifthenelse notice that cont is passed to two different recursive calls to check, each of which will perhaps call cont. Or perhaps not; the symbolic execution of Sexit n (to break out of n nested blocks) does not call cont at all, but looks up the nth item in BR.

The miracle of termination. In Coq, a **Fixpoint** function must have a structurally inductive parameter, such that in every recursive call the actual parameter is a substructure of the formal parameter. Here the structural parameter is the statement c. Most of the recursive calls are buried in continuations (lambda-expressions passed to the cont parameter)—and may not actually occur until much later, inside other calls to check. The miracle is that Coq still recognizes this function as structurally recursive.

10 Ghost Variables

A variable mentioned in an assertion but not in the program is a *ghost variable*. In a Hoare logic with ghost variables, one has rules capable of proving such derivations as,

$$\frac{\{a = x\} \quad a \leftarrow a+1 \ \{a = x+1\}}{\{a = x-1\} \ a \leftarrow a+1 \ \{a = x\}}$$

That is, taking advantage of the fact that x is not free in the command $a := a + 1$, substitute for x in both the pre- and postcondition.

Our underlying Hoare logic does not handle ghost variables directly. We could add such a rule, as it is provable from the underlying operational model of C minor. But instead we find that our Concurrent Separation Logic is expressive enough to derive a new Separation Logic *with* a ghost-variable rule; its rules are proved sound as derived lemmas from the underlying rules. In the new logic, we add a separate namespace of *logical variables* (or ghost variables) visible to semantic assertions but not to ordinary commands. (Also, the underlying Separation Logic has variables-as-resources [5], but the top layer has a conventional (nonresource) treatment of variables; the underlying layer has fractional ownership shares [9], but the top layer is a conventional all-or-nothing separation logic.)

The Smallfoot algorithm would like to think that there's just one namespace of variables, so our *syntactic* separation logic (Section 3) has just one namespace. Let the variable ghost be the first one beyond the highest-numbered variabled used in the program. In our interpretation of Hoare triples during symbolic execution, all the variables beyond ghost in the syntactic Hoare logic will be interpreted as logical variables.

Then we do some predicate translation, as follows. Let P be a predicate on states. We define the *ghostly denotation* of P as a predicate on worlds:

$$[\![P]\!]_{\mathsf{ghost}} = \lambda w.\ P\,(\mathsf{State}\ (\mathsf{mix\_envs}\ 1\ \mathsf{ghost}\ (\mathrm{w\_rho}\, w)\ (\mathrm{w\_aux}\, w)\ )\ (\mathrm{w\_m}\, w))$$

where mix_envs lo hi ρ a is the environment that consults the "real" local-variable environment ρ on variables lo $\leq i <$ hi, otherwise consults the ghost environment a.

At the start of the symbolic execution, the check0 function computes the ghost boundary x for the given program by taking the max of all variable names in use:

```
Definition check0 (P: assertion) (c: stmt) (Q: assertion) : bool :=
 let x := Pmax (Pmax (freshmax_assertion P) (freshmax_stmt c)) (freshmax_assertion Q)
  in check P nil c x (fun Q' _ ⇒ oracle (Entailment Q' Q)).
```

11 Soundness of Symbolic Execution

```
Theorem check_sound: ∀G P c Q,
    check0 P c Q = true →
    semax G (assertion2wpred P) (erase_stmt c) (RET1 (assertion2wpred Q)).
```

Theorem [check_sound]. If the symbolic executor checks a Hoare triple (check0 P c Q) then that triple is semantically sound, according to our axiomatic semantics semax. Since check0 takes syntactic assertions and semax takes semantic assertions, in the statement of this theorem we must do world-to-state translations and take assertion-denotations, which is what assertion2wpred does.

Proof. By induction on the height of commands, using the following induction scheme.

```
Definition check_sound_scheme (c: stmt) :=
 ∀ghost G P BR x cont
 (GHOST: Zpos ghost <= Zpos x)
 (H: check P BR c x cont = true)
 (FRESHc: fresh freshmax_stmt c ghost)
 (FRESHP: fresh freshmax_assertion P x)
 (FRESHBR: fresh (freshmax_list freshmax_assertion) BR x),
 semax G [assertion_denote P]ghost (erase_stmt c)
     (RET [cont_assert x cont]ghost [map assertion_denote BR]ghost).
```

```
Lemma check_sound_helper: ∀n c, (height c < n) → check_sound_scheme c.
```

We could *almost* do the induction directly on the structure of commands, except for one case involving C minor's block command.

The most difficult and annoying issues in the proof involve the treatment of fresh variables, especially in the case where some postcondition will hold provided that the fresh variable has an appropriate value.

The treatment of the continuation argument of the check function is one of the interesting parts of the proof. By the time symbolic execution calls cont with postcondition Q and a fresh variable x, that means variables <x may be used in Q. The semantic meaning of this cont function is the disjunction of all the arguments Q for which cont could return true, but the instantiation of fresh variables must also be taken into account, as follows:

```
Definition agree_except_range (lo hi: var) (σ σ': state) : Prop :=
 match σ, σ' with State s h , State s' h' ⇒
 (∀ x, lo <= x < hi ∨ env_get s (Some x) = env_get s (Some x)) ∧ h = h' end.
```

```
Definition existsvs (lo hi: var) (P: spred) : spred :=
   fun σ ⇒ ∃σ', agree_except_range lo hi σ σ' ∧ P σ'.
```

```
Definition cont_assert (lo: positive) (cont: assertion → var → bool) : spred :=
 fun σ ⇒ ∃(Q : assertion) (hi: var),
             cont Q hi = true ∧ lo <= hi ∧ fresh freshmax_assertion Q hi
             ∧ existsvs lo hi (assertion_denote Q) σ.
```

12 Conclusion

Yang et al. [16] improve on Smallfoot with a symbolic join operator $\sqcup$ that reduces or eliminates the need to symbolically re-execute the same statements. The *bi-abduction* algorithm infers frames and anti-frames for function calls [7]. Both $\sqcup$ and bi-abduction should be implementable and provable in Coq; even if not, an unverified static analyzer using them can generate annotations checkable by Smallfoot (and the extra $\sqcup$-joined annotations would avoid symbolic re-execution in Smallfoot).

The SLAyer program analysis [4] infers loop invariants using external theorem provers; at this scale of software we might skip proving SLAyer correct, and just have it generate assertions checkable by our proved-correct tool.

Our implementation and verification is not much more than just competent engineering. Computer Science has reached the point where one can take a result from the literature, use conventional functional programming to write a purely functional program, and then use Coq or Isabelle (etc.) to prove correctness. The proofs were accomplished by the expedient of having a reasonably competent Coq hacker (the author) slog away in the tactical theorem prover for a few weeks; the paramodulation implementation and proofs are the work of Gordon Stewart, Lennart Beringer, and the author.

The soundness proofs are complete except for one issue: our C minor Hoare logic requires for the command M[p]:=q that q be an initialized variable; neither the original Smallfoot nor our implementation tracks initialized variables, but this would be trivial to add and prove sound.

Component	*Program Lines*	*Proof Lines*
Paramodulation	1096	~4000
Isolate/exorcize	125	798
Symbolic execution	149	3606

We will not report performance benchmarks in this paper, as they depend critically on the performance of the entailment oracle (paramodulation), and our implementation of paramodulation is not yet tuned. However, the check function—when extracted to Caml—is about as clean and efficient as one could imagine an implementation of Smallfoot to be in any language, and even our preliminary untuned implementation of paramodulation is competitive with Smallfoot on Navarro's benchmark suite [11].

Here we proved Smallfoot sound w.r.t. a separation logic; the deeper and more difficult scientific results are in the soundness of that logic w.r.t. an optimizing compiler [1]. That logic is very highly expressive, with many features (concurrency, impredicativity, indirection) that are entirely unneeded by the Smallfoot soundness proof. But if static analyzers such as this one are to be connected to other automatic or semiautomatic program analyses and program-proof systems, it would be helpful to have them all proved sound w.r.t. the same logic—hence the desire for "expressiveness overkill" in the logic.

Furthermore, there are difficult results in the specification of optimizing compilers for thread-concurrent languages (such as C with Pthreads) or for programs that do shared-memory interaction with an operating system—and in the connection of these specifications to the Hoare logic [8]. But having all that infrastructure in place means that our soundness result is not just "if the shape checker says true then there's a Separation Logic proof;" it means that *and* "if there's a Separation Logic proof then the assembly code that results from the optimizing compilation in CompCert behaves according to the specification checked by the shape checker." Connecting all the components end-to-end gives a much more valuable result.

Acknowledgments. This research was supported in part by the Air Force Office of Scientific Research (grant FA9550-09-1-0138) and the National Science Foundation (grant CNS-0910448).

References

1. Appel, A.W.: Verified Software Toolchain. In: Barthe, G. (ed.) ESOP 2011. LNCS, vol. 6602, pp. 1–17. Springer, Heidelberg (2011)
2. Berdine, J., Calcagno, C., O'Hearn, P.W.: Symbolic Execution with Separation Logic. In: Yi, K. (ed.) APLAS 2005. LNCS, vol. 3780, pp. 52–68. Springer, Heidelberg (2005)
3. Berdine, J., Calcagno, C., O'Hearn, P.W.: Smallfoot: Modular Automatic Assertion Checking with Separation Logic. In: de Boer, F.S., Bonsangue, M.M., Graf, S., de Roever, W.-P. (eds.) FMCO 2005. LNCS, vol. 4111, pp. 115–137. Springer, Heidelberg (2006)
4. Berdine, J., Cook, B., Ishtiaq, S.: SLAYER: Memory Safety for Systems-Level Code. In: Gopalakrishnan, G., Qadeer, S. (eds.) CAV 2011. LNCS, vol. 6806, pp. 178–183. Springer, Heidelberg (2011)
5. Bornat, R., Calcagno, C., Yang, H.: Variables as resource in separation logic. Electronic Notes in Theoretical Computer Science 155, 247–276 (2006); Proc. 21st Annual Conf. on Mathematical Foundations of Programming Semantics (MFPS XXI)
6. Cachera, D., Pichardie, D.: Comparing techniques for certified static analysis. In: Proc. 1st NASA Formal Methods Symposium (NFM 2009), pp. 111–115. NASA Ames Research Center (2009)
7. Calcagno, C., Distefano, D., O'Hearn, P., Yang, H.: Compositional shape analysis by means of bi-abduction. In: POPL 2009: Proc. 36th Annual ACM SIGPLAN-SIGACT Symposium on Principles of Programming Languages, pp. 289–300. ACM (2009)
8. Dockins, R., Appel, A.W.: Behavioral refinement for unsafe languages (submitted for publication, 2011)
9. Dockins, R., Hobor, A., Appel, A.W.: A Fresh Look at Separation Algebras and Share Accounting. In: Hu, Z. (ed.) APLAS 2009. LNCS, vol. 5904, pp. 161–177. Springer, Heidelberg (2009)
10. Leroy, X.: A formally verified compiler back-end. Journal of Automated Reasoning 43(4), 363–446 (2009)
11. Pérez, J.A.N., Rybalchenko, A.: Separation logic + superposition calculus = heap theorem prover. In: Proceedings of the 32nd ACM SIGPLAN Conference on Programming Language Design and Implementation, PLDI 2011, New York, NY, USA, pp. 556–566. ACM (2011)
12. Nieuwenhuis, R., Rubio, A.: Paramodulation-based theorem proving. In: Robinson, J.A., Voronkov, A. (eds.) Handbook of Automated Reasoning, vol. I, ch. 7, pp. 371–443. Elsevier (2001)
13. Stewart, G., Beringer, L., Appel, A.W.: Verified heap theorem prover by paramodulation (in preparation, 2011)
14. Tuerk, T.: A Separation Logic Framework for HOL. PhD thesis, Univ. of Cambridge (June 2011)
15. Wu, D., Appel, A.W., Stump, A.: Foundational proof checkers with small witnesses. In: 5th ACM SIGPLAN International Conference on Principles and Practice of Declarative Programming (August 2003)
16. Yang, H., Lee, O., Berdine, J., Calcagno, C., Cook, B., Distefano, D., O'Hearn, P.W.: Scalable Shape Analysis for Systems Code. In: Gupta, A., Malik, S. (eds.) CAV 2008. LNCS, vol. 5123, pp. 385–398. Springer, Heidelberg (2008)

Verification of *Scalable Synchronous Queue*★

Jinjiang Lei[1,2] and Zongyan Qiu[1,2]

[1] LMAM and Department of Informatics, School of Math., Peking University, Beijing, China
[2] State Key Laboratory of Computer Science, ISCAS, China

Abstract. Lock-free algorithms are extremely hard to be built correct due to their fine-grained concurrency natures. Formal techniques for verifying them are crucial. We present a framework for verification of CAS-based lock-free algorithms, and prove a nontrivial lock-free algorithm *Scalable Synchronous Queue* that is practically adopted in Java 6. The strength of our approach lies on that it relieves the dependence on auxiliary variables/commands, thus is relatively easier to conduct and comprehend, comparing to existing works.

1 Introduction

Implementations of concurrent programs usually rely on locks, semaphores, or other similar mechanisms. Although lock-based synchronization approaches are widely used, they also incur many defects, including deadlock, livelock, starvation, priority inversion, convoy effect, etc.

Lock-free algorithms are immune to deadlock and achieve progress even if some threads are descheduled or fail. Various primitives for these algorithms have developed, e.g., CAS, DCAS, LL/SC, etc. As an example, a CAS instruction, $\mathtt{cas}(loc, old, new)$, takes a location *loc*, an *old* value, and a *new* value as arguments. In the execution, if location *loc* holds value *old* as expected, it is replaced by *new* and **true** is returned, otherwise the cas returns **false** and changes nothing. To develop a lock-free algorithm, designers need to arrange the instructions in the algorithm, especially the CASs (or other primitives), skillfully to achieve the goal. Due to the concurrent and fine-grained natures, to guarantee that such a algorithm meets its requirements by regular means is a hard work. CAS-based lock free algorithms have been widely used in practice. As a recent important example, the *Scalable Synchronous Queue* [11] (SSQ) is adopted by Java 6 concurrency libraries, thus becomes one corner stone for large amount of real applications. However, no one gives arguments for the correctness of this algorithm till now, that leaves all those applications in a dangerous situation. Proving this important (and other similar industry-strength) algorithm(s) is one motivation of our work.

Possible theoretical weapons for reasoning lock-free programs include Concurrent Separation Logic (CSL) [15, 2], rely-guarantee technique [13, 14], etc. CSL adapts a form of local reasoning in which one needs only to focus on the footprint of (the part of memory actually touched by) a thread. This facilitates modular reasoning based on the resource invariant. Rely-guarantee reasoning is super in tackling fine-grained interferences among threads and effects of intertwined atomic commands. Every specification

★ Supported by NNSFC grant no. 60718002 and Open Foundation of State Key Laboratory of Computer Science, ISCAS grant no. SYSKF1103.

J.-P. Jouannaud and Z. Shao (Eds.): CPP 2011, LNCS 7086, pp. 247–263, 2011.

in rely-guarantee is accompanied with a pair of rely ($\mathcal{R}$) and guarantee ($\mathcal{G}$) conditions, where $\mathcal{R}$ is the thread's expectations for the state transitions made by environment, and $\mathcal{G}$ denotes the promise of the transitions made by the thread itself.

However, so far, most proof techniques developed based on either CSL or rely guarantee for lock-free algorithms heavily rely on auxiliary variables/commands, e.g. [19, 22]. On one side, auxiliary variables/commands would usually obscure the proofs; on the other side, the layout of auxiliary variables/commands itself is also a tricky work. A clearer abstract level for verification is worth to seek.

We report here a new approach for proving CAS-based lock-free algorithms, which is easier to conduct and more intuitive comparing with existing works. It can be viewed as a development of RGSep [19], in which we package CASs into control flow commands, thus rule out the heavy dependence on the auxiliary variables and commands. Based on the framework, we successfully prove that SSQ is correct and find some informative issues. The main contributions of our work are as follows:

- We propose a method for reasoning programs under garbage collection (GC). It allow primitives touch the shared state of concurrent programs, and also capture the semantics that is implicitly guaranteed by GC.
- We package CAS primitives into control flow commands. This choice is reasonable for proving lock-free algorithms. A set of rules for packaged-CAS commands are developed, and the soundness of rules are proved.
- We generalize rules for control flow commands; and the new rules are more expressive due to the relaxed premise. This and above techniques are important for practically proving concurrent programs, and also relatively intuitive because they don't require for any auxiliary variables and commands.
- We prove the SSQ algorithm, and find some semantically equivalent variations of it. We also conduct an investigation on its fairness and liveness properties.

The rest of the paper is organized as follows. In Section 2, we propose rules for reasoning with garbage collection, rules for packaged-CAS commands, and generalized rules for control flow commands. The SSQ algorithm is introduced in Section 3 with its proof in Section 4. We discuss some related work in Section 5 then conclude.

2 The Inference Rules

To prove SSQ, we build our foundations first. We give basic concepts and some definitions and then the inference rules. We don't claim these rules are more general than existing methods. These rules are proposed in the aim of concise and direct proofs of our target algorithms.

2.1 Notations and Definitions

Now we introduce some notations and concepts that used in this paper, which are developed from RGSep [19]. The techniques are general for most heap-stack models.

In our model, the heap is partitioned into one part shared by all threads in the program, and several disjunct local portions where each is privately owned by a thread. By shared state, we mean all the resources accessible to all threads. In Separation Logic originally, variables are viewed as shared resource. This is not suitable for concurrent programs, thus comes the *variables-as-resource* [18, 1]. For simplicity, we do not explicitly specify variables' ownership, but still distinguish shared/private variables. Finally, we take "shared state = shared heap + shared variables".

The assertion language is a variant of SL, where P stands for any SL assertion:

$p, q, r ::= P$	Local assertion	$\mid \boxed{P}$	Shared assertion
$\mid\ p * q$	Separating conjunction	$\mid p \wedge q$	Conjunction
$\mid\ p \vee q$	Disjunction	$\mid \forall x.p$	Universal quant.
$\mid\ \exists x.p$	Existential quant.		

Here we have a form of boxed assertion, $\boxed{P}$, named *shared assertion*. An unboxed assertion, P, is a *local assertion*. Local assertions specify states of private portions (i.e., *private states*), while the shared $\boxed{P}$ specifies *shared state*. Comparing with standard Separation Logic, $*$ splits only the local states, but not the shared state. In this case, every shared assertion targets to the whole shared part, in particular, $\boxed{P} * \boxed{Q} \Leftrightarrow \boxed{P \wedge Q}$. The conjunction of local and shared assertions, $P * \boxed{Q}$, describes the whole state.

We use *action* to depict transitions on the shared state. An action $P \leadsto Q$ means replacing a portion of the shared state satisfying P to a portion satisfying Q. The semantics of an action and a set of actions are a set of pre- and post-states pairs:

$$\begin{array}{l} [\![P \leadsto Q]\!] \ \widehat{=}\ \{(\sigma, \sigma') | (\sigma = \sigma_1 * \sigma_0) \wedge (\sigma' = \sigma_2 * \sigma_0) \wedge \sigma_1 \models P \wedge \sigma_2 \models Q\} \\ [\![\{P_i \leadsto Q_i \mid i \in 1..n\}]\!] \ \widehat{=}\ (\cup_{i=1}^{n} [\![P_i \leadsto Q_i]\!])^* \end{array}$$

Here σ denotes a state, $\sigma \models P$ means P holds in state σ, and δ^* is the reflective transitive closure of transition δ.

Assertion P is *precise* [20], if for any state σ, there is at most one sub-state of σ satisfying P. Action $P \leadsto Q$ is *precise* if both P and Q are precise. Precise actions ensure precise state transitions.

In rely-guarantee reasoning, a thread expresses its interaction with the environment by a pair of rely/guarantee conditions. The rely condition $\mathcal{R}$ specifies all possible mutations on shared state caused by the environment, and the guarantee condition $\mathcal{G}$ specifies what caused by the thread itself. The thread ensures its guarantee condition as long as its environment respects its rely condition. In this paper, we express $\mathcal{R}$ and $\mathcal{G}$ as two sets of actions. Each action in rely/guarantee conditions are required to be *precise*.

A predicate u is *stable* under an action if it keeps valid across the action. A predicate u is *stable* under a set $\mathcal{S}$ of actions, if it is stable under every element of $\mathcal{S}$. A *compatible* concurrent program requires that the guarantee of a thread ensures what all the other threads rely on. Formal definitions are given in [22].

Hoare rules in rely-guarantee reasoning take the form of $\mathcal{R}, \mathcal{G} \models \{P\}\, C\, \{Q\}$, where in the triple we require P and Q to be stable under $\mathcal{R}$, and $[\![P \leadsto Q]\!] \subseteq [\![\mathcal{G}]\!]$. In addition, we introduce a function Fd such that $Fd(x)$ returns the set of fields according to the type of x, and then the rule takes the form of $\mathcal{R}, \mathcal{G}, Fd \models \{P\}\ C\ \{Q\}$. For example, if x denotes a node of binary trees, then we may have $Fd(x) = \{\texttt{lchild}, \texttt{rchild}\}$. Introduction of Fd is reasonable because Fd can be statically determined at compile-time. The usage of Fd can be seen in Section 2.2. We may write $Fd \models \bullet$, or $\mathcal{R}, \mathcal{G} \models \bullet$, when the other components are not important.

In order to prove algorithms involving mutable data structures, e.g. SSQ, we propose some new rules for commands running in an environment with *garbage collection* (GC). The existence of GC and no explicit memory release guarantee, as long as some thread holds a reference to a dynamic node, the node will not be reclaimed.

For most algorithms under GC, we need to specify the reachable part from an object in the heap. We use $\mathsf{rset}(x, s)$ to represent *reachable set* of nodes from x wrt. a given set s of field names. When a node is in $\mathsf{rset}(x, s)$, we say it is *reachable* from x via s:

$$\mathsf{rset}(x, s) \stackrel{def}{=} \begin{cases} \mathbf{emp} & \text{when } x \text{ is non-pointer or null} \\ cell(x) * (\underset{\forall n \in s}{\circledast} \mathsf{rset}(x.n, s)) & \text{when } s \subseteq Fd(x) \\ \mathbf{false} & \text{otherwise} \end{cases}$$

Here $\circledast$ is the iterative version of "$*$", and $cell(x)$ denotes the object x points to, where $cell$ is some object type. For example, if x is `node` which has a link field `next`, then $\mathsf{rset}(x, \{\mathtt{next}\})$ represents $\mathtt{node}(x) * \mathsf{rset}(x.\mathtt{next}, \{\mathtt{next}\})$; if x is a `tnode` for binary trees and s is $\{\mathtt{lchild}, \mathtt{rchild}\}$, $\mathsf{rset}(x, s)$ is $\mathtt{tnode}(x) * \mathsf{rset}(x.\mathtt{lchild}, s) * \mathsf{rset}(x.\mathtt{rchild}, s)$. We allow s contain some (but not all) fields of object x. If x refers to a `tnode`, $\mathsf{rset}(x, \{\mathtt{lchild}\})$ is fine.

Some words for rset need to be added. The above definition is valid when the references forms a DAG, while for general case, rset can be defined as a fixpoint. Sometimes we write rset without its second argument, to mean the full reachable set from x. However, usually we need, in real proof, the second argument to make proofs concise.

2.2 Reasoning Algorithms under Garbage Collection

Some people may wonder how a correct garbage collection (GC) implementation may influence the correctness of an algorithm. Indeed, we reason about GC is to explore the semantics that provided by GC, that is the "reachability" among nodes in the heap.

The SSQ algorithm we will prove runs under the effect of garbage collection. Generally speaking, two different levels may be considered when reasoning an algorithm with GC. On the low-level, the behavior and effects of GC are taken into account. We can view GC as a thread concurrently running with other threads, and it would affect the state of a program. However, on the high-level, we can abstract the explicit effect of GC, when all assertions are GC-insensitive. Intuitively, assertion P is GC-insensitive if $\{P\}\, \mathtt{gc()}\, \{P\}$. Hur *et al.* [10] discussed separation logic in the presence of GC. They formally constructed the relation between low- and high-levels. Here we only reason programs in the high-level, where all assertions are GC-insensitive.

The idea for reasoning under GC is clear: because GC will not reclaim any node that is accessible to some thread, $\mathsf{rset}(x, s)$ is a GC-insensitive assertion. We can add $\mathsf{rset}(x, s)$ in the pre- and post-conditions to extend the expressiveness of a specification. Taken as an example, the rule for assignments under GC is as follows, where s is any given attribute set, x and y are program variables:

$$Fd \vdash_{\mathrm{SL}} \{\mathsf{rset}(y, s)\}\ x := y\ \{x = y \wedge \mathsf{rset}(x, s)\}$$

This rule is valid for any given type of x. It says if all nodes reachable from y via s are in the heap before the assignment, then afterward, $x = y$ and all those nodes from x via s are still in the heap (they will not be reclaimed).

In concurrent programs, situations become more complicated: $\mathsf{rset}(x, s)$ is volatile when x is shared, because it may be modified by other threads. If x or y is in shared state, assertion $x = y$ is not stable under $\mathcal{R}$. To solve this problem, we require that all assignments to shared variables are implemented by `cas` commands, then we only need to consider assignment to private variables. When x is private variable, we define:

$$\mathcal{R}, \mathcal{G}, Fd \vdash \{\boxed{\mathsf{rset}(Y, s)}\} x := Y \{\boxed{\mathsf{rset}(x, s) * \mathbf{true}} * \boxed{\mathsf{rset}(Y, s) * \mathbf{true}}\} \quad \text{(ASS-GC)}$$

Note, since Y is a shared variable, $x = Y$ is not guaranteed after the assignment, but the post-assertion is still stable under $\mathcal{R}$.

It is true that the expanded nodes of rset can be specified by auxiliary variables, but obviously that will most prolix and obscure. Proper usage of rset can alleviate the abuse of auxiliaries. Besides, (ASS-GC) is intuitively sound for any $\mathcal{R}, \mathcal{G}$ and Fd under GC. The validity of its pre- and post-conditions are guaranteed by GC. Because reading a shared variable into a private variable does not change the shared state, it satisfies any $\mathcal{G}$. When x and Y are both private variables; the explanation is similar.

2.3 Rules for Packaged-CAS Commands

One crux of this work is to formalize the `cas` primitive. A `cas` gives a boolean result, and is often used as guards in **if** or **while** statements. However, a `cas` may also change the state as a side effect. If we formalized it as an expression, that would force us to introduce side-effect feature into all expressions, and make things very complicated.

Based on an analysis of various lock-free algorithms, we decide to package `cas` primitive into control statements, that is, always considering **if** $\mathtt{cas}(l, o, n)$ **then** `C1` **else** `C2` or **do** `C` **while** $\mathtt{cas}(l, o, n)$ as whole structures, and define inference rules directly for them. Clearly, a stand alone `cas` can be encoded into the semantical equivalent form **if** $\mathtt{cas}(l, o, n)$ **then** `skip` **else** `skip`. Comparing with RGSep, although packaging `cas` commands don't extend the capability, it helps us to rule out auxiliary variables and commands considerably. The situation can be seen below.

Rule for packaged-CAS `if` commands is:

$$\frac{\begin{array}{c}\emptyset, \mathcal{G}, Fd \vdash \{p * \boxed{l \mapsto o * \mathbf{true}}\}\ \langle [l] := n \rangle\ \{r_1\} \quad \mathcal{R}, \mathcal{G}, Fd \vdash \{r_1\}\ C_1\ \{q\} \\ p * \boxed{l \mapsto o' * \mathbf{true}} \wedge o' \neq o \Rightarrow r_2 \quad \mathcal{R}, \mathcal{G}, Fd \vdash \{r_2\}\ C_2\ \{q\} \\ p \Rightarrow \boxed{l \mapsto - * \mathbf{true}} * \mathbf{true} \quad \mathbf{stable}(\{p, q, r_1, r_2\}, \mathcal{R})\end{array}}{\mathcal{R}, \mathcal{G}, Fd \vdash \{p\}\ \mathbf{if}\ \mathtt{cas}(l, o, n)\ \mathbf{then}\ C_1\ \mathbf{else}\ C_2\ \{q\}} \quad \text{(IF-CAS)}$$

Note that $[l] := n$ means the assignment of value n to location l, which is called *mutation* in Separation Logic.

Because the location accessed by $\mathtt{cas}(l, o, n)$ must be in the shared state (which may be modified by other threads), any CAS-based command must rely on the nondeterministic value at location l. Specifically, in rule (IF-CAS), we require $\mathcal{R}$ to be $\emptyset$ to ensure the stability of $p * \boxed{l \mapsto o * \mathbf{true}}$. Another difference between ordinary **if** and **if**

$\texttt{cas}(l, o, n)$ **then** C_1 **else** C_2 is that, the guard of the latter one will access the heap location l, while in ordinary **if**, guard B is *pure* without any effect to/from the heap. Therefore, here premise $p \Rightarrow \boxed{l \mapsto - * \textbf{true}} * \textbf{true}$ is necessary.

Similarly, there is a symmetrical rule:

$$\frac{\begin{array}{c}\varnothing, \mathcal{G}, Fd \vdash \{p * \boxed{l \mapsto o * \textbf{true}}\}\ \langle [l] := n \rangle\ \{r_2\} \\ \mathcal{R}, \mathcal{G}, Fd \vdash \{r_2\}\ C_2\ \{q\} \quad p * \boxed{l \mapsto o' * \textbf{true}} \wedge o' \neq o \Rightarrow r_1 \\ \mathcal{R}, \mathcal{G}, Fd \vdash \{r_1\}\ C_1\ \{q\} \quad p \Rightarrow \boxed{l \mapsto - * \textbf{true}} * \textbf{true} \\ \textbf{stable}(\{p, q, r_1, r_2\}, \mathcal{R})\end{array}}{\mathcal{R}, \mathcal{G}, Fd \vdash \{p\}\ \textbf{if}\ !\mathrm{cas}(\mathrm{l}, \mathrm{o}, \mathrm{n})\ \textbf{then}\ \mathrm{C}_1\ \textbf{else}\ \mathrm{C}_2\ \{q\}} \quad \text{(IF-N-CAS)}$$

We also define the packaged cas command with **do-while**. Taking here **do-while** but not **while-do**, since the former reflects a pattern of optimistic programming, which is used more frequently in lock-free algorithms.

$$\frac{\begin{array}{c}\mathcal{R}, \mathcal{G}, Fd \vdash \{p\}\ C\ \{r\} \quad \textbf{stable}(\{p, q, r\}, \mathcal{R}) \\ \varnothing, \mathcal{G}, Fd \vdash \{r * \boxed{l \mapsto o * \textbf{true}}\}\ \langle [l] := n \rangle\ \{q\} \\ r * \boxed{l \mapsto o' * \textbf{true}} \wedge o' \neq o \Rightarrow p \quad p \Rightarrow \boxed{l \mapsto - * \textbf{true}} * \textbf{true}\end{array}}{\mathcal{R}, \mathcal{G}, Fd \vdash \{p\}\ \textbf{do}\ C\ \textbf{while}\ !\texttt{cas}(l, o, n)\ \{q\}} \quad \text{(WHILE-N-CAS)}$$

2.4 Improved Rules for Control Flow Commands

We also propose improved rules for control statements. The naive rules given below are not useful in many practical cases, if we do not introduce auxiliary variables:

$$\frac{p \Rightarrow (B = B) * \textbf{true} \quad \mathcal{R}, \mathcal{G} \vdash \{p \wedge B\}\ C_1\ \{q\} \quad \mathcal{R}, \mathcal{G} \vdash \{p \wedge \neg B\}\ C_2\ \{q\}}{\mathcal{R}, \mathcal{G} \vdash \{p\}\ \textbf{if}\ B\ \textbf{then}\ C_1\ \textbf{else}\ C_2\ \{q\}} \quad \text{(N-IF)}$$

$$\frac{p \Rightarrow (B = B) * \textbf{true} \quad \mathcal{R}, \mathcal{G} \vdash \{p \wedge B\}\ C\ \{q\}}{\mathcal{R}, \mathcal{G} \vdash \{p\}\ \textbf{while}\ B\ \textbf{do}\ C\ \{p \wedge \neg B\}} \quad \text{(N-WHILE)}$$

These rules are sound, but not usable when B relies on shared variables. Take (N-IF) for example, rely guarantee requires the pre-condition of the specification to be stable under $\mathcal{R}$. However, if B accesses shared variables, $\{p \wedge B\}$ is not stable under $\mathcal{R}$. To tackle the problem, people (in existing works) suggested to introduce auxiliary variables and commands, to make sure B stable. Coleman [3] proposed to decompose B into a stable part and an unstable part, so that B is allowed to be unstable as long as its unstable part has one unstable variable at most. In this paper, as a development of Coleman, we propose the following rules which we find very effective in the reasoning.

$$\frac{\begin{array}{c}p \wedge B \Rightarrow r_1 \quad p \wedge \neg B \Rightarrow r_2 \quad \textbf{stable}(\{r_1, r_2, p\}, \mathcal{R}) \\ \mathcal{R}, \mathcal{G}, Fd \vdash \{r_1\}\ C_1\ \{q\} \quad \mathcal{R}, \mathcal{G}, Fd \vdash \{r_2\}\ C_2\ \{q\}\end{array}}{\mathcal{R}, \mathcal{G}, Fd \vdash \{p\}\ \textbf{if}\ B\ \textbf{then}\ C_1\ \textbf{else}\ C_2\ \{q\}} \quad \text{(IF)}$$

$$\frac{p \wedge B \Rightarrow r \quad p \wedge \neg B \Rightarrow q \quad \mathcal{R}, \mathcal{G}, Fd \vdash \{r\}\ C\ \{p\} \quad \textbf{stable}(\{p, q\}, \mathcal{R})}{\mathcal{R}, \mathcal{G}, Fd \vdash \{p\}\ \textbf{while}\ B\ \textbf{do}\ C\ \{q\}} \quad \text{(WHILE)}$$

Table 1. Scalable Dual Queue — Enqueue

```
void enqueue(e) {
  offer = new Node(e, DATA);
  while true {
    t = Tail; h = Head;
    if (h == t || t.state == DATA) {
      n = t.next;
      if (t == Tail){
        if (n != null) {
          casTail(t, n);
        }
        else
          if(t.casNext(n, offer)) {
            casTail(t, offer);
            while (offer.data != null))   /* spin */;
            h = Head;
            if (offer == h.next)
              casHead(h, offer);
            return;
          }
    }else{
      n = h.next;
      if(t != Tail || h != Head || n == null)
        continue;
      if(n.casData(null, e)) {
        casHead(h, n);
        return;
      }else casHead(h, n);
    }
  }
}
```

These rules still keep the information of guard B, but do not require B to be stable under $\mathcal{R}$. Besides, (IF) and (WHILE) can be seen as general cases of (N-IF) and (N-WHILE), but doesn't rely on auxiliaries as (N-IF) and (N-WHILE) in many cases.

The soundness proof of these rules can be found in [12].

3 Scalable Synchronous Queue

The SSQ algorithm we are going to prove comes from [11]. It is used in Java 6 concurrency libraries because of its remarkable performance. There are two modes of the implementation: the (FIFO) queue and the (LIFO) stack. We give here only proof for the queue mode, while leave the proof for stack mode in [12]. As mentioned in [11], `dequeue` is symmetric with `enqueue` except for the direction of data transfer, therefore we only present prove of `enqueue` in this paper and proof of `dequeue` is symmetric. The code is listed in Table 1.

The data structure is a single linked list held by global variables `Head` and `Tail` (Figure 1), whose first node is dummy. A node contains three fields: `data`, `state` (either `DATA` or `REQ`), `next`. In Figure 1, `N` denotes a null field, and `!N` for not null. `D` denotes that the `state` field is `DATA`, while `R` says it is `REQ`. A *matched data node* is a node where `state` field is `DATA` but its `data` field is null in the same time. A *matched request node* is a node whose `state` is `REQ` and its `data` field is not null. We say a *data node* x is *matched*, when another thread "withdraws" x.`data` and sets x.`data` `null`; for a request node, vice versa.

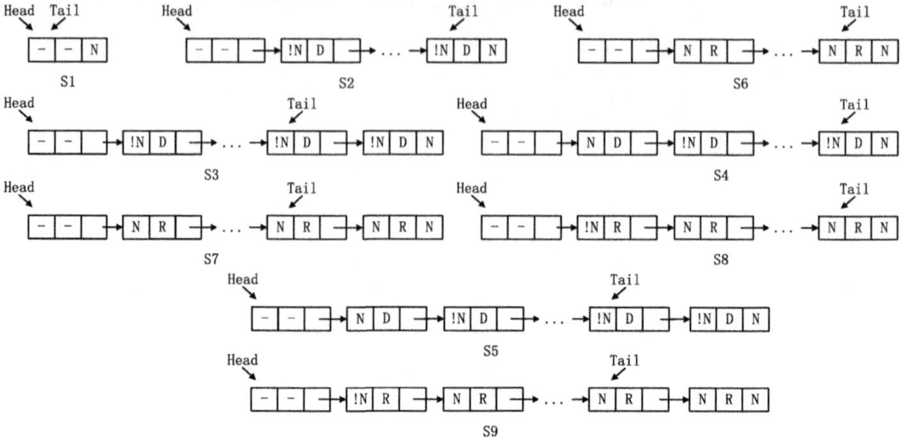

Fig. 1. The Link List's States

For the code, the first step is to read shared variables `Head` and `Tail` into local variables `h` and `t` (line 04). Then there are two branches: if the list is empty or has at least one *data node*, it tries to attach the newly created *data node* pointed by local variable `offer` to the end of the list (line 05-19); otherwise, when the list has at least one *request node*, it tries to match the data with the *unmatched request node* that behinds `Head` (line 20-27). In addition, the call to enqueue will not return until the `offer` is matched (line 13-18) or a *request node* is matched (line 24-26).

4 Proof for Synchronous Queue

Now we prove the SSQ algorithm. We give here the fundamental work, and put proof listing in Appendix A.

4.1 Adapted Rules for casHead, casTail, n.casData, and n.casNext

Primitives `casHead`, `casTail`, n.`casData` and n.`casNext` are used in SSQ. Their semantics is a bit different from plain `cas` discussed before: `casHead`(o, n) (alt., `casTail`(o, n)) compares the value of `Head` (or `Tail`) with o, and probably assigns the new value n to it. On the other hand, v.`casData`(o, n) (alt., v.`casNext`(o, n)) compares (and probably assigns) the value stored in the `data` (`next`) field of v with o. However, the essence of these CAS-based primitives is identical, and the adapted rules are similar with rules introduced in Section 2.3:

$$\frac{\begin{array}{c}\varnothing, \mathcal{G} \vdash \{p * \boxed{\texttt{Head} = o * \textbf{true}}\}\ \langle \texttt{Head} := n \rangle\ \{r_1\} \\ \mathcal{R}, \mathcal{G} \vdash \{r_1\}\ C_1\ \{q\} \quad \mathcal{R}, \mathcal{G} \vdash \{r_2\}\ C_2\ \{q\} \\ p * \boxed{\texttt{Head} = o' * \textbf{true}} \wedge o' \neq o \Rightarrow r_2 \quad \textbf{stable}(\{p, q, r_1, r_2\}, \mathcal{R})\end{array}}{\mathcal{R}, \mathcal{G} \vdash \{p\}\ \textbf{if}\ \texttt{casHead}(o, n)\ \textbf{then}\ C_1\ \textbf{else}\ C_2\ \{q\}} \quad \text{(IF-casHead)}$$

Rules for `casTail`, v.`casData`(o, n) and v.`casNext`(o, n) are similar.

$$S_1 = \mathsf{Nd}(\texttt{Head}, -, -, \texttt{null}) \wedge \texttt{Head} = \texttt{Tail}$$
$$S_2 = \mathsf{Nd}(\texttt{Head}, -, -, \texttt{Head.next}) * \mathsf{Dls}(\texttt{Head.next}, \texttt{Tail.next}) \wedge \texttt{Tail.next} = \texttt{null}$$
$$S_3 = \exists x.\mathsf{Nd}(\texttt{Head}, -, -, -) * \mathsf{Dls}(\texttt{Head.next}, \texttt{Tail.next}) * \mathsf{Nd}(x, !\texttt{null}, \texttt{DATA}, \texttt{null}) \wedge \texttt{Tail.next} = x$$
$$S_4 = \exists x, y.\mathsf{Nd}(\texttt{Head}, -, -, x) * \mathsf{Nd}(x, \texttt{null}, \texttt{DATA}, y) * \mathsf{Dls}(y, \texttt{Tail.next}) \wedge \texttt{Tail.next} = \texttt{null}$$
$$S_5 = \exists x, y.\mathsf{Nd}(\texttt{Head}, -, -, x) * \mathsf{Nd}(x, \texttt{null}, \texttt{DATA}, -) * \mathsf{Dls}(x.\texttt{next}, \texttt{Tail.next}) * \mathsf{Nd}(y, !\texttt{null}, \texttt{DATA}, \texttt{null}) \wedge \texttt{Tail.next} = y$$
$$S_6 = \mathsf{Nd}(\texttt{Head}, -, -, \texttt{Head.next}) * \mathsf{Rls}(\texttt{Head.next}, \texttt{Tail.next}) \wedge \texttt{Tail.next} = \texttt{null}$$
$$S_7 = \exists x.\mathsf{Nd}(\texttt{Head}, -, -, -) * \mathsf{Rls}(\texttt{Head.next}, \texttt{Tail.next}) * \mathsf{Nd}(x, \texttt{null}, \texttt{REQ}, \texttt{null}) \wedge \texttt{Tail.next} = x$$
$$S_8 = \exists x, y.\mathsf{Nd}(\texttt{Head}, -, -, x) * \mathsf{Nd}(x, !\texttt{null}, \texttt{REQ}, y) * \mathsf{Rls}(y, \texttt{Tail.next}) \wedge \texttt{Tail.next} = \texttt{null}$$
$$S_9 = \exists x, y.\mathsf{Nd}(\texttt{Head}, -, -, x) * \mathsf{Nd}(x, !\texttt{null}, \texttt{REQ}, -) * \mathsf{Rls}(x.\texttt{next}, \texttt{Tail.next}) * \mathsf{Nd}(y, \texttt{null}, \texttt{REQ}, \texttt{null}) \wedge \texttt{Tail.next} = y$$

Fig. 2. Shared States

4.2 Predicates for State Assertions

Some predicates are defined first. Predicate Nd asserts a node with specific contents:

$$\mathsf{Nd}(x, v, s, y) \mathrel{\widehat{=}} x \mapsto \{\texttt{data} \mapsto v, \texttt{state} \mapsto s, \texttt{next} \mapsto y\}$$

where `state` could be DATA or REQ.

Predicates for asserting list segments, and special types of list segments are defined:

$$\mathsf{ls}(x, y) \mathrel{\widehat{=}} (\mathbf{emp} \wedge x = y) \vee (\mathsf{Nd}(x, -, -, -) * \mathsf{ls}(x.\texttt{next}, y))$$
$$\mathsf{Dls}(x, y) \mathrel{\widehat{=}} (\mathbf{emp} \wedge x = y) \vee (\mathsf{Nd}(x, !\texttt{null}, \texttt{DATA}, -) * \mathsf{Dls}(x.\texttt{next}, y))$$
$$\mathsf{Rls}(x, y) \mathrel{\widehat{=}} (\mathbf{emp} \wedge x = y) \vee (\mathsf{Nd}(x, \texttt{null}, \texttt{REQ}, -) * \mathsf{Rls}(x.\texttt{next}, y))$$

Here $\mathsf{ls}(x, y)$ asserts a list segment that begins with x and ends with a node whose `next` field is y. $\mathsf{Dls}(x, y)$ asserts additional that each node in the list is *unmatched data node*, i.e., whose `state` field is DATA and `data` field is not null; $\mathsf{Rls}(x, y)$ asserts that each node is *unmatched request node*, whose `state` field is REQ and `data` field is null.

With these predicates, we specify the possible states of the linked list in the queue, totally nine sets of them as given in Figure 1. Figure 2 lists the specifications. Two properties are worth to mention: $S_1 = S_2 \cap S_6$; and there is at most one *matched node* in each of the states. These states are crucial for our proof, because they cover and partition all possible shared states of the algorithm.

4.3 The Rely/Guarantee Conditions

The essence of rely-guarantee logic is using rely/guarantee conditions, $\mathcal{R}$ and $\mathcal{G}$, to describe the interferences between a thread and its environment. We construct these conditions for the algorithm in this subsection.

The Rely Condition. Figure 3 depicts all the possible state transitions, it shows out a very beautiful graph: regular, and symmetric. The regularity should be the real foundation of the algorithm's correctness. Note that we do not depict state set S_1 as other sets in the graph, because S_1 belongs to both S_2 and S_6. The transition between S_2 and S_6 is not depicted in $\mathcal{R}$ and $\mathcal{G}$, but via the common regression state of both sets.

We specify the state transitions of Figure 3 by a set of actions, as given in Figure 4. In order to distinguish values of shared variables in pre- and post-states of an *action*, we use the hat form to denote the pre-values. We substitute all shared variables in the pre-state with their hat form, and use additional boolean expressions to depict the relation. Note that this relation can also be depicted using existential quantification, as in RGSep.

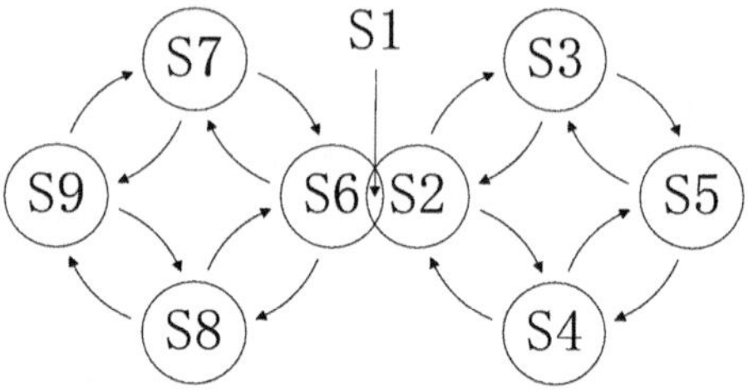

Fig. 3. Transition Graph of Shared States

Take the 7th action in Figure 4 as an example, which makes a transition from $S4$ to $S2$. After substitution, the left hand side becomes $\exists x, y.\mathsf{Nd}(\widehat{\texttt{Head}}, -, -, x) * \mathsf{Nd}(x, \texttt{null}, \texttt{DATA}, y) * \mathsf{Dls}(y, \widehat{\texttt{Tail}}.\texttt{next}) \wedge \widehat{\texttt{Tail}}.\texttt{next} = \texttt{null}$, depicting that the head of the queue is followed by a matched data node and then some unmatched data nodes. The right hand side says the head node is followed by some unmatched data node and there exists a node, say $\mathsf{Nd}(z_0, -, \texttt{DATA}, \texttt{Head})$ (use z_0 instead of z to remove the $\exists$), which is the pre-node of Head. In order to precisely confine the state transition to an atomic action, $\widehat{\texttt{Head}} = z_0$ and $\widehat{\texttt{Tail}} = \texttt{Tail}$ are indispensable, which confine the footprint's transition caused by the action: $\widehat{\texttt{Head}} = z_0$ and $\mathsf{Nd}(z_0, -, \texttt{DATA}, \texttt{Head})$ implies Head moves one step ahead; $\widehat{\texttt{Tail}} = \texttt{Tail}$ implies Tail does not move.

The state transitions among $S6, S7, S8, S9$ are not given, because they are symmetric with the ones listed in Figure 4. We define our rely condition, $\mathcal{R}$, as the reflexive transitive closure of all the actions above.

In addition, all the actions in $\mathcal{R}/\mathcal{G}$ defined in Figure 2 are *precise*. The portion of state that remain unchanged is not depicted in the actions, which satisfy our requirement at Section 2.1. We use an invariant I to describe the whole shared state: $I = \boxed{S_2 \cup S_3 \ldots \cup S_9 * \mathbf{true}}$.

In fact, the detailed state of the shared memory may be affected by GC, because GC may periodically (or by other manners) reclaim nodes that are not accessible to any thread. However, this makes no trouble, every assertion in the proof is GC-insensitive. In fact, GC may shrink some part of the heap originally covered by the **true** part in invariant I, but make no effect to our $\mathcal{R}$ condition. For this algorithm, GC may only reclaim nodes that previously in the queue but not still. No thread will notice the state modification caused by GC, and the definition of $\mathcal{R}$ is thus reasonable for SSQ.

$S2[\widehat{\texttt{Head}}/\texttt{Head}, \widehat{\texttt{Tail}}/\texttt{Tail}] \leadsto S3 \wedge \widehat{\texttt{Head}} = \texttt{Head} \wedge \widehat{\texttt{Tail}} = \texttt{Tail}$
Attach a *data node* to the tail of a list when the list is in $S2$

$S3[\widehat{\texttt{Head}}/\texttt{Head}, \widehat{\texttt{Tail}}/\texttt{Tail}] \leadsto S2 \wedge \widehat{\texttt{Head}} = \texttt{Head} \wedge \widehat{\texttt{Tail}}.\texttt{next} = \texttt{Tail}$
Move `Tail` to its next node when the list is in $S3$

$S3[\widehat{\texttt{Head}}/\texttt{Head}, \widehat{\texttt{Tail}}/\texttt{Tail}] \leadsto S5 \wedge \widehat{\texttt{Head}} = \texttt{Head} \wedge \widehat{\texttt{Tail}} = \texttt{Tail}$
Match the first *unmatched data node* when the list is in $S3$

$S5[\widehat{\texttt{Head}}/\texttt{Head}, \widehat{\texttt{Tail}}/\texttt{Tail}] \leadsto \exists z.\mathsf{Nd}(z, -, D, \texttt{Head}) * S3 \wedge \widehat{\texttt{Head}} = z \wedge \widehat{\texttt{Tail}} = \texttt{Tail}$
Move `Head` to its next node when the list is in $S5$

$S5[\widehat{\texttt{Head}}/\texttt{Head}, \widehat{\texttt{Tail}}/\texttt{Tail}] \leadsto S4 \wedge \widehat{\texttt{Head}} = \texttt{Head} \wedge \widehat{\texttt{Tail}}.\texttt{next} = \texttt{Tail}$
Move *Tail* to its next node when the list is in $S5$

$S4[\widehat{\texttt{Head}}/\texttt{Head}, \widehat{\texttt{Tail}}/\texttt{Tail}] \leadsto S5 \wedge \widehat{\texttt{Head}} = \texttt{Head} \wedge \widehat{\texttt{Tail}} = \texttt{Tail}$
Attach a *data node* to the tail of a list when the list is in $S4$

$S4[\widehat{\texttt{Head}}/\texttt{Head}, \widehat{\texttt{Tail}}/\texttt{Tail}] \leadsto \exists z.\mathsf{Nd}(z, -, D, \texttt{Head}) * S2 \wedge \widehat{\texttt{Head}} = z \wedge \widehat{\texttt{Tail}} = \texttt{Tail}$
Move `Head` to its next node when the list is in $S4$

$S2[\widehat{\texttt{Head}}/\texttt{Head}, \widehat{\texttt{Tail}}/\texttt{Tail}] \leadsto S4 \wedge \widehat{\texttt{Head}}.\texttt{next} = \texttt{Head} \wedge \widehat{\texttt{Tail}} = \texttt{Tail}$
Match the first *unmatched data node* when the list is in $S3$

Fig. 4. State Transitions Rules

Deduced Properties. Since the rely condition depicts all *actions* in the shared state (except for the nodes reclaimed by GC), the rest of the shared state will remain the same on the assertion level. The following facts trivially hold:

Property 1. In SSQ, the `state` field of a node will not change until reclaimed by GC.

Property 2. In SSQ, if the `next` field of a node is not null, it will not change until reclaimed by GC .

Property 3. In SSQ, the `value` field of a matched data or request node will not change until reclaimed by GC.

These properties are used in our proof. For example, as mentioned in Section 2.2, $\mathsf{rset}(x, \{\texttt{next}\})$ is volatile when x refers to some shared variable; however, Property 2 rules out this possibility: the `next` field will not redirect to other node in this algorithm. Two rules for assignment in the algorithm are proposed based on that consideration.

Proposition 1 (Assignment Rules for SSQ). *Under GC and the rely condition* $\mathcal{R}$ *defined in* Section 4.3, *the following tuples hold for* SSQ algorithm*:*

$$\mathcal{R}, \mathcal{G}, Fd \vdash \{\boxed{\mathsf{ls}(\texttt{Head}, \texttt{Tail}) * \mathbf{true}}\}\ h := \texttt{Head}$$
$$\{\boxed{\mathsf{ls}(h, \texttt{Head}) * \mathbf{true}} * \boxed{\mathsf{ls}(\texttt{Head}, \texttt{Tail}) * \mathbf{true}}\}$$

$$\mathcal{R}, \mathcal{G}, Fd \vdash \{\boxed{\mathsf{ls}(\texttt{Head}, \texttt{Tail}) * \mathbf{true}}\}\ t := \texttt{Tail}$$
$$\{\boxed{\mathsf{ls}(t, \texttt{Tail}) * \mathbf{true}} * \boxed{\mathsf{ls}(\texttt{Head}, \texttt{Tail}) * \mathbf{true}}\}$$

Proof. Based on the assignment rule under GC in Section 2.2 and Property 2.

Note that in (ASS-GC), rset is translated into ls here. It is obvious that many auxiliary variables will be introduced to specify the relation between h (t) and `Head` (`Tail`), if not using this rule.

The Guarantee Condition. The behavior of the `enqueue` method includes: attaching a *data node* to `Tail` when `Tail.next` $=$ `null`; moving `Tail` to its next node when `Tail.next` $\neq$ `null`; matching an *unmatched request node* that referred by `Head.next`; .... More importantly, each of these *actions* caused by `enqueue` is also an action included in $\mathcal{R}$ defined in Section 4.3. Based on this, we define $\mathcal{G} = \mathcal{R}$, thus we have sure a pair of *compatible* rely guarantee conditions.

4.4 Thread's Shared and Private States

In our model, each thread in a system has a private state and can access the shared state. As discussed in Section 4.3, now the shared state is the list of nodes in queue, which is depicted by invariant I. Particularly, the state is composed with three parts: list segment ls(`Head`, `Tail`); the node that is about to be taken into the queue when `Tail.next` refers to it; the nodes that once in the queue but do not in the segment ls(`Head`, `Tail`) as `Head` and `Tail` move and not reclaimed by GC yet. The first two parts are depicted by $S_2 \cup S_3 \ldots \cup S_9$ in I, and the third part is depicted by **true** in I.

The private part of the thread is the node that referred by `offer` before it is attached to the queue in the shared part. Note the 1st and 6th actions in Figure 4, the post assertion's footprint of them is one node larger than their pre counterpart. This is because the node that originally belongs to the private part, is attached to the queue and belongs to the shared part after these transitions.

4.5 The Proof, and What We Find

Appendix A lists the whole proof. We use Hoare rules for restricted forms of jumps. For a loop with invariant I and exit condition q, the pre-condition and postcondition for `break` are q and `false` respectively; and for `continue`, they are I and `false`.

Variants. In the proof, we find that the pre/post conditions of lines 7 and 15 in Table 1 are the same. These mean that, if we deleted either or both of these lines, the algorithm remains its semantic meaning. Take line 15 in Table 1 as example. This is the second read for the shared variable `Head`. Clearly, `Head` may be modified between line 5 and line 15 by other threads, but it may also be modified after line 15 and before line 16. As a result, semantically, the read primitive in line 15 provides no useful information, neither for our proof, nor for real execution.

Via emails, one of the algorithm's designer, Doug Lea, agrees with us that the algorithm is still correct if the two lines are deleted, but he thinks the deletions will sacrifice performance. However, in our mind, this fact can only be confirmed by extensive test running on benchmarks, and may depend on the circumstance. Anyway, Lea's feedback demonstrates the power of our techniques from another side.

Linearizability. Linearizability requires a method call take effect instantaneously at some moment between its invocation and response. Every linearizable method has a well-defined linearization point. We find the linearization point of SSQ either. Because there are two branches in figure 1, there are two linearization points, line 12 and 24 respectively. Take line 12 as an example, the node `offer` is attached to the list at the instant when `casNext` succeed. Therefore, SSQ is linearizable.

Now we discuss some other issues related to the algorithms.

Fairness. For SSQ, *fairness* can be defined as, every call to `enqueue/dequeue` will eventually return when there exist corresponding calls. Unfortunately, SSQ itself is clearly unfair. Suppose a thread calls `enqueue`, but each time before it executes line 8 in Table 1, it is preempted by other one with a successful `enqueue` call. When the thread is rescheduled the guard of line 8 is not true. In this scenario, the thread will loop forever, and its call to `enqueue` will never return.

Liveness. For the whole system, *liveness* means: if ls(`Head`, `Tail`) has at least one unmatched data node, at least one call of `dequeue` will return; for calls to `enqueue`, it is similar definition. Note, since SSQ implements a synchronous queue, for any return from `dequeue/enqueue`, its corresponding `enqueue/dequeue` returns too. Based on our rely/guarantee conditions, this property holds for SSQ. However, strict proof may need temporal logic, and a transition model of the rely/guarantee conditions.

Modularity. Our proof is modular in the sense that it has no restriction on the number of threads. The shared state can only be accessed by `enqueue/dequeue`. All calls to these procedures share the same pair of rely/guarantee conditions. The compatibility thus remains and is independent of the number of threads.

5 Related Work and Conclusion

People have worked on reasoning concurrent programs for decades. [16] is a pioneer paper in this area. [17] defined the semantics of concurrent programs as a set of *execution sequences*; [9] proposed a way for modular verification of concurrent programs, and viewed a programs as sets of modules that interact via procedure calls.

A new story in this area began around the turn of this century, after the seminal work of Reynolds *et al.* on the Separation Logic (SL) [20]. Many variants of SL and relating theories were blossomed. S. Brookes [2] discussed a semantics of concurrent SL. Bornat *et al.* [1] and Parkinson *et al.* [18] proposed the view of *variable-as-resource*.

As a most important tool for reasoning concurrent programs, the rely/guarantee method [13] has a long history. It is one of the best studied techniques for compositional concurrent program verification. Rely/guarantee method has been combined with SL [6, 23]. Recent theories intend to improve the approaches in aspects of modularity and expressiveness, as concurrent abstract predicate [5] and theories of refinement [21].

One character of these works, including ours, is that they require one to possess a deep understanding about how the algorithm operates, as "explore the algorithm thoroughly, and then construct a proof correspondingly". Most recent works take the similar way, e.g., Colvin [4], Parkinson [19]. There are also some works on automate verification of lock-free algorithms, e.g., Yahav [24]. We would like to think about the combination of both approaches in the future.

On the another side, in this paper we present strict proof for the safety properties, and discuss its fairness and liveness. Strict proof of these properties need additional techniques. A. Gotsman *et al.* [8] and M. Fu *et al.* [7] incorporated rely guarantee logic with temporal logic and facilitate reasoning of liveness property to some extend. It is possible to incorporate their ideas with our approach too.

In this paper, we developed an approach based on reply-guarantee reasoning for verifying CAS-based lock-free algorithms, and verified a nontrivial industry-respected lock-free algorithm adopted by Java 6 concurrency libraries, the *Scalable Synchronous Queue*. The contributions of the work lie in some aspects.

First, we developed rules for verifying programs with GC, which captures more information guaranteed by GC. Second, by packaging CAS commands into control commands, we relieved the reliance of auxiliaries that heavily used in existing approaches, thus made the inference more direct and clear. Although we do not have a proof about whether or not we can rule out completely all auxiliaries, the experiences tell us that the improvement is significant, that gives the possibility to prove larger and complicated algorithms. Third, we developed some new rules for control flow commands that can be used in more circumstances. The soundness of our inference rules are proved in [12].

By the proof of Scalable Synchronous Queue, the power of our framework is illustrated. During the work, we found also some semantical equivalent variations of the algorithm, which shows further the value of our work. These discoveries have been confirmed by the developers of the algorithms.

For future work, we plan to enhance our framework, and to exercise it by proving more lock-free algorithms. We also plan an implementation on some existing proof-assistant systems, to support the development of CAS-based algorithms.

References

1. Bornat, R., Calcagno, C., Yang, H.: Variables as resource in separation logic. Electr. Notes Theor. Comput. Sci. 155, 247–276 (2006)
2. Brookes, S.D.: A Semantics for Concurrent Separation Logic. In: Gardner, P., Yoshida, N. (eds.) CONCUR 2004. LNCS, vol. 3170, pp. 16–34. Springer, Heidelberg (2004)
3. Coleman, J.W.: Expression Decomposition in a Rely/Guarantee Context. In: Shankar, N., Woodcock, J. (eds.) VSTTE 2008. LNCS, vol. 5295, pp. 146–160. Springer, Heidelberg (2008)
4. Colvin, R., Groves, L.: A scalable lock-free stack algorithm and its verification. In: SEFM, pp. 339–348 (2007)
5. Dinsdale-Young, T., Dodds, M., Gardner, P., Parkinson, M.J., Vafeiadis, V.: Concurrent Abstract Predicates. In: D'Hondt, T. (ed.) ECOOP 2010. LNCS, vol. 6183, pp. 504–528. Springer, Heidelberg (2010)
6. Feng, X., Ferreira, R., Shao, Z.: On the Relationship Between Concurrent Separation Logic and Assume-Guarantee Reasoning. In: De Nicola, R. (ed.) ESOP 2007. LNCS, vol. 4421, pp. 173–188. Springer, Heidelberg (2007)
7. Fu, M., Li, Y., Feng, X., Shao, Z., Zhang, Y.: Reasoning about Optimistic Concurrency using a Program Logic for History. In: Gastin, P., Laroussinie, F. (eds.) CONCUR 2010. LNCS, vol. 6269, pp. 388–402. Springer, Heidelberg (2010)
8. Gotsman, A., Cook, B., Parkinson, M.J., Vafeiadis, V.: Proving that non-blocking algorithms don't block. In: POPL, pp. 16–28 (2009)
9. Hailpern, B., Owicki, S.S.: Modular verification of concurrent programs. In: POPL, pp. 322–336 (1982)
10. Hur, C.-K., Dreyer, D., Vafeiadis, V.: Separation logic in the presence of garbage collection. In: LICS (2011)
11. Scherer III, W.N., Lea, D., Scott, M.L.: Scalable synchronous queues. Commun. ACM 52(5), 100–111 (2009)
12. Lei, J., Qiu, Z.: Verification of Scalable Synchronous Queue. Technical Report 2011-32, School of Math., Peking University (September 2011), `http://www.mathinst.pku.edu.cn/index.php?styleid=2`

13. Jones, C.B.: Specification and design of (parallel) programs. In: IFIP Congress, pp. 321–332 (1983)
14. Jones, C.B.: Tentative steps toward a development method for interfering programs. ACM Trans. Program. Lang. Syst. 5(4), 596–619 (1983)
15. O'Hearn, P.W.: Resources, concurrency, and local reasoning. Theor. Comput. Sci. 375(1-3), 271–307 (2007)
16. Owicki, S.S., Gries, D.: Verifying properties of parallel programs: an axiomatic approach. Commun. ACM 19(5), 279–285 (1976)
17. Owicki, S.S., Lamport, L.: Proving liveness properties of concurrent programs. ACM Trans. Program. Lang. Syst. 4(3), 455–495 (1982)
18. Parkinson, M.J., Bornat, R., Calcagno, C.: Variables as resource in hoare logics. In: LICS, pp. 137–146 (2006)
19. Parkinson, M.J., Bornat, R., O'Hearn, P.W.: Modular verification of a non-blocking stack. In: POPL, pp. 297–302 (2007)
20. Reynolds, J.C.: Separation logic: A logic for shared mutable data structures. In: LICS, pp. 55–74 (2002)
21. Turon, A.J., Wand, M.: A separation logic for refining concurrent objects. In: POPL, pp. 247–258 (2011)
22. Vafeiadis, V.: Modular fine-grained concurrency verification. Technical Report UCAM-CL-TR-726, University of Cambridge, Computer Laboratory (July 2008)
23. Vafeiadis, V., Parkinson, M.: A Marriage of Rely/Guarantee and Separation Logic. In: Caires, L., Vasconcelos, V.T. (eds.) CONCUR 2007. LNCS, vol. 4703, pp. 256–271. Springer, Heidelberg (2007)
24. Yahav, E., Sagiv, S.: Automatically verifying concurrent queue algorithms. Electr. Notes Theor. Comput. Sci. 89(3) (2003)

A Proof Listing of SSQ Algorithm

Based on $\mathcal{R}$ and $\mathcal{G}$ defined in Section 4.3, and rules for packaged-CAS-command and assignment under GC, we list the whole proof, where the invariant is $I = \boxed{S_2 \cup S_3 \ldots \cup S_9 * \textbf{true}}$.

```
{I ∧ e ≠ null}
offer = new Node(e, DATA);
{I ∗ Nd(offer, !null, DATA, null)}
while true {
   {I ∗ Nd(offer, !null, DATA, null)}
   t = Tail; h = Head;
   {I ∗ Nd(offer, !null, DATA, null) ∗ [ls(t, Tail) ∗ true] ∗ [ls(h, Head) ∗ true]}   ≡ G
   if (h == t || t.state == DATA) {
      {G ∧ (h = t ∨ t.state = DATA)}                                                ≡ H
      n = t.next;
      {H ∗ (n = null ∨ [Nd(t, −, −, n) ∗ Nd(n, −, −, null) ∗ true])}                 (1)
      if (t == Tail){                                                                 (†)
         if (n != null) {
            {H ∗ [Nd(t, −, −, n) ∗ Nd(n, −, −, null) ∗ true]}
            casTail(t, n);                                                            (2)
            {I ∗ Nd(offer, !null, DATA, null)}
```

```
      } else {
        {H ∧ n = null}
        if(t.casNext(n, offer)) {
          {I ∗ |ls(h, Head) ∗ true|
            ∗ |Nd(t, −, −, offer) ∗ Nd(offer, −, DATA, −) ∗ true|}    (3)
          casTail(t, offer);
          {I ∗ |Nd(offer, −, DATA, −) ∗ true| ∗ |ls(h, Head) ∗ true|}
          while (offer.data != null)); /* spin */;
          {I ∗ |Nd(offer, null, DATA, −) ∗ true| ∗ |ls(h, Head) ∗ true|}  (4)
          h = head;                                                     (††)
          {I ∗ |Nd(offer, null, DATA, −) ∗ true| ∗ |ls(h, Head) ∗ true|}
          if (offer == h.next)
            {I ∗ |Nd(h, −, −, offer) ∗ Nd(offer, null, DATA, −) ∗ true|}
            casHead(h, offer);
          {I}
          return;
        }}
  } else{
    {G ∧ h ≠ t ∧ t.state = REQ}                                         ≡ J
    n = h.next;
    {J ∗ (n = null ∨ |Nd(h, −, −, n) ∗ Nd(n, −, −, −) ∗ true|)}
    if(t != Tail || h != Head || n == null) continue;                  (5)
    {J ∗ |Nd(h, −, −, n) ∗ Nd(n, −, REQ, −) ∗ true|}
    if(n.casData(null, e)) {
      {J ∗ |Nd(h, −, −, n) ∗ Nd(n, e, REQ, −) ∗ true|}                 (6)
      casHead(h, n);
      {I}
      return;
    }else
      {J ∗ |Nd(t, −, −, n) ∗ Nd(n, !null, REQ, −) ∗ true|}
      casHead(h, n);
      {I ∗ Nd(offer, !null, DATA, null)}
  }}
```

The following are discussions to the numbered lines here:

(1) $\texttt{t.next} = \texttt{n}$ is stable under $\mathcal{R}$ (Property 2).

(2) `casTail(t, n)` is translated to `if(casTail(t,n)) then skip; else skip;` and rule `IF-casTail` is applied. Note:

$\{H * \boxed{\mathsf{Nd}(\texttt{t}, -, -, \texttt{n}) * \mathsf{Nd}(\texttt{n}, -, -, \texttt{null}) * \textbf{true}} \wedge \texttt{Tail} = \texttt{t}\}$
`Tail`:= n;
$\{I * \mathsf{Nd}(\texttt{offer}, e, \texttt{DATA}, \texttt{null})\}$

(3) Here, the node referred by `offer` was transformed from private state to shared state. Note:

$\{H \wedge \texttt{n} = \texttt{null} \wedge \texttt{t.next} = \texttt{n}\}$
$\{H \wedge \texttt{t} = \texttt{Tail} \wedge \texttt{Tail.next} = \texttt{null}\}$

t.next= offer;
$\{I * \boxed{\mathsf{ls}(\mathtt{h}, \mathtt{Head}) * \textbf{true}} * \boxed{\mathsf{Nd}(\mathtt{t}, -, -, \mathtt{offer}) * \mathsf{Nd}(\mathtt{offer}, -, \mathtt{DATA}, -) * \textbf{true}}\}$
offer.state = DATA is stable under $\mathcal{R}$ (Property 1); t.next = offer is stable under $\mathcal{R}$ (Property 2).

(4) The offer.data = null is stable under $\mathcal{R}$ (Property 3).

(5) Only n = null is stable under $\mathcal{R}$ because n is private. Below, n is a REQ node, this is because h $\neq$ t, t.state = REQ and $n \neq h$ (indicated by n = h.next). Note that all the nodes in the list, except the dummy, have the same state, so t and n have the same state.

(6) The n.data = e is stable under $\mathcal{R}$ (Property 3).

Besides, the algorithm is semantically equivalent if the lines that marked (†) (with corresponding close-brace) and (††) are deleted, because each of them has identical pre- and post-conditions.

Coq Mechanization of Featherweight Fortress with Multiple Dispatch and Multiple Inheritance

Jieung Kim and Sukyoung Ryu

Computer Science Department, KAIST
{gbali,sryu.cs}@kaist.ac.kr

Abstract. In object-oriented languages, overloaded methods with multiple dispatch extend the functionality of existing classes, and multiple inheritance allows a class to reuse code in multiple classes. However, both multiple dispatch and multiple inheritance introduce the possibility of ambiguous method calls that cannot be resolved at run time. To guarantee no ambiguous calls at run time, the overloaded method declarations should be checked statically.

In this paper, we present a core calculus for the Fortress programming language, which provides both multiple dispatch and multiple inheritance. While previous work proposed a set of static rules to guarantee no ambiguous calls at run time, the rules were parametric to the underlying programming language. To implement such rules for a particular language, the rules should be instantiated for the language. Therefore, to concretely realize the overloading rules for Fortress, we formally define a core calculus for Fortress and mechanize the calculus and its type safety proof in Coq.

Keywords: Coq, Fortress, overloading, multiple dispatch, multiple inheritance, type system, proof mechanization.

1 Introduction

Most object-oriented programming languages support method *overloading*: a method may have multiple declarations with different parameter types. Multiple method declarations with the same name can make the program logic clear and simple. When several of the overloaded methods are applicable to a particular call, the most specific applicable declaration is selected by the *dispatch mechanism*.

Several dispatch mechanisms exist for various object-oriented languages. For example, the Java™ programming language [11] uses a single-dispatch mechanism, where the dynamic type of only a single argument (the receiver of the method) and the static types of the other arguments are considered for method selection. CLOS [9] uses asymmetric multiple dispatch, where the dynamic type of each argument is considered in a specified order (usually left to right) for method selection. Fortress [3] uses *symmetric multiple dispatch*, where the dynamic types of all the arguments are equally considered. Because previous work [24,26,4,13] observed that using static types of arguments or a particular order of method parameters for method selection often produces confusing results, we focus on symmetric multiple dispatch throughout this paper.

Multiple inheritance lets a type have multiple super types, which allows the type to reuse code in its multiple super types, and permits more type hierarchies than what

J.-P. Jouannaud and Z. Shao (Eds.): CPP 2011, LNCS 7086, pp. 264–279, 2011.

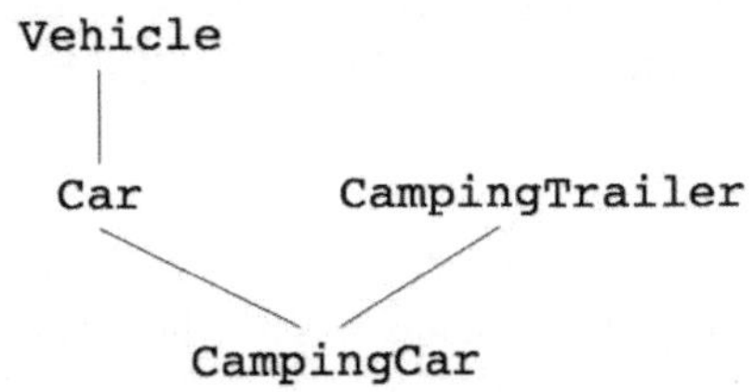

Fig. 1. Example type hierarchy

are allowed in single inheritance. While multiple inheritance provides high expressive power, it has well-known problems such as "name conflicts" and "state conflicts" [28]. Several object-oriented languages support multiple inheritance by addressing these problems in different ways. For example, C++ [29] requires programmers specify how to resolve conflicts between inherited fields. Scala [27] supports multiple inheritance via *traits* [28], where the order of super traits resolves any conflicts. Fortress [3] also provides multiple inheritance via traits, but the order of super traits does not affect the language semantics. Instead, Fortress traits do not include any fields, which removes the possibility of state conflicts. Similarly to the dispatch mechanism, we focus on symmetric multiple inheritance in this paper.

However, both multiple dispatch and multiple inheritance introduce the possibility of ambiguous method calls that cannot be resolved at run time. Consider a type hierarchy illustrated in Figure 1 in a language with multiple dispatch and multiple inheritance. The following overloaded method declarations:

```
collide(Car c, CampingCar cc)
collide(CampingCar cc, Car c)
```

introduce a possibility of an ambiguous method call due to multiple dispatch. For a method call `collide(cc1, cc2)` where both `cc1` and `cc2` have the `CampingCar` type at run time, we cannot select the best method to call because none of the `collide` method declarations is more specific than the other. Likewise, the following overloaded method declarations:

```
lightOn(Car c)
lightOn(CampingTrailer ct)
```

introduce a possibility of an ambiguous method call due to multiple inheritance. For a method call `lightOn(cc)` where `cc` has the `CampingCar` type at run time, we cannot select the best method to call because none of the `lightOn` method declarations is more specific than the other.

To break ties between ambiguous method declarations to a call, there should exist a disambiguating method declaration that is more specific than the ambiguous declarations and also applicable to the call. For example, if we add the following declaration to the above set of `collide` method declarations:

```
collide(CampingCar cc1, CampingCar cc2)
```

the set would be a valid overloading, and the set of `lightOn` method declarations would be a valid overloading if it includes the following declaration:

```
lightOn(CampingCar cc)
```

Finding such a disambiguating method declaration is not always trivial. While the following set of method declarations seems to be valid because the third declaration is more specific than the first and second:

```
tow(Vehicle v, Car c)
tow(Car c, Vehicle v)
tow(CampingCar cc1, CampingCar cc2)
```

it is not a valid overloading: For a method call `tow(c1, c2)` where both `c1` and `c2` have the `Car` type at run time, we cannot select the best method to call because none of the first and the second declarations is more specific than the other, and the third declaration is not even applicable to `tow(c1, c2)`.

Previous work proposed a set of rules to check overloaded method declarations statically to guarantee no ambiguous calls at run time. The Fortress team designed such overloading rules and proved that the rules ensure that there exist no ambiguous calls at run time [4]. While the overloading rules were designed in the context of Fortress, they were not closely tied to a particular programming language. To make the rules more concrete enough to be clearly implementable, the team defined a calculus, *Core Fortress with Overloading* (CFWO) [3, Appendix A.2], but the calculus was not proved type sound.

In this paper, we present a core calculus for the Fortress programming language, *Featherweight Fortress with Multiple Dispatch and Multiple Inheritance* (FFMM), which provides both multiple dispatch and multiple inheritance. Unlike CFWO, FFMM does not support generic types which are orthogonal to the overloading rules as we discuss in Section 2. We formally define FFMM and mechanize it and its type safety proof in COQ. While proving the type safety of FFMM in COQ, we found a bug in CFWO and proposed a fix to it. Our COQ implementation is available online [20].

The remainder of this paper is organized thus. In Section 2, we discuss the overloading rules that are statically checked to guarantee no ambiguous calls at run time. We formally define such rules in the context of FFMM in Section 3, describe our mechanization of the calculus using COQ in Section 4, and present its type safety proof using COQ in Section 5. We share our experience of using COQ in the development of FFMM in Section 6 and discuss the related work in Section 7. Section 8 discusses future work of our research and concludes.

2 Overloading Rules

To make sure that one can always pick the best method to call among overloaded methods at run time, the Fortress team devised a set of overloading rules to check statically [4]. The rules determine whether a set of overloaded declarations is valid by considering each pair of declarations in the set independently. A pair of declarations is a valid overloading if it satisfies one of the following rules:

1. **The Exclusion Rule**
   If the parameter types of the declarations are disjoint types, then the pair is a valid overloading.
2. **The Subtype Rule**
   If the parameter type of one declaration is a strict subtype of the parameter type of the other declaration, and the return type of the former is a subtype of the return type of the latter, then the pair is a valid overloading.
3. **The Meet Rule**
   If the parameter types of the declarations are not in the subtype relation, then the pair is a valid overloading if there is a declaration whose parameter type is an intersection type of the parameter types of the declarations.

We refer interested readers to the work of the Fortress team for detailed explanation of the rules [4]. They proved that the static rules placed on overloaded declarations are sufficient to guarantee no undefined nor ambiguous calls at run time. While the overloading rules are clearly described and the overloading resolution is proved to be unambiguous at run time, there still exists a gap between the rules and a particular programming language, Fortress.

Because the rules are specified independently for the underlying language, the Fortress team designed a calculus, CFwO [3, Appendix A.2], to describe the overloading rules more closely to the Fortress programming language. However, they did not prove the type safety of the calculus, and we found a bug in CFwO while we were working on our calculus as we discuss in Section 3.

To solve the problems arose from the previous approaches, we define a core calculus for the Fortress programming language, FFMM, which provides both multiple dispatch and multiple inheritance. Because CFwO is a strict extension of the *Basic Core Fortress* calculus [3, Appendix A.1], it includes generic types and top-level functions. While generic types can permit various interesting overloadings as the team presents in their recent work [5], the previous approaches assume that overloaded declarations should have type parameters that are identical (up to α-equivalence). Therefore, generic types and top-level functions are orthogonal to the overloading rules presented in [4] and FFMM does not support generic types nor top-level functions for simplicity. We formally define FFMM in Section 3, and we mechanize the definition and its type safety proof in Coq as we present in Sections 4 and 5.

3 FFMM: Featherweight Fortress with Multiple Dispatch and Multiple Inheritance

In this section, we formally define our calculus FFMM that we mechanize using Coq in the next section. Due to the space limitation, we describe only the rules that are closely related to overloaded methods in this paper. The full syntax, static semantics, and dynamic semantics of FFMM are available from a companion report [21].

p	$::=$	$\overrightarrow{d}\ e$	program
d	$::=$	$\texttt{trait}\ T\ \texttt{extends}\ \{\overrightarrow{T}\}\ \overrightarrow{md}\ \texttt{end}$	trait declaration
	$\mid$	$\texttt{object}\ O(\overrightarrow{f\colon\tau})\ \texttt{extends}\ \{\overrightarrow{T}\}\ \overrightarrow{md}\ \texttt{end}$	object declaration
md	$::=$	$m(\overrightarrow{x\colon\tau})\colon\tau = e$	method declaration
e	$::=$	x	parameter
	$\mid$	$\texttt{self}$	self
	$\mid$	$O(\overrightarrow{e})$	object creation
	$\mid$	$e.f$	field access
	$\mid$	$e.m(\overrightarrow{e})$	method invocation
τ	$::=$	T	trait type
	$\mid$	O	object type

Fig. 2. Syntax of FFMM

3.1 Syntax

The syntax of FFMM is provided in Figure 2. The metavariables T ranges over trait names; O ranges over object names; m ranges over method names; f ranges over field names; and x ranges over method parameter names. We write $\overrightarrow{x}$ for a (possibly empty) sequence $x_1, \cdots, x_n$.

A program consists of a sequence of trait and object declarations followed by a single top-level expression. Following the precedent set by prior core calculi such as Featherweight Java (FJ) [19], we have abided by the restriction that all valid FFMM programs are valid Fortress programs except that certain simple syntactic abbreviations such as commas and semicolons must be expanded.

Trait and object declarations in a program may include method declarations. Object declarations may include field declarations, which are shown as value parameters. Both traits and objects may extend multiple traits; they inherit the methods provided by the extended traits. Method declarations in a trait or an object may have the same name; method declarations may be *overloaded*.

Valid expressions are parameter references, references to the special identifier `self` which is like `this` in Java, constructor calls, field accesses, and method invocations. Types are trait types including the top trait Object and object types; note that object types are leaves of any FFMM type hierarchy.

For brevity, we make several assumptions that are easily checked syntactically: (1) every trait or object declaration declares a unique name, (2) every trait or object extends at least one trait, (3) extended traits in every trait or object are different, (4) every field has a unique name in its defining object, (5) no trait nor object declares Object, (6) type hierarchies are acyclic, and (7) every variable in type environments is unique.

3.2 Overloading Rules

In this section, we formally define the overloading rules described in prose in Section 2. The static semantics of FFMM describes how to type-check a given program at compile time. Type-checking a program consists of checking its trait and object declarations and the top-level expression:

Getting visible methods: $\boxed{\mathit{defined}_p(C)\,/\,\mathit{inherited}_p(C)\,/\,\mathit{visible}_p(C) = \{\overrightarrow{(m, \overrightarrow{\tau_p} \rightarrow \tau_r, \overrightarrow{x}\,.\,e)}\}}$

$$\begin{array}{lll}
\mathit{defined}_p(C) & = & \{(m, \overrightarrow{\tau_p} \rightarrow \tau_r, \overrightarrow{x}\,.\,e) \mid m(\overrightarrow{x\colon \tau_p})\colon \tau_r = e \in \overrightarrow{md}\} \\
& & \text{where } _\ C\ _\ \overrightarrow{md}\ \texttt{end} \in p \\
\mathit{inherited}_p(C) & = & \{(m, \overrightarrow{\tau_p} \rightarrow \tau_r, \overrightarrow{x}\,.\,e) \mid (m, \overrightarrow{\tau_p} \rightarrow \tau_r, \overrightarrow{x}\,.\,e) \in \mathit{visible}_p(T), \\
& & \qquad \textit{there is no } \tau' \textit{ such that } (m, \overrightarrow{\tau_p} \rightarrow \tau', _) \in \mathit{defined}_p(C)\} \\
& & \text{where } _\ C\ \texttt{extends}\ \{\overrightarrow{T}\}\ _ \in p \\
\mathit{visible}_p(C) & = & \mathit{defined}_p(C) \cup \mathit{inherited}_p(C)
\end{array}$$

Fig. 3. $\mathit{visible}_p$ function

$$[\text{T-PROGRAM}]\ \frac{p = \overrightarrow{d}\ e \qquad p \vdash \overrightarrow{d}\ \mathsf{ok} \qquad p; \emptyset \vdash e : \tau}{\vdash p : \tau}$$

Type-checking a trait, for example, includes checking a set of method declarations visible from the trait:

$$[\text{T-TRAITDEF}]\ \frac{p \vdash \overrightarrow{T'}\ \mathsf{ok} \qquad p; \texttt{self} : T \vdash \overrightarrow{md}\ \mathsf{ok} \qquad p \vdash \mathit{validMeth}(T)}{p \vdash \texttt{trait}\ T\ \texttt{extends}\ \{\overrightarrow{T'}\}\ \overrightarrow{md}\ \texttt{end}\ \mathsf{ok}}$$

via the *validMeth* judgment:

$$[\text{VALIDMETH}]\ \frac{\begin{array}{l}\forall\{(md, C), (md', C')\} \subseteq \mathit{visible}_p(C^o). \\ \quad md \neq md', \ \text{(not same declaration)} \\ \quad md = m(\overrightarrow{\_ : \tau^a}) : \tau^r\ \_, \qquad md' = m(\overrightarrow{\_ : \tau^{a'}}) : \tau^{r'}\ \_, \\ \quad p \vdash \mathit{valid}(m, C, \overrightarrow{\tau^a} \rightarrow \tau^r, C', \overrightarrow{\tau^{a'}} \rightarrow \tau^{r'}, \mathit{visible}_p(C^o))\end{array}}{p \vdash \mathit{validMeth}(C^o)}$$

It checks a set of method declarations visible from a trait or object C^o, $\mathit{visible}_p(C^o)$ presented in Figure 3, to see whether each pair in the set is a valid overloading. The metavariable C ranges over both trait and object names.

A pair of declarations md and md' is a valid overloading if it satisfies one of the overloading rules: Exclusion Rule, Subtype Rule, and Meet Rule. The pair is checked by the *valid* judgment described in Figure 4. The [VALIDEXC] rule describes the Exclusion Rule. While Fortress allows programmers to declare disjoint types with `excludes` clauses, FFMM does not support such clauses for brevity because they are largely orthogonal to multiple dispatch. Therefore, a pair of method declarations satisfies the Exclusion Rule when they have different number of parameters. The [VALIDSUBTYR] and [VALIDSUBTYL] rules describe the Subtype Rule. If the parameter type of one declaration is a strict subtype of the parameter type of the other declaration, and the return type of the former is a subtype of the return type of the latter, then the pair satisfies the Subtype Rule. Finally, the [VALIDMEET] rule describes the Meet Rule. If there is

[VALIDEXC]
$$\frac{|\overrightarrow{\tau^a}| \neq |\overrightarrow{\tau^{a'}}|}{p \vdash valid(m, C, \overrightarrow{\tau^a} \rightarrow \tau^r, C', \overrightarrow{\tau^{a'}} \rightarrow \tau^{r'}, S)}$$

[VALIDSUBTYR]
$$\frac{|\overrightarrow{\tau^a}| = |\overrightarrow{\tau^{a'}}| \quad C\,\overrightarrow{\tau^a} \neq C'\,\overrightarrow{\tau^{a'}} \quad p \vdash \overrightarrow{\tau^{a'}} <: \overrightarrow{\tau^a} \\ p \vdash \tau^{r'} <: \tau^r \quad p \vdash C' <: C}{p \vdash valid(m, C, \overrightarrow{\tau^a} \rightarrow \tau^r, C', \overrightarrow{\tau^{a'}} \rightarrow \tau^{r'}, S)}$$

[VALIDSUBTYL]
$$\frac{|\overrightarrow{\tau^a}| = |\overrightarrow{\tau^{a'}}| \quad C\,\overrightarrow{\tau^a} \neq C'\,\overrightarrow{\tau^{a'}} \quad p \vdash \overrightarrow{\tau^a} <: \overrightarrow{\tau^{a'}} \\ p \vdash \tau^r <: \tau^{r'} \quad p \vdash C <: C'}{p \vdash valid(m, C, \overrightarrow{\tau^a} \rightarrow \tau^r, C', \overrightarrow{\tau^{a'}} \rightarrow \tau^{r'}, S)}$$

[VALIDMEET]
$$\frac{\begin{array}{l} l = |\overrightarrow{\tau^a}| = |\overrightarrow{\tau^{a'}}| \quad C\,\overrightarrow{\tau^a} \neq C'\,\overrightarrow{\tau^{a'}} \quad \tau_0^a = C \quad \tau_0^{a'} = C' \\ \exists (m(\overrightarrow{\_ : \tau^{a''}}) : \_, \tau_0^{a''}) \in S. \\ \text{where } (l = |\overrightarrow{\tau^{a''}}|) \wedge (\forall\, 0 \leq i \leq l.\, p \vdash meet(\{\tau_i^a, \tau_i^{a'}, \tau_i^{a''}\}, \tau_i^{a''})) \end{array}}{p \vdash valid(m, C, \overrightarrow{\tau^a} \rightarrow \tau^r, C', \overrightarrow{\tau^{a'}} \rightarrow \tau^{r'}, S)}$$

Fig. 4. Overloading rules

Meet type: $\boxed{p \vdash meet(\{\overrightarrow{\tau}\}, \tau)}$

[MEET]
$$\frac{\tau' \in \{\overrightarrow{\tau}\} \quad p \vdash \tau' <: \overrightarrow{\tau} \quad p \vdash \cap\{\overrightarrow{\tau}\} <: \tau'}{p \vdash meet(\{\overrightarrow{\tau}\}, \tau')}$$

Intersection type: $\boxed{\tau \cap \tau = \tau}$

$$\tau_1 \cap \tau_2 = \begin{cases} \tau_3 & \text{if } \tau_3 <: \tau_1 \wedge \tau_3 <: \tau_2 \wedge \tau_1 \not<: \tau_2 \wedge \tau_2 \not<: \tau_1 \\ & \quad \wedge (\forall \tau_4, (\tau_4 <: \tau_1 \wedge \tau_4 <: \tau_2) \rightarrow \tau_4 <: \tau_3) \\ \tau_1 & \text{if } \tau_1 <: \tau_2 \\ \tau_2 & \text{if } \tau_2 <: \tau_1 \end{cases}$$

Fig. 5. Meet and intersection of types

a disambiguating declaration whose parameter type is the *meet* of the parameter types of the declarations in a pair, the pair satisfies the Meet Rule .

Definitions of the meet and the intersection of types are presented in Figure 5. The meet of a set of types is the most specific type in the set, and the intersection of two types is the greatest lower bound of them. These definitions serve a key role in the Meet Rule. As the `tow` example in Section 1 shows, finding a tie-breaking meet is not trivial. Actually, the bug we found in CFWO was in its definition of the Meet Rule: CFWO incorrectly specifies the definition of the meet type and the Meet Rule for multiple inheritance.

Applicable definitions: $\boxed{applicable_p(m(\overrightarrow{\tau}), \{\overrightarrow{(md, C)}\}) = \{\overrightarrow{(md, C)}\}}$

$$applicable_p(m(\overrightarrow{\tau}), S) = \{(md, C) \mid (md, C) \in S,\ md = m(\overrightarrow{x : \tau'}) : \_,\ p \vdash \overrightarrow{\tau} <: \overrightarrow{\tau'}\}$$

Most specific definitions: $\boxed{mostspecific_p(\{\overrightarrow{(md, C)}\}) = \{\overrightarrow{md}\}}$

$$mostspecific_p\ (\{\overrightarrow{(md, C)}\}) = \begin{cases} \{md_i\} & \text{if } |\overrightarrow{md}| = n \\ & \overrightarrow{md} = m((\overrightarrow{\_ : \tau^a})_1) : \tau_1^r\ \_ \quad \cdots \quad m((\overrightarrow{\_ : \tau^a})_n) : \tau_n^r\ \_ \\ & (md_i, C_i) \in \{\overrightarrow{(md, C)}\} \\ & \forall\, 1 \leq j \leq n.\, (p \vdash \overrightarrow{\tau^a}_i <: \overrightarrow{\tau^a}_j \land p \vdash C_i <: C_j) \\ \emptyset & \text{Otherwise} \end{cases}$$

Fig. 6. $applicable_p$ and $mostspecific_p$functions

3.3 Overloading Resolution

When several of the overloaded methods are applicable to a particular call, the most specific applicable declaration is selected by the dispatch mechanism. The following rule describes how FFMM evaluates a method invocation at run time:

$$[\text{R-METHOD}]\ \frac{\texttt{object}\ O\ \_\ \texttt{end} \in p \qquad \overrightarrow{type(v')} = \overrightarrow{\tau} \qquad mostspecific_p(applicable_p(m(\overrightarrow{\tau}), visible_p(O))) = \{m(\overrightarrow{x : \_}) : \_ = e\}}{p \vdash E[O(\overrightarrow{v}).m(\overrightarrow{v'})] \longrightarrow E[[O(\overrightarrow{v})/\texttt{self}][\overrightarrow{v'}/\overrightarrow{x}]e]}$$

Among all the visible methods from the receiver O, $applicable_p(m(\overrightarrow{\tau}), visible_p(O))$ selects the applicable declarations to a method call of name m with arguments of type $\overrightarrow{\tau}$. Note that because our dispatch mechanism uses symmetric multiple dispatch, it considers the dynamic types of all the arguments equally: $\overrightarrow{type(v')} = \overrightarrow{\tau}$. Finally, among the applicable declarations, select the single most specific declaration via the $mostspecific_p$ function.

The definitions of the $applicable_p$ and $mostspecific_p$ functions are presented in Figure 6. For a given method name m, the dynamic types of all the arguments $\overrightarrow{\tau}$, and a set of visible methods $\{\overrightarrow{(md, C)}\}$, the $applicable_p$ function collects all the methods that have the given name and whose parameter types are super types of the given arguments' dynamic types from the set of visible methods. The $mostspecific_p$ function selects the best method to call from a given set of applicable methods.

When a program is well typed under the static semantics of FFMM, there are no undefined nor ambiguous calls at run time. The type safety of FFMM described in Section 5 guarantees this property. Thanks to the type safety, the $mostspecific_p$ function always picks the single most specific method to call at run time.

```
Definition validmeet' (mn: mname) (tys : list typ) (ty : typ) (mS : mSet) : Prop :=
  exists2 mdt, mdt \in mS &
        ((tys = (getartys mdt)) ∧ (ty = (snd mdt)) ∧ (mn = (getmname mdt))).
Inductive valid (mdt1 mdt2 : mdttype) (mS : mSet) : Prop :=
  | valid_same :
      (getmid mdt1) = (getmid mdt2) →
      (snd mdt1) = (snd mdt2) →
      valid mdt1 mdt2 mS
  | valid_diff_name :
      (getmname mdt1) ≠ (getmname mdt2) →
      valid mdt1 mdt2 mS
  | valid_diff_arg_len :
      (getmid mdt1) ≠ (getmid mdt2) ∨ (snd mdt1) ≠ (snd mdt2) →
      (getmname mdt1) = (getmname mdt2) →
      (getenvlen mdt1) ≠ (getenvlen mdt2) →
      valid mdt1 mdt2 mS
  | valid_sub_ty_r :
      (getmid mdt1) ≠ (getmid mdt2) ∨ (snd mdt1) ≠ (snd mdt2) →
      (getartys mdt1) ≠ (getartys mdt2) ∨ (snd mdt1) ≠ (snd mdt2) →
      (getmname mdt1) = (getmname mdt2) →
      (getenvlen mdt1) = (getenvlen mdt2) →
      sub_tys (getartys mdt2) (getartys mdt1) →
      sub_ty (getrty mdt2) (getrty mdt1) →
      sub_ty (snd mdt2) (snd mdt1) →
      valid mdt1 mdt2 mS
  | valid_sub_ty_l :
      (getmid mdt1) ≠ (getmid mdt2) ∨ (snd mdt1) ≠ (snd mdt2) →
      (getartys mdt1) ≠ (getartys mdt2) ∨ (snd mdt1) ≠ (snd mdt2) →
      (getmname mdt1) = (getmname mdt2) →
      (getenvlen mdt1) = (getenvlen mdt2) →
      sub_tys (getartys mdt1) (getartys mdt2) →
      sub_ty (getrty mdt1) (getrty mdt2) →
      sub_ty (snd mdt1) (snd mdt2) →
      valid mdt1 mdt2 mS
  | valid_meet : ∀ tys ty,
      (getmid mdt1) ≠ (getmid mdt2) ∨ (snd mdt1) ≠ (snd mdt2) →
      (getartys mdt1) ≠ (getartys mdt2) ∨ (snd mdt1) ≠ (snd mdt2) →
      (getmname mdt1) = (getmname mdt2) →
      (getenvlen mdt1) = (getenvlen mdt2) →
      ~(sub_tys (getartys mdt1) (getartys mdt2)) →
      ~(sub_tys (getartys mdt2) (getartys mdt1)) →
      is_tys (getartys mdt1) (getartys mdt2) tys →
      is_ty (snd mdt1) (snd mdt2) ty →
      validmeet' (getmname mdt1) tys ty mS→
      valid mdt1 mdt2 mS.
```

Fig. 7. Overloading rules in COQ

4 FFMM in COQ

To mechanically prove the type safety of FFMM in Section 5, we describe our implementation of FFMM using COQ 8.3 in this section. Our implementation is largely based on Cast-Free Featherweight Java (CFFJ) by Fraine *et al.* [18]. Among others, the `Metatheory` library in CFFJ provides auxiliary constructs and properties of atoms [2] and environments including membership tests, accessors, and uniqueness guarantee. For our convenience, we extend the library with utility functions mostly for list manipulation. The full implementation is available online [20].

We implement the seven assumptions described in Section 3.1 in three ways:

- The uniqueness assumptions are implemented by the existing `Metatheory` library: (1), (4), and (7)
- We extend the `Metatheory` library to implement the assumptions on traits and objects: (2) and (3)
- To separate the concerns of the well-formed programs assumptions, we use a module system of COQ [14, Chapter 5]: (5) and (6)

While the COQ implementation of FFMM is very close the FFMM calculus, there are small differences in the implementation:

1. Unlike Java-like languages which provide only classes, FFMM provides both traits and objects as we discuss in Section 3. While the calculus does not distinguish between traits and objects by using the metavariable C in most cases, the implementation maintains two separate class tables for traits and objects. With two class tables, we could reuse the existing `Metatheory` library instead of forking a variant of it to handle both traits and objects in one class table.
2. While the calculus identifies each method by its name and parameter types, the implementation introduces a unique identity of type `nat` for each method. With the unique identity for every method, we could again reuse the existing `Metatheory` library as it is instead of forking a variant of it to use a pair of method name and parameter types as keys of environments.
3. As the [VALIDMETH] in Section 3 describes, the calculus does not check the overloading rules for a pair of method declarations if the pair is the same method or the pair has different names. However, for simplicity, the implementation checks the overloading rules for such declarations as Figure 7 illustrates. Unlike the overloading rules of the calculus in Figure 4, the implementation includes two extra cases: valid_same and valid_diff_name. The valid_same constructor specifies that if we check the overloading rules with a single method declaration and itself, the rules are vacuously satisfied. The valid_diff_name constructor specifies that if we check the overloading rules with a pair of method declarations with different names, again the overloading rules hold vacuously.
4. While the overloading rules in the calculus are not disjoint, its corresponding implementation is. The [VALIDMEET] rule in the calculus could be satisfied by a pair of method declarations whose parameter types and return types are in the subtype relation. However, the valid_meet case is not satisfied by such a pair. To make the

implementation of the overloading rules deterministic, the **valid_meet** constructor specifies that two method declarations are not in the subtype relation:

```
~ (sub_tys (getartys mdt1) (getartys mdt2)) →
~ (sub_tys (getartys mdt2) (getartys mdt1)) →
```

Because the differences in the calculus and the implementation are minor implementation details, we believe that we faithfully implement FFMM in Coq.

5 Type Safety Proof

We prove the type safety of the FFMM calculus in Coq. Among approximately 150 facts, lemmas, and theorems in our proof, we describe only those that are closely related to multiple dispatch and multiple inheritance.

First, the following lemma guarantees that, in a well-typed program, every non-empty set of applicable methods always includes the most specific method:

Lemma 1. *Suppose that p is well typed. If* $applicable_p(m(\overrightarrow{\tau}), visible_p(C))=\{\overrightarrow{md, C}\}$ *and* $\{\overrightarrow{md, C}\} \neq \emptyset$, *then there exists* md' *such that* $mostspecific_p(\{\overrightarrow{md, C}\}) = \{md'\}$.

It implies that there are *no ambiguous method calls* at run time in FFMM. In a well-typed program, each method call has at least one applicable method. In other words, the set of applicable methods to a call is not empty as guaranteed by the typing rule for method invocations as follows:

$$[\text{T-METHOD}]\quad \frac{p;\Gamma \vdash e_0 : \tau_0 \qquad p;\Gamma \vdash \overrightarrow{e} : \overrightarrow{\tau} \qquad mostspecific_p(applicable_p(m(\overrightarrow{\tau}), visible_p(\tau_0))) = \{m(\_) : \tau^r \_\}}{p;\Gamma \vdash e_0 . m(\overrightarrow{e}) : \tau^r}$$

It implies that there are *no undefined method calls* at run time in FFMM. This lemma plays an important role in the proofs of the following two lemmas:

Lemma 2. *Let p be well typed. If* $mostspecific_p(applicable_p(m(\overrightarrow{\tau'}), visible_p(C))) = \{m(\overrightarrow{x' \colon \tau^{a'}}) \colon \tau^{r'}{=}e'\}$ *and* $p \vdash \overrightarrow{\tau} <: \overrightarrow{\tau'}$ *for some* $\overrightarrow{\tau}$, *then there exists* $m(\overrightarrow{x \colon \tau^a}) \colon \tau^r{=}e$ *such that* $mostspecific_p(applicable_p(m(\overrightarrow{\tau}), visible_p(C))) = \{m(\overrightarrow{x \colon \tau^a}) \colon \tau^r = e\}$ *and* $p \vdash \tau^r <: \tau^{r'}$.

Lemma 3. *Let p be well typed. If* $mostspecific_p(applicable_p(m(\overrightarrow{\tau}), visible_p(C'))) = \{m(\overrightarrow{x' \colon \tau^{a'}}) \colon \tau^{r'} = e'\}$ *and* $p \vdash C <: C'$ *for some C, then there exists* $m(\overrightarrow{x \colon \tau^a}) \colon \tau^r{=}e$ *such that* $mostspecific_p(applicable_p(m(\overrightarrow{\tau}), visible_p(C))) = \{m(\overrightarrow{x \colon \tau^a}) \colon \tau^r = e\}$ *and* $p \vdash \tau^r <: \tau^{r'}$.

These lemmas state that dynamically selecting a method that is more specific than the statically chosen method is type safe. They serve an important role in proving that term substitutions preserve typing:

Table 1. COQ mechanization of CFFJ, FBCF, and FFMM

Language	Metatheory		Calculus	Type Safety Proof		Total	
	Spec	Proofs	Spec	Spec	Proofs	Spec	Proofs
CFFJ	114	158	164	249	338	527	496
FBCF	114	158	226	233	348	573	506
FFMM	136	203	402	742	1786	1280	1989

Lemma 4. *Suppose that p is well typed. If* $p; \Gamma\overrightarrow{x:\tau} \vdash e : \tau$ *and* $p; \Gamma \vdash \overrightarrow{e'} : \overrightarrow{\tau'}$, *and* $p \vdash \overrightarrow{\tau'} <: \overrightarrow{\tau}$, *then* $p; \Gamma \vdash [\overrightarrow{e'/x}]e : \tau'$ *for some* τ' *such that* $p \vdash \tau' <: \tau$.

Finally, the type safety proof of FFMM follows the traditional technique:

Theorem 1 (Progress). *Suppose that p is well typed. If* $p; \emptyset \vdash e : \tau$, *then e is a value or there exists some* e' *such that* $p \vdash e \longrightarrow e'$.

The Progress theorem is proved by induction on the derivation of $p; \emptyset \vdash e : \tau$. The most interesting part is the [T-METHOD] case where we find a witness e' using the previous lemmas.

Theorem 2 (Preservation). *Suppose that p is well typed. If* $p; \Gamma \vdash e : \tau$ *and* $p \vdash e \longrightarrow e'$, *then* $p; \Gamma \vdash e' : \tau'$ *where* $p \vdash \tau' <: \tau$.

The type safety theorem is immediate from Theorem 1 and Theorem 2:

Theorem 3 (Type Safety). *Suppose that p is well-typed. If* $p; \emptyset \vdash e : \tau$ *and* $p \vdash e \longrightarrow^* v$, *then* $p; \emptyset \vdash v : \tau'$ *and* $p \vdash \tau' <: \tau$.

The full proof of all the facts, lemmas, and theorems in COQ is available online [20].

6 Lessons

In this section, we share our experience and lessons from mechanizing the type soundness of FFMM.

6.1 Extensibility of COQ Mechanization

Before FFMM, we mechanized Featherweight Basic Core Fortress (FBCF) [22], a very small core of the Fortress programming language, in COQ. It supports both traits and objects like FFMM, but it does not provide multiple dispatch nor multiple inheritance. Its mechanization heavily relies on the implementation of CFFJ. Table 1 compares the line numbers of the COQ implementations of CFFJ, FBCF, and FFMM. While the size of the FBCF implementation is similar to that of the CFFJ implementation, the size of the FFMM implementation is almost three times bigger than them.

While FBCF provides method overriding and single inheritance, FFMM supports method overloading and multiple inheritance. Similarly to CFFJ, FBCF uses the $mtype_p$ and $mbody_p$ functions for method lookup: it traverses up the type hierarchy one by one until it finds the intended method. On the contrary, FFMM uses the $visible_p$ function for method lookup: it collects all the visible methods first instead of traversing up the type hierarchy.

We observed that adding multiple inheritance to FFMM was much easier and natural than adding it to FBCF. We first tried to extend FBCF by generalizing the $mtype_p$ and $mbody_p$ functions to support multiple inheritance, but it became easily exponential. Extending FFMM with multiple inheritance amounts to collecting all the super types before collecting the visible methods via the $visible_p$ function, which does not require much code changes.

6.2 Witness Finding

When a rule in FFMM is not algorithmic, finding a witness to mechanize the rule in COQ could be not trivial. For example, in the following subtype transitivity rule:

$$[\text{S-TRANS}]\ \frac{p \vdash \tau_1 <: \tau_2 \qquad p \vdash \tau_2 <: \tau_3}{p \vdash \tau_1 <: \tau_3}$$

the premise uses τ_2 that does not appear in the conclusion. Therefore, when we prove a statement including subtype relations, we should be able to find a witness for the [S-TRANS] case. Consider the following fact:

```
Fact sub_ty_implies_visible_super_set: ∀ ty ty' mS mS',
  sub_ty ty ty' →
  visible ty' mS' →
  visible ty mS →
  (∀ mdt, mdt \in mS' → mdt \in mS).
```

which states that "If *ty* is a subtype of *ty'*, $visible_p$(*ty'*) = *mS'*, and $visible_p$(*ty*) = *mS*, then *mS'* is a subset of *mS*." Proving the above fact by induction on the derivation of **sub_ty** *ty ty'* fails to find a witness for the [S-TRANS] case: while COQ has to find a witness of $visible_p$(*ty"*) for some *ty"* that is a subtype of *ty'* and a supertype of *ty*, two method sets, *mS'* and *mS*, are already bound to two types, *ty'* and *ty*, respectively. Moreover, the fact statement does not specify how to find a set of visible methods for a certain type that is not mentioned in the statement such as *ty"*.

Instead, we need to restate the above fact as follows:

```
Fact sub_ty_implies_visible_super_set: ∀ ty ty' mS',
  sub_ty ty ty' →
  visible ty' mS' →
  exists2 mS, visible ty mS & (∀ mdt, mdt \in mS' → mdt \in mS).
```

By using `exists2` in the induction hypothesis, COQ can find a witness to prove the [S-TRANS] case.

7 Related Work

Several object-oriented languages provide multiple dispatch. As we discussed in Section 1, there are two camps in multiple dispatch: asymmetric multiple dispatch and symmetric multiple dispatch. Languages supporting asymmetric multiple dispatch such as

CLOS [9] and Dylan [1] distinguish method arguments to eliminate the possibility of ambiguous method calls. However, languages with symmetric multiple dispatch such as Nice [10] and Fortress [3] treat all the arguments equally, and they provide some restrictions to guarantee no ambiguous calls at run time.

Some languages support symmetric multiple dispatch with static rules on overloaded method declarations. Castagna *et al.* [12] proposed the λ&-calculus, an extension of the typed lambda calculus with overloaded functions, and presented constraints to ensure that for each call site, there exists a unique best method to call at run time. Similarly, the Fortress team [4] proposed static rules on overloaded methods in the context of the Fortress type system. Millstein and Chambers designed the Dubious [24] language and restrictions to ensure the *modular* type safety in the presence of symmetric multiple dispatch.

Researchers have proposed extensions of the Java programming language with symmetric multiple dispatch. Clifton *et al.* [13] presented MultiJava which adds symmetric multiple dispatch and open classes to Java. Lorenzo *et al.* [7,8] proposed Featherweight Java with Multi-methods (FJM) and proved its type soundness. Lievens and Harrison [23] proposed a very similar approach to FJM but they included casting expressions that are omitted in FJM. None of these extensions support multiple inheritance.

Type safety proofs of several languages are mechanized in COQ. Dubois [17] proved type soundness of ML [25] using COQ, Fraine *et al.* [18] proved the type safety of CFFJ, Delaware *et al.* [16] verified the type soundness of Lightweight Feature Java, a subset of Java extended with support for features, and Cremet and Altherr [6,15] mechanized the type safety of FGJ_{Ω}, an extension of FGJ with variables representing type constructors. However, none of these mechanizations of Java-like languages provide multiple dispatch nor multiple inheritance.

8 Conclusion and Future Work

We present a core calculus for the Fortress programming language with multiple dispatch and multiple inheritance. The calculus formally specifies the static restrictions on valid overloaded method declarations. For a well-typed program, which satisfies the overloading rules statically, there are no undefined nor ambiguous calls at run time. We mechanize the calculus and prove its type safety in COQ. As far as we know, our work is the first mechanized calculus of multiple dispatch in the presence of multiple inheritance, and we believe that our work is adaptable to any object-oriented languages with multiple dispatch and multiple inheritance.

We are planning to extend the calculus with more features in Fortress and mechanize the extended calculus and its type safety proof. First, we are planning to add `excludes` clauses to the calculus so that the Exclusion Rule can allow more methods as valid overloadings. Secondly, using the Fortress team's recent work on valid overloadings of parametrically polymorphic methods [5], we will extend the calculus to permit overloadings on generic methods. Finally, we will support the Fortress module system so that the calculus can faithfully capture the core expressive power of the Fortress overloading mechanism.

Acknowledgments. This work is supported in part by the Engineering Research Center of Excellence Program of Korea Ministry of Education, Science and Technology (MEST) / National Research Foundation of Korea(NRF) (Grant 2011-0000974).

References

1. Dylan, `http://www.opendylan.org/`
2. Metatheory Library: Atom, `http://www.cis.upenn.edu/ plclub/popl08-tutorial/code/coqdoc/Atom.html`
3. Allen, E., Chase, D., Hallett, J.J., Luchangco, V., Maessen, J.-W., Ryu, S., Steele Jr., G.L., Tobin-Hochstadt, S.: The Fortress Language Specification Version 1.0 (March 2008)
4. Allen, E., Hallett, J.J., Luchangco, V., Ryu, S., Steele Jr., G.L.: Modular Multiple Dispatch with Multiple Inheritance. In: Proceedings of the 2007 ACM Symposium on Applied Computing, New York, NY, USA, pp. 1117–1121. ACM (2007)
5. Allen, E., Hilburn, J., Kilpatrick, S., Ryu, S., Chase, D., Luchangco, V., Steele Jr., G.L.: Type-checking Modular Multiple Dispatch with Parametric Polymorphism and Multiple Inheritance. In: Proceedings of the 26th ACM SIGPLAN Conference on Object-Oriented Programming Systems Languages and Applications. ACM (2011)
6. Altherr, P., Cremet, V.: Adding Type Constructor Parameterization to Java. In: Formal Techniques for Java-like Programs (2007)
7. Bettini, L., Capecchi, S., Venneri, B.: Featherweight Java with Multi-methods. In: Proceedings of the 5th International Symposium on Principles and Practice of Programming in Java, New York, NY, USA, pp. 83–92. ACM (2007)
8. Bettini, L., Capecchi, S., Venneri, B.: Featherweight Java with Dynamic and Static Overloading. Science of Computer Programming 74, 261–278 (2009)
9. Bobrow, D.G., DiMichiel, L.G., Gabriel, R.P., Keene, S.E., Kiczales, G., Moon, D.A.: Common Lisp Object System Specification. ACM SIGPLAN Notices, 23 (September 1988)
10. Bonniot, D., Keller, B., Barber, F.: The Nice user's manual (2003), `http://nice.sourceforge.net/NiceManual.pdf`
11. Bracha, G., Steele, G., Joy, B., Gosling, J.: Java^{TM} Language Specification, 3rd edn. Java Series. Addison-Wesley Professional (July 2005)
12. Castagna, G., Ghelli, G., Longo, G.: A Calculus for Overloaded Functions with Subtyping. SIGPLAN Lisp Pointers 1, 182–192 (1992)
13. Clifton, C., Leavens, G.T., Chambers, C., Millstein, T.: MultiJava: Modular Open Classes and Symmetric Multiple Dispatch for Java. In: Proceedings of the 15th ACM SIGPLAN Conference on Object-Oriented Programming, Systems, Languages, and Applications, New York, NY, USA, pp. 130–145. ACM (2000)
14. The Coq Development Team. The Coq Proof Assistant, `http://coq.inria.fr/`
15. Cremet, V., Altherr, P.: FGJ-ω in Coq (2007), `http://lamp.epfl.ch/~cremet/FGJ-omega`
16. Delaware, B., Cook, W.R., Batory, D.: Fitting the Pieces Together: a Machine-checked Model of Safe Composition. In: Proceedings of the the 7th Joint Meeting of the European Software Engineering Conference and the ACM SIGSOFT Symposium on The Foundations of Software Engineering, ESEC/FSE 2009, New York, NY, USA, pp. 243–252. ACM (2009)
17. Dubois, C.: Proving ML Type Soundness within Coq. In: Aagaard, M.D., Harrison, J. (eds.) TPHOLs 2000. LNCS, vol. 1869, pp. 126–144. Springer, Heidelberg (2000)
18. De Fraine, B., Ernst, E., Südholt, M.: Cast-Free Featherweight Java (2008), `http://soft.vub.ac.be/~bdefrain/featherj`

19. Igarashi, A., Pierce, B., Wadler, P.: Featherweight Java: A Minimal Core Calculus for Java and GJ. In: Meissner, L. (ed.) Proceedings of the 1999 ACM SIGPLAN Conference on Object-Oriented Programming, Systems, Languages & Applications (OOPSLA 1999), vol. 34(10), pp. 132–146. NY (1999)
20. Kim, J.: FFMM in Coq (2011), http://plrg.kaist.ac.kr/_media/research/software/ffmm_in_coq.tar.gz
21. Kim, J., Ryu, S.: FFMM: Featherweight Fortress with Multiple Dispatch and Multiple Inheritance. Technical report, KAIST (June 2011)
22. Kim, J., Ryu, S.: Coq Mechanization of Featherweight Basic Core Fortress for Type Soundness. Technical Report ROSAEC-2011-011, KAIST (May 2011)
23. Lievens, D., Harrison, W.: Symmetric Encapsulated Multi-methods to Abstract over Application Structure. In: Proceedings of the 2009 ACM Symposium on Applied Computing, New York, NY, USA, pp. 1873–1880. ACM (2009)
24. Millstein, T., Chambers, C.: Modular Statically Typed Multimethods. In: Information and Computation, pp. 279–303 (2002)
25. Milner, R., Tofte, M., Harper, R., MacQueen, D.: The Definition of Standard ML (Revised). The MIT Press (1997)
26. Muschevici, R., Potanin, A., Tempero, E., Noble, J.: Multiple Dispatch in Practice. In: Proceedings of the 23rd ACM SIGPLAN Conference on Object-Oriented Programming Systems Languages and Applications, pp. 563–582. ACM (2008)
27. Odersky, M., Spoon, L., Venners, B.: Programming in Scala: A Comprehensive Step-by-Step Guide, 2nd edn. Artima Inc. (2011)
28. Schärli, N., Ducasse, S., Nierstrasz, O., Black, A.P.: Traits: Composable Units of Behaviour. In: Cardelli, L. (ed.) ECOOP 2003. LNCS, vol. 2743, pp. 248–274. Springer, Heidelberg (2003)
29. Stroustrup, B.: The C++ Programming Language. Addison-Wesley (1985)

Mechanizing the Metatheory of mini-XQuery

James Cheney[1] and Christian Urban[2]

[1] University of Edinburgh
[2] TU Munich

Abstract. We present a Nominal Isabelle formalization of an expressive core fragment of XQuery, a W3C standard functional language for querying XML documents. Our formalization focuses on results presented in the literature concerning XQuery's operational semantics, typechecking, and optimizations. Our core language, called mini-XQuery, omits many complications of XQuery such as ancestor and sibling axes, recursive types and functions, node identity, and unordered processing modes, but does handle distinctive features of XQuery including monadic comprehensions, downward XPath steps and regular expression types. To our knowledge no language with similar features has been mechanically formalized previously. Our formalization is a first step towards a complete formalization of full XQuery.

1 Introduction

The long-term vision of research on mechanized metatheory is to develop practical computer-assisted techniques for designing new programming languages, validating implementations and optimization techniques, and improving the reliability and efficiency of existing languages. To realize this vision, it is important to apply mechanized metatheory tools to real programming languages, not just well-studied core calculi [1,4,27]. In this paper, we take first steps toward formalizing and verifying properties of the XQuery language [6]. Formalizing XQuery's semantics and verifying optimization techniques will both stretch the capabilities of mechanized metatheory tools and improve confidence in XQuery-based programs.

Over the last two decades the World Wide Web Consortium (W3C) has promulgated many key standards such as Hypertext Markup Language (HTML) used for Web pages, and the more general Extensible Markup Language (XML) that can be used to exchange data and documents. These standard protocols and data formats help ensure cross-compatibility for Web browsers, servers, and other applications, catalyzing the rapid growth of the Web over the past decades. More recently, the W3C has put a great deal of effort into standardizing languages for querying, transforming, or processing XML data — particularly XPath and XQuery.

XQuery is a flagship W3C standard language for querying XML databases that manage efficient access to large amounts of data in XML form. XQuery can be used to write high-level programs; in fact, some Web applications can be written entirely in XQuery. XQuery is considered particularly suitable for integrating loosely-structured data from diverse sources. Moreover, unlike many calculi used for mechanized metatheory tools to date, the commercial value of XQuery is recognized by industry. Relational database

J.-P. Jouannaud and Z. Shao (Eds.): CPP 2011, LNCS 7086, pp. 280–295, 2011.

vendors such as IBM and Oracle view XML support as required functionality; over 50 commercial products use XQuery. Another vendor, MarkLogic, has over 180 government and publishing industry clients for its native XML database software. There are also several popular open-source XQuery implementations, such as Galax, MonetDB/XQuery, BaseX and Saxonica.

XML databases offer great potential, but also pose new challenges. Efficient XQuery implementations perform sophisticated optimizations based on equational reasoning about programs [14]. Testing these optimizations can detect bugs, but can never guarantee that all bugs have been eliminated. Moreover, most equational reasoning has been validated at the level of a simplified, purely functional form of XQuery. Equational reasoning about these core languages does not necessarily hold for the full language, in part because full XQuery is not *really* pure: node identity allocation is an effect that can be observed by identity tests or duplicate elimination. Consider the following two XQuery expressions:

```
e1:   e union e           e2:   let $x := e in ($x union $x)
```

Here, e2 performs less work than e1 by evaluating e only once, so it is tempting to rewrite e1 to e2. This is an example of *common subexpression elimination*, an important optimization technique. However, naive use of common subexpression elimination for XQuery can produce wrong answers, because e1 and e2 are not always equivalent. For example, suppose e is an expression such as <a/> that constructs a new node. In e1, two nodes are created because <a/> is evaluated twice yielding two element nodes with distinct node identifiers, while in e2, only one new node identifier is created. The union operation eliminates duplicate nodes, so count(<a/> union <a/>) = 2 while count(let $x := <a/> in $x union $x) = 1. Such corner cases are unusual. Common subexpression elimination and its inverse, inlining, are important for optimizing XQuery programs, but care is needed to ensure they are used safely.

A great deal of research about XQuery (e.g. [14,12,8]) has focused on simpler core languages that do not exhibit the above pathological behavior. Thus, as a starting point for formalizing full XQuery, our strategy is to first formalize this well-understood core and conduct mechanically-checked proofs of the main results about it. In this paper, we focus on a Turing-incomplete core language called mini-XQuery that exhibits many of the issues needed to handle full XQuery, but whose semantics is much cleaner and easier to deal with because it omits features such as node identity. Mini-XQuery is nevertheless rich enough to study several previously-developed type systems, equivalence laws, and static analyses, including those of Fernandez et al. [14], Colazzo et al. [12] and Cheney [8]. The formalization is available online [9].

The structure of the rest of this paper is as follows: Section 2 presents the mini-XQuery language we will formalize. Section 3 discusses the basic metatheory, including the operational semantics, type soundness, determinacy and totality. Section 4 presents formalizations of operational equivalences about XQuery, including the laws presented by Fernandez et al. [14]. Section 5 presents the formalization of laws and properties of regular expression subtyping. Section 6 reflects on the formalization process itself; Section 7 presents related and future work and Section 8 concludes.

2 Background

Values We use a simplified model of XML values, in common with previous work that focuses on the element and text node structure of trees and ignores attributes and other leaf node types. While these details matter in implementations, the main challenges lie in formalizing the handling of elements and text nodes.

$$\hat{v} ::= \texttt{text}\{w\} \mid \texttt{elem}\,l\,\{v\} \qquad v ::= \hat{v}, v \mid (\,)$$

Here, $w, l \in \Sigma^*$ are strings and values v are lists of atomic values $\hat{v}$ which can be text nodes $\texttt{text}\{w\}$ or element nodes $\texttt{elem}\ l\ \{v\}$. We define functions such as $[\hat{v}]$ (which makes an atomic value into a singleton list $\hat{v}$, $(\,)$) and $v \mathbin{@} v'$ (which concatenates two sequences).

Types. In XQuery, the type system is based on regular expression types and the related notion of subtyping is language inclusion [19]. The syntax of types is as follows:

$$\alpha ::= \texttt{text} \mid \texttt{elem}\,l\,\{\tau\} \mid \texttt{item} \qquad \tau ::= \alpha \mid (\,) \mid \tau_1, \tau_2 \mid \tau_1|\tau_2 \mid \tau^*$$

Note that we distinguish syntactically between atomic types α that match atomic values, versus general types τ that represent values v. We also omit recursive types; although star types provide a limited form of recursion, our types will always have a bounded nesting depth of element constructors. This is not an essential limitation; however, we chose to focus on non-recursive types in this paper to focus attention on the new issues arising for regular expression types.

The meaning of types is defined using the judgments $\hat{v} \mathrel{\hat{:}} \alpha$ and $v : \tau$ indicating when a value matches a type. (Equivalently, we could write this denotationally as $v \in [\![\tau]\!]$). This judgment is defined as follows:

$$\frac{}{\texttt{text}\{w\} \mathrel{\hat{:}} \texttt{text}} \qquad \frac{v : \tau}{\texttt{elem}\,l\,\{v\} \mathrel{\hat{:}} \texttt{elem}\,l\,\{\tau\}} \qquad \frac{}{\hat{v} \mathrel{\hat{:}} \texttt{item}} \qquad \frac{}{(\,) : (\,)} \qquad \frac{\hat{v} \mathrel{\hat{:}} \alpha}{[\hat{v}] : \alpha}$$

$$\frac{v_1 :: \tau_1 \quad v_2 : \tau_2}{v_1 \mathbin{@} v_2 : \tau_1, \tau_2} \qquad \frac{v : \tau_1}{v : \tau_1|\tau_2} \qquad \frac{v : \tau_2}{v : \tau_1|\tau_2} \qquad \frac{}{(\,) : \tau^*} \qquad \frac{v : \tau}{v : \tau^*} \qquad \frac{v_1 : \tau^* \quad v_2 : \tau^*}{v_1 \mathbin{@} v_2 : \tau^*}$$

XPath steps and tests. We handle the downward XPath axis steps and basic node tests:

$$ax ::= \texttt{self} \mid \texttt{child} \mid \texttt{dos} \qquad \phi ::= \texttt{node}() \mid \texttt{text}() \mid l$$

Axis steps include `self`, which corresponds to the identity relation; `child`, which corresponds to the parent-child relation; and the `dos` or *descendant-or-self* axis which corresponds to the transitive, reflexive closure of `child`. We also consider the basic node tests `node()` which selects any node, `text()` which selects text nodes only, and l which selects element nodes with label l. Note that the (seemingly superfluous) parentheses on `node()` and `text()` are part of the syntax of XPath/XQuery. Also note that the `node()` test is often abbreviated $*$, and $/\texttt{child} :: \phi$ is often abbreviated as just $/\phi$.

Expressions. The expressions of mini-XQuery are as follows:

$$e ::= \texttt{()} \mid e, e' \mid \texttt{elem}\, l\, \{e\} \mid w \mid x \mid x/ax :: \phi \mid \texttt{let}\; x := e\; \texttt{in}\; e'$$
$$\mid\; \texttt{if}\; e\; \texttt{then}\; e_1\; \texttt{else}\; e_2 \mid \texttt{for}\; x \in e\; \texttt{return}\; e'$$

where again $w, l \in \Sigma^*$ are strings. These include expressions for constructing values, such as the empty sequence `()`, sequential composition e_1, e_2, element nodes $\texttt{elem}\ l\ \{e\}$ and string literals $w \in \Sigma^*$, as well as standard variables, let-bindings $\texttt{let}\ x\ :=\ e_1\ \texttt{in}\ e_2$, and conditionals $\texttt{if}\ e\ \texttt{then}\ e_1\ \texttt{else}\ e_2$. Furthermore, the expression $x/ax\ ::\ \phi$ denotes the result of taking an XPath step from a variable and $\texttt{for}\ x \in e\ \texttt{return}\ e'$ denotes iteration over a value viewed as a sequence. Note that in XQuery source programs it is typical to write programs with compound XPath steps and abbreviations.

$$\texttt{for}\; x \in y/a//b\; \texttt{return}\; e$$

However, in XQuery this is desugared to

$$\texttt{for}\; z_1 \in y/a\; \texttt{return}\; \texttt{for}\; z_2 \in z_1/*\; \texttt{return}\; \texttt{for}\; x \in z_2/\texttt{dos}::b\; \texttt{return}\; e$$

where the extra $*$-step is due to the fact that $//b$ really abbreviates the descendant step, which is irreflexive.

Value and typing contexts. We write γ for value contexts mapping variables to values v, and Γ for typing contexts mapping variables to types τ. These are represented as lists of pairs of variables with values or types, such that no variable is repeated. As usual, we need to define validity for such contexts and prove a number of routine properties to ensure that e.g. a variable bound in a valid context has a unique value. We also build validity assumptions into the evaluation and typing judgments (for example, by requiring validity in the rules for base cases such as variables) to decrease the number of explicit validity hypotheses we need to state the main results. We omit the details of this part of the formalization. A value context γ is considered well-formed with respect to a typing context Γ if they bind the same variables and for each x in their common domain, we have $\gamma(x) : \Gamma(x)$. We write $\gamma : \Gamma$ to indicate that this is the case. Again, in the formalization we need to be more pedantic (e.g. we also require that γ and Γ bind the same variables in the same order) but for the purposes of exposition these details are omitted.

Evaluation. The XQuery standard gives an operational semantics, while several papers give a denotational semantics for mini-XQuery-like core languages. Although denotational semantics is attractive for a purely functional, terminating core language such as mini-XQuery, we expect that operational techniques will scale better to handling the full language, so we will use a simplified operational semantics for mini-XQuery.

We define the operational semantics judgments via inference rules as shown in Figure 1. The semantics uses two judgments, one for ordinary expression evaluation $\gamma \models e \Rightarrow v$ and one for iterated evaluation $\gamma;\ x \in v \models^* e \Rightarrow v'$. Intuitively, the iteration judgment does a list comprehension over the input value list v, binding x to each atomic $\hat{v}$, evaluating e, and then concatenating the resulting value sequences. It is an important fact about mini-XQuery that these iterations are completely independent, that

$\boxed{\gamma \vDash e \Rightarrow v}$

$$\frac{}{\gamma \vDash () \Rightarrow ()} \qquad \frac{\gamma \vDash e \Rightarrow v}{\gamma \vDash \texttt{elem}\, l\, \{e\} \Rightarrow \texttt{elem}\, l\, \{v\}} \qquad \frac{\gamma \vDash e_1 \Rightarrow v_1 \quad \gamma \vDash e_2 \Rightarrow v_2}{\gamma \vDash e_1, e_2 \Rightarrow v_1 \mathbin{@} v_2}$$

$$\frac{}{\gamma \vDash w \Rightarrow w} \qquad \frac{\gamma(x) = v}{\gamma \vDash x \Rightarrow v} \qquad \frac{}{\gamma \vDash x/ax :: \phi \Rightarrow (\gamma(x)/ax) :: \phi}$$

$$\frac{\gamma \vDash e_0 \Rightarrow \hat{v}, v \quad \gamma \vDash e_1 \Rightarrow v_1}{\gamma \vDash \texttt{if}\ e_0\ \texttt{then}\ e_1\ \texttt{else}\ e_2 \Rightarrow v_1} \qquad \frac{\gamma \vDash e_0 \Rightarrow () \quad \gamma \vDash e_2 \Rightarrow v_2}{\gamma \vDash \texttt{if}\ e_0\ \texttt{then}\ e_1\ \texttt{else}\ e_2 \Rightarrow v_2}$$

$$\frac{\gamma \vDash e_1 \Rightarrow v_1 \quad \gamma, x \mapsto v_1 \vDash e_2 \Rightarrow v_2}{\gamma \vDash \texttt{let}\ x := e_1\ \texttt{in}\ e_2 \Rightarrow v_2} \qquad \frac{\gamma \vDash e_1 \Rightarrow v_1 \quad \gamma;\ x \in v_1 \vDash^* e_2 \Rightarrow v_2}{\gamma \vDash \texttt{for}\ x \in e_1\ \texttt{return}\ e_2 \Rightarrow v_2}$$

$\boxed{\gamma;\ x \in v \vDash^* e \Rightarrow v'}$

$$\frac{}{\gamma;\ x \in () \vDash^* e \Rightarrow ()}\ \textsc{Nil}^* \qquad \frac{\gamma, x \mapsto \hat{v} \vDash e \Rightarrow v'}{\gamma;\ x \in [\hat{v}] \vDash^* e \Rightarrow v'}\ \textsc{Sng}^*$$

$$\frac{\gamma;\ x \in v_1 \vDash^* e \Rightarrow v_1' \quad \gamma;\ x \in v_2 \vDash^* e \Rightarrow v_2'}{\gamma;\ x \in v_1 \mathbin{@} v_2 \vDash^* e \Rightarrow v_1' \mathbin{@} v_2'}\ \textsc{Seq}^*$$

Fig. 1. Large-step operational semantics

is, the sub-computations can be evaluated in any order or in parallel (as long as they are reassembled in the correct order). Also note that conditionals test emptiness.

We use auxiliary functions v/ax and $v :: \phi$ to define the behavior of axis steps and node tests on values. Their definitions are:

$$\begin{aligned} \hat{v}/\texttt{self} &= [\hat{v}] \\ \texttt{elem}\, l\, \{v\}/\texttt{child} &= v \\ \texttt{text}\{w\}/\texttt{child} &= () \\ \texttt{elem}\, l\, \{v\}/\texttt{dos} &= \texttt{elem}\, l\, \{v\}, (v/\texttt{dos}) \\ \texttt{text}\{w\}/\texttt{dos} &= \texttt{text}\{w\} \\ ()/ax &= () \\ (\hat{v}, v)/ax &= \hat{v}/ax \mathbin{@} v/ax \end{aligned} \qquad \begin{aligned} \hat{v} :: \texttt{node}() &= [\hat{v}] \\ \texttt{elem}\, l\, \{v\} :: l &= \texttt{elem}\, l\, \{v\} \\ \texttt{elem}\, l\, \{v\} :: l' &= () \\ \texttt{text}\{w\} :: l &= () \\ \texttt{elem}\, l\, \{v\} :: \texttt{text}() &= () \\ \texttt{text}\{w\} :: l &= \texttt{text}\{w\} \\ () :: \phi &= () \\ (\hat{v}, v) :: \phi &= \hat{v} :: \phi \mathbin{@} v :: \phi \end{aligned}$$

Observe that in both cases, the behavior over value sequences is uniform. In the case of the descendant-or-self axis, we always return the value itself followed by the result of evaluating the descendant-or-self axis on the sequence of children of the node (if any). For example:

$$\begin{aligned} &\texttt{elem}\, a\, \{\texttt{elem}\, b\, \{\texttt{text}\{\text{"}x\text{"}\}\}, \texttt{elem}\, c\, \{\texttt{elem}\, b\, \{\texttt{text}\{\text{"}y\text{"}\}\}\}\}/\texttt{dos} :: b/\texttt{text}() \\ = &\ \texttt{text}\{\text{"}x\text{"}\}, \texttt{text}\{\text{"}y\text{"}\} \end{aligned}$$

This example also illustrates another (minor) simplification in mini-XQuery: we do not normalize values to merge adjacent text nodes.

Substitution. We define a form of substitution adapted to mini-XQuery as follows:

$$
\begin{aligned}
x[e/x] &= e \\
y[e/x] &= y \\
(x/ax :: \phi)[e/x] &= \texttt{let}\ x := e\ \texttt{in}\ x/ax :: \phi \\
(y/ax :: \phi)[e/x] &= y/ax :: \phi \\
w[e/x] &= w \\
()\,[e/x] &= () \\
(e_1, e_2)[e/x] &= (e_1[e/x], e_2[e/x]) \\
\texttt{elem}\ l\ \{e_0\}[e/x] &= \texttt{elem}\ l\ \{e_0[e/x]\} \\
(\texttt{if}\ e_0\ \texttt{then}\ e_1\ \texttt{else}\ e_2)[e/x] &= \texttt{if}\ e_0[e/x]\ \texttt{then}\ e_1[e/x]\ \texttt{else}\ e_2[e/x] \\
(\texttt{let}\ y := e_1\ \texttt{in}\ e_2)[e/x] &= \texttt{let}\ y := e_1[e/x]\ \texttt{in}\ e_2[e/x] \qquad (y \notin FV(x, e, e_1)) \\
(\texttt{for}\ y \in e_1\ \texttt{return}\ e_2)[e/x] &= \texttt{for}\ y \in e_1[e/x]\ \texttt{return}\ e_2[e/x]\ \ (y \notin FV(x, e, e_1))
\end{aligned}
$$

Note that for variable occurrences in axis steps, we cannot always substitute for the variable, so we simply re-bind it locally (following [12] and other core calculi on which mini-XQuery is based). This function can be defined as a total function in Nominal Isabelle as described for similar languages in [25,27]. It is worth pointing out that this substitution function is not the one that comes "for free" with higher-order abstract syntax, because of the rules for XPath step expressions $x/ax :: \phi$.

Type system. The typing rules used for mini-XQuery are shown in Figure 2. These rules include the ordinary expression typing judgment $\Gamma \vdash e : \tau$ and iteration typing judgment $\Gamma;\ x \in \tau \vdash^* e : \tau'$.

Auxiliary rules for axis and node test typechecking are given in Figure 3. In most cases these simply follow the operational behavior or the structure of the regular expression type. Because we omitted recursive types, we can employ a more precise rule for typechecking descendant-or-self steps: specifically, we symbolically evaluate the descendant-or-self step on the regular expression type. This level of precision is not possible in the presence of recursion, because the resulting language is not necessarily regular. Instead, Colazzo et al.'s μXQ system [12] simply approximates the descendant-or-self step as $(\alpha_1|\cdots|\alpha_n)^*$, where $\alpha_1, \ldots, \alpha_n$. We believe that the nonrecursive case is common enough to warrant special handling for this increased precision.

Our rules are not the same as the original W3C type system either. The W3C system does not use rules similar to the iteration judgment; instead, when an expression of the form $\texttt{for}\ x \in e_1\ \texttt{return}\ e_2$ is typechecked, the type of e_1 is split into a *prime* type $\alpha_1|\cdots|\alpha_n$ and a *quantifier* $q \in \{1, +, *, ?\}$. The body of the loop is then checked with x bound to $\alpha_1|\cdots|\alpha_n$ and the return type is adjusted using q.

We believe it is more interesting to prove type soundness for the more precise approach; soundness for the W3C type system can then be proved easily by showing that the type inferred by our system is always a subtype of that inferred by the W3C system. We have not formalized the W3C system or this proof, but this appears straightforward (Colazzo and Sartiani [13] present such a result and discuss a number of related issues concerning the expressiveness of the two systems).

$\boxed{\Gamma \vdash e : \tau}$

$$\frac{}{\Gamma \vdash \texttt{()} : \texttt{()}} \qquad \frac{\Gamma \vdash e : \tau}{\Gamma \vdash \texttt{elem}\, l\, \{e\} : \texttt{elem}\, l\, \{\tau\}} \qquad \frac{\Gamma \vdash e : \tau \quad \Gamma \vdash e' : \tau'}{\Gamma \vdash e, e' : \tau, \tau'}$$

$$\frac{w \in \Sigma^*}{\Gamma \vdash \texttt{text}\{w\} : \texttt{text}} \qquad \frac{x{:}\tau \in \Gamma}{\Gamma \vdash x : \tau} \qquad \frac{\Gamma \vdash e_1 : \tau_1 \quad \Gamma, x{:}\tau_1 \vdash e_2 : \tau_2}{\Gamma \vdash \texttt{let}\, x = e_1\, \texttt{in}\, e_2 : \tau_2}$$

$$\frac{\Gamma \vdash c : \tau_0 \quad \Gamma \vdash e_1 : \tau_1 \quad \Gamma \vdash e_2 : \tau_2}{\Gamma \vdash \texttt{if}\, c\, \texttt{then}\, e_1\, \texttt{else}\, e_2 : \tau_1 | \tau_2} \qquad \frac{x : \tau \in \Gamma \quad \tau/ax \Rightarrow \tau' \quad \tau' :: \phi \Rightarrow \tau''}{\Gamma \vdash x/ax :: \phi : \tau''}$$

$$\frac{\Gamma \vdash e_1 : \tau_1 \quad \Gamma;\, x \in \tau_1 \vdash^* e_2 : \tau_2}{\Gamma \vdash \texttt{for}\, x \in e_1\, \texttt{return}\, e_2 : \tau_2} \qquad \frac{\Gamma \vdash e : \tau \quad \tau <: \tau'}{\Gamma \vdash e : \tau'}$$

$\boxed{\Gamma;\, x \in \tau \vdash^* e : \tau'}$

$$\frac{}{\Gamma;\, x \in \texttt{()} \vdash^* e : \texttt{()}} \qquad \frac{\Gamma, x{:}\alpha \vdash e : \tau}{\Gamma;\, x \in \alpha \vdash^* e : \tau} \qquad \frac{\Gamma;\, x \in \tau_1 \vdash^* e : \tau_2}{\Gamma;\, x \in \tau_1^* \vdash^* e : \tau_2^*}$$

$$\frac{\Gamma;\, x \in \tau_1 \vdash^* e : \tau_1' \quad \Gamma;\, x \in \tau_2 \vdash^* e : \tau_2'}{\Gamma;\, x \in \tau_1, \tau_2 \vdash^* e : \tau_1', \tau_2'} \qquad \frac{\Gamma;\, x \in \tau_1 \vdash^* e : \tau_1' \quad \Gamma;\, x \in \tau_2 \vdash^* e : \tau_2'}{\Gamma;\, x \in \tau_1 | \tau_2 \vdash^* e : \tau_1' | \tau_2'}$$

Fig. 2. Query well-formedness rules

3 Basic Metatheory

In this section we present some of the basic properties of evaluation needed for the main results in the later sections. Values and value typing have a number of properties that are often needed:

Lemma 1. *1. For any v, we have $v/\texttt{self} = v = v :: \texttt{node}()$.*
2. For all τ, there exists $v : \tau$.
3. For all v, we have $v : \texttt{item}^*$.

Theorem 1 (Weakening). *Assume $\gamma \subseteq \gamma'$. Then:*
1. If $\gamma \models e \Rightarrow v$ holds and $\gamma \subseteq \gamma'$ then $\gamma' \models e \Rightarrow v$ holds.
2. If $x \notin \mathrm{dom}(\gamma')$ and $\gamma;\, x \in v \models^* e \Rightarrow v'$ holds then $\gamma';\, x \in v \models^* e \Rightarrow v'$ holds.

Theorem 2 (Strengthening). *Suppose $x \notin \mathrm{dom}(\gamma_1, \gamma_2)$ and $x \notin FV(e)$. Then:*
1. If $\gamma_1, x \mapsto v_0, \gamma_2 \models e \Rightarrow v$ then $\gamma_1, \gamma_2 \models e \Rightarrow v$.
2. If $y \notin \mathrm{dom}(\gamma_1, x \mapsto v_0, \gamma_2)$ and $\gamma_1, x \mapsto v_0, \gamma_2;\, y \in v_1 \models^* e \Rightarrow v_2$ then $\gamma_1, \gamma_2;\, y \in v_1 \models^* e \Rightarrow v_2$.

Theorem 3 (Exchange)
1. If $\gamma_1, x \mapsto v_1, y \mapsto v_2, \gamma_2 \models e \Rightarrow v$ then $\gamma_1, y \mapsto v_2, x \mapsto v_1, \gamma_2 \models e \Rightarrow v$.
2. If $\gamma_1, x \mapsto v_1, y \mapsto v_2, \gamma_2;\, z \in v \models^* e \Rightarrow v'$ then $\gamma_1, y \mapsto v_2, x \mapsto v_1, \gamma_2;\, z \in v \models^* e \Rightarrow v'$.

Another important property is that the iteration rules are invertible, despite their rather nondeterministic flavor. In particular:

$\boxed{\tau :: \phi \Rightarrow \tau'}$

$$\frac{}{\alpha :: \texttt{node()} \Rightarrow \alpha} \qquad \frac{}{\texttt{text} :: \texttt{text()} \Rightarrow \texttt{text}} \qquad \frac{}{\texttt{elem}\, l\, \{\tau\} :: \texttt{text()} \Rightarrow ()}$$

$$\frac{}{\texttt{text} :: l \Rightarrow ()} \qquad \frac{}{\texttt{elem}\, l\, \{\tau\} :: l \Rightarrow \texttt{elem}\, l\, \{\tau\}} \qquad \frac{l \neq l'}{\texttt{elem}\, l\, \{\tau\} :: l' \Rightarrow ()}$$

$$\frac{}{\texttt{item} :: \phi \Rightarrow \texttt{item}^*} \qquad \frac{}{() :: \phi \Rightarrow ()} \qquad \frac{\tau_1 :: \phi \Rightarrow \tau_1' \quad \tau_2 :: \phi \Rightarrow \tau_2'}{\tau_1, \tau_2 :: \phi \Rightarrow \tau_1', \tau_2'}$$

$$\frac{\tau_1 :: \phi \Rightarrow \tau_1' \quad \tau_2 :: \phi \Rightarrow \tau_2'}{\tau_1 | \tau_2 :: \phi \Rightarrow \tau_1' | \tau_2'} \qquad \frac{\tau_1 :: \phi \Rightarrow \tau_2}{\tau_1^* :: \phi \Rightarrow \tau_2^*}$$

$\boxed{\tau / \phi \Rightarrow \tau'}$

$$\frac{}{\alpha/\texttt{self} \Rightarrow \alpha} \qquad \frac{}{\texttt{text}/\texttt{child} \Rightarrow ()} \qquad \frac{}{\texttt{elem}\, l\, \{\tau\}/\texttt{child} \Rightarrow \tau} \qquad \frac{}{\texttt{text}/\texttt{dos} \Rightarrow \texttt{text}}$$

$$\frac{\tau/\texttt{dos} \Rightarrow \tau'}{\texttt{elem}\, l\, \{\tau\}/\texttt{dos} \Rightarrow \texttt{elem}\, l\, \{\tau\}, \tau'} \qquad \frac{}{\texttt{item}/ax \Rightarrow \texttt{item}^*} \qquad \frac{}{()/ax \Rightarrow ()}$$

$$\frac{\tau_1/ax \Rightarrow \tau_1' \quad \tau_2/ax \Rightarrow \tau_2'}{\tau_1, \tau_2/ax \Rightarrow \tau_1', \tau_2'} \qquad \frac{\tau_1/ax \Rightarrow \tau_1' \quad \tau_2/ax \Rightarrow \tau_2'}{\tau_1|\tau_2/ax \Rightarrow \tau_1'|\tau_2'} \qquad \frac{\tau_1/ax \Rightarrow \tau_2}{\tau_1^*/ax \Rightarrow \tau_2^*}$$

Fig. 3. Auxiliary judgments

Theorem 4 (Inversion)
1. *If* $\gamma;\ x \in () \models^* e \Rightarrow v$ *then* $v = ()$.
2. *If* $\gamma;\ x \in [\hat{v}] \models^* e \Rightarrow v'$ *then* $\gamma, x \mapsto \hat{v} \models e \Rightarrow v'$.
3. *If* $\gamma;\ x \in v_1 \ @\ v_2 \models^* e \Rightarrow v$ *then there exist* v_1', v_2' *such that* $v = v_1' \ @\ v_2'$ *and* $\gamma;\ x \in v_1 \models^* e \Rightarrow v_1'$ *and* $\gamma;\ x \in v_2 \models^* e \Rightarrow v_2'$.

Proof. Of these, the inversion of SEQ* is the most complex. We need to reason carefully by induction on the structure of the derivation, using parts (1) and (2) as well as a number of facts about lists and @. □

A final useful property of the value typing rules is that we can "lift" functions on atomic values to functions on values, preserving typing. Let f be a function from atomic values to values, and let *lift* f be the natural extension of f to a function on values, given by *lift* $f\ () = ()$ and *lift* $f\ (\hat{v}_1, v_2) = f(\hat{v}_1)\ @\ (\textit{lift}\ f\ v_2)$. Then:

Lemma 2 (Star lifting). *Assume that for all* $v' : \tau_1$ *we have* *lift* $f\ v' : \tau_2$. *Then if* $v : \tau_1^*$ *then* $(\textit{lift}\ f)\ v : \tau_2^*$.

3.1 Type Soundness

We now have enough infrastructure to show type soundness. First, we need to establish soundness properties for variables, axis steps, and iterations:

Lemma 3. *1. (Variable soundness) If* $\gamma : \Gamma$ *then* $\gamma(x) : \Gamma(x)$.

2. *(Axis soundness) If* $\tau/ax \Rightarrow \tau'$ *and* $v : \tau$ *then* $v/ax : \tau'$.
3. *(Test soundness) If* $\tau :: \phi \Rightarrow \tau'$ *and* $v : \tau$ *then* $v :: \phi : \tau'$.
4. *(Star soundness) Assume that for all* v_1, v_2 *such that* $v_1 : \tau_1$ *and* $\gamma;\ x \in v_1 \models^* e \Rightarrow v_2$ *we have* $v_2 : \tau_2$. *Then if* $v_1' : \tau_1^*$ *and* $\gamma;\ x \in v_1' \models^* e \Rightarrow v_2'$ *then* $v_2' : \tau_2^*$.

Proof. Part (1) is immediate. Parts (2) and (3) are by induction on axis or test typing derivations, using Star Lifting for the case involving τ^*. Part (4) is by induction on value typing judgments, using evaluation inversion principles. □

Theorem 5 (Type soundness)

1. *If* $\Gamma \vdash e : \tau$ *and* $\gamma : \Gamma$ *and* $\gamma \models e \Rightarrow v$ *then* $v : \tau$.
2. *If* $\Gamma;\ x \in \tau \vdash^* e : \tau'$ *and* $\gamma;\ x \in v \models^* e \Rightarrow v'$ *and* $\gamma : \Gamma$ *and* $v : \tau$, *then* $v' : \tau'$.

Proof. By induction on the typing derivations, using the previous lemma and evaluation inversion principles for the iteration cases. □

In fact, we can also show that all well-formed programs evaluate to a value. Although mini-XQuery has iteration, the iteration is always bounded. Naturally, this property does not carry over to the full language, but it is useful here since it means we can eliminate some termination side-conditions on equivalence laws.

Lemma 4 (Star convergence). *Suppose that for all* v_1, *if* $v_1 : \tau_1$ *then there exists* v_2 *such that* $\gamma;\ x \in v_1 \models^* e \Rightarrow v_2$. *Then if* $v_1' : \tau_1^*$ *then there exists* v_2' *such that* $\gamma;\ x \in v_1' \models^* e \Rightarrow v_2'$.

Theorem 6 (Convergence)

1. *If* $\gamma \vdash e : \tau$ *and* $\gamma : \Gamma$ *then there exists* $v : \tau$ *such that* $\gamma \models e \Rightarrow v$.
2. *If* $\gamma;\ x \in \tau \vdash^* e : \tau'$ *and* $\gamma : \Gamma$ *and* $v : \tau$ *then there exists* v' *such that* $\gamma;\ x \in v \models^* e \Rightarrow v'$.

3.2 Determinacy

The evaluation relation is also deterministic. This is not easy to show directly. Instead, we introduce an alternative evaluation relation that is easy to prove deterministic, and show that it is equivalent to the first presentation. This alternative presentation replaces the SNG* and SEQ* rules with the following CONS* rule:

$$\frac{\gamma, x \mapsto \hat{v_1} \models e \Rightarrow v_1' \quad \gamma;\ x \in v_2 \models^* e \Rightarrow v_2'}{\gamma;\ x \in \hat{v_1}, v_2 \models^* e \Rightarrow v_1' \mathbin{@} v_2'}\ \text{CONS}^*$$

Theorem 7 (Equivalence of presentations). *The* CONS* *rule is derivable from* NIL*, SNG* *and* SEQ*; *conversely,* SNG* *and* SEQ* *are admissible using* NIL* *and* CONS*.

Theorem 8 (Determinacy)

1. *If* $\gamma \models e \Rightarrow v$ *and* $\gamma \models e \Rightarrow v'$ *then* $v = v'$.
2. *If* $\gamma;\ x \in v_0 \models^* e \Rightarrow v$ *and* $\gamma;\ x \in v_0 \models^* e \Rightarrow v'$ *then* $v = v'$.

Proof. We first show this for the evaluation relation defined using NIL* and CONS*, which is straightforward. We then use the previous lemma to transfer the result to the NIL*, SNG*, SEQ* presentation of the rules. □

It is probably not strictly necessary to introduce the second presentation based on NIL* and CONS*, but it is sometimes convenient to use it since there are fewer cases to cover. Proving the systems equivalent was also useful as a sanity check.

4 Operational Equivalences

We define operational equivalence as follows:

$$\Gamma \models e \cong e' \iff \forall \gamma : \Gamma.\gamma \models e \Rightarrow v \iff \gamma \models e' \Rightarrow v$$

This definition suffices for our semantics in the absence of side-effects such as node identifier generation. Note that the context Γ is necessary to track the variables of e and e' that are needed for evaluation to be sensible. Without this constraint, proving even simple equivalences such as FOREMPTY is nontrivial, since we have to explicitly rename the bound name to avoid names already present in the context. Nevertheless, using an explicit typing context for this is just for convenience (we could also have just used a list of variables), since we can always use the top type `item`* for all the variables. Using a typing context Γ to impose nontrivial type constraints on the free variables of e also means we can take type information into account when reasoning about equivalence. We give an example at the end of the next section.

One important lemma is that we can drop let-bindings if the value of the bound variable is never used:

Lemma 5 (Let weakening). *Suppose $x \notin FV(e_2)$. Then $\Gamma \models \texttt{let}\ x := e\ \texttt{in}\ e_2 \cong e_2$.*

To prove operational equivalence laws involving commuting `let` and `for`, we need additional properties of evaluation.

Lemma 6 (Variable iteration). *Assume $x \notin \mathrm{dom}(\gamma)$. Then $\gamma;\ x \in v \models^* x \Rightarrow v'$ holds if and only if $v = v'$.*

Lemma 7 (Let iteration). *Assume $\gamma \models e_1 \Rightarrow v_1$. Then $\gamma, x \mapsto v_1;\ y \in v \models^* e_2 \Rightarrow v_2$ iff $\gamma;\ y \in v \models^* \texttt{let}\ x := e_1\ \texttt{in}\ e_2 \Rightarrow v_2$.*

Lemma 8 (For iteration). *Assume $x \notin FV(e_1) \cup \mathrm{dom}(\gamma)$ and $y \notin FV(e_2) \cup \mathrm{dom}(\gamma)$ and $x \neq y$. Then $\gamma;\ y \in v \models^* \texttt{for}\ x \in e_1\ \texttt{return}\ e_2 \Rightarrow v_2$ holds iff there exists a v_1 such that $\gamma;\ y \in v \models^* e_1 \Rightarrow v_1$ and $\gamma;\ x \in v_1 \models^* e_2 \Rightarrow v_2$.*

Using these laws, we can verify all of the standard properties of XQuery expressions discussed in for example [14], and a number of others.

Theorem 9. *The operational equivalences and laws listed in Figure 4 are valid for mini-XQuery.*

Proof. Most of the laws are straightforward using inversion, weakening, and strengthening. The FORVAR, FORLET and FORFOR laws require variable, let, and for-iteration respectively. □

We can also verify congruence laws for all of the expression forms. These are shown in Figure 5. Note that in the congruence rules for `let` and `for`, we use the `item`* type for the type of x in the extended context, meaning that we must verify that e_2 is equivalent to e_2' under all possible values for x. The congruence rule for `for` also requires a lemma:

$\Gamma \models \texttt{()}, e \cong e$ SEQUNITL
$\Gamma \models e, \texttt{()} \cong e$ SEQUNITR
$\Gamma \models (e_1, e_2), e_3 \cong e_1, (e_2, e_3)$ SEQASSOC
$\Gamma \models \texttt{for}\ x \in \texttt{()}\ \texttt{return}\ e \cong \texttt{()}$ FOREMPTY
$\Gamma \models \texttt{for}\ x \in e\ \texttt{return}\ x \cong e$ FORVAR
$\Gamma \models \texttt{for}\ x \in (e_1, e_2)\ \texttt{return}\ e \cong (\texttt{for}\ x \in e_1\ \texttt{return}\ e, \texttt{for}\ x \in e_2\ \texttt{return}\ e)$ FORSEQ
$\Gamma \models \texttt{for}\ x \in \texttt{text}\{w\}\ \texttt{return}\ e \cong \texttt{let}\ x := w\ \texttt{in}\ e$ FORSTRING
$\Gamma \models \texttt{for}\ x \in \texttt{elem}\ l\ \{e_1\}\ \texttt{return}\ e_2 \cong \texttt{let}\ x := \texttt{elem}\ l\ \{e_1\}\ \texttt{in}\ e_2$ FORELEM
$\Gamma \models \texttt{for}\ x \in (\texttt{if}\ e\ \texttt{then}\ e_1\ \texttt{else}\ e_2)\ \texttt{return}\ e_0$
$\quad \cong \texttt{if}\ e\ \texttt{then}\ (\texttt{for}\ x \in e_1\ \texttt{return}\ e_0)\ \texttt{else}\ \texttt{for}\ x \in e_2\ \texttt{return}\ e_0$ FORCOND
$\Gamma \models \texttt{for}\ x \in (\texttt{let}\ y := e_1\ \texttt{in}\ e_2)\ \texttt{return}\ e \cong \texttt{let}\ y := e_1\ \texttt{in}\ (\texttt{for}\ x \in e_2\ \texttt{return}\ e)$ FORLET
$\Gamma \models \texttt{for}\ x \in (\texttt{for}\ y \in e_1\ \texttt{return}\ e_2)\ \texttt{return}\ e$
$\quad \cong \texttt{for}\ y \in e_1\ \texttt{return}\ (\texttt{for}\ x \in e_2\ \texttt{return}\ e)$ FORFOR
$\Gamma \models \texttt{if ()}\ \texttt{then}\ e_1\ \texttt{else}\ e_2 \cong e_2$ CONDEMPTY
$\Gamma \models \texttt{if}\ \texttt{text}\{w\}\ \texttt{then}\ e_1\ \texttt{else}\ e_2 \cong e_1$ CONDSTRING
$\Gamma \models \texttt{if}\ \texttt{elem}\ l\ \{e\}\ \texttt{then}\ e_1\ \texttt{else}\ e_2 \cong e_1$ CONDELEM
$\Gamma \models \texttt{if}\ (e_1, e_2)\ \texttt{then}\ e_1'\ \texttt{else}\ e_2' \cong \texttt{if}\ e_1\ \texttt{then}\ e_1'\ \texttt{else}\ (\texttt{if}\ e_2\ \texttt{then}\ e_1'\ \texttt{else}\ e_2')$ CONDCOND
$\Gamma \models x/\texttt{self} :: \texttt{node}() \cong x$ SELFID
$\Gamma \models \texttt{let}\ x := e\ \texttt{in}\ x \cong e$ LETVAR
$\Gamma \models \texttt{let}\ x := e\ \texttt{in}\ \texttt{elem}\ l\ \{e_0\} \cong \texttt{elem}\ l\ \{\texttt{let}\ x := e\ \texttt{in}\ e_0\}$ LETELEM
$\Gamma \models \texttt{let}\ x := e\ \texttt{in}\ (e_1, e_2) \cong (\texttt{let}\ x := e\ \texttt{in}\ e_1, \texttt{let}\ x := e\ \texttt{in}\ e_2)$ LETSEQ
$\Gamma \models \texttt{let}\ x := e\ \texttt{in}\ (\texttt{if}\ e_0\ \texttt{then}\ e_1\ \texttt{else}\ e_2)$
$\quad \cong \texttt{if}\ (\texttt{let}\ x := e\ \texttt{in}\ e_0)\ \texttt{then}\ (\texttt{let}\ x := e\ \texttt{in}\ e_1)\ \texttt{else}\ (\texttt{let}\ x := e\ \texttt{in}\ e_2)$ LETCOND
$\Gamma \models \texttt{let}\ x := e\ \texttt{in}\ (\texttt{let}\ y := e_1\ \texttt{in}\ e_2)$
$\quad \cong \texttt{let}\ y := (\texttt{let}\ x := e\ \texttt{in}\ e_1)\ \texttt{in}\ (\texttt{let}\ x := e\ \texttt{in}\ e_2)$ LETLET
$\Gamma \models \texttt{let}\ x := e\ \texttt{in}\ (\texttt{for}\ y \in e_1\ \texttt{return}\ e_2)$
$\quad \cong \texttt{for}\ y \in (\texttt{let}\ x := e\ \texttt{in}\ e_1)\ \texttt{return}\ (\texttt{let}\ x := e\ \texttt{in}\ e_2)$ LETFOR

Fig. 4. Verified operational equivalence laws for mini-XQuery

Lemma 9 (For-iteration congruence). *Assume* $\Gamma, x : \texttt{item}^* \models e \cong e'$ *and* $\gamma : \Gamma$ *where* $x \notin \mathrm{dom}(\gamma)$. *Then* $\gamma;\ x \in v \models^* e \Rightarrow v'$ *holds if and only if* $\gamma;\ x \in v \models^* e' \Rightarrow v'$ *holds.*

Theorem 10 (Evaluation is a congruence). *The congruence laws of Figure 5 all hold of operational equivalence.*

Finally, using the equivalence laws, congruences, and the definition of substitution we can prove that inlining is sound in mini-XQuery. Of course, inlining is not sound for full XQuery but nevertheless it is an important program transformation, and its soundness proof in mini-XQuery sheds some light on what is needed for specific instances of inlining to be sound in XQuery.

Theorem 11 (Inlining). $\Gamma \models \texttt{let}\ x = e_1\ \texttt{in}\ e_2 \cong e_2[e_1/x]$

Proof. By induction on the structure of e, using many of the rules in Figure 4, congruence rules, and the definition of substitution. □

$$\frac{}{\Gamma \models e \cong e} \quad \frac{\Gamma \models e' \cong e}{\Gamma \models e \cong e'} \quad \frac{\Gamma \models e \cong e' \quad \Gamma \models e' \cong e''}{\Gamma \models e \cong e''} \quad \frac{\Gamma \models e \cong e'}{\Gamma \models \texttt{elem}\, l\, \{e\} \cong \texttt{elem}\, l\, \{e'\}}$$

$$\frac{\Gamma \models e_1 \cong e_1' \quad \Gamma \models e_2 \cong e_2'}{\Gamma \models (e_1, e_2) \cong (e_1', e_2')} \quad \frac{\Gamma \models e \cong e' \quad \Gamma \models e_1 \cong e_1' \quad \Gamma \models e_2 \cong e_2'}{\Gamma \models \texttt{if}\, e\, \texttt{then}\, e_1\, \texttt{else}\, e_2 \cong \texttt{if}\, e'\, \texttt{then}\, e_1'\, \texttt{else}\, e_2'}$$

$$\frac{\Gamma \models e_1 \cong e_1' \quad \Gamma, x : \texttt{item}^* \models e_2 \cong e_2'}{\Gamma \models \texttt{let}\, x := e_1\, \texttt{in}\, e_2 \cong \texttt{let}\, x := e_1'\, \texttt{in}\, e_2'}$$

$$\frac{\Gamma \models e_1 \cong e_1' \quad \Gamma, x : \texttt{item}^* \models e_2 \cong e_2'}{\Gamma \models \texttt{for}\, x \in e_1\, \texttt{return}\, e_2 \cong \texttt{for}\, x \in e_1'\, \texttt{return}\, e_2'}$$

Fig. 5. Congruence laws for mini-XQuery equivalence

$$\frac{}{\tau <: \tau} \quad \frac{\tau_1 <: \tau_2 \quad \tau_2 <: \tau_3}{\tau_1 <: \tau_3} \quad \frac{\tau_1 <: \tau_2}{\texttt{elem}\, l\, \{\tau_1\} <: \texttt{elem}\, l\, \{\tau_2\}} \quad \frac{\tau_1 <: \tau_1' \quad \tau_2 <: \tau_2'}{\tau_1, \tau_2 <: \tau_1', \tau_2'}$$

$$\frac{\tau_1 <: \tau_1' \quad \tau_2 <: \tau_2'}{\tau_1|\tau_2 <: \tau_1'|\tau_2'} \quad \frac{\tau_1 <: \tau_2}{\tau_1^* <: \tau_2^*} \quad \frac{() <: \tau_2 \quad \tau_1 <: \tau_2 \quad \tau_2, \tau_2 <: \tau_2}{\tau_1^* <: \tau_2}$$

Fig. 6. Subtyping congruence rules and $*$-induction

$$\begin{aligned} \tau, () &\equiv \tau & \tau_1|(\tau_2|\tau_3) &\equiv (\tau_1|\tau_2)|\tau_3 \\ (), \tau &\equiv \tau & \tau, (\tau_1|\tau_2) &\equiv (\tau, \tau_1)|(\tau, \tau_2) \\ \tau, (\tau', \tau'') &\equiv (\tau, \tau'), \tau'' & (\tau_1|\tau_2), \tau &\equiv (\tau_1, \tau)|(\tau_2, \tau) \\ \tau_1|\tau_2 &\equiv \tau_2|\tau_1 & \texttt{elem}\, l\, \{\tau_1|\tau_2\} &\equiv \texttt{elem}\, l\, \{\tau_1\}|\texttt{elem}\, l\, \{\tau_2\} \\ \tau_1 &<: \tau_1|\tau_2 & () &<: \tau^* \\ \tau_2 &<: \tau_2|\tau_2 & \tau &<: \tau^* \\ ()^* &\equiv () & \tau^*, \tau^* &<: \tau^* \end{aligned}$$

Fig. 7. Subtyping and type equivalence laws

5 Subtyping

Subtyping is based on containment of regular expression types, that is,

$$\tau <: \tau' \iff \forall v. v : \tau \implies v : \tau'$$

Moreover, we often employ type equivalence $\tau \equiv \tau'$ defined as the symmetric closure of $<:$ (or equivalently, as $\forall v. v : \tau \iff v : \tau'$).

We first establish a number of routine properties of regular expression types, including pre-congruence and $*$-induction rules shown in Figure 6, and equivalence or subtyping laws shown in Figure 7.

Since we do not include an empty type or recursive types that could be used to define empty types, we can show that any subtype of `()` is equivalent to `()`. We first need a few auxiliary properties:

Lemma 10. *1. Neither* `()` $<: \alpha$ *nor* $\alpha <:$ `()` *holds for any atomic type* α.

2. *If* $\tau_1, \tau_2 <: ()$ *then* $\tau_1 <: ()$ *and* $\tau_2 <: ()$.
3. *If* $\tau_1|\tau_2 <: ()$ *then* $\tau_1 <: ()$ *and* $\tau_2 <: ()$.
4. *If* $\tau^* <: ()$ *then* $\tau <: ()$.

Theorem 12. *If* $\tau <: ()$ *then* $\tau \equiv ()$.

Proof. Proof is by induction on the structure of τ, using the previous lemma to rule out the case $\tau = \alpha$ and to bridge the gap between the assumption and induction hypotheses. Consider the case for τ^*. We can assume that $\tau^* <: ()$ holds and that $\tau <: ()$ implies $\tau \equiv ()$. By part (4) of the lemma, we have that $\tau <: ()$, which implies $\tau \equiv ()$, which is equivalent to $()^*$ as shown in Figure 7. □

To conclude the section, we show how types can be used to help optimize programs.

Theorem 13 (Statically dead code elimination). *If* $\Gamma \vdash e : ()$ *then* $\Gamma \models e \cong ()$.

Proof. Let $\gamma : \Gamma$ be given; we must show that $\gamma \models e \Rightarrow v$ holds if and only if $\gamma \models () \Rightarrow v$. For the forward direction of the equivalence, we just use soundness and the fact that $()$ is the only value of type $()$. For the reverse direction, we use Convergence to show that e must evaluate to some value, then use soundness again to show that it must be $()$, and finally use inversion to show that the only value $()$ can evaluate to is also $()$. □

This suggests the following optimization technique: traverse the term, typecheck each subterm and replace each subterm whose type is $()$ with $()$.

6 Discussion

Table 1 provides some summary information about the formalization (see also [9]), including the number of lines of proof and number of lemmas for each theory (corresponding to the previous three sections of the paper). We have not attempted to be rigorous about blank lines or comments; the merit of raw lines of proof as a metric is unclear to us (especially across different systems), but these figures could at least provide a rough comparison with other possible formalizations. Many of the more subtle proofs turn out to be short, while longer proofs tend to be mostly "brute force" induction steps for which most cases follow a similar reasoning pattern. We have made no attempt to shorten proofs by leveraging Isabelle's automation beyond the basics, because it makes the behavior (and termination) of proof search tactics much harder to control, but it is plausible that more sophisticated use of Isabelle's existing automation (or use of a different technique altogether) could lead to much shorter proofs.

We have begun to formalize some more complex results such as the admissibility of subsumption (a simplified version of the main result of [8]). We have a partial formalization of the syntactic part of the proof from [8], but the semantic aspects of the proof are proving tricky to formalize — the original proof involves reasoning about various regular expression homomorphisms, some of which are partial. We have already found one minor bug in the proof, and we are investigating workarounds.

Table 1. Overview of the formalization

Theory	Description	Lines	Lemmas
XQuery	Basic definitions, evaluation metatheory, and type soundness.	1740	95
Equivalence	Operational equivalence laws and congruences	965	45
Subtyping	Properties of subtyping; type-based equivalences	459	42

7 Related Work

Among W3C standards, XQuery is distinctive in that formalization of its semantics was integrated into the standardization process from an early stage. Fernandez et al. [14] presented a core XML query language that served as one starting point for XQuery, which included several features not present in XQuery such as pattern matching, while excluding other features such as node identity and schema validation. Siméon and Wadler [24] studied and formalized the behavior of validation in XML Schema and XQuery, identifying some formal properties that helped influence the final design.

The only previous work we are aware of on mechanically checking properties of XML query languages is by Genevés and Vion-Dury [17]. They formalize the XML tree model in Coq and define the semantics of XPath axes (including ancestor and sibling axes that we do not handle), and they formalize some equivalence laws for XPath. However, they do not consider XQuery constructs involving name-binding (`for`, `let`) or construction of new XML document values (`elem` l $\{e\}$). We view their formalization as complementary; ultimately a formalization of full XQuery will have to handle all of the features in their work, all those in this paper, and more.

Malecha et al. [20] formalize an implementation of a core SQL-like relational query language in Coq, including a formalization of the B-tree data structure, and leveraging proof automation techniques available in Coq, as also documented by Chlipala et al. [11]. XQuery can be implemented by translating XML trees and queries to algebraic or relational languages (see e.g. Grust et al. [18] or Ré et al. [22]) and it would be interesting to verify such translations. Rose [23] is developing a rewriting-based compiler for full XQuery. While the aim of this system is to simplify compiler development and experimentation with optimization rules, it is also an attractive starting point for verification.

There are also a number of other results from papers on XQuery that we would like to formalize, including the *path-error analysis* of Colazzo et al. [12]. Also, there are refinements to the typechecking algorithm that we have not verified, including some discussed further in more recent work by Colazzo and Sartiani [13]. More ambitious would be to formalize the W3C XQuery Update Facility [7]. Its semantics is defined informally, but Benedikt and Cheney [3] give a candidate operational semantics for a core language based on the standard.

Naturally it would also be interesting to extend the regular expression type system to handle recursion; it would be even more interesting to formalize the syntax-oriented algorithm for deciding subtyping of Hosoya et al. [19]. Another interesting direction for future work is extending mini-XQuery with XQuery 3.0 features such as higher-order functions, exceptions, and grouping or aggregation constructs in order to understand how they interact with XQuery's distinctive approach to typechecking.

Nominal Isabelle is an implementation of the nominal abstract syntax approach pioneered by Gabbay and Pitts [15]. Our work employs mature aspects of the Nominal Isabelle infrastructure [25], particularly *strong induction principles* [26] and *strong inversion principles* [5] that enable reasoning about name-binding syntax in a way that parallels on-paper reasoning, and which have been used in a number of other case studies, including various lambda-calculi such as LF [27] and the π-calculus [4]. More recent work on Nominal Isabelle has aimed at supporting additional name-binding constructs [28], such as simultaneous binding in function definitions, and these features should be very useful in scaling our formalization up to XQuery; conversely, the formalization needs of full XQuery may help motivate further investigation of mechanized metatheory techniques, much as the POPLMark challenge has helped spur research on such tools [1].

While some of the properties we proved are essentially syntactic and ought (in principle) to be formalizable in any mechanized metatheory system, others such as operational equivalence and type equivalence involve a mixture of syntactic and semantic methods, which seems to require the expressiveness of first-order logical definitions, which are not available in certain systems. We expect that it would be straightforward (albeit possibly labor-intensive) to formalize mini-XQuery in Coq using standard techniques [2,10]. It is less clear how suitable tools such as Twelf [21] would be since they lack support for first-order logical definitions. It would be particularly interesting to consider Abella [16], which combines higher-order abstract syntax with first-order logical definitions.

8 Conclusions

XQuery is a compelling target for formalization because it is commercially relevant, there is a growing literature on optimization techniques for XQuery, and there is a detailed (albeit not fully mechanized) formal semantics for XQuery already. In this paper, we have taken an important step towards a complete mechanical formalization of XQuery, by formalizing an expressive core language called mini-XQuery. Our formalization captures some of the distinctive features of XQuery but additional work is needed to scale this approach (or alternative mechanization methodologies) up to full XQuery, starting with the impure features discussed in the introduction.

References

1. Aydemir, B.E., Bohannon, A., Fairbairn, M., Foster, J.N., Babu, C.S., Sewell, P., Vytiniotis, D., Washburn, G., Weirich, S., Zdancewic, S.: Mechanized Metatheory for the Masses: The POPLMARK Challenge. In: Hurd, J., Melham, T. (eds.) TPHOLs 2005. LNCS, vol. 3603, pp. 50–65. Springer, Heidelberg (2005)
2. Aydemir, B.E., Charguéraud, A., Pierce, B.C., Pollack, R., Weirich, S.: Engineering formal metatheory. In: POPL, pp. 3–15 (2008)
3. Benedikt, M., Cheney, J.: Semantics, Types and Effects for XML Updates. In: Gardner, P., Geerts, F. (eds.) DBPL 2009. LNCS, vol. 5708, pp. 1–17. Springer, Heidelberg (2009)
4. Bengtson, J., Parrow, J.: Formalising the pi-calculus using nominal logic. Logical Methods in Computer Science 5(2) (2008)

5. Berghofer, S., Urban, C.: Nominal Inversion Principles. In: Mohamed, O.A., Muñoz, C., Tahar, S. (eds.) TPHOLs 2008. LNCS, vol. 5170, pp. 71–85. Springer, Heidelberg (2008)
6. Boag, S., Chamberlin, D., Fernández, M.F., Florescu, D., Robie, J., Siméon, J.: XQuery 1.0: An XML query language. W3C Recommendation (January 2007), `http://www.w3.org/TR/xquery`
7. Chamberlin, D., Robie, J.: XQuery update facility 1.0. W3C Candidate Recommendation (August 2008), `http://www.w3.org/TR/xquery-update-10/`
8. Cheney, J.: Regular Expression Subtyping for XML Query and Update Languages. In: Gairing, M. (ed.) ESOP 2008. LNCS, vol. 4960, pp. 32–47. Springer, Heidelberg (2008)
9. Cheney, J., Urban, C.: Formalization of mini-XQuery in Nominal Isabelle, `http://homepages.inf.ed.ac.uk/jcheney/projects/XQuery`
10. Chlipala, A.: Parametric higher-order abstract syntax for mechanized semantics. In: ICFP, pp. 143–156 (2008)
11. Chlipala, A., Malecha, J.G., Morrisett, G., Shinnar, A., Wisnesky, R.: Effective interactive proofs for higher-order imperative programs. In: ICFP, pp. 79–90 (2009)
12. Colazzo, D., Ghelli, G., Manghi, P., Sartiani, C.: Static analysis for path correctness of XML queries. J. Funct. Program. 16(4-5), 621–661 (2006)
13. Colazzo, D., Sartiani, C.: Precision and complexity of XQuery type inference. In: PPDP (to appear, 2011); Preliminary version in ICTCS 2010
14. Fernandez, M., Siméon, J., Wadler, P.: A Semi-Monad for Semi-Structured Data. In: Van den Bussche, J., Vianu, V. (eds.) ICDT 2001. LNCS, vol. 1973, pp. 263–300. Springer, Heidelberg (2000)
15. Gabbay, M.J., Pitts, A.M.: A new approach to abstract syntax with variable binding. Formal Aspects of Computing 13, 341–363 (2002)
16. Gacek, A.: The Abella Interactive Theorem Prover (System Description). In: Armando, A., Baumgartner, P., Dowek, G. (eds.) IJCAR 2008. LNCS (LNAI), vol. 5195, pp. 154–161. Springer, Heidelberg (2008)
17. Genevès, P., Vion-Dury, J.-Y.: XPath formal semantics and beyond: A Coq-based approach. In: TPHOLs Emerging Trends, Salt Lake City, Utah, United States, August 2004, pp. 181–198. University Of Utah (2004)
18. Grust, T., Rittinger, J., Teubner, J.: Pathfinder: XQuery off the relational shelf. IEEE Data Eng. Bull. 31(4) (2008)
19. Hosoya, H., Vouillon, J., Pierce, B.C.: Regular expression types for XML. ACM Trans. Program. Lang. Syst. 27(1), 46–90 (2005)
20. Malecha, J.G., Morrisett, G., Shinnar, A., Wisnesky, R.: Toward a verified relational database management system. In: POPL, pp. 237–248 (2010)
21. Pfenning, F., Schürmann, C.: System Description: Twelf - A Meta-Logical Framework for Deductive Systems. In: Ganzinger, H. (ed.) CADE 1999. LNCS (LNAI), vol. 1632, pp. 202–206. Springer, Heidelberg (1999)
22. Ré, C., Siméon, J., Fernández, M.F.: A complete and efficient algebraic compiler for XQuery. In: ICDE, p. 14 (2006)
23. Rose, K.: CRSX - combinatory reduction systems with extensions. In: RTA (2011)
24. Siméon, J., Wadler, P.: The essence of XML. In: POPL, New York, NY, USA, pp. 1–13. ACM (2003)
25. Urban, C.: Nominal techniques in Isabelle/HOL. J. Autom. Reasoning 40(4), 327–356 (2008)
26. Urban, C., Berghofer, S., Norrish, M.: Barendregt's Variable Convention in Rule Inductions. In: Pfenning, F. (ed.) CADE 2007. LNCS (LNAI), vol. 4603, pp. 35–50. Springer, Heidelberg (2007)
27. Urban, C., Cheney, J., Berghofer, S.: Mechanizing the metatheory of LF. ACM Trans. Comput. Log. 12(2), 15 (2011)
28. Urban, C., Kaliszyk, C.: General Bindings and Alpha-Equivalence in Nominal Isabelle. In: Barthe, G. (ed.) ESOP 2011. LNCS, vol. 6602, pp. 480–500. Springer, Heidelberg (2011)

Automatically Verifying Typing Constraints for a Data Processing Language

Michael Backes[1,2], Cătălin Hriţcu[1,3], and Thorsten Tarrach[1,4,5]

[1] Saarland University, Saarbrücken, Germany
[2] MPI-SWS, Saarbrücken, Germany
[3] University of Pennsylvania, Philadelphia, USA
[4] Atomia AB, Västerås, Sweden
[5] Troxo DOO, Niš, Serbia

Abstract. In this paper we present a new technique for automatically verifying typing constraints in the setting of Dminor, a first-order data processing language with refinement types and dynamic type-tests. We achieve this by translating Dminor programs into a standard while language and then using a general-purpose verification tool. Our translation generates assertions in the while program that faithfully represent the sophisticated typing constraints in the original program. We use a generic verification condition generator together with an SMT solver to prove statically that these assertions succeed in all executions. We formalise our translation algorithm using an interactive theorem prover and provide a machine-checkable proof of its soundness. We provide a prototype implementation using Boogie and Z3 that can already be used to efficiently verify a large number of test programs.

1 Introduction

Dminor [7] is a first-order data processing language with *refinement types* (types qualified by Boolean expressions) and *dynamic type-tests* (Boolean expressions testing whether a value belongs to a type). The combination of refinement types and dynamic type-tests seems to be very useful in practice [2]. However, the increased expressivity allowed by this combination makes statically type-checking programs very challenging.

In this paper we present a new technique for statically checking the typing constraints in Dminor programs by translating these programs into a standard while language. The sophisticated typing constraints in the original program are faithfully encoded as assertions in the generated program and we use a general-purpose verification tool to show statically that these assertions succeed in all executions. This opens up the possibility to take advantage of the huge amount of proven techniques and ongoing research done on general-purpose verification tools.

We have proved that if all assertions succeed in the translated program then the original Dminor program does not cause typing errors when executed. This proof was done in the Coq interactive theorem prover [5], based on a formalisation of our translation algorithm. We thus show formally that, for the language we are considering, a generic verification tool can check the same properties as a sophisticated type-checker. To the best of our knowledge, this is the first machine-checked proof of a translation to an intermediate verification language (IVL).

J.-P. Jouannaud and Z. Shao (Eds.): CPP 2011, LNCS 7086, pp. 296–313, 2011.

Finally, we provide a prototype implementation using Boogie and Z3 that can already be used to verify a large number of test programs and we report on an experimental evaluation against the original Dminor type-checker.

1.1 Related Work

Bierman et al. [7] were the first to study the combination of refinement types and dynamic type-tests. They introduce a first-order functional language called Dminor, which captures the essence of the Microsoft code name M language [2], but which is simple enough to express formally. They show that the combination of refinement types and dynamic type-tests is highly expressive; for instance intersection, union, negation, singleton, dependent sum, variant and algebraic types are all derivable in Dminor. This expressivity comes, however, at a cost: statically type-checking Dminor programs is very challenging, since the type information can be "hidden" deep inside refinements with arbitrary logical structure. For instance intersection types $T\&U$ are encoded in Dminor as refinement types $(x : \mathsf{Any}\ \mathsf{where}\ (x\ \mathsf{in}\ T\ \&\&\ x\ \mathsf{in}\ U))$, where the refinement formula is the boolean conjunction of the results of two dynamic type-tests. Syntax-directed typing rules cannot deal with such "non-structural" types, so Bierman et al. [7] propose a solution based on semantic subtyping. They formulate a semantics in which types are interpreted as first-order logic formulae, subtyping is defined as a valid implication between the semantics of types and they use an SMT solver to discharge such logical formulae efficiently.

The idea of using an SMT solver for type-checking languages with refinement types is quite well established and was used in languages such as SAGE [17], F7 [6], Fine [30] and Dsolve [29]. Bierman et al. [7] show that, in the setting of a first-order language, the SMT solver can play a more central role: They successfully use the SMT solver to check semantic subtyping, not just the refinement constraints. However, while in Dminor [7] subtyping is semantic and checked by the SMT solver, type-checking is still specified by syntax-directed typing rules, and implemented by bidirectional typing rules. In the current work we show that we can achieve very similar results to those of the Dminor type-checker without relying on *any* typing rules, by using the logical semantics of Dminor types directly to generate assertions in a while program, and then verifying the while program using standard verification tools.

Relating type systems and software model-checkers is a topic that has received attention recently from the research community [18, 25, 15]. Our approach is different since we enforce typing constraints using a verification condition generator. Our implementation uses Boogie [20], the generic verification condition generation back-end used by the Verified C Compiler (VCC) [10] and Spec# [4].

There is previous work on integrating verification tools such as Boogie [8] and Why [13] with proof assistants, for the purpose of manually aiding the verification process or proving the correctness of background theories with respect to natural models. However, we are not aware of other machine-checked correctness proofs for translations from surface programming languages into IVLs, even for a language as simple as the one described in this paper. A translation from Java bytecode into Boogie was proved correct in the Mobius project [1, 19], but we are not aware of any mechanised formalisation of this proof.

1.2 Overview

In §2 we provide a brief review of Dminor and in §3 we give a short introduction to our intermediate verification language. §4 and §5 describe our translation algorithm and its implementation. In §6 we compare our work to the Dminor type-checker [7]. Finally, in §7 we conclude and discuss some future work. Further details, our implementation, our Coq formalisation and proofs are all available online at:
`http://www.infsec.cs.uni-saarland.de/projects/dverify`.

2 Review of Dminor (Data Processing Language)

Dminor is a first-order functional language for data processing. We will briefly review this language below; full details are found in the paper by Bierman et al. [7].

Values in Dminor can be simple values (integers, strings, Booleans or null), collections (multi-sets of values) or entities (records). Dminor types include the top type (Any), scalar types (Integer, Text, Logical), collection types ($T*$) and entity types ($\{\ell\colon T\}$). More interestingly, Dminor has *refinement types*: the refinement type ($x : T$ where e) consists of the values x of type T satisfying the arbitrary Boolean expression e.

Syntax of Dminor Expressions:

$e ::=$	Dminor expression
$x \mid k$	variables and scalar constants
$\oplus(e_1, \ldots, e_n)$	primitive operator application
$e_1 ? e_2 : e_3$	conditional (if-then-else)
let $x = e_1$ in e_2	let-expression (x bound in e_2)
e in T	dynamic type-test
$\{\ell_i \Rightarrow e_i \;{}^{i \in 1..n}\}$	entity (record with n fields $\ell_i \ldots \ell_n$)
$e.\ell$	selects field ℓ of entity e
$\{v_1, \ldots, v_n\}$	collection (multiset)
$e_1 :: e_2$	adding element e_1 to collection e_2
from x in e_1 let $y = e_2$ accumulate e_3	collection iteration (x, y bound in e_3)
$f(e_1, \ldots, e_n)$	function application

Refinement types can be used to express pre- and postconditions of functions, as shown in the type of removeNulls below, where the postcondition states that the resulting collection has at most as many elements as the original collection.

Refinement type used to encode pre- and postconditions

e.Count $\triangleq$ from x in e let $y = 0$ accumulate $y + 1$

NullableInteger $\triangleq$ x : Any where (x in Integer $||$ $x ==$ null)

removeNulls(c : NullableInteger$*$) : (x : Integer$*$ where x.Count $\leq$ c.Count) {
  from x in c let $y = \{\}$ accumulate (($x \neq$ null)?($x :: y$) : y)
}

The dynamic type-test expression e in T matches the result of expression e against the type T; it returns **true** if e has type T and **false** otherwise. While dynamic type-tests are useful on their own in a data processing language (e.g. for pattern-matching an XML document against a schema represented as a type [14]), they can also be used inside refinement types, which greatly increases the expressivity of Dminor (e.g. it allows encoding union, intersection, negation types, etc., as seen in the example above, where NullableInteger is an encoded union between type Integer and the singleton type containing only the value null).

Bierman et al. [7] define a big-step operational semantics for Dminor, in which evaluating an expression can return either a value or "error". An error can for instance arise if a non-existing field is selected from an entity. In Dminor such errors are avoided by the type system, but in this work we rule them out using standard verification tools. The type system by Bierman et al. uses semantic subtyping: they formulate a logical semantics (denotational) in which types are interpreted as first-order logic formulae and subtyping is defined as the valid implication between such formulae. More precisely, they define a function $\mathbf{F}[\![T]\!](v)$ that returns a first-order logic formula testing if the value v belongs to a type T. Since (pure) expressions can appear inside refinement types, $\mathbf{F}[\![T]\!]$ is defined by mutual recursion together with two other functions: $\mathbf{R}[\![e]\!]$ returns a first-order logic term denoting the result[1] of an expression e; and $\mathbf{W}[\![T]\!](v)$ a formula that tests if checking whether v is in type T causes an execution error. The reason for the existence of **W** is that **F** is total and has to return a boolean even when evaluating the expression inside a refinement type causes an error. Our translation makes use of the functions **F** and **W** to faithfully encode the typing constraints in Dminor as assertions in the generated while program.

3 Bemol (Intermediate Verification Language)

We define a simple intermediate verification language (IVL) we call Bemol. Bemol is much simplified compared to a generic IVL: the number of language constructs has been reduced and some Dminor-specific constructs that would normally be encoded were added as primitives. We use Bemol to simplify the presentation, the formalisation of our translation and the soundness proof. In our implementation we use Boogie [20, 3, 11] as the IVL and we encode all Bemol constructs that do not have a direct correspondent in Boogie.

3.1 Syntax and Operational Semantics

Bemol is a while language with collections, records, asserts, mutually recursive procedures, variable scoping and evaluation of logical formulae. The syntax of Bemol is separated into two distinct classes: expressions e, which are side-effect free, and commands c, which have side-effects.

[1] Bierman et al. [7] show that $\mathbf{R}[\![e]\!]$ coincides with the big-step operational semantics on pure expressions – i.e., expressions without side-effects such as non-determinism (accumulate) and non-termination (recursive functions).

Bemol expressions allow basic operations on values, most of which directly correspond to the operations in Dminor. Also the available primitive operators $\oplus$ are the same as in Dminor. The only significant difference is the expression formula f which "magically" evaluates the logical formula f and returns a boolean encoding the validity or invalidity of the formula – such a construct is standard in most IVLs. We use the notation $[\![e]\!]_{st}$ for the evaluation of expression e under state st. In case of a typing error (such as selecting a non-existing field from an entity) $\bot$ is returned.

Syntax of Bemol Expressions:

$e ::=$	Bemol expression
x	variable
v	Dminor value (scalar, collection or entity)
$\oplus(e_1, \ldots, e_n)$	primitive Dminor operator application
$e.\ell$	selects field ℓ of entity e
$e_1 [\ell := e_2]$	updates field ℓ in entity e_1 with e_2 (produces new entity)
$e_1 :: e_2$	adds element e_1 to collection e_2 (produces new collection)
$e_1 \backslash \{e_2\}$	removes one instance of e_2 from e_1 (produces new collection)
is_empty e	returns **true** if e is the empty collection; **false** otherwise
formula f	returns **true** if formula f is valid in the current state

Syntax of Bemol Commands:

$c ::=$	Bemol command
skip	does nothing
$c_1;\ c_2$	executes c_1 and then c_2
$x\ :=\ e$	assigns the result of e to x
if e then c_1 else c_2	conditional
while e inv a do c end	while loop with invariant a
assert f	expects that formula f holds, causes error otherwise
$x\ :=$ pick e	puts an element of e in x (non-deterministic)
call P	calls the procedure P
backup x in c	backs up the current state

Bemol commands manipulate the current global state, which is a total function that maps variables to values. The invariant specified in the while command does not affect evaluation; its only goal is to aid the verification condition generator. The pick command chooses non-deterministically an element from collection e and assigns its value to variable x. The call P command transfers control to procedure P, which will also operate on the same global state. The backup x in c command backs up the current state, executes c and once this is finished restores all variables to their former value except for x. This is useful for simulating a call-stack for procedures, and we also introduce it during the translation to simplify the soundness proof. A similar technique is employed by Nipkow [26] for representing local variables. We use this in our encoding of procedure calls below. The encoding uses an entity to pass multiple arguments.

Encoding of Procedure Calls

x := call P$(e_1, \ldots, e_n) \triangleq$
  backup x in (
    arg := {}; arg := arg$[\ell_1 := e_1]; \ldots;$ arg := arg$[\ell_n := e_n]$; call P; x := ret)

procedure P$(x_1, \ldots, x_n)\{ c \} \triangleq$ proc P $\{ x_1$:= arg.$\ell_1; \ldots; x_n$:= arg.ℓ_n; $c \}$

3.2 Operational Semantics

We define the big-step semantics of Bemol as a relation $st_{init} \xrightarrow{c} r$, where r can be either a final state st_{final} or **Error**. The only command that can cause an **Error** is the assert command; all the other commands simply "bubble up" the errors produced by failed assertions. If an expression evaluates to $\perp$ it will lead to the divergence of the command that contains it, but this does not cause an error.[2]

Operational Semantics

(Eval Skip)
$$\frac{}{st \xrightarrow{\mathsf{skip}} st}$$

(Eval Assign)
$$\frac{[\![e]\!]_{st} = v}{st \xrightarrow{x := e} st[x := v]}$$

(Eval Seq)
$$\frac{st \xrightarrow{c_1} st' \quad st' \xrightarrow{c_2} r}{st \xrightarrow{c_1;\, c_2} r}$$

(Eval Seq Error)
$$\frac{st \xrightarrow{c_1} \mathbf{Error}}{st \xrightarrow{c_1;\, c_2} \mathbf{Error}}$$

(Eval If True)
$$\frac{[\![e]\!]_{st} = \mathbf{true} \quad st \xrightarrow{c_1} r}{st \xrightarrow{\mathsf{if}\ e\ \mathsf{then}\ c_1\ \mathsf{else}\ c_2} r}$$

(Eval If False)
$$\frac{[\![e]\!]_{st} = \mathbf{false} \quad st \xrightarrow{c_2} r}{st \xrightarrow{\mathsf{if}\ e\ \mathsf{then}\ c_1\ \mathsf{else}\ c_2} r}$$

(Eval While End)
$$\frac{[\![e]\!]_{st} = \mathbf{false}}{st \xrightarrow{\mathsf{while}\ e\ \mathsf{inv}\ a\ \mathsf{do}\ c\ \mathsf{end}} st}$$

(Eval While Loop)
$$\frac{[\![e]\!]_{st} = \mathbf{true} \quad st \xrightarrow{c} st' \quad st' \xrightarrow{\mathsf{while}\ e\ \mathsf{inv}\ a\ \mathsf{do}\ c\ \mathsf{end}} r}{st \xrightarrow{\mathsf{while}\ e\ \mathsf{inv}\ a\ \mathsf{do}\ c\ \mathsf{end}} r}$$

(Eval While Error)
$$\frac{[\![e]\!]_{st} = \mathbf{true} \quad st \xrightarrow{c} \mathbf{Error}}{st \xrightarrow{\mathsf{while}\ e\ \mathsf{inv}\ a\ \mathsf{do}\ c\ \mathsf{end}} \mathbf{Error}}$$

(Eval Assert)
$$\frac{[\![\mathsf{formula}\ f]\!]_{st} = \mathbf{true}}{st \xrightarrow{\mathsf{assert}\ f} st}$$

(Eval Assert Error)
$$\frac{[\![\mathsf{formula}\ f]\!]_{st} = \mathbf{false}}{st \xrightarrow{\mathsf{assert}\ f} \mathbf{Error}}$$

(Eval Pick)
$$\frac{v \in [\![e]\!]_{st}}{st \xrightarrow{x := \mathsf{pick}\ e} st[x := v]}$$

(Eval Call)
$$\frac{st \xrightarrow{c} r \quad \text{given } P\{c\}}{st \xrightarrow{\mathsf{call}\ P} r}$$

(Eval Backup)
$$\frac{st \xrightarrow{c} st'}{st \xrightarrow{\mathsf{backup}\ x\ \mathsf{in}\ c} st[x := st'x]}$$

(Eval Backup Error)
$$\frac{st \xrightarrow{c} \mathbf{Error}}{st \xrightarrow{\mathsf{backup}\ x\ \mathsf{in}\ c} \mathbf{Error}}$$

3.3 Hoare Logic and Verification Condition Generation

We define a Hoare logic for our commands, based on the Software Foundations lecture notes [27] and the ideas of Nipkow [26].

[2] Since we only reason about partial correctness, diverging programs are considered correct. This makes the assumptions on our encoding of Bemol into Boogie be minimal: we only assume that the asserts and successful evaluations of the other commands are properly encoded in Boogie. In §4 we prove that we add enough asserts to capture all errors in the original Dminor program, even under these conservative assumptions we make in the Bemol semantics.

Definition 1 (Hoare Triple). *We say that a Hoare triple* $\models \{P\}\, c\, \{Q\}$ *holds semantically if* $\forall st\, r.\ st \xrightarrow{c} r \implies \forall z.\, P\ z\ st = \mathbf{true} \implies (\exists st'.\, r = st' \wedge Q\ z\ st' = \mathbf{true})$.

By requiring that the result of the command is not $\mathbf{Error}$ but an actual state st' we ensure that correct programs do not cause assertions to fail. The meta-variable z is an addition to the traditional Hoare triple and models auxiliary variables, which need to be made explicit in the presence of recursive procedures. Our treatment of auxiliary variables and procedures follows the one of Nipkow [26], who formalises an idea by Morris [24] and Kleymann [16]. We also use Nipkow's definition of extended Hoare triples and Hoare judgements, which are needed for recursive and mutually recursive procedures respectively.

Definition 2 (Extended Hoare Triple). *We say that an extended Hoare triple* $C \models \{P\}\, c\, \{Q\}$ *holds semantically if and only if* (C *valid* $\implies \models \{P\}\, c\, \{Q\}$), *where* C *is valid means* $\forall P\, c\, Q.\ \{P\}c\{Q\} \in C \implies \models \{P\}\, c\, \{Q\}$.

Definition 3 (Hoare Judgement). *We say that a Hoare judgement* $C_1 \models C_2$ *holds semantically if and only if* C_1 *valid* $\implies$ C_2 *valid.*

We give a complete list of the Hoare rules for our commands. Except for backup, pick and assert they are the same as in Nipkow's work [26].

Hoare Rules for Bemol

(Hoare Assign)

$$\frac{}{C \models \{Q\{v/x\}\}\, x := v\, \{Q\}}$$

(Hoare Sequence)

$$\frac{C \models \{P\}\, c_1\, \{Q\} \quad C \models \{Q\}\, c_2\, \{R\}}{C \models \{P\}\, c_1;\ c_2\, \{R\}}$$

(Hoare If)

$$\frac{C \models \{\lambda z\ st.\ P\ z\ st \wedge [\![e]\!]_{st}\}\, c_1\, \{Q\} \quad C \models \{\lambda z\ st.\ P\ z\ st \wedge \neg [\![e]\!]_{st}\}\, c_2\, \{Q\}}{C \models \{P\}\ \mathsf{if}\ e\ \mathsf{then}\ c_1\ \mathsf{else}\ c_2\, \{Q\}}$$

(Hoare Skip)

$$\frac{}{C \models \{Q\}\, \mathsf{skip}\, \{Q\}}$$

(Hoare While)

$$\frac{C \models \{\lambda z\ st.\ P\ z\ st \wedge [\![e]\!]_{st}\}\, c\, \{P\}}{C \models \{P\}\ \mathsf{while}\ e\ \mathsf{inv}\ P\ \mathsf{do}\ c\ \mathsf{end}\, \{\lambda z\ st.\ P\ z\ st \wedge \neg [\![e]\!]_{st}\}}$$

(Hoare Context)

$$\frac{\{P\}\, c\, \{Q\} \in C}{C \models \{P\}\, c\, \{Q\}}$$

(Hoare Pick)

$$\frac{}{C \models \{\lambda z\ st.\ \forall v \in [\![e]\!]_{st}, P\{v/x\}\ z\ st\}\, x := \mathsf{pick}\ e\, \{P\}}$$

(Hoare Consequence)

$$\frac{C \models \{P'\}\, c\, \{Q'\} \qquad \forall st\ st'.\ (\forall z.\ P'\ z\ st \implies Q'\ z\ st') \implies (\forall z.\ P\ z\ st \implies Q\ z\ st')}{C \models \{P\}\, c\, \{Q\}}$$

(Hoare Triple)

$$\frac{\forall P\, c\, Q.\ \{P\}\, c\, \{Q\} \in C_2 \implies C_1 \models \{P\}\, c\, \{Q\}}{C_1 \models C_2}$$

(Hoare Call MutRec)

$$\frac{\begin{array}{c}\forall P\,c\,Q.\,\{P\}\,c\,\{Q\} \in C_2 \implies \exists S.\ c = \mathsf{call}\ S\\ \forall P\,Q\,S.\ \{P\}\,\mathsf{call}\ S\,\{Q\} \in C_2 \implies C_1 \cup C_2 \models \{P\}\,c\,\{Q\}\quad \text{given } S\{c\}\end{array}}{C_1 \models C_2}$$

(Hoare Assert)

$$\frac{}{C \models \{Q \wedge a\}\,\mathsf{assert}\ a\,\{Q\}}$$

(Hoare Call Simple)

$$\frac{\{P\}\,\mathsf{call}\ S\,\{Q\} :: C \models \{P\}\,c\,\{Q\}\quad \text{given } S\{c\}}{C \models \{P\}\,\mathsf{call}\ S\,\{Q\}}$$

(Hoare Backup)

$$\frac{\forall st'.\, C \models \{\lambda z\ st.\ P\ z\ st \wedge st' = st\}\,c\,\{\lambda z\ st.\ Q\{st\ x/x\}\ z\ st'\}}{C \models \{P\}\,\mathsf{backup}\ x\ \mathsf{in}\ c\,\{Q\}}$$

The backup x in c command requires that the Hoare triple for c has the same state for the pre- and postcondition, except for variable x which is updated. We "transfer" the state from the pre- to the postcondition by quantifying over a new state st' that we require to be equal to the state in the precondition.

For our semantics of the Hoare triples it is possible to define a weakest precondition, but not a strongest postcondition function. This is because if c evaluates to **Error** no postcondition is strong enough to make the triple valid. Corresponding to the Hoare rules, we define a verification condition generator (VCgen c Q), which takes a command c and a postcondition Q as arguments and generates a precondition. We have proved that this is sound; however, the VCgen is not guaranteed to return the weakest precondition, because the used loop invariants are not necessarily the best. The soundness proof of the VCgen crucially relies on the soundness of the Hoare logic rules above.

More importantly for our application, we have proved as a corollary that the programs deemed correct by our VCgen do not cause errors when executed.

Theorem 1 (Soundness of VCgen)
If VCgen c Q *returns a valid formula, then* $\nexists st.\ st \xrightarrow{c} \mathbf{Error}$.

4 Translation from Dminor to Bemol

4.1 Translation Function

Our translation algorithm is a function $\langle\!\langle\, e \,\rangle\!\rangle_x$ that takes a Dminor program and a variable name x as input and outputs a Bemol program. The variable x is where the generated Bemol program should store the result after it executes. Variables beginning with z are freshly chosen, meaning no variable with that name exists in the program.

Translation from Dminor to Bemol

$$\langle\!\langle\, e \,\rangle\!\rangle_x = \mathsf{backup}\ x\ \mathsf{in}\ \langle e\rangle_x$$

$$\langle x\rangle_{\mathsf{out}} = \mathsf{out} := x$$

$$\langle v\rangle_{\mathsf{out}} = \mathsf{out} := v$$

$$\begin{array}{lcl}
\langle \oplus(e_1,\ldots,e_n)\rangle_{\mathsf{out}} & = & \langle\!\langle e_1 \rangle\!\rangle_{z_1};\ldots;\langle\!\langle e_n \rangle\!\rangle_{z_n};\\
\text{where operands of } \oplus & & \mathsf{assert}\ \mathbf{F}[\![T_1]\!](z_1);\ldots;\mathsf{assert}\ \mathbf{F}[\![T_n]\!](z_n);\\
\text{need to have types } T_1,\ldots,T_n & & \mathsf{out} := \oplus(z_1,\ldots,z_n)\\
\langle e_1 ? e_2 : e_3\rangle_{\mathsf{out}} & = & \langle\!\langle e_1 \rangle\!\rangle_{z_1};\ \mathsf{assert}\ \mathbf{F}[\![\mathsf{Logical}]\!](z_1);\\
 & & \mathsf{if}\ z_1\ \mathsf{then}\ \langle\!\langle e_2 \rangle\!\rangle_{\mathsf{out}}\ \mathsf{else}\ \langle\!\langle e_3 \rangle\!\rangle_{\mathsf{out}}\\
\langle \mathsf{let}\ x = e_1\ \mathsf{in}\ e_2\rangle_{\mathsf{out}} & = & \langle\!\langle e_1 \rangle\!\rangle_{x};\ \langle\!\langle e_2 \rangle\!\rangle_{\mathsf{out}}\\
\langle e\ \mathsf{in}\ T\rangle_{\mathsf{out}} & = & \langle\!\langle e \rangle\!\rangle_{z_1};\ \mathsf{assert}\ (\neg\mathbf{W}[\![T]\!](z_1));\\
 & & \mathsf{out} := \mathsf{formula}\ \mathbf{F}[\![T]\!](z_1)\\
\langle \{\ell_i \Rightarrow e_i\ ^{i\in 1..n}\}\rangle_{\mathsf{out}} & = & \langle\!\langle e_1 \rangle\!\rangle_{z_1};\ldots;\langle\!\langle e_n \rangle\!\rangle_{z_n};\ \mathsf{out} := \{\};\\
 & & \mathsf{out} := \mathsf{out}[\ell_1 := z_1];\ldots;\mathsf{out} := \mathsf{out}[\ell_n := z_n]\\
\langle e.\ell\rangle_{\mathsf{out}} & = & \langle\!\langle e \rangle\!\rangle_{z_1};\ \mathsf{assert}\ \mathbf{F}[\![\{\ell : \mathsf{Any}\}]\!](z_1);\\
 & & \mathsf{out} := z_1.\ell\\
\langle \{v_1,\ldots,v_n\}\rangle_{\mathsf{out}} & = & \mathsf{out} := \{v_1,\ldots,v_n\}\\
\langle e_1 :: e_2\rangle_{\mathsf{out}} & = & \langle\!\langle e_1 \rangle\!\rangle_{z_1};\ \langle\!\langle e_2 \rangle\!\rangle_{z_2};\ \mathsf{assert}\ \mathbf{F}[\![\mathsf{Any}*]\!](z_2);\\
 & & \mathsf{out} := z_1 :: z_2\\
\langle \mathsf{from}\ x\ \mathsf{in}\ e_1 & = & \langle\!\langle e_1 \rangle\!\rangle_{z_c};\ \langle\!\langle e_2 \rangle\!\rangle_{y};\\
\quad \mathsf{let}\ y = e_2 & & \mathsf{assert}\ \mathbf{F}[\![\mathsf{Any}*]\!](z_c);\\
\quad \mathsf{accumulate}\ e_3\rangle_{\mathsf{out}} & & \mathsf{while}\ !(\mathsf{is_empty}\ z_c)\ \mathsf{inv}\ i(z_c,\ldots)\ \mathsf{do}\\
 & & \quad x := \mathsf{pick}\ z_c;\\
 & & \quad z_c := z_c \backslash \{x\};\\
 & & \quad \langle\!\langle e_3 \rangle\!\rangle_{y}\\
 & & \mathsf{end};\\
 & & \mathsf{out} := y\\
\langle f(e_1,\ldots,e_n)\rangle_{\mathsf{out}} & = & \langle\!\langle e_1 \rangle\!\rangle_{z_1};\ldots;\langle\!\langle e_n \rangle\!\rangle_{z_n};\\
 & & \mathsf{out} := \mathsf{call}\ f(z_1,\ldots,z_n)
\end{array}$$

Our translation directly uses function $\mathbf{F}$ of the Dminor semantics of types [7] to generate logical formulae that check whether a value has a certain Dminor type or not. These formulae are used in asserts to make sure that the variables involved in a command have the right types before actually executing the command. For type-tests we first use an assert $(\neg\mathbf{W}[\![T]\!](z_1))$ to make sure that the type-test itself does not cause an error in Dminor, before actually checking whether z_1 has type T using $\mathbf{F}$. To translate procedure calls we use the encoding of call given in §3.1. In our translation of accumulate the loop invariant i needs to be provided as an input, either directly in the translated code, or in the original Dminor program as a type annotation T on the accumulator (as required by the Dminor type-checker). In the later case we can use $\mathbf{F}[\![T]\!]$ to obtain an invariant; however, below we show an example where the necessary invariant cannot be expressed this way. In §5.1 we discuss how we use the Dminor infrastructure for automatically inferring invariants for a special class of accumulate expressions corresponding to LINQ queries [23].

4.2 Examples

In the examples below we consider out to be the variable where the result is put. In Example 1 we show how the removeNulls example from §2 is translated to a while loop that picks and removes elements from the collection until it is empty.

Example 1: Accumulate filtering null **values**

removeNulls(c : NullableInteger$*$) :
  (x : (Integer$*$) where (x.Count $\leq$ c.Count))
{
  from x in c let $y = \{\}$
    accumulate (($x \neq$ null)?($x :: y$) : y)
}
	procedure removeNulls(c) {
	  assert $\mathbf{F}[\![$NullableInteger$*]\!](c)$;
	  $y := \{\}$;
	  $c' := c$;
	  while !is_empty c' inv $i(c, c', y)$ do
	    $x :=$ pick c';
	    $c' := c' \backslash \{x\}$;
	    if $x \neq$ null then
	      $y := x :: y$
	    else
	      $y := y$
	  end;
	  ret $:= y$;
	  assert ($\mathbf{F}$ $[\![x$: Integer$*$
	    where y.Count $\leq$ c.Count) ret$]\!]$
	}

$i(c, c', y) \triangleq \mathbf{F}[\![y$: (Integer$*$) where (c'.Count $+$ y.Count $\leq$ c.Count)$]\!]$ y

The loop invariant specifies that the sum of the number of elements in the intermediate collection c' and the resulting collection y is less than or equal to the number of elements in the original collection c. It is not sufficient for the invariant to just reason over y and c as this would be too weak. In this case the invariant is provided by hand on the generated code because this loop invariant is not expressible as a Dminor type. Loop invariant inference on the Dminor side is in general deemed to fail for global properties of collections. Our implementation successfully verifies this example with the provided invariant. In the future we hope to infer such invariants automatically using the Boogie infrastructure for this task.

As seen in Example 2 for type-tests we first use an assert to check that the type-test does not cause a typing error and then perform the actual type-test which returns a Logical. Note that **F** is total and would also return a value on a wrongly typed argument.

Example 2: Type-test

x in (y : Integer where $y > 5$)	assert ($\neg(\mathbf{W}[\![y$: Integer where $y > 5]\!]$ x)); out := formula ($\mathbf{F}[\![y$: Integer where $y > 5]\!]$ x)

For illustration, we expand **W** and **F** in the example above; please see the paper by Bierman et al. [7] for the precise definition of these functions of the logical semantics.

$$
\begin{aligned}
&\mathbf{W}[\![y : \mathsf{Integer\ where}\ y > 5]\!]\ x\\
\models\ &\mathbf{W}[\![\mathsf{Integer}]\!]\ x \vee \mathsf{let}\ y = x\ \mathsf{in}\\
&\qquad \neg(\mathbf{R}[\![y > 5]\!] = \mathbf{Return(false)} \vee \mathbf{R}[\![y > 5]\!] = \mathbf{Return(true)})\\
\models\ &\mathbf{false} \vee \neg((\mathsf{if}\ \mathbf{F}[\![\mathsf{Integer}]\!]\ x\ \mathsf{then}\ \mathbf{Return}(x > 5)\ \mathsf{else}\ \mathbf{Error}) = \mathbf{Return(false)}\\
&\qquad\qquad \vee (\mathsf{if}\ \mathbf{F}[\![\mathsf{Integer}]\!]\ x\ \mathsf{then}\ \mathbf{Return}(x > 5)\ \mathsf{else}\ \mathbf{Error}) = \mathbf{Return(true)})\\
\models\ &\neg\mathbf{F}[\![\mathsf{Integer}]\!]\ x \models \neg(\mathsf{In\_Integer}\ x)
\end{aligned}
$$

$$
\begin{aligned}
&\mathbf{F}[\![y : \mathsf{Integer\ where}\ y > 5]\!]\ x\\
\models\ &\mathbf{F}[\![\mathsf{Integer}]\!]\ x \wedge \mathsf{let}\ y = x\ \mathsf{in}\ \mathbf{R}[\![y > 5]\!] = \mathbf{Return(true)}\\
\models\ &\mathsf{In\_Integer}\ x \wedge (\mathsf{if\ In\_Integer}\ x\ \mathsf{then}\ \mathbf{Return}(x > 5)\ \mathsf{else}\ \mathbf{Error}) = \mathbf{Return(true)}\\
\models\ &\mathsf{In\_Integer}\ x \wedge x > 5
\end{aligned}
$$

In case x is not an integer the formula $\mathsf{In\_Integer}\ x \wedge x > 5$ is logically equivalent to **false**. Our translation asserts that x is an integer before calling formula in order to match the semantics of Dminor, in which $x > 5$ causes an error when x is not an integer.

We construct another example using type-tests where our technique is more complete than the Dminor type system.

Example 3: Valid type-test

$4\ \mathsf{in}\ (x : \mathsf{Any\ where}\ x > 5)$	$\mathsf{assert}\ (\neg\mathbf{W}[\![x : \mathsf{Any\ where}\ x > 5]\!](4));$ $\mathsf{out} := \mathsf{formula}\ (\mathbf{F}[\![x : \mathsf{Any\ where}\ x > 5]\!]\ 4)$

The Dminor type system rejects Example 3 as ill-typed because x is typed as Any, which is not a valid type for an operand of the greater operator. The big-step semantics of Dminor, however, evaluates this expression successfully to **false** because the x always evaluates to 4, which is a valid operand of the greater operator. The translated program is accepted by Boogie, since our translation aims to be complete with respect to the operational semantics, whereas Dminor implements an inherently incomplete type system.

4.3 Soundness

We have proved in Coq that if a Dminor program e can evaluate to **Error**[3], then the translated program $\langle\!\langle\, e \,\rangle\!\rangle_x$ can evaluate to **Error** in Bemol. The contrapositive of this is: if the translated program cannot evaluate to **Error**, then the original Dminor program cannot evaluate to **Error** either. We have proved this theorem in Coq by induction over the big-step semantics of Dminor $\Downarrow$.

Theorem 2 (Soundness of the Translation). *If* $e \Downarrow \mathbf{Error}$ *then* $\forall st.\ st \xrightarrow{\langle\!\langle e \rangle\!\rangle_x} \mathbf{Error}$.

As an immediate consequence of Theorem 1 and Theorem 2 we obtain the soundness of our whole technique.

Corollary 1 (Soundness). *If* $\mathsf{VCgen}\ \langle\!\langle\, e \,\rangle\!\rangle_x\ \mathsf{true}$ *is a valid formula, then* $e \not\Downarrow \mathbf{Error}$.

[3] Because of non-determinism a program could evaluate to **Error** only in some of the possible executions.

4.4 Formalisation and Machine-Checked Proofs

We have proved Theorem 2 in Coq, by mutual induction together with Lemma 1.

Lemma 1. *If* $e \Downarrow v$ *then* $\forall st.\ \exists st'.\ st \xrightarrow{\langle\!\langle e \rangle\!\rangle_x} st'$ *and* $st'x = v$.

To prove this we require three additional assumptions: The expression we want to translate must not contain impure refinements, none of the functions contains impure refinements and the only free variable a function may have is the argument[4]. An impure refinement is when an impure expression (possibly non-deterministic or non-terminating) is used in a refinement type.

We prove this theorem by induction on the Dminor big-step semantics [7], which gives us 42 cases. In each case we have to prove that the generated Bemol code evaluates to the same result (value or error) as the original Dminor expression. On the Bemol side we use the big-step semantics we have defined in §3.2.

The first step in the proof is always to remove the backup command that is added by every translation step. For this we use two lemmas, one for the error case and one for successful evaluation. In the case of successful evaluation only the output variable out changed after executing the backup command. This fact significantly simplifies the proof and is the main reason we add a backup command at every step of the translation.

Lemma 2 (Backup Error). $st \xrightarrow{\mathsf{backup}\ x\ \mathsf{in}\ c}$ **Error** *if and only if* $st \xrightarrow{c}$ **Error**

Lemma 3 (Backup). *If* $st \xrightarrow{c} st'$ *and* $st'\ x = v$ *then* $st \xrightarrow{\mathsf{backup}\ x\ \mathsf{in}\ c} st[x := v]$

In our proof we had to strengthen the induction hypotheses in several different ways. First, as already mentioned above, we needed to prove Theorem 2 together with Lemma 1 by mutual induction. Second, we needed to generalize the statements of Theorem 2 and Lemma 1 to open expressions. Since the big-step semantics of Dminor is only defined for closed expressions we needed to substitute the free variables with the values from the Bemol state before evaluating Dminor expressions. Finally, we also needed to strengthen the inner induction hypotheses of a number of cases, such as the accumulate and the entity creation cases.

Our formal development consists of 5000 lines of Coq and our proofs are done in full detail. Our development is made on top of the Dminor formalisation consisting of 4000 lines [7], which makes the total size of the formal development approach 9000 lines of Coq. The soundness proof alone consists of 1300 lines of Coq code and the Coq proof checker takes more than $2\frac{1}{2}$ minutes to check the proof. Three custom Coq tactics were defined to solve steps that are commonly used in the translation proof.

5 Implementation

Our implementation is called DVerify and translates a Dminor program into a Boogie program. DVerify is written in F# 2.0 [22] and consists of more than 1200 lines of code,

[4] For a formal statement see the third author's Master's thesis [31], or the Coq formalization, where Theorem 2 is named `translation_closed_sound`.

as well as a 700 line axiomatisation that defines the Dminor types and functions in Boogie. The Boogie tool then takes the translated Boogie program as input and outputs either an error message that describes points in the program where certain postconditions or assertions may not hold [21] or otherwise prints a message indicating that the program has been verified successfully.

5.1 High-Level Overview

The heart of our translation algorithm consists of a recursive function that goes over a Dminor expression and translates it into Boogie code (corresponding to the $\langle\!\langle\, e \,\rangle\!\rangle_x$ function in §4.1). This function is called once per Dminor function and produces a Boogie procedure. Types in Dminor are translated into Boogie function symbols returning bool, using another recursive function in our implementation.

Our translation uses the type annotations on accumulate expressions in the original Dminor program to generate invariants for while loops, so that very often the user does not have to provide loop invariants for the generated Boogie program. However, as illustrated by Example 1, there are also cases for which the invariant needed to verify the program is not expressible as a Dminor type. Such loop invariants are completely out of reach for the Dminor type system, and currently can be provided manually in DVerify. In the future we intend to infer such invariants automatically using the Boogie infrastructure for this task.

The Dminor implementation allows for one more construct to define a loop, a from-where-select as in LINQ [23]. In theory from-where-select can be encoded using accumulate, but in the Dminor implementation it is considered primitive in the interest of efficiency and to reduce the type annotation burden. Since from-where-select does not carry a type annotation, we have to find one during translation so that we can use it as a loop invariant. For that we use a modified version of the type-synthesis from Dminor that does not call the type-checking algorithm and therefore never fails to synthesise a type for an expression.

We use the Dminor implementation as a library so that we do not have to reimplement existing functionality. This is mainly the parser for Dminor files, the purity checking and a weak form of type-synthesis for from-where-select.

5.2 Axiomatisation

Our implementation also comprises an axiomatisation of Dminor values and functions in Boogie. This is necessary because Boogie as such understands only two sorts, bool and int, whereas Dminor and Bemol have a number of primitive and composite values, such as collections and entities. Our axiomatisation is similar to the axiomatisation the Dminor type-checker feeds to Z3 [7]. In Dminor this axiomatisation is written in SMT-LIB 1.2 syntax [28] and directly fed to Z3 with the proof obligation. Our axiomatisation is in the Boogie language and Boogie translates it to Simplify syntax [12] and feeds it to Z3 along with the verification conditions it generates. Dminor makes heavy usage of the theories Z3 offers, such as extensional arrays and datatypes. We use the weak arrays provided by Boogie by default and encode datatypes by hand.

Table 1 Precision Comparison

	well-typed	ill-typed
Test suite	76	33
DMinor accepts	66	0
DVerify accepts	62	2 (correct programs)

Chart 1 Speed Comparison (average times for 66 well-typed samples)

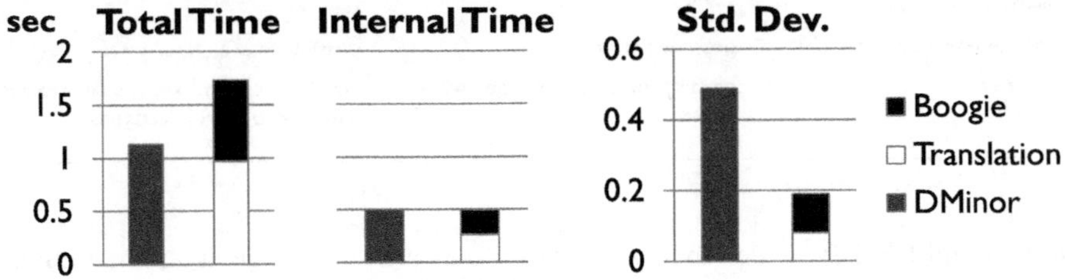

6 Comparison between Dminor and DVerify

We have tested our implementation against Dminor 0.1.1 from September 2010. Microsoft Research gave us access to their Dminor test suite that contains 109 sample programs. Out of these 109 tests 76 are well-typed Dminor programs and 33 are ill-typed. Out of the 76 well-typed programs, the Dminor type-checker cannot verify 10 tests because of its imprecision.

As shown in Table 1, from the 66 cases on which Dminor succeeds, DVerify manages to verify 62 as correct. Out of the 33 that Dminor rejects, DVerify rejects 31. The other two are correct operationally, but are ill-typed with respect to the (inherently incomplete) Dminor type system. Overall this means that DVerify succeeds on 94% of the cases Dminor succeeds on and is able to verify two correct programs Dminor cannot verify. For the 4 correct programs that DVerify cannot verify the most common problem is that type-synthesis generates too complicated loop invariants and Z3 cannot handle the resulting proof obligations. Giving explicit type annotations on the Dminor side (instead of relying on Dminor type-synthesis) makes DVerify also accept these programs.

In order to compare efficiency, we first measured the overall wall-clock time that is needed by the two tools, which includes the time the operating system requires to start the process. Because we are dealing with a large number of small test files and both tools are managed .NET assemblies, initialisation dominates the total running times of both tools. Since initialisation is a constant factor that becomes negligible on bigger examples, we also measured the time excluding initialisation and parsing, which we call "internal time". Chart 1 shows both times (averaged over the 62 well-typed samples accepted by both tools) on a 2.1 GHz laptop. The internal time is 0.5s on average for both

Table 2 Qualitative Comparison of Dminor and DVerify

Area	Our verification approach (DVerify)	Type-checking approach (DMinor)
Verification cond. generation	Weakest precondition	Bidirectional type-checking (type synthesis ≈ strongest postcondition)
Formulae discharged	One per postcondition/assertion (larger, but less obligations)	One per subtyping test (smaller, but more obligations)
Backend	Boogie + SMT-Solver (Z3)	SMT-Solver (Z3)
Loop invariants	In principle Boogie could infer some (even for accumulates) (even for global properties)	For from-where-select (but not for accumulates) (but not for global properties)
Speed	similar	similar
Precision (practise)	similar	similar
Completeness (theory)	possibly better	possibly worse (type system)
Theories	equality, integers, datatypes, weak arrays	equality, integers, extensional arrays and native encoding of datatypes

Dminor and DVerify, which means that both tools are very efficient and that our combination of a translation and an off-the-shelf verification condition generator matches the average speed of a well-optimised type-checker on its own test suite. One should still keep in mind that all examples in this test suite are relatively small, the biggest one consisting of 176 lines.

7 Conclusion

In this paper we have presented a new technique for statically checking the typing constraints in Dminor programs by translating these programs into a standard while language and then using a general-purpose verification tool. We have formalised our translation algorithm using an interactive theorem prover and provide a machine-checkable proof of its soundness. We also provide a prototype implementation using Boogie and Z3, which can already be used to verify a large number of test programs and which is able to match and on some examples even surpass the precision and efficiency of the Dminor type-checker.

Future Work

Using a general verification tool for checking the types of Dminor programs should allow us to increase the expressivity of the Dminor language more easily. For example, adding support for *mutable state* would be easy in DVerify: Bemol already supports state, moreover Boogie is used mainly for imperative programming languages [9]. An interesting consequence is that it should be easier to support strong updates in DVerify (i.e. updates that change the type of variables), which is usually quite hard to achieve with a type-checker.

Another very interesting extension is *inferring loop invariants*. Dminor requires that each accumulate expression is annotated with a type for the accumulator which constitutes the invariant of the loop, whereas Boogie has build-in support for abstract interpretation for automatically inferring such invariants [3]. While the invariant inference support in Boogie seems currently very much focused on integer domains, it seems possible to extend it to include support for our Dminor types.

Completeness. A theoretical goal would be to prove the completeness of our technique, rather than just soundness. Completeness would ensure that if our VCgen for Bemol generates an invalid formula, then the original program indeed evaluates to an error. This would guarantee that the only source of false positives (i.e., programs that are rejected by our technique, but are actually correct with respect to the Dminor big-step semantics) is the tool used to discharge the verification conditions (i.e., Z3). A crucial step in this direction would be to show the translation complete.

Claim (Completeness of translation). If $\forall st.\ st \xrightarrow{\langle\!\langle e \rangle\!\rangle_x}$ **Error** then $e \Downarrow$ **Error** and if $\forall st, st'.\ st \xrightarrow{\langle\!\langle e \rangle\!\rangle_x} st'$ then $e \Downarrow st'\ x$.

As for soundness, the completeness of the translation can probably be combined with the completeness of the Hoare logic. We expect our Hoare logic to be complete because Nipkow proved completeness for a similar set of Hoare rules [26]. The verification condition generator is, however, inherently incomplete, because of the user-provided annotations for loop invariants and procedure pre- and postconditions. However, for loop-and-procedure-free programs a completeness proof should be possible even for the verification condition generator. Even more, one should be able to prove the *expressive completeness* of the verification condition generator: for every operationally correct program without user annotations, there exists a set of annotations that makes the verification condition generator output a valid formula.

Certified Implementation. We have proved in Coq that our translation is sound and we have implemented this translation in DVerify and tested it to be sound on a considerable number of samples. However, there is no proof that our implementation in F# is sound or indeed implements our proven translation. Coq has the ability to extract OCaml code from Coq source files [5]. This feature could be used to create a certified implementation by extracting our Coq translation function as OCaml code and using it as part of our F# project. To make this extracted code produce proper Boogie programs in practise, we would have to deal with a large number of implementation details we have ignored so far. For example, we would need to deal with the shallow embedding of the logic in Coq and relate our formalisation of Bemol to real Boogie.

Acknowledgements. We thank Andrew D. Gordon for his helpful comments as well as the BOOGIE 2011 and CPP 2011 reviewers for their very useful feedback. Microsoft Research made our work much easier by making the Dminor source code and test suite available to us. Cătălin Hriţcu was supported by a fellowship from Microsoft Research and the International Max Planck Research School for Computer Science.

References

1. Bytecode level specification language and program logic. Mobius Project, Deliverable D3.1 (2006)
2. The Microsoft code name "M" modeling language specification (October 2009), http://msdn.microsoft.com/en-us/library/dd548667.aspx
3. Barnett, M., Chang, B.-Y.E., DeLine, R., Jacobs, B., Leino, K.R.M.: Boogie: A Modular Reusable Verifier for Object-Oriented Programs. In: de Boer, F.S., Bonsangue, M.M., Graf, S., de Roever, W.-P. (eds.) FMCO 2005. LNCS, vol. 4111, pp. 364–387. Springer, Heidelberg (2006)
4. Barnett, M., Leino, K.R.M., Schulte, W.: The Spec# Programming System: An Overview. In: Barthe, G., Burdy, L., Huisman, M., Lanet, J.-L., Muntean, T. (eds.) CASSIS 2004. LNCS, vol. 3362, pp. 49–69. Springer, Heidelberg (2005)
5. Barras, B., Boutin, S., Cornes, C., Courant, J., Coscoy, Y., Delahaye, D., de Rauglaudre, D., Filliâtre, J., Giménez, E., Herbelin, H., et al.: The Coq proof assistant reference manual, version 8.2. INRIA (2009)
6. Bengtson, J., Bhargavan, K., Fournet, C., Gordon, A.D., Maffeis, S.: Refinement types for secure implementations. ACM Transactions on Programming Languages and Systems 33(2), 8 (2011)
7. Bierman, G.M., Gordon, A.D., Hriţcu, C., Langworthy, D.: Semantic subtyping with an SMT solver. In: 15th ACM SIGPLAN International Conference on Functional programming (ICFP 2010), pp. 105–116. ACM Press (2010)
8. Böhme, S., Leino, K.R.M., Wolff, B.: HOL-Boogie — An Interactive Prover for the Boogie Program-Verifier. In: Mohamed, O.A., Muñoz, C., Tahar, S. (eds.) TPHOLs 2008. LNCS, vol. 5170, pp. 150–166. Springer, Heidelberg (2008)
9. Cohen, E., Moskal, M., Tobies, S., Schulte, W.: A precise yet efficient memory model for C. Electronic Notes in Theoretical Computer Science 254, 85–103 (2009)
10. Dahlweid, M., Moskal, M., Santen, T., Tobies, S., Schulte, W.: VCC: Contract-based modular verification of concurrent C. In: 31st International Conference on Software Engineering (ICSE), pp. 429–430. IEEE (2009)
11. DeLine, R., Leino, K.: BoogiePL: A typed procedural language for checking object-oriented programs. Technical Report MSR-TR-2005-70, Microsoft Research (2005)
12. Detlefs, D., Nelson, G., Saxe, J.: Simplify: A theorem prover for program checking. Journal of the ACM (JACM) 52(3), 473 (2005)
13. Filliâtre, J.-C., Marché, C.: The Why/Krakatoa/Caduceus Platform for Deductive Program Verification. In: Damm, W., Hermanns, H. (eds.) CAV 2007. LNCS, vol. 4590, pp. 173–177. Springer, Heidelberg (2007)
14. Hosoya, H., Pierce, B.: XDuce: A statically typed XML processing language. ACM Transactions on Internet Technology 3(2), 117–148 (2003)
15. Jhala, R., Majumdar, R., Rybalchenko, A.: HMC: Verifying Functional Programs Using Abstract Interpreters. In: Gopalakrishnan, G., Qadeer, S. (eds.) CAV 2011. LNCS, vol. 6806, pp. 470–485. Springer, Heidelberg (2011)
16. Kleymann, T.: Hoare logic and auxiliary variables. Formal Aspects of Computing 11(5), 541–566 (1999)
17. Knowles, K., Tomb, A., Gronski, J., Freund, S., Flanagan, C.: Sage: Unified hybrid checking for first-class types, general refinement types and Dynamic. Technical report, UCSC (2007)
18. Kobayashi, N., Ong, C.-H.L.: A type system equivalent to the modal mu-calculus model checking of higher-order recursion schemes. In: 24th Annual IEEE Symposium on Logic in Computer Science (LICS), pp. 179–188. IEEE Computer Society (2009)

19. Lehner, H., Müller, P.: Formal translation of bytecode into BoogiePL. Electronic Notes in Theoretical Computer Science 190(1), 35–50 (2007)
20. Leino, K.R.M.: This is Boogie 2. TechReport (2008)
21. Leino, K.R.M., Millstein, T., Saxe, J.: Generating error traces from verification-condition counterexamples. Science of Computer Programming 55(1-3), 209–226 (2005)
22. Marinos, C.: An introduction to functional programming for.NET developers. MSDN Magazine (April 2010)
23. Meijer, E., Beckman, B., Bierman, G.M.: LINQ: reconciling object, relations and XML in the .NET framework. In: ACM SIGMOD International Conference on Management of Data (SIGMOD), page 706. ACM (2006)
24. Morris, J.: Comments on "procedures and parameters". Undated and unpublished
25. Naik, M., Palsberg, J.: A type system equivalent to a model checker. ACM Transactions on Programming Languages and Systems (TOPLAS) 30(5), 29 (2008)
26. Nipkow, T.: Hoare logics in Isabelle/HOL. In: Proof and System-Reliability, pp. 341–367. Kluwer (2002)
27. Pierce, B., Casinghino, C., Greenberg, M., Sjöberg, V., Yorgey, B.: Software Foundations (2010), `http://www.cis.upenn.edu/~bcpierce/sf/`
28. Ranise, S., Tinelli, C.: The satisfiability modulo theories library, SMT-LIB (2006), `http://www.SMT-LIB.org`
29. Rondon, P.M., Kawaguchi, M., Jhala, R.: Liquid types. In: ACM SIGPLAN 2008 Conference on Programming Language Design and Implementation (PLDI), pp. 159–169 (2008)
30. Swamy, N., Chen, J., Chugh, R.: Enforcing Stateful Authorization and Information Flow Policies in FINE. In: Gordon, A.D. (ed.) ESOP 2010. LNCS, vol. 6012, pp. 529–549. Springer, Heidelberg (2010)
31. Tarrach, T.: Automatically verifying "M" modeling language constraints. Master's thesis, Saarland University (2010)

Hardware-Dependent Proofs of Numerical Programs*

Thi Minh Tuyen Nguyen[1,2] and Claude Marché[1,2]

[1] INRIA Saclay – Île-de-France, Orsay, F-91893
[2] Lab. de Recherche en Informatique, Univ Paris-Sud, CNRS, Orsay, F-91405

Abstract. We present an approach for proving behavioral properties of numerical programs by analyzing their compiled assembly code. We focus on the issues and traps that may arise on floating-point computations. Direct analysis of the assembly code allows us to take into account architecture- or compiler-dependent features such as the possible use of extended precision registers.

The approach is implemented on top of the generic Why platform for deductive verification, which allows us to perform experiments where proofs are discharged by combining several back-end automatic provers.

1 Introduction

The C language is the first choice for embedded systems or critical software from domains such as simulation of physical systems, control-command programs in transportation, etc. For such systems, floating-point (FP for short) computations are involved and precision of calculations is an important issue. The IEEE-754 standard [1] enforces a precise definition on how the basic arithmetic operations (+, -, *, /, and also absolute value, square root, etc.) must be computed on a given FP format (32 bits, 64 bits, etc.) and w.r.t a given rounding mode. This standard is currently supported by most of the processor chips. However, this does not imply that a given C program must produce exactly the same results whatever is the compiler and the underlying architecture. There are for that several possible reasons, e.g. the x87 floating-point unit (FPU) uses 80-bit internal FP registers, FMA instructions compute $xy \pm z$ with a single rounding, or the compiler may optimize the assembly code by changing the order of operations. Such issues have been extensively analyzed by D. Monniaux [29]. A small example that illustrates such an issue is as follows.

```
double doublerounding() {
  double x = 1.0;
  double y = 0x1p-53 + 0x1p-64;
  double z = x + y;
  return z;
}
```

* This work was partly funded by the U3CAT project (ANR-08-SEGI-021, `http://frama-c.com/u3cat/`) of the French national research organization (ANR), and the Hisseo project, funded by Digiteo (`http://hisseo.saclay.inria.fr/`)

J.-P. Jouannaud and Z. Shao (Eds.): CPP 2011, LNCS 7086, pp. 314–329, 2011.

If computations follow the IEEE-754 standard, the result should be $1 + 2^{-52}$, but if compiled using the x87 FPU, a *double rounding* happens and the result is 1. The latter compilation does not *strictly* follows the standard[1].

In the context of static verification, FP computations have been considered in part. In analyses based on the abstract interpretation framework, support for FP computations is proposed in tools like Fluctuat [22], Astrée [19] and the Value analysis of Frama-C [20]. Generally speaking, FP arithmetic has been formalized since 1989 to formally prove hardware components or algorithms [17,26,34].

However, there are very few attempts to analyze FP programs in the so-called *extended static checking* techniques, or *deductive verification* techniques, where verification is typically performed by producing proof obligations, i.e. formulas to be shown valid using theorem provers. In this context, complex behavioral properties are formally specified using specification languages such as JML [15] for Java, ACSL [7] for C, Spec#[4] for C#. The support for FP computations in such approaches is poorly studied. In 2006, Leavens [27] enumerates a set of possible traps when one attempts to specify FP programs. In 2007, Boldo and Filliâtre [10] propose both a specification language to specify FP programs and an approach to generate proof obligations to be proved in the Coq proof assistant. In 2010, Ayad and Marché [2,11] extended this to the support of special values and to the use of automated theorem provers. However, the former approaches assume that the compiler strictly follows the IEEE-754 standard. In other words, on the example above they can prove that the result is $1 + 2^{-52}$.

In 2010, Boldo and Nguyen [12,13] proposed a deductive verification approach which is *compiler- and architecture-independent*, is the sense that the behavioral properties that can be proved valid on a FP program are true whatever does the compiler (up to some extent). On the same example, the only property that can be proved is that the result is between 1 and $1 + 2^{-52}$. In this paper, we propose an approach which is compiler- and architecture- *dependent*: the requirements are proved valid with respect to the assembly code generated by the compiler. At the level of the assembly, all architecture-dependent information is known, such as the precision of each operation.

This paper is organized as follows. Section 2 presents the approach from a user's point of view, and is largely illustrated by examples. Section 3 explains the technicalities of our approach, which consists in translating annotations and assembly instructions into the Why intermediate language [25]. Conclusions and future work will be presented in the last section.

2 Overview of the Approach

Fig. 1 presents all the steps to prove an annotated C program by analyzing its assembly code. The specification language we consider is ACSL [7], where annotations are put in comments. To transport these annotations into the

[1] The term *strict* here refers to the `-fp-model strict` or `/fp:strict` options of C compilers, or the `strictfp` keyword of Java, which explicitly requires the compilation to strictly conform to the standard.

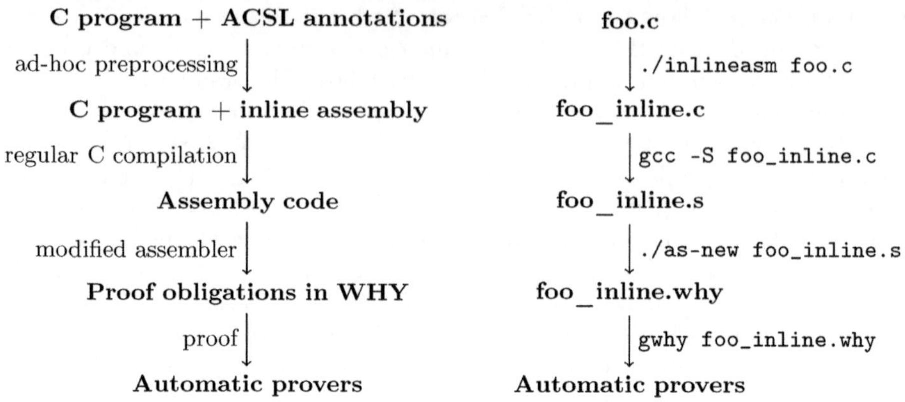

Fig. 1. Step-by-step from C program to WHY proof obligations

generated assembly, we have a preprocessing step in which we rewrite annotations as inline assembly. Assembly code is then generated from this new C source by the regular `gcc` compiler to generate assembly code, with precise architecture-related options, e.g. `-mfpmath=387` to generate x87 assembly or `-O`n to optimize at level n. The main original step is then a translation of the assembly code to a Why program. This is implemented by modifying the GNU assembler so as to produce Why source instead of binary object. The Why environment is finally invoked to generate proof obligations, and prove them by automatic provers such as Gappa [28], Alt-Ergo [9], CVC3 [5], Z3 [21] or interactive provers like Coq [8].

2.1 Example: Double-Rounding

To illustrate this process, let's add a specification in the short program given in introduction under the form of an ACSL *assertion*:

```
double doublerounding() {
  double x = 1.0;
  double y = 0x1p-53 + 0x1p-64;
  double z = x + y;
  //@ assert z == 1.0;
  return z;
}
```

The assembly code generated with our tools in the x87 mode is shown in Fig. 2. Notice on line 16 how the inline assembly preprocessing allowed to replace the occurrence of `z` in the assertion by its assembly counterpart. Notice also that the first addition is computed at compile-time (line 9), whereas the second one is compiled into x87 instructions (lines 11-14). When this assembly code is fed into our translator to Why, and the result analyzed by Why, three proof obligations are produced. One is naturally for proving the assertion, the two others are

```
1  .globl doublerounding
2          .type   doublerounding, @function
3  doublerounding:
4  .LFB0:
5          .cfi_startproc
6  ....
7          movabsq $4607182418800017408, %rax
8          movq    %rax, -16(%rbp)
9          movabsq $4368493837572636672, %rax
10         movq    %rax, -8(%rbp)
11         fldl    -8(%rbp)
12         fld1
13         faddp   %st, %st(1)
14         fstpl   -24(%rbp)
15 #APP
16         /* assert #double#-24(%rbp)# == 1.0;*/
17 #NO_APP
18         movq    -24(%rbp), %rax
19         movq    %rax, -40(%rbp)
20         movsd   -40(%rbp), %xmm0
21         leave
22         ret
23         .cfi_endproc
```

Fig. 2. Assembly code of the double-rounding example in x87 mode

required to prove the absence of overflow, once at line 13 of Fig. 2, corresponding to the addition `x+y` in the source code, and once at line 14, which amount to store the 80-bit value of the x87 stack into a 64-bit memory cell. These three obligations are proved valid using the Gappa automatic prover.

If we compile the program in SSE2 mode (Streaming SIMD Extensions, 64-bits precision arithmetic) then the generated proof obligation corresponding to the assertion cannot be proved anymore. The modified assertion `z == 1.0 + 0x1p-52` can be proved instead. As seen on this example, our approach produces proof obligations to show that the program satisfies its specification, but also to show the absence of overflow in FP computations.

2.2 Example: Architecture Dependent Overflow

Monniaux [29] considers the following program to illustrate differences between architectures with respect to overflows.

```
double foo() {
  double v = 1e308;
  double y = v * v;
  return y/v;
}
```

No optimization

```
1  movabsq  $9214871658872686752, %rax
2  movq     %rax, -8(%rbp)
3  fldl     -8(%rbp)
4  fmull    -8(%rbp)
5  fstpl    -16(%rbp)
6  fldl     -16(%rbp)
7  fdivl    -8(%rbp)
8  fstpl    -24(%rbp)
9  movsd    -24(%rbp), %xmm0
10 ....
```

Optimized, level 1

```
1     fldl  .LC0(%rip)
2     fld   %st(0)
3     fmul  %st(1), %st
4     fdivp %st, %st(1)
5     fstpl -8(%rsp)
6     movsd -8(%rsp), %xmm0
7  ....
8  .LC0:
9     .long 2246822048
10    .long 2145504499
```

Fig. 3. Optimized versus non-optimized assembly

Excerpts of the generated assembly code are shown on Fig. 3. The left part corresponds to non-optimized x87 code where $v * v$ is rounded in 64 bits whereas the right part is optimized (-O1) where $v * v$ is store in 80-bit registers and then the division is also done in 80-bit register. This is the reason why with no-optimization, overflow occurs but with optimization, it does not.

For the non-optimized version, 5 obligations are generated to check absence of overflow at lines 4, 5, 7 and 8, and to check that divisor is not null at line 7. All are proved by Gappa except the overflow at line 5, where the content of the 80-bit register holding the result of the multiplication is moved into a 64-bit memory cell, which indeed overflows. On the other hand, 4 obligations are generated on the optimized code at lines 3, 4 and 5 and all are proved by Gappa. Indeed there is no overflow in this version because the result of multiplication is not temporarily stored into a 64-bit register. Finally, notice that we can also analyze the code compiled in the SSE2 mode, resulting in 3 obligations: overflows for the multiplication and division and check divisor is not null. As expected, it cannot be proved that multiplication does not overflow.

2.3 Example: KB3D

Our next example illustrates the handling of function calls, and the way we express properties on rounding errors across functions. This example is an excerpt of the KB3D collision detection and resolution system developed by Dowek and Munoz [23] and formally proved in PVS, but using exact calculations on real numbers. An analysis of the same code but with FP calculations was done by Boldo and Nguyen [13] using their architecture-independent approach. The annotated C source is given on Fig. 4. The logical symbol `l_sign` returns the sign of a real number: 1 for positive and -1 for negative (sign of zero is not pertinent). The C function `sign` returns the sign of a FP number x. To make sure that the result is correct, a precondition requires that the rounding error on previous computation on x (written as $x - \texttt{\textbackslash exact}(x)$) is between bounds e_1 and e_2 given as arguments. The C function `eps_line` then attempts to decide whether a aircraft at position sx, sy with velocity vx, vy should avoid the point (0,0) on the left or on the right. The decision is taken from the sign of some quantities, for

```
#define E 0x1p-45

//@ logic integer l_sign(real x) = (x >= 0.0) ? 1 : -1;

/*@ requires e1<= x-\exact(x) <= e2;
  @ ensures  (\result != 0 ==> \result == l_sign(\exact(x))) &&
  @           \abs(\result) <= 1 ; */
int sign(double x, double e1, double e2) {
  if (x > e2) return 1;
  if (x < e1) return -1;
  return 0;
}

/*@ requires sx == \exact(sx)  && sy == \exact(sy) &&
  @          vx == \exact(vx)  && vy == \exact(vy) &&
  @          \abs(sx) <= 100.0 && \abs(sy) <= 100.0 &&
  @          \abs(vx) <= 1.0   && \abs(vy) <= 1.0;
  @ ensures \result != 0 ==>
  @   \result == l_sign(\exact(sx)*\exact(vx)+\exact(sy)*\exact(vy))
  @              * l_sign(\exact(sx)*\exact(vy)-\exact(sy)*\exact(vx)); */
int eps_line(double sx, double sy,double vx, double vy){
  int s1,s2;
  s1=sign(sx*vx+sy*vy, -E, E);
  s2=sign(sx*vy-sy*vx, -E, E);
  return s1*s2;
}
```

Fig. 4. Excerpt of KB3D program

which rounding errors must be taken into account, here in function of a constant E declared at the beginning. Our goal is to analyze what should be the value of E depending on the architecture.

Feeding this annotated source code in our assembly analyser in SSE2 mode, each VC is automatically proved valid using either Gappa or one of the SMT solvers Alt-Ergo or CVC3. The bound E is indeed in that case exactly the same as the one found by Boldo and Nguyen [13] in a strict IEEE-754 mode. At least on this example, this shows that SSE2 assembly conforms strictly to the standard. The table below shows the value of E that are proved correct using various architecture-dependent settings.

Architecture Optim. level	SSE2	x87 -O0	x87 -O2	FMA -O2
E	2048×2^{-56}	1025×2^{-56}	1025×2^{-56}	1536×2^{-56}

The FMA setting[2] asks to use the *fused-multiply-add* operation, which computes expressions of the form $x * y \pm z$ with only one rounding [1]. As expected, using FMA improves over SSE2 (25% less) since fewer roundings occur. The extended precision of x87 is even better (around 50% less whatever the optimization level).

[2] Obtained by options `-mfma4` of `gcc-4.5`, requires `-O2`.

```
#define NMAX 10
#define NMAXR 10.0
#define B 0x1.1p-50

/*@ requires 0 <= n <= NMAX;
  @ requires \valid_range(x,0,n-1) && \valid_range(y,0,n-1) ;
  @ requires \forall integer i; 0 <= i < n ==>
  @              \abs(x[i]) <= 1.0 && \abs(y[i]) <= 1.0 ;
  @ ensures \abs(\result - exact_scalar_product(x,y,n)) <= n * B;  */
double scalar_product(double x[], double y[], int n) {
  double p = 0.0;
  /*@ loop invariant 0 <= i <= n ;
    @ loop invariant \abs(exact_scalar_product(x,y,i)) <= i;
    @ loop invariant \abs(p - exact_scalar_product(x,y,i)) <= i * B;
    @ loop variant n-i;        */
  for (int i=0; i < n; i++) {
    //@ assert \abs(x[i]) <= 1.0 && \abs(y[i]) <= 1.0;
    //@ assert \abs(p) <= NMAXR*(1+B) ;
  L: p = p + x[i]*y[i];
    //@ assert \abs(p - (\at(p,L) + x[i]*y[i])) <= B;
    /*@ assert \abs(p - exact_scalar_product(x,y,i+1)) <=
         \abs(p - (\at(p,L) + x[i]*y[i])) + \abs((\at(p,L) + x[i]*y[i]) -
         (exact_scalar_product(x,y,i) + x[i]*y[i])) ; */
    /*@ assert \abs(exact_scalar_product(x,y,i+1)) <=
          \abs(exact_scalar_product(x,y,i)) + \abs(x[i]) * \abs(y[i]); */
    //@ assert \abs(x[i]) * \abs(y[i]) <= 1.0;
  }
  return p;
}
```

Fig. 5. Scalar product: annotated code

Of course, all these bounds are smaller than the one found by Boldo and Nguyen for *any* architecture, which was 0x1.90641p-45$\approx 3203 \times 2^{-56}$ [13], that is more than 50% higher than the SSE2 one.

2.4 Example: Scalar Product

Our last example illustrates how we combine FP analysis with other features such as loops and arrays. The annotated C program on Fig. 5 computes the scalar product of two vectors represented as arrays of doubles. Similarly as the l_sign function of previous example, exact_scalar_product(x, y, n) is defined to denote the scalar product $\sum_{0 \leq i < n} x_i y_i$ computed in real numbers. The post-condition expresses a bound B on the accumulated rounding error in function of a bound NMAX on the size of the vectors. We also assume a bound, here 1.0, on each component of the vectors. Several extra assertions are added in the body of the loop: these are needed to help the automatic provers to solve the generated VCs. In particular, to make Gappa solve the VCs on the accumulated rounding error, it is necessary to guarantee that p remains bounded: it appears to be bounded by NMAX$(1 + B)$.

The table below displays the value of `B` in function of `NMAX` and the architecture-dependent settings.

Architecture / `NMAX`	SSE2	x87 -O0	x87 -O2	FMA
10	$2^{-50}+2^{-54}$	$2^{-50}+\ 17\times 2^{-65}$	17×2^{-65}	2^{-50}
100	$2^{-47}+2^{-54}$	$2^{-47}+\ 33\times 2^{-63}$	129×2^{-65}	2^{-47}
1000	$2^{-44}+2^{-54}$	$2^{-44}+513\times 2^{-64}$	1025×2^{-65}	2^{-44}

The SSE2 mode, supposed to be strictly compliant with the standard, is worse than FMA and x87 without optimization, because the roundings are, as expected, slightly more precise. However the improvement with x87 with optimization is impressive: around $2^{11} \simeq 2000$ times better. The reason is that optimization makes the value of p stored into the x87 stack thus with extended 80-bit precision for the complete execution of the loop: no intermediate rounding to 64-bit is done.

3 Underlying Technique

The core of our technique is to interpret the assembly code into the input language of Why. First we describe the general principles to follow, then we show how we interpret the various assembly statements. Due to the lack of space and the large number of different assembly instructions to handle, we only present a few of them. We focus on the support of FP arithmetic which is the point of interest in this paper. We refer to our technical report [32] for more details.

3.1 Principles of Interpretation in Why

In the input language of Why, one can define a pure model in the logic world by declaring abstract sort names, declaring logic symbols operating on these sorts and posing first-order axioms to axiomatize the behavior of these symbols. Equality and both integer and real arithmetic are built-in in the logic. One can then declare a set of *references* which are mutable variables denoting logic values. Finally, one can define procedures which can modify these references. The body of such a procedure is made of statements in a while-style language. Procedures are also equipped with pre- and post-conditions. The Why VC generator then produces the necessary VCs to ensures that the body respects the post-condition.

One can alternatively just *declare* procedures by only giving pre- and post-conditions, but also declaring the set of modified references. This feature allows us to declare how the atomic operations on a given data type behave. It is exemplified below.

3.2 Model of Data

Machine Integers and Floating-Point Numbers. The Why logic has unbounded mathematical integers and reals only. We reuse the modeling of machine integers provided by the Jessie plug-in of Frama-C [30]. This is done as follows for 32-bit integers; the type `int64` is modelled similarly.

```
type int32
logic integer_of_int32: int32 -> int
predicate is_int32(x:int) = -2147483648 <= x and x <= 2147483647
axiom int32_coerce: forall x:int32. is_int32(integer_of_int32(x))
```

An abstract type `int32` for 32-bit integers is declared, together with a function `integer_of_int32` returning the value such a machine integer denotes. The predicate `is_int32` checks whether an integer is in the range of a 32-bit word or not, and we pose an axiom to specify that the value denoted by an `int32` is always in this range.

To model FP numbers, we reuse the modeling of 32- and 64-bit floats defined by Ayad and Marché [2], which introduces the corresponding abstract types `single` and `double`. We also complete this modeling by the type `binary80` for handling 80-bit floats. Here are the main parts of this model for 64-bit floats

```
type mode = nearest_even | to_zero | up | down | nearest_away
type double
logic double_value : double -> real
logic round_double : mode, real -> real
predicate no_overflow_double(m:mode,x:real) =
      abs(round_double(m,x)) <= 0x1.FFFFFFFFFFFFFp1023
```

An enumerated type `mode` is defined for the 5 possible rounding modes. The abstract type `double` for 64-bit floats is declared, together with the function `double_value` returning the real value it denotes. `round_double`(m, x) returns the closest to x representable number with *unbounded exponent* [2], w.r.t mode m. It is declared and partially axiomatized [2]. Gappa has this function built-in, that's why it is able to solve the VCs about rounding. The predicate `no_overflow_double` checks if overflow occurred when computing x.

Registers. A central feature of our approach is how we model the CPU registers on which the assembly instructions operate. The issue is that a given register only stores a sequence of bits, that can be interpreted either as an integer, a FP number or a memory address. Moreover, for a given register one can either consider it as a 64-bit value or as the 32-bit value stored in its lower part, e.g. the `rax` versus `eax` register of the x86 chip.

To model that behavior, we introduce an abstract type `register` equipped with several access symbols, each of them denotes a different "view" of the value stored in the register. The symbol `sel_exact` is the view for calculations in infinite precision, to model the `\exact` construct of Fig. 4.

```
type register
logic sel_int32  : register -> int32
logic sel_int64  : register -> int64
logic sel_single : register -> single
logic sel_double : register -> double
logic sel_80     : register -> binary80
logic sel_exact  : register -> real
```

Then, for each register, we introduce a Why variable of type `register`, e.g. for the code of Fig. 2 two variables `xmm0` and `rax` are declared with type `register ref` (but not for `%rip` and `%st` which have a special meaning).

Model of the Memory. Interpreting memory access and update at the level of assembly is a major issue, since unlike high-level languages, we have no type information to help interpretation of raw data. In particular a given 64-bit word can be indifferently interpreted as an integer or a memory address. The memory is thus interpreted as a large array of data indexed by integers. However, without type information we would need to know how to encode and decode structured data, like FP numbers, into sequences of bits. Encoding and decoding are defined by complex computations that cannot be handle easily in a purely logical context: the generated VCs would be largely polluted with decoding and encoding hence would unlikely be proved by automatic provers.

We thus decide to keep a typed model of memory instead. This implies that we cannot handle C sources which non type-safe operations: pointer casts and union types. Our ad-hoc preprocessor keeps track of the C type of variables. The memory is then represented not only by one but by several arrays which contains different types of data. Each of these arrays is represented by a Why variable, e.g. a variable `int32M` holds an array of int32 indexed by integers, another variable `doubleM` holds an array of `double` indexed by integers, etc. The Why type for such an array is declared as a polymorphic type:

```
type 'v memory
logic select: 'v memory -> int -> 'v
```

where `select` is the function to access the element at the given index. In assembly, a *memory reference* is an operand of the general form $disp(base, index, scale)$ where *base* and *index* are registers, and *disp* and *scale* are integer constants. *index* defaults to 0 and *scale* defaults to 1. A memory reference $mem = d(b, i, s)$ is thus interpreted as the integer address $b + d + i \times s$.

3.3 Interpretation of Assembly Instructions

Operands. An operand is either an immediate constant, a register or a memory reference. Simple instructions for copying (with name typically starting with `mov`) and arithmetic operations have an output operand called *destination* and one or more input operands called *sources*. There are indeed 6 different interpretations of a source operand depending on the type of the expected value. We denote by $[\![opr]\!]_{int32}$, $[\![opr]\!]_{int64}$, $[\![opr]\!]_{single}$, $[\![opr]\!]_{double}$ and $[\![opr]\!]_{binary80}$ the interpretation of a source operand, respectively as a 32-bit, 64-bit integers and a 32-bit, 64-bit and 80-bit FP number. We also denote by $[\![opr]\!]_{exact}$ its abstract `\exact` value.

$$\begin{aligned}
[\![imm]\!]_{int32} &= imm \\
[\![imm]\!]_{int64} &= imm \\
[\![imm]\!]_{single} &= decode_float32(imm) \\
[\![imm]\!]_{double} &= decode_float64(imm) \\
[\![reg]\!]_{int32} &= \texttt{integer_of_int32(sel_int32(!}reg\texttt{))}
\end{aligned}$$

$$
\begin{aligned}
[\![reg]\!]_{int64} &= \texttt{integer\_of\_int64(sel\_int64(!}reg\texttt{))}\\
[\![reg]\!]_{single} &= \texttt{single\_value(sel\_single(!}reg\texttt{))}\\
[\![reg]\!]_{double} &= \texttt{double\_value(sel\_double(!}reg\texttt{))}\\
[\![reg]\!]_{binary80} &= \texttt{binary80\_value(sel\_80(!}reg\texttt{))}\\
[\![reg]\!]_{exact} &= \texttt{sel\_exact(!}reg\texttt{)}\\
[\![d(b,i,s)]\!]_{addr} &= [\![b]\!]_{int64}\texttt{+d+s*}[\![i]\!]_{int64}\\
[\![mem]\!]_{int32} &= \texttt{integer\_of\_int32(select(int32M,}[\![mem]\!]_{addr}\texttt{))}\\
[\![mem]\!]_{int64} &= \texttt{integer\_of\_int64(select(int64M,}[\![mem]\!]_{addr}\texttt{))}\\
[\![mem]\!]_{single} &= \texttt{single\_value(select(singleM,}[\![mem]\!]_{addr}\texttt{))}\\
[\![mem]\!]_{double} &= \texttt{double\_value(select(doubleM,}[\![mem]\!]_{addr}\texttt{))}\\
[\![mem]\!]_{exact} &= \texttt{select(exactM,}[\![mem]\!]_{addr}\texttt{)}
\end{aligned}
$$

Notations *decode_float32* and *decode_float64* are *not* Why logic functions but denote the operations of transforming a decimal literal into the real it represents respectively in single and double format. This decoding is done "at compile-time" in our translator from assembly to Why.

Move Instructions. The move instructions have a mnemonic prefixed with `mov` and their suffix details the size of source and destination[3]. Their interpretation depends on whether the destination is a register or a memory reference. Moving to a 64-bit register and to memory respectively are interpreted as procedure calls:

$$
\begin{aligned}
[\![\ \text{movq imm, reg}\ ]\!]_i &= \texttt{move\_cte64}\ [\![imm]\!]_{int64}\ [\![imm]\!]_{double}\ [\![imm]\!]_{double}\ \text{reg}\\
[\![\ \text{movq src, reg}\ ]\!]_i &= \texttt{move\_cte64}\ [\![src]\!]_{int64}\ [\![src]\!]_{double}\ [\![src]\!]_{exact}\ \text{reg}\\
[\![\ \text{movq reg, } mem\ ]\!]_i &= \texttt{move\_reg\_to\_mem64}\ \texttt{!reg}\ [\![mem]\!]_{addr}
\end{aligned}
$$

where the Why procedure `move_cte64` is declared as

```
parameter move_cte64: a:int -> b:real -> exact:real -> r:register ref ->
{ } unit writes r
{ integer_of_int64(sel_int64(r)) = a and
     double_value(sel_double(r)) = b and sel_exact(r) = exact }
```

It reads as follows: calling this procedure modifies the register `r` (and nothing else) and after the call, the new content of `r` denotes both a 64-bit integer representing `a`, the FP double representing `b` and an exact real value `exact`. Notice how this interpretation abstracts away from bitwise representation details. Similarly, the Why procedure `move_reg_to_mem64` is declared as

```
parameter move_reg_to_mem64: r:register -> addr:int ->
{ } unit writes int64M, doubleM, exactM
{ integer_of_int64(select(int64M, addr)) = integer_of_int64(sel_int64(r))
  and double_value(select(doubleM, addr)) = double_value(sel_double(r))
  and select(exactM, addr) = sel_exact(r) and
  forall a:int. not (addr <= a <= addr+7) ->
    integer_of_int64(select(int64M,a))=integer_of_int64(select(int64M@,a))
    and double_value(select(doubleM,a))=double_value(select(doubleM@,a))
    and select(exactM,a) =select(exactM@,a)) }
```

[3] All the instructions we use in this paper are written in AT&T assembly syntax.

The `@` sign in a Why post-condition denotes the value of a variable before the call. Thus, the quantified part of the post-condition above amounts to specifying that the rest of the memory is unmodified.

SSE2 Scalar Arithmetic Instructions. Instructions for arithmetic operations of the SSE family operate on 32- or 64-bit integers or floats, depending on the suffix. The destination is always a register. Here is the interpretation of the multiplication, other operations being similar (with an additional precondition for division to check the divisor is not zero).

$$\begin{aligned} [\![\text{ mull src, reg }]\!]_i &= \texttt{set\_int32 } ([\![dest]\!]_{int32} * [\![src]\!]_{int32})\ reg \\ [\![\text{ mulsd src, reg }]\!]_i &= \texttt{set\_double } ([\![dest]\!]_{double} * [\![src]\!]_{double}) \\ &\qquad ([\![dest]\!]_{exact} * [\![src]\!]_{exact})\ reg \end{aligned}$$

```
parameter set_int32: imm:int -> dest: register ref ->
  { is_int32(imm) } unit writes dest
  { integer_of_int32(sel_int32(dest)) = imm }
parameter set_double : a:real -> exact:real -> b:register ref ->
  { no_overflow_double(nearest_even,a) } unit writes b
  { double_value(sel_double(b)) = round_double(nearest_even,a) and
    sel_exact(b) = exact }
```

Notice that these procedures have pre-conditions to check for overflow. Moreover, the post-condition of `set_double` applies FP rounding: this is the way we interpret the IEEE-754 standard: "the result of multiplication should be the same as it was first computed in infinite precision then rounded to the result format". (We hardwire the "nearest even" rounding mode here for simplicity.)

x87 Arithmetic Instructions. The x87 FPU has 8 FP registers to hold FP numbers in extended 80-bit precision. These registers are organized as a stack ST0–ST7 and the current top of stack is identified internally with a special register. In assembly code, `%st` or `%st(0)` denote the top of stack, whereas `%st(`i`)` denote the i-th register below the top.

We could have represented this stack by an array in Why with an additional integer variable. However we can indeed identify statically the current top of stack while translating into Why, so we just represent the stack by 8 variables `st0`, ..., `st7` of type `register ref`. The only assumption we make in this optimization is that the stack is empty at functions entrance and exit. Our translator statically computes the value of the top-of-stack pointer at each instruction. This value must be unique whatever is the path of the control-flow graph to reach the instruction. We thus translate x87 registers `%st(`i`)` into Why variable `st`$\bar{i}$ where $\bar{i} = top\_of\_stack - i$.

For example, in optimized x87 assembly code compiled from Fig. 5, the top-of-stack pointer has value 1 at the loop entrance, because `p` is stored in the stack. Note that this way of interpreting the x87 stack, instead of considering the stack as an array, greatly improves the verification of VCs by Gappa back-end, since it does not know the theory of arrays as SMT solvers do.

f:	$\longrightarrow$	`let f() =`
`.cfi_startproc`		
`/*@ requires` P`; */`	$\longrightarrow$	`assumes {`$[\![P]\!]_{annot}$`};`
(body of the function f)	$\longrightarrow$	$[\![$ (body of the function f) $]\!]_i$
`/*@ ensures` Q`; */`	$\longrightarrow$	`assert {`$[\![Q]\!]_{annot}$`};`
`leave`		`void`
`ret`		
`.cfi_endproc`		`parameter f: unit ->`
		`{` $[\![P]\!]_{annot}$ `} unit writes` w `{` $[\![Q]\!]_{annot}$ `}`

Fig. 6. Translation of a function in assembly to Why

Instructions for loading in the stack and storing from the stack are interpreted as follows.

$$
\begin{aligned}
[\![\text{ fldl src }]\!]_i &= \texttt{set\_80 } [\![src]\!]_{double}\ [\![src]\!]_{exact}\ \texttt{st}\overline{0} \\
[\![\text{ fld st(\%i) }]\!]_i &= \texttt{set\_80 } [\![\texttt{st}\overline{i}]\!]_{binary80}\ [\![\texttt{st}\overline{i}]\!]_{exact}\ \texttt{st}\overline{0} \\
[\![\text{ fldl1 }]\!]_i &= \texttt{set\_80 (1.0) (1.0) st}\overline{0} \\
[\![\text{ fstl reg }]\!]_i &= \texttt{set\_double } [\![\texttt{st}\overline{0}]\!]_{binary80}\ [\![\texttt{st}\overline{0}]\!]_{exact}\ \texttt{!reg} \\
[\![\text{ fstl mem }]\!]_i &= \texttt{set\_double\_mem } [\![\texttt{st}\overline{0}]\!]_{binary80}\ [\![\texttt{st}\overline{0}]\!]_{exact}\ [\![mem]\!]_{addr}
\end{aligned}
$$

where `set_80` is the analogous of `set_double` for 80-bit FP numbers.

Arithmetic instructions in the stack are interpreted as follows (for `fmul`, `fadd` and `fsub` being similar).

$$
\begin{aligned}
[\![\text{ fmull src }]\!]_i &= \texttt{set\_80 (}[\![\texttt{st}\overline{0}]\!]_{binary80}\texttt{*}[\![src]\!]_{double}\texttt{)} \\
&\qquad \texttt{(}[\![\texttt{st}\overline{0}]\!]_{exact}\texttt{*}[\![src]\!]_{exact}\texttt{) st}\overline{0} \\
[\![\text{ fmul \%st(i), \%st(j) }]\!]_i &= \texttt{set\_80 (}[\![\texttt{st}\overline{j}]\!]_{binary80}\texttt{*}[\![\texttt{st}\overline{i}]\!]_{binary80}\texttt{)} \\
&\qquad \texttt{(}[\![\texttt{st}\overline{j}]\!]_{exact}\texttt{*}[\![\texttt{st}\overline{i}]\!]_{exact}\texttt{) st}\overline{j}
\end{aligned}
$$

3.4 Translation of Annotated Functions

Assume that we have a function with preconditions, post-conditions and assertions. The translation of this function in assembly language to Why is illustrated in Fig. 6. Our preprocessing moved the post-condition to the end of the function. More generally, each annotation is preprocessed into an inline assembly instruction of the form

```
asm("/* <keyword> P */"::"X"(x0),..,"X"(xn));
```

where each variable x_i of type τ in proposition P is replaced by #τ#%i#. The compiler thus transforms this inline assembly into lines between *#APP* and *#NO_APP* as on Fig. 2, where each %i is replaced by any appropriate memory reference or register, even the 80-bit ones[4]. The Why interpretation of an annotation A is denoted by $[\![A]\!]_{annot}$. It is defined by a straightforward structural recursion, where the variables are interpreted as follows. in Why.

[4] As specified by the constraint "X", see
http://gcc.gnu.org/onlinedocs/gcc/Simple-Constraints.html

$$
\begin{array}{ll}
[\![\#int\#v\#]\!]_{annot} & = [\![v]\!]_{int32} \\
[\![\#float\#v\#]\!]_{annot} & = [\![v]\!]_{single} \\
[\![\#double\#v\#]\!]_{annot} & = [\![v]\!]_{double} \\
[\![\texttt{\textbackslash exact}(\#\tau\#v\#)]\!]_{annot} & = [\![v]\!]_{exact} \text{ where } \tau \in \{\text{float, double}\} \\
[\![\texttt{\textbackslash valid\_range}(\#\tau\#v\#, a, b)]\!]_{annot} & = \texttt{forall i, } a\texttt{<=i<}b \texttt{ ->} \\
& \quad [\![e]\!]_{int64}\texttt{+i*}\mathit{sizeof}(\tau) \texttt{ >= !rsp}
\end{array}
$$

The interpretation of `\valid_range` allows to provide *separation* information for pointers [32], e.g. in Fig. 5, we need to know that local variable `p` is separated from `x[0..9]` and `y[0..9]`. Finally, the translation of code assertions and function calls are

$$
\begin{array}{ll}
[\![\ \texttt{assert P}\ ]\!]_i & = \texttt{assert } [\![P]\!]_{annot} \\
[\![\ \texttt{call } f\ ]\!]_i & = f\texttt{\_parameter ()}
\end{array}
$$

About Translation of Compound Statements. When the source contains compound statements like conditional and loops, then the assembly code contains conditional jumps to labels. We cannot interpret such arbitrary jumps directly into Why since its programming language only support structured statements. We proceed differently by interpreting the arbitrary control flow graph of the assembly into a finite set of small pieces of codes, where loop invariants play the role of pre- or post-conditions. This technique is not the purpose of this paper and indeed is not original: we refer to our report [32] and earlier work done above Boogie [3] or Why [24] for more details.

4 Conclusion

An early work on verification of machine code is due to Boyer and Yu in 1992 [14]. They formalize the assembly language of a particular micro-processor and its operational semantics in Nqthm, and were able to verify a few programs, specified in Nqthm too. Their approach provides a deep embedding of assembly code, whereas ours is based on a shallow embedding: assembly code is simulated in Why. In our approach, behaviors are specified in the general-purpose ACSL language, and the proofs can be conducted with a large set of automated provers. Former studies on the verification of assembly code are in the context of the so-called *proof-carrying-code* [18], where proof obligations for *safety* (of memory dereferencing, absence of overflow, etc.) are generated on the object code. However these do not consider any behavioral specification language to specify deeper properties than safety. Although, it is worth noting that there is an identified need to generate loop invariants in the target code to explicitate compilation choices [18,33]. In 2006, Burdy and Pavlova [16] consider a specification language on Java bytecode. Barthe et al. [6] showed how proofs of VCs at source level can be reused for proving VCs at bytecode level, but they do not admit compiler optimizations. In 2008, Myreen [31] proposes to compile assembly code into functions in the HOL4 system. Our approach is somewhat close to this, using Why as target language instead of HOL4. Where Myreen must perform the proofs within HOL4, we can use various automatic provers thanks to the multi-prover feature of Why.

As far as we know, nobody ever considered any aspect of FP computations behavioral verification at the level of assembly. We believe that what we present in this paper is the first method being able to prove architecture- and compiler-dependent behavioral properties of FP programs.

Our approach and our prototype implementation demonstrate that handling architecture-dependent aspects is indeed possible. However it is clearly not mature enough for a non-expert user, because there are a lot of open issues. First, some languages features are not supported at the C level (like pointer casts) and also at the assembly level. Second, we are not always able to interpret all the compiler optimizations. For example we do not support inlining of functions. We believe that to go further, we should integrate our approach into the compiler itself, following the ideas of proof-carrying-code: the optimizations made by the compiler should also produce annotations of the generated assembly (assertions, loop invariants) to make the optimizations explicit.

References

1. IEEE standard for floating-point arithmetic. Technical report (2008), `http://dx.doi.org/10.1109/IEEESTD.2008.4610935`
2. Ayad, A., Marché, C.: Multi-Prover Verification of Floating-Point Programs. In: Giesl, J., Hähnle, R. (eds.) IJCAR 2010. LNCS (LNAI), vol. 6173, pp. 127–141. Springer, Heidelberg (2010)
3. Barnett, M., Leino, K.R.M.: Weakest-precondition of unstructured programs. In: 6th PASTE, New York, NY, USA, pp. 82–87. ACM (2005)
4. Barnett, M., Leino, K.R.M., Schulte, W.: The Spec# Programming System: An Overview. In: Barthe, G., Burdy, L., Huisman, M., Lanet, J.-L., Muntean, T. (eds.) CASSIS 2004. LNCS, vol. 3362, pp. 49–69. Springer, Heidelberg (2005)
5. Barrett, C.W., Tinelli, C.: CVC3. In: Damm, W., Hermanns, H. (eds.) CAV 2007. LNCS, vol. 4590, pp. 298–302. Springer, Heidelberg (2007)
6. Barthe, G., Rezk, T., Saabas, A.: Proof Obligations Preserving Compilation. In: Dimitrakos, T., Martinelli, F., Ryan, P.Y.A., Schneider, S. (eds.) FAST 2005. LNCS, vol. 3866, pp. 112–126. Springer, Heidelberg (2006)
7. Baudin, P., Filliâtre, J.-C., Marché, C., Monate, B., Moy, Y., Prevosto, V.: ANSI/ISO C Specification Language (2009), `http://frama-c.cea.fr/acsl.html`
8. Bertot, Y., Castéran, P.: Interactive Theorem Proving and Program Development. Springer, Heidelberg (2004)
9. Bobot, F., Conchon, S., Contejean, E., Iguernelala, M., Lescuyer, S., Mebsout, A.: The Alt-Ergo automated theorem prover (2008), `http://alt-ergo.lri.fr/`
10. Boldo, S., Filliâtre, J.-C.: Formal Verification of Floating-Point Programs. In: 18th ARITH, pp. 187–194 (June 2007)
11. Boldo, S., Marché, C.: Formal verification of numerical programs: from C annotated programs to mechanical proofs. In: Mathematics in Computer Science (2011)
12. Boldo, S., Nguyen, T.M.T.: Hardware-independent proofs of numerical programs. In: 2nd NASA Formal Methods Symposium, pp. 14–23 (April 2010)
13. Boldo, S., Nguyen, T.M.T.: Proofs of numerical programs when the compiler optimizes. Innovations in Systems and Software Engineering 7, 151–160 (2011)
14. Boyer, R.S., Yu, Y.: Automated proofs of object code for a widely used microprocessor. J. ACM 43(1), 166–192 (1996)

15. Burdy, L., Cheon, Y., Cok, D., Ernst, M., Kiniry, J., Leavens, G.T., Leino, K.R.M., Poll, E.: An overview of JML tools and applications. Technical Report NIII-R0309, Dept. of Computer Science, University of Nijmegen (2003)
16. Burdy, L., Pavlova, M.: Java bytecode specification and verification. In: SAC, pp. 1835–1839. ACM (2006)
17. Carreño, V.A., Miner, P.S.: Specification of the IEEE-854 floating-point standard in HOL and PVS. In: HOL 1995 (September 1995)
18. Colby, C., Lee, P., Necula, G., Blau, F., Plesko, M., Cline, K.: A certifying compiler for Java. In: PLDI, pp. 95–107. ACM (2000)
19. Cousot, P., Cousot, R., Feret, J., Mauborgne, L., Miné, A., Monniaux, D., Rival, X.: The ASTREÉ Analyzer. In: Sagiv, M. (ed.) ESOP 2005. LNCS, vol. 3444, pp. 21–30. Springer, Heidelberg (2005)
20. Cuoq, P., Prevosto, V.: Value Plugin Documentation, *Carbon* version. In: CEA-List (2011), `http://frama-c.com/download/frama-c-value-analysis.pdf`
21. de Moura, L., Bjørner, N.S.: Z3: An Efficient SMT Solver. In: Ramakrishnan, C.R., Rehof, J. (eds.) TACAS 2008. LNCS, vol. 4963, pp. 337–340. Springer, Heidelberg (2008)
22. Delmas, D., Goubault, E., Putot, S., Souyris, J., Tekkal, K., Védrine, F.: Towards an Industrial use of FLUCTUAT on Safety-Critical Avionics Software. In: Alpuente, M., Cook, B., Joubert, C. (eds.) FMICS 2009. LNCS, vol. 5825, pp. 53–69. Springer, Heidelberg (2009)
23. Dowek, G., Muñoz, C.: Conflict detection and resolution for 1,2,.,N aircraft. In: 7th AIAA Aviation, Technology, Integration, and Operations Conference (2007)
24. Filliâtre, J.-C.: Formal Verification of MIX Programs. In: Journées en l'honneur de Donald E. Knuth (2007), `http://knuth07.labri.fr/exposes.php`
25. Filliâtre, J.-C., Marché, C.: The Why/Krakatoa/Caduceus Platform for Deductive Program Verification. In: Damm, W., Hermanns, H. (eds.) CAV 2007. LNCS, vol. 4590, pp. 173–177. Springer, Heidelberg (2007)
26. Harrison, J.: Formal Verification of Floating Point Trigonometric Functions. In: Johnson, S.D., Hunt Jr., W.A. (eds.) FMCAD 2000. LNCS, vol. 1954, pp. 217–233. Springer, Heidelberg (2000)
27. Leavens, G.T.: Not a number of floating point problems. Journal of Object Technology 5(2), 75–83 (2006)
28. Melquiond, G.: Proving Bounds on Real-Valued Functions with Computations. In: Armando, A., Baumgartner, P., Dowek, G. (eds.) IJCAR 2008. LNCS (LNAI), vol. 5195, pp. 2–17. Springer, Heidelberg (2008)
29. Monniaux, D.: The pitfalls of verifying floating-point computations. Transactions on Programming Languages and Systems 30(3), 12 (2008)
30. Moy, Y., Marché, C.: Jessie Plugin Tutorial, Beryllium version. INRIA (2009), `http://www.frama-c.cea.fr/jessie.html`
31. Myreen, M.O.: Formal verification of machine-code programs. PhD thesis, University of Cambridge (2008)
32. Nguyen, T.M.T., Marché, C.: Proving floating-point numerical programs by analysis of their assembly code. Research Report 7655, INRIA (2011), `http://hal.inria.fr/inria-00602266/en/`
33. Rival, X.: Abstract Interpretation-Based Certification of Assembly Code. In: Zuck, L.D., Attie, P.C., Cortesi, A., Mukhopadhyay, S. (eds.) VMCAI 2003. LNCS, vol. 2575, pp. 41–55. Springer, Heidelberg (2002)
34. Russinoff, D.M.: A mechanically checked proof of IEEE compliance of the floating point multiplication, division and square root algorithms of the AMD-K7 processor. LMS Journal of Computation and Mathematics 1, 148–200 (1998)

Coquet: A Coq Library for Verifying Hardware

Thomas Braibant*

LIG, UMR 5217, INRIA

Abstract. We propose a new library to model and verify hardware circuits in the Coq proof assistant. This library allows one to easily build circuits by following the usual pen-and-paper diagrams. We define a deep-embedding: we use a (dependently typed) data-type that models the architecture of circuits, and a meaning function. We propose tactics that ease the reasoning about the behavior of the circuits, and we demonstrate that our approach is practicable by proving the correctness of various circuits: a text-book divide and conquer adder of parametric size, some higher-order combinators of circuits, and some sequential circuits: a buffer, and a register.

Introduction

Formal methods are widely used in the verification of circuit design, and appear as a necessary alternative to test and simulation techniques. Among them, model checking methods have the advantage of being fully automated but can only deal with circuits of fixed size and suffers from combinatorial explosion. On the other hand, circuits can be formally specified and certified using theorem provers [10,9,14]. For instance, the overall approach introduced in [9,17] to model circuits in higher-order logic is to use predicates of the logic to express the possible behaviour of devices.

We present a study for specifying and verifying circuits in Coq. Our motivations are two-fold. First, there has been a lot of work describing and verifying circuits in logic in the HOL and ACL2 family of theorem provers. However, Coq features dependent types that are more expressive. The Veritas language experiment [10] hinted that these allow for specifications that are both clearer and more concise. We also argue that dependent types are invaluable for developing circuits reliably: some errors can be caught early, when type-checking the circuits or their specifications. Second, most of these works model circuits using a shallow-embedding: circuits are defined as predicates or functions in the logic of the theorem prover, with seldom, if any, way to reason about the devices inside the logic: for instance, functions that operate on circuits must be built at the meta-level [21], which precludes one from proving their correctness. We define a data-type for circuits and a meaning function: we can write (and reason about) Coq functions that operate on the structure of circuits.

* The author has been partially funded by the French projects "Choco", ANR-07-BLAN-0324 and "PiCoq", ANR-10-BLAN-0305.

J.-P. Jouannaud and Z. Shao (Eds.): CPP 2011, LNCS 7086, pp. 330–345, 2011.

Circuit diagrams describe the wire connections between gates and have nice algebraic properties [5,15]. While we do not prove algebraic laws, our library features a set of basic blocks and combinators that allows one to describe such diagrams in a hierarchic and modular way. We make precise the interconnection of circuits, yet, we remain high-level because we make implicit the low-level diagram constructs such as wires and ports. Circuit diagrams are also used to present recursive or parametric designs. We use Coq recursive definitions to *generate* circuits of parametric size, e.g., to generate an n-bit adder for a given n. Then, we reason about these functions rather than on the tangible (fixed-size) instantiations of such generators. Circuits modelled by recursion have already been verified in other settings [14,17]. The novelty of our approach is that we derive circuit designs in a systematic manner: we structure circuits generators by mimicking the usual circuit diagrams, using our combinators. Then, the properties of these combinators allow us to prove the circuits correct.

We are interested in two kinds of formal dependability claims. First, we want to capture some properties of well-formedness of the diagrams. Second, we want to be able to express the *functional correctness* of circuits – the fact that a circuit complies to some specification, or that it implements a given function. Obviously, the well-formedness of a circuit is a prerequisite to its functional correctness. We will show that using dependent types, we can get this kind of verification for free. As an example, the type-system of Coq will preclude the user to make invalid compositions of circuits. Hence, we can focus on what is the intrinsic *meaning* of a circuit, and prove that the meaning of some circuits entails a high-level specification, e.g., some functional program.

Our contributions can be summarized as follows: we propose a new framework to model and verify circuits in Coq that allows us to define circuits in a systematic manner by following the usual diagrams; we provide tactics that allow to reason about circuits; we demonstrate that our approach is practicable on practical examples: text-book n-bit adders, high-level combinators, and sequential circuits.

Outline. In §1, we give a small overview of all the basic concepts underlying our methodology to present how the various pieces fit together. We present the actual definitions we use in §2. Then, in §3 and §4, we demonstrate the feasibility of our approach on some examples. We analyse some benefits of using a deep-embedding in §5. Finally, we compare our study to other related work in §6.

1 Overview of Our System

We give a global overview of the basic concepts of our methodology first, before giving a formal Coq definition to these notions in the next section. We take this opportunity to illustrate the use of our system to represent parametrized systems through the example of a simple n-bit adder: it computes an n-bit sum and a 1-bit carry-out from two n-bit inputs and a 1-bit carry-in. The recursive construction scheme of this adder is presented in Fig. 1 (data flows from left to right), using a *full-adder*, i.e., a 1-bit adder, as basic building block.

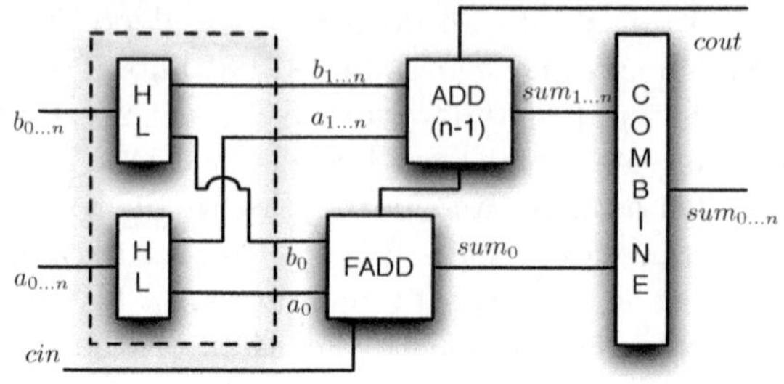

Fig. 1. A recursive n-bit ripple-carry adder

Circuit Interfaces. Informally, we want to build circuits that relate n input wires to m output wires, where n, m are integers. For instance, the door `AND` has two inputs and one output. However, using integers to number the wires does not give much structure: how are grouped the $2n+1$ input wires of the n-bit adder? Hence, we use arbitrary *finite-types* as indexes for the wires rather than integers [11]. A circuit that relates inputs indexed by n to outputs indexed by m has type $\mathbb{C}\ n\ m$, where n and m are types. For instance, the full-adder, a circuit with three inputs and one output, has type $\mathbb{C}\ (\mathbf{1} \oplus (\mathbf{1} \oplus \mathbf{1}))\ (\mathbf{1} \oplus \mathbf{1})$, where $\oplus$ is the disjoint sum (associative to the left) and $\mathbf{1}$ is a singleton type. Hence, the n-bit adder has type $\mathbb{C}\ (\mathbf{1} \oplus \texttt{sumn}\ \mathbf{1}\ \texttt{n} \oplus \texttt{sumn}\ \mathbf{1}\ \texttt{n})\ (\texttt{sumn}\ \mathbf{1}\ \texttt{n} \oplus \mathbf{1})$, where `sumn` $A\ n$ is a n-ary disjoint sum.

Circuits Combinators. The n-bit adder is made of several sub-components that are composed together. We use *circuit combinators* (or combining forms [19]) to specify the connection layout of circuits. For instance, in Fig. 1, the dashed-box is built by composing in parallel two `HL` circuits that are then composed serially with a combinator that reorders the wires. These combinators leave implicit the connection points in the circuits and focus on how information flow through the circuit: the wire names given in Fig. 1 do not correspond to variables, and are provided for the sake of readability.

In our "nameless" setting, wires have to be forked and reordered using *plugs*: a plug is a circuit of type $\mathbb{C}\ n\ m$, defined using a map from m to n that defines how to connect an output wire (indexed by m) to an input wire (indexed by n). Since we use functions rather than relations, this definition naturally forbids short-circuits (two input wires connected to the same output wire).

Meaning of a Circuit. We now depart from the syntactic definitions of circuits to give an overview of their semantics. We assume a type $\mathbb{T}$ of what is carried by the wires, for instance booleans ($\mathbb{B}$) or streams of booleans ($\texttt{nat} \to \mathbb{B}$). Let x be a circuit of type $\mathbb{C}\ n\ m$. The *inputs* (resp. *outputs*) of x are a finite function *ins* of type $n \to \mathbb{T}$ (resp. *outs* of type $m \to \mathbb{T}$). The *meaning* of x is a relation $x \vdash_m^n ins \bowtie outs$ between *ins* and *outs* that we define by induction on x.

This is an abstract mathematical characterization, which may or may not be computational (we will come back to this point later).

Abstracting from the Implementation. The meaning of a circuit is defined by induction on its structure: this relation may be complex and may give information about the internal implementation of a circuit. Thus, we want to move from the definition of this relation, for instance, to give high-level specifications, or to abstract their behavior. These abstractions can be expressed through the following kind of entailment [17]:

$$\forall ins, \forall outs, \texttt{RIPPLE n} \vdash ins \bowtie outs \rightarrow R\ outs\ ins$$

We use *data-abstraction* [17] to be more elegant. Indeed, a value of type $\mathbf{1} \oplus \texttt{sumn 1 n} \oplus \texttt{sumn 1 n} \rightarrow \mathbb{B}$ is isomorphic to a value of type $\mathbb{B} \times \mathbb{W}_n \times \mathbb{W}_n$ (where $\mathbb{W}_n$ is the type of integers from 0 to 2^n). We use *type-isomorphisms* to give tractable specifications for circuits: we prove that the parametric n-bit adder depicted in fig 1 implements the addition with carry function on $\mathbb{W}_n$.

2 Formal Development

We now turn to define formally the concepts that were overviewed in the previous section. We use Coq type-classes to structure and parametrize our development.

2.1 Circuit Interfaces

We use arbitrary finite types (types with finitely many elements) as interfaces for the circuits, i.e., as indexes for the wires. One can create such finite types by using the disjoint-sum operator $\oplus$ and the one-element type $\mathbf{1}$. This construction can be generalized to n-ary disjoint sums written `sumn A n`, for a given `A`. However, using a single singleton type for all wires can be confusing: there is no way to distinguish one $\mathbf{1}$ from another, except by its position in the type (which is frustrating). Hence, we use an infinite family of singleton types $\mathbf{1}_x$ where x is a *tag*. Circuits are parametrised by some tags, which allows the Coq type-system to rule out some ill-formed combinations. This tagging discipline makes it easy to follow circuit diagrams to define circuits in Coq, without much room for mistakes.

```
Inductive tag (t : string) : Type := _tag : tag t. (** we write 1_t for tag t*)
```

Finite types are defined as a class `Fin A` that packages a duplicate-free list of all the elements of the type `A`, defined along the lines of [8].

2.2 Type Isomorphisms

We use type-isomorphisms as "lenses" to express the specification of circuits in user-friendly types, without loss of information. In a nutshell, we define in Coq an isomorphism between two types A and B as a pair of functions $iso : A \to B$ and $uniso : B \to A$ that are proved to be inverse of each other. We use the notation $A \cong B$ for an isomorphism between A and B, and we define some notations for operations (or instances) that allow us to build such isomorphisms in Fig. 2. The most important instance states the duality between disjoint-sums in the domain of the finite functions to a cartesian product.

```
Class Iso (A B : Type) :={
 iso : A → B;
 uniso : B → A}.
```

```
Class Iso_Props {A B: Type} (I : Iso A B):= {
 iso_uniso : ∀ (x : B), iso (uniso x) = x;
 uniso_iso : ∀ (x : A), uniso (iso x) = x}.
```

2.3 Plugs

Rewiring circuits of type $\mathbb{C}\ n\ m$ are defined by mapping output wires indexed by m to input wires indexed by n. We define plugs using usual Coq functions to get small and computational definition of maps. (Note that, since we map the indexes of the wires, there is no way to embed an arbitrary function inside our circuits to compute, e.g., the addition of the value carried by two input wires.)

We give three examples: (a) is a circuit that forgets its first input (types must be read bottom-up on diagrams); (b) is a circuit that duplicates its inputs; (c) implements some re-ordering and duplication of the wires. (We leave implicit the associativity of wires on the diagrams.)

(a) (b) (c)

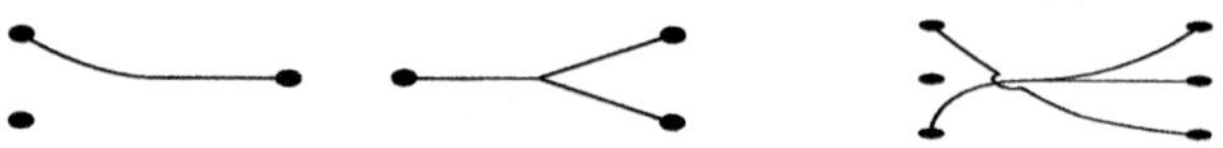

$\mathbb{C}\ (n \oplus m)\ m$ $\quad$ $\mathbb{C}\ n\ (n \oplus n)$ $\quad$ $\mathbb{C}\ (n \oplus m \oplus p)\ (p \oplus (n \oplus n))$

A possible implementation for (a) is `fun x ⇒ inr x` and (b) can be implemented as `fun x ⇒ match x with inl e ⇒ e | inr e ⇒ e end`. If the type of the circuit gives enough information, like the examples above, it is possible to define such plugs using proof-search. Indeed, plugs that deal with the associativity of the wires, or even re-orderings, are completely defined by their type, and we use tactics to write the map between wires (it amounts to some case splitting and little automation). Hence, in the formal definition of circuits, we omit the plugs that deal with associativity or re-orderings of the wires, not only for the sake of readability, but also because we do so in the actual Coq code: we leave holes in the code (thanks to the Coq Program feature) that will be filled automatically.

$$\cdot\bullet\cdot\ \frac{A \to \mathbb{T} \cong \sigma \qquad B \to \mathbb{T} \cong \tau}{A \oplus B \to \mathbb{T} \cong (\sigma \times \tau)} \qquad \iota_x\ \frac{}{\mathbf{1}_x \to \mathbb{T} \cong \mathbb{T}} \qquad \frac{A \to \mathbb{T} \cong \sigma}{\texttt{sumn}\ A\ n \to \mathbb{T} \cong \texttt{vector}\ \sigma\ \texttt{n}}$$

Fig. 2. Isomorphisms between types

```
Context {tech : Techno}
Inductive ℂ : Type → Type → Type :=
| Atom : ∀ {n m : Type} {Hfn : Fin n} {Hfm : Fin m}, techno n m → ℂ n m
| Plug : ∀ {n m : Type} {Hfn : Fin n} {Hfm : Fin m} (f : m → n), ℂ n m
| Ser : ∀ {n m p : Type}, ℂ n m → ℂ m p → ℂ n p
| Par : ∀ {n m p q : Type}, ℂ n p → ℂ m q → ℂ (n ⊕ m) (p ⊕ q)
| Loop : ∀ {n m p : Type}, ℂ (n ⊕ p) (m ⊕ p) → ℂ n m.
```

Fig. 3. Syntax

2.4 Abstract Syntax

In the following, we use some basic gates from which all other circuits are defined. Hence, we parametrize the definition of circuits by the type of the gates:

```
Class Techno := techno : Type → Type → Type.
```

Fig. 3 presents the dependent type that models circuits, as defined in Coq. This abstract syntax is strongly typed: it ensures that circuits built using the provided combinators are well-formed: dimensions have to agree, and it is not possible to connect circuits in the wrong direction. (Note that this is not anecdotal: if we were to describe circuits with ports and wires, ensuring these properties would require some boilerplate.) We denote serial composition (`Ser`) with the infix ▷ symbol, and parallel composition (`Par`) with &. (Note that these definitions do not deal with what transits in the wires.)

2.5 Structural Specifications

Let $\mathbb{T}$ be the type of what is carried in the wires. We now define the meaning relation for circuits. For a given circuit of type $\mathbb{C}\ n\ m$, we build a relation between two functions of type $n \to \mathbb{T}$ and $m \to \mathbb{T}$. We define several operations on such functions, in order to express the meaning relation in a legible manner:

```
Context {𝕋 : Type}.
Definition left {n} {m} (x : (n ⊕ m) → 𝕋) : n → 𝕋 := fun e ⇒ (x (inl _ e)).
Definition right {n} {m} (x : (n ⊕ m) → 𝕋) : m → 𝕋 := fun e ⇒ (x (inr _ e)).
Definition lift {n} {m} (f : m → n) (x : n → 𝕋) : m → 𝕋 := fun e ⇒ x (f e).
Definition app {n m} (x : n → 𝕋) (y : m → 𝕋) : n ⊕ m → 𝕋 :=
  fun e ⇒ match e with inl e ⇒ x e | inr e ⇒ y e end.
```

We define the semantics of a given set of basic gates `tech: Techno` by defining instances of the following type-class, (typically, one instance for the boolean setting, and one instance for the boolean stream setting):

```
Class Technology_spec (tech : Techno) 𝕋 :=
   spec : ∀ {a b: Type}, tech a b → (a → 𝕋) → (b → 𝕋) → Prop.
```

The meaning relation for circuits is generated by this parameter and rules for each combinator. These rules are presented on Fig. 4 using inference rules rather than the corresponding Coq inductive, for the sake of readability.

2.6 Abstractions

The meaning relation defines precisely the behavior of a circuit, but cannot be used as it is. First, it may be too precise, e.g., by leaking some internal details or imposing constraints between the inputs and the outputs of a circuit that are not relevant from an external point of view. Second, it defines a constraint between the inputs and outputs of a circuit as a relation between two functions $n \to \mathbb{T}$ and $m \to \mathbb{T}$, which is not user-friendly. In his book [17], Melham defines two kinds of abstractions that are relevant here: behavioral abstraction (expressed through the logical entailment of a weak specification R by the meaning relation) and data-abstraction (when the specification is expressed in terms of higher-level types than the above function types).

We combine these two notions to specify that a given circuit realises a specification R up to two type isomorphisms, and to get more concise specifications, we also define the fact that a circuit implements a function `f` up to isomorphisms:

```
Context {n m N M : Type} (Rn : (n→𝕋) ≅ N) (Rm : (m→𝕋) ≅ M).
Class Realise (c : ℂ n m) (R : N → M → Prop) := realise: ∀ ins outs,
    c ⊢ⁿₘ ins ⋈ outs → R (iso ins) (iso outs).
Class Implement (c : ℂ n m) (f : N → M) := implement: ∀ ins outs,
    c ⊢ⁿₘ ins ⋈ outs → iso outs = f (iso ins).
```

2.7 Atoms and Modular Proofs

We develop circuits in a modular way: to build a complex circuit, we define a functor that takes as an argument a module that packages the implementations of the sub-components, and the proofs that they meet some specification. This means that our proofs are hierarchical: we do not inspect the definition of the sub-components when we prove a circuit. These functors can then be applied to a module that contains a set of basic doors (of type `Techno`) and its meaning relation (of type `Technology_spec`).

$$\textsc{KSer}\ \frac{x \vdash^n_m \mathit{ins} \bowtie \mathit{middle} \qquad y \vdash^m_p \mathit{middle} \bowtie \mathit{outs}}{x \rhd y \vdash^n_p \mathit{ins} \bowtie \mathit{outs}}$$

$$\textsc{KPar}\ \frac{x \vdash^n_p \texttt{left}\ \mathit{ins} \bowtie \texttt{left}\ \mathit{outs} \qquad y \vdash^m_q \texttt{right}\ \mathit{ins} \bowtie \texttt{right}\ \mathit{outs}}{x \& y \vdash^{n \oplus m}_{p \oplus q} \mathit{ins} \bowtie \mathit{outs}}$$

$$\textsc{KPlug}\ \frac{}{\texttt{Plug}\ f \vdash^n_m \mathit{ins} \bowtie \texttt{lift}\ f\ \mathit{ins}} \qquad \textsc{KLoop}\ \frac{x \vdash^{n \oplus p}_{m \oplus p} \texttt{app}\ \mathit{ins}\ r \bowtie \texttt{app}\ \mathit{outs}\ r}{\texttt{Loop}\ x \vdash^n_m \mathit{ins} \bowtie \mathit{outs}}$$

Fig. 4. Meaning of circuits (omitting the rule for Atom)

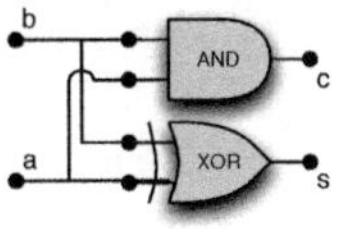


```
Context a b s c : string. (*section variables*)
Definition HADD : ℂ (1ₐ ⊕ 1_b) (1ₛ ⊕ 1_c) :=
Fork2 (1ₐ ⊕ 1_b) ▷ (XOR a b s & AND a b c).
```

Fig. 5. Definition of a half-adder

3 Proving Some Combinatorial Circuits

In this section, we focus on acyclic combinational circuits, and implement some arithmetic circuits. We assume a set of basic gates (AND, XOR among others, that can all be defined and proved correct starting from NOR only). Wires carry booleans, i.e., the meaning relation is defined on booleans for the basic gates. We first illustrate our proof methodology on a half-adder. Then, we present operations on n-bits integers that will be used to specify n-bit adders.

3.1 Proving a Half-Adder

A half-adder adds two 1-bit binary numbers together, producing a 1-bit number and a carry-out. However, they cannot be chained together since they have no carry in-input. We present a diagram of this circuit, along with its formal definition, in Fig. 5. The left-hand side of the following Coq excerpt is the statement we prove: the circuit HADD implements the function hadd on booleans (defined as λ(a,b).(a $\oplus$ b, a $\wedge$ b), where $\oplus$ is the boolean exclusive-or, and $\wedge$ is the boolean and) up to isomorphisms (we use the notations from Fig. 2 for isos). The Coq system ask us to give evidence of the right-hand side.

```
Instance HADD_Spec : Implement
(ιa • ιb) (* iso on inputs *)
(ιs • ιc) (* iso on outputs *)
HADD hadd.
```

```
I : 1a ⊕ 1b → 𝔹, O : 1s ⊕ 1c → 𝔹
H : HADD ⊢^{1a⊕1b}_{1s⊕1c} ins ⋈ outs
======================
@iso (ιs • ιc) O = hadd (@iso (ιa • ιb) I)
```

We have developed several tactics that help to prove these kind of goals. First, we automatically invert the derivation of the meaning relation in the hypothesis H, following the structure of the circuit, to get rid of parallel and serial combinators. This leaves the user with one meaning relation hypothesis per sub-component in the circuit (plugs included). Second, we use the type-class Implement as a dictionary of interesting properties. We use it to make fast-forward reasoning by applying implements in any hypothesis stating a meaning relation for a sub-component. The type-class resolution mechanism will look for an instance of Implement for this sub-component, and transform the "meaning relation" hypothesis into an equation. (Note that at this point, the user may have to interact with the proof-assistant, e.g., to choose other Implement instances than the ones that are picked automatically, but in many cases, this step is automatic.) At this point, the goal looks like the left-hand side of the following excerpt:

```
I : 1a ⊕ 1b → 𝔹, O : 1s ⊕ 1c → 𝔹
M : (1a ⊕ 1b) ⊕ (1a ⊕ 1b) → 𝔹
H0: iso M = (fun x ⇒ (x,x)) (iso I)
H1: iso (left O) = uncurry ⊕ (iso (left M))
H2: iso (right O)= uncurry ∧ (iso (right M))
============================
iso O = hadd (iso I)
```

```
I: 𝔹 * 𝔹, O: 𝔹 * 𝔹,
M : (𝔹 * 𝔹) * (𝔹 * 𝔹),
H0: M = (fun x ⇒ (x,x)) I
H1: fst O = uncurry ⊕ (fst M)
H2: snd O = uncurry ∧ (snd M)
===================
O = hadd I
```

Third, we move to the right-hand side of the excerpt: we massage the goal to make some `iso` commute with the `left`, `right` and `app` operations, in order to generalize the goal w.r.t. the isos. (Note that the user may be required to interact with Coq if different isos are applied to the same term in different equations.) Finally, the proof context deals only with high-level data-types, and functions operating on these. The user may then prove the "interesting" part of the lemma.

3.2 n-Bits Integers

From now, we use a dependently typed definition of n-bits integers, along the lines of the fixed-size machine integers of [16]. We omit the actual definitions of functions when they can be infered from the type. In the following, we prove that various (recursive) circuits implement the `carry_add` function (that adds two n-bit numbers and a carry).

```
Record word (n:nat) := mk_word {val : Z; range: 0 ≤ val < 2^n}.   (* W_n *)

Definition repr n (x : Z) : W_n := ...
Definition high n m (x : W_(n+m)) : W_m := ...
Definition low n m (x : W_(n+m)) : W_n := ...
Definition combine n m (low : W_n) (high : W_m) : W_(n+m) := ...
Definition carry_add n (x y : W_n) (b : 𝔹) : W_n * 𝔹 :=
    let e := val x + val y + (if b then 1 else 0) in (e mod 2^n,2^n ≤ e)

Definition Φ_x^n : Iso (sumn 1_x n → 𝔹) (W_n) := ...
```

3.3 Two Specifications of a 1-Bit Adder

A full-adder adds two 1-bit binary numbers with a carry-in, producing a 1-bit number and a carry-out, and is built from two half-adders. We present a diagram of this circuit, along with its formal definition in Fig. 6.

From this circuit, we can derive two specifications of interest. First, the meaning of the full-adder can be expressed in terms of a boolean function that mimics the truth-table of the circuit. Second, we can prove that this circuit actually implements the `carry_add` function up-to isomorphism. Both these specifications are proved using the aforementioned tactics, only the interesting parts differ.

```
Instance FADD_1 : Implement
(ι_cin • (ι_a • ι_b)) (* iso on inputs *)
(ι_sum • ι_cout) (* iso on outputs *)
FADD (fun (c,(x,y)) ⇒
  (x ⊕ (y ⊕ c),(x ∧ y) ∨ c ∧ (x ⊕ y))).
```

```
Instance FADD_2 : Implement
(ι_cin • (Φ_a^1 • Φ_b^1)) (* iso on inputs *)
(Φ_sum^1 • ι_cout) (* iso on outputs *)
FADD (fun (c,(x,y)) ⇒
  carry_add 1 x y c).
```

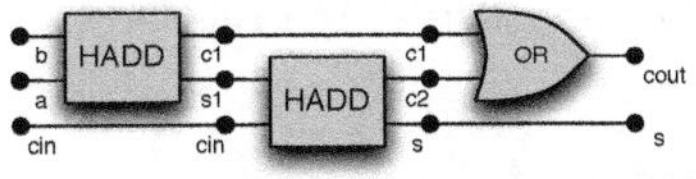


```
Context a b cin sum cout : string.
Program Definition FADD :
 ℂ (1_cin ⊕ (1_a ⊕ 1_b)) (1_sum ⊕ 1_cout) :=
  (ONE 1_cin & HADD a b "s1" "c1")
  ▷ ... (* associativity plug *)
  ▷ (HADD cin "s1" sum "c2" & ONE 1_"c1")
  ▷ ... (* associativity plug *)
  ▷ (ONE 1_sum & OR "c2" "c1" cout).
```

Fig. 6. Definition of a full-adder

```
Program Fixpoint RIPPLE cin a b cout sum n :
  ℂ (1_cin ⊕ sumn 1_a n ⊕ sumn 1_b n) (sumn 1_sum n ⊕ 1_cout) :=
match n with
  | O ⇒ ... (* Associativity *)
  | S p ⇒ ... ▷ (ONE (1_cin) & HIGHLOWS a b 1 p)
     ▷ ... ▷ (FADD a b cin sum "c" & ONE (sumn 1_a p ⊕ sumn 1_b p))
     ▷ ... ▷ (ONE (sumn 1_sum 1) & RIPPLE "c" a b cout s p)
     ▷ ... ▷ COMBINE sum 1 p & ONE (1_cout)
end.
```

Fig. 7. Implementation of the ripple-carry-adder from Fig. 1

3.4 Ripple-Carry Adder

We present in Fig. 7 the formal definition of the ripple-carry adder from Fig. 1 (we omit the rewiring plugs). This definition is based on two new circuits to split wires, and combine them. Indeed, to build a $1 + n$-bit adder, the lowest-order wire of each parameter is connected to a full-adder, while the n high-order wires of each parameter are connected to another ripple-carry adder. Conversely, the wires corresponding to the sum must be combined together. We use two plugs to define the HL and the COMBINE circuits.

```
Definition HL x n p : ℂ (sumn 1_x (n + p)) (sumn 1_x n ⊕ sumn 1_x p):= Plug ...
Definition COMBINE x n p : ℂ (sumn 1_x n ⊕ sumn 1_x p) (sumn 1_x (n + p)):= Plug ...
```

Then, we prove that these functions on wires implement their counterparts on words. These gates are then easily combined two-by-two to build HIGHLOWS and COMBINES that work with two sets of wires at the same time to get more economical designs (i.e., designs with fewer sub-components).

```
Lemma HL_Spec x n p: Implement
(Φ_x^{n+p}) (Φ_x^n • Φ_x^p) (HL x n p)
(fun x ⇒ (low n p x, high n p x)).
```

```
Lemma COMBINE_Spec x n p: Implement
(Φ_x^n • Φ_x^p) (Φ_x^{n+p}) (COMBINE x n p)
(fun x ⇒ (combine n p (fst x) (snd x))).
```

Finally, we prove by induction on the size of the circuit that it implements the high-level carry_add addition function on words. (Note that this is a high-level specification of the circuit: the carry_add function is not recursive and discloses nothing of the internal implementation of the device.) This boils down to the proof of lemma add_parts.

```
Lemma add_parts n m (xH yH: word m) (xL yL : word n) cin:
  let (sumL,middle) := carry_add n xL yL cin in
  let (sumH,cout) := carry_add m xH yH middle in
  let sum := combine n m sumL sumH in
  carry_add (n + m) (combine n m xL xH)(combine n m yL yH) cin = (sum,cout).
```

Instance RIPPLE_Spec cin a b cout sum n : Implement (RIPPLE cin a b cout s n)
  $(\iota_{\text{cin}} \bullet (\Phi^n_a \bullet \Phi^n_b))$ $(\Phi^n_{sum} \bullet \iota_{\text{cout}})$ (fun (c,(x,y)) ⇒ carry_add c x y).

This design is simple (a linear chain of 1-bit adders) and slow (each full-adder must wait for the carry-in bit from the previous full-adders). In the next subsection, we address the case of a more efficient adder, which is incidentally more complicated, and a better benchmark for our methodology.

3.5 Divide and Conquer Adder

A text-book [1] solution to improve on the delay of the previous ripple-carry adder is to use a divide and conquer scheme, and to compute both the sum when there is a carry-in, and the sum when there is no carry-in. It is then possible to compute at the same time the sum for the high-order bits, and the sum for the low order bits. Hence, we build a circuit that computes four pieces of data: s (resp. t), the n-bit sum of the inputs, assuming that there is no carry-in (resp. assuming that there is a carry-in); p the *carry-propagate* bit (resp. g the *carry-generate* bit), which is true when there is a carry-out of the circuit, assuming that there is a carry-in (resp. that there is no carry-in).

We provide a diagram in Fig. 8 that depicts the base case and the recursive case, but we omit the actual Coq implementation for the sake of readability. We prove that this circuit implements the following Coq function:

Definition dc n : $\mathbb{W}_{2^n} * \mathbb{W}_{2^n} \rightarrow \mathbb{B} * \mathbb{B} * \mathbb{W}_{2^n} * \mathbb{W}_{2^n}$:= fun (x,y) ⇒
let (s,g) := carry_add 2^n x y false in
let (t,p) := carry_add 2^n x y true in (g,p,s,t).

Again, this is a high-level specification w.r.t. the definition of the circuit: it does not disclose how the circuit compute its results (for instance, the dc function is not recursive). In a nutshell, the circuit computes in parallel the 4-uple of results for the high-order and low-order part of the inputs. Then, the propagate and generate bits for both parts can be combined by the PG circuit to compute the propagate and generate bits for the entire circuit. In parallel, the FIX circuit is made of two 2^{n-1}-bit multiplexers (easily defined with a fixpoint using 1-bit multiplexers), and updates the high-order parts of the sum, w.r.t. the propagate and generate carry-bits of the low-order adder.

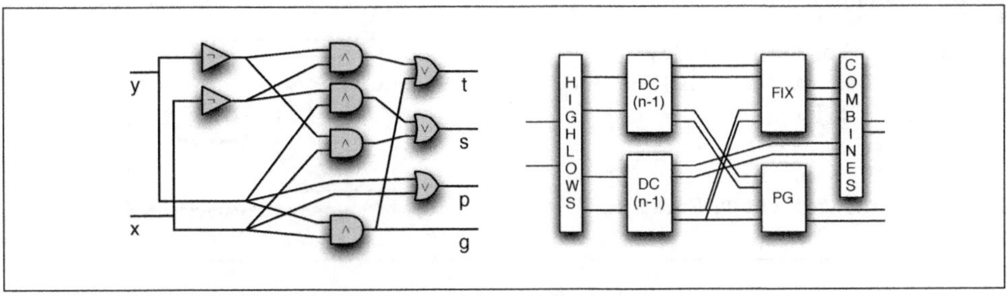

Fig. 8. Divide and conquer adder

4 Sequential Circuits: Time and Loops

While we have focused our case studies on combinational circuits, our methodology can be applied to sequential circuits, with or without the loops that were allowed in the syntax of circuits in §2.4. In this section, the wires carry streams of booleans (of type `nat` → 𝔹), and we assume a basic gate `DFF` that implements the `pre` function (in the particular case of booleans):

```
Definition pre {A} (d : A):
stream A → stream A := fun f t ⇒
match t with | 0 ⇒ d | S p ⇒ f p end.
```

```
Instance DFF_Realise_stream {a out}:
Implement (DFF a out) (ι_a) (ι_out)
(pre false).
```

A Buffer. A `DFF` delays one wire by one unit of time; a `FIFO` buffer generalizes this behavior in two dimensions, by chaining layers of `DFF` one after another. This circuit is simple, but is a good example for the use of high-level combinators.

These combinators capture the underlying regularity in some common circuit pattern, for instance replicating a sub-component in a serial or parallel manner.

```
Variable CELL : ℂ n n.
Fixpoint COMPOSEN k : ℂ n n :=
match k with
| 0 ⇒ Plug id
| S p ⇒ CELL ▷ (COMPOSEN p)
end.
```

```
Variable CELL : ℂ n m.
Fixpoint MAP k : ℂ (sumn 1_n k) (sumn 1_m k):=
match k with
| 0 ⇒ Plug id
| S p ⇒ CELL & (MAP p)
end.
```

We prove that the `COMPOSEN` combinator implements a higher-order iteration function, up-to isomorphism: if `CELL` implements a given function `f`, then `COMPOSEN k` implements the iteration of `f`. Respectively, we prove that the `MAP` circuit implements the higher-order `map` function on vectors. Hence, we define a `FIFO` buffer in one-line, and we prove that it implements the function below.

```
Definition FIFO x n k : ℂ (sumn 1_x k) (sumn 1_x k) := COMBINEN (MAP (DFF x x) k) n.
Definition fifo n k (v : stream (vector 𝔹 k)) : stream (vector 𝔹 k) :=
  fun t ⇒ if n < t then v (t − n) else Vector.repeat k false.
Remark useful_iso : sumn 1 n → stream 𝔹 ≅ stream (vector 𝔹 n) := ...
```

The proof of this specification relies on the above useful isomorphism between groups of wires that carries streams of booleans, and a stream of vectors of

```
Context a load out : string.
Program Definition REGISTER:
ℂ (1_load ⊕ 1_a) 1_out :=
@Loop (1_load ⊕ 1_a) 1_out 1_out
(... ▷ MUX2 a out load "in_dff"
▷ DFF "in_dff" out ▷ Fork2 1_out).
```

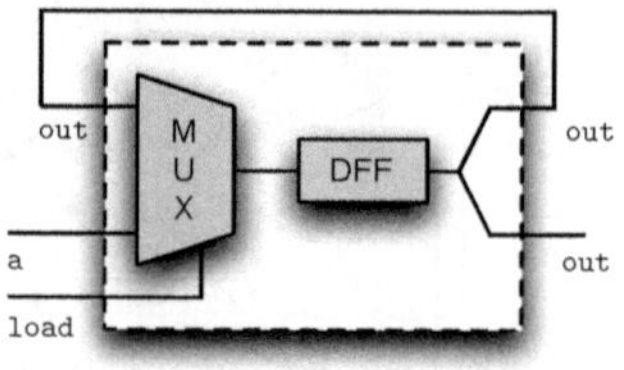

Fig. 9. A memory element

booleans. The proof that the circuit implements a function on streams is done in the same fashion as the proofs from the previous section.

A Memory Element. Our next goal is to demonstrate how we deal with state-holding structures. Hence, we turn to the implementation of a 1-bit memory element, as implemented in Fig. 9. The register is meant to hold 1-bit of information through time, which does not fit nicely in the `Implement` framework (we cannot easily express the meaning of the register in terms of a stream transformer). Hence, we use a relational specification through the use of `Realise`:

```
Instance Register_Spec : Realise
(... : 1_load ⊕ 1_a → stream 𝔹 ≅ stream 𝔹 * 𝔹) (ι_out) REGISTER
(fun (ins : stream (𝔹 * 𝔹)) (outs : stream 𝔹) ⇒
  outs = pre false (fun t ⇒ if  fst (ins t) then snd (ins t) else outs t)).
```

Here, the state of the register is stored inside the history of the stream (the previous values that were taken by the output). While we do not advocate that this is the nicest way to reason about state holding devices, we were able to prove this specification in the same fashion as the previous combinatorial devices. We leave a more thorough investigation of state-holding devices to future work.

5 Interesting Corollaries

We now turn on to investigating some interesting consequences of the use of a concrete data-type to represent circuits. First, we prove that the behavior of combinatorial circuits without delay can be lifted to the stream setting. Second, we build some functions (or interpretations [2]) that operates on circuits.

Lifting Combinatorial Circuits. The meaning relation is parametrized by the semantics of the basic gates. This can be put to good use to prove the functional correctness of some designs in the boolean setting, and then, to mechanically lift this proof of functional correction to the boolean stream setting (for the same set of gates). For instance, if a loop-less and delay-less circuit implements a function `f` in the boolean setting, we can prove that the very same circuit implements the function `Stream.map f` in the boolean stream setting.

Simulating and Checking Designs. One feature of our first-order encoding of circuits in Coq is that designs can be checked by simulation before attempting to prove them. This verification is done on the definition that will be proved later, allowing a seamless approach, and remains a valuable help to avoid dead-ends even if we cannot simulate circuit parametrized by a size. We define a simulation function `sim` that works on loop-free circuits, if the user provides a computational interpretation of each basic gate. For instance, we can simulate the adders of §3.

Delay and Pretty-Printing. Using the same ideas, we can build functions that compute the list of gates of circuits (with or without loops), or compute the length of the critical path in combinatorial circuits. While this is more anecdotal, and less directly useful than the previous simulation function, these functions are still interesting: one could, for instance, prove that some complex designs meet some time (or gate-count) complexity properties. (Note that is the only place where we exploit the finiteness of types.)

6 Comparisons with Related Work

Verifying Circuits with Theorem Provers. There has been a substantial amount of work on specification and verification of hardware in HOL. In [9,17], HOL is used as a hardware description language and as a formalism to prove that a design meets its specification. They model circuits as predicates in the logic, using a shallow-embedding that merges the architecture of a circuit and its behavior. Building on the former methodology, [21] defines a compiler from a synthetisable subset of HOL that creates correct-by-construction clocked synchronous hardware implementations of mathematical functions specified in HOL. This methodology allows the designer to focus on high-level abstraction instead of reasoning and verifying at the gate level, admitting the existence of some base low-level circuits (like the addition on words [13]). By contrast, our work complements their behavioral "correct by design" synthesis from a subset of the high-level language of the theorem-prover with structural verification of circuits.

In the Boyer Moore theorem prover (untyped, quantifier-free and first-order), Brock and Hunt proved the correctness of functions that generate correct hardware designs. They studied the correctness of an arithmetic and logic unit, parametrised by a size [14]. This verified synthesis approach was used to verify a microprocessor design [4]. While our proofs are not as automated, and our examples are less ambitious, we are able to prove higher-order circuits. Moreover, the dependent-types we use are helpful when defining complex circuits.

In Coq, Paulin-Mohring [18] proved the correctness of a multiplier unit, using a shallow-embedding similar to the methodology used in HOL: circuits are modelled as functions of the Coq language. More recently, [6] investigated how to take advantage of dependent types and co-inductive types in hardware verification: they use a shallow embedding of Mealy automata to describe sequential circuits. By contrast with both works, we use a deep-embedding of circuits in Coq, that makes explicit the definition of circuits. We still need to investigate the examples of sequential circuits studied in these papers.

Algebraic Definitions of Circuits. Circuit diagrams have nice algebraic properties. Lafont [15] studied the algebraic theory of boolean circuits and Hinze [12] studied the algebra of parallel prefix circuits. Both settings are close to ours: however the former focused on the algebraic structure of circuits, while the latter defined combinators that allows one to model (and prove correct using algebraic reasoning) all standard designs of a restricted class of circuits.

Functional Languages in Hardware Design. Sheeran [20] made a thorough review of the use of functional languages in hardware design, and of the challenges to address. Our work is a step towards one of them: the design and verification of parametrized designs, through the use of circuit combinators. Lava [2] is a language embedded in Haskell to describe circuits, allowing one to define parametric circuits or higher-order combinators. While much of our goals are common, one key difference is that our encoding of circuits in Coq avoids the use of bound variables (we use only combinators). Moreover, we use dependent types, that are required to deal precisely with parametric circuits. Finally, we prove the correctness of these parametric circuits in Coq, while verification in Lava is reduced to the verification of finite-size circuits.

7 Conclusion and Future Works

We have presented a deep-embeding of circuits in the Coq proof-assistant that allows to build and reason about circuits, proving high-level specifications through the use of type-isomorphism. We have demonstrated that dependent types are useful to prove automatically some well-formedness conditions on the circuits, and help to avoid time consuming mistakes. Then, we proved by induction the correctness of some arithmetic circuits of parametric size: this could not have been possible without mimicking the structure of the usual circuit diagrams to define circuit generators in Coq. The formal development accompanying this paper is available from the author's web-page [3].

In the immediate future, we plan to continue the case studies described in §3. In particular, we would like to investigate how to construct parallel prefix circuits in our framework [12,20], and to investigate combinational multipliers. In the more distant future, it would be interesting to study some front-ends to automatically generate some circuits: this could range to the reduction of the boilerplate inherent to the definition of plugs, to the compilation of circuits from automaton. A major inspiration on behavioral synthesis is the work of Ghica [7]. We also look forward to studying how our methodology applies to other settings than booleans or streams of booleans. For instance, if we move from booleans to the three-valued Scott's domain (unknown, true, false), we may interpret circuits in the so-called constructive semantics. We also hope that some of our methods could be applied to the probabilistic setting.

Acknowledgements. We thank D. Pous for his precious supervision, the reviewers who made useful comments, and J. Alglave, P. Boutillier, J. Planul who commented on a draft. We also thank J. Vuillemin for his encouragements.

References

1. Aho, A.V., Ullman, J.D.: Foundations of Computer Science. Computer Science Press, W. H. Freeman and Company (1992)
2. Bjesse, P., Claessen, K., Sheeran, M., Singh, S.: Lava: Hardware Design in Haskell. In: Proc. ICFP, pp. 174–184. ACM Press (1998)
3. Braibant, T.: `http://sardes.inrialpes.fr/~braibant/coquet` (June 2011)
4. Brock, B., Hunt Jr., W.A.: The DUAL-EVAL Hardware Description Language and Its Use in the Formal Specification and Verification of the FM9001 Microprocessor. Formal Methods in System Design 11(1), 71–104 (1997)
5. Brown, C., Hutton, G.: Categories, allegories and circuit design. In: Proc. LICS, pp. 372–381. IEEE Computer Society (1994)
6. Coupet-Grimal, S., Jakubiec, L.: Certifying circuits in type theory. Formal Asp. Comput. 16(4), 352–373 (2004)
7. Ghica, D.R.: Geometry of synthesis: a structured approach to VLSI design. In: Proc. POPL, pp. 363–375 (2007)
8. Gonthier, G., Mahboubi, A.: An introduction to small scale reflection in Coq. Journal of Formalized Reasoning 3(2), 95–152 (2010)
9. Gordon, M.: Why Higher-Order Logic is a Good Formalism for Specifying and Verifying Hardware. Technical Report UCAM-CL-TR-77, Cambridge Univ., Computer Lab (1985)
10. Hanna, F.K., Daeche, N., Longley, M.: Veritas$^+$: A Specification Language Based on Type Theory. In: Leeser, M., Brown, G. (eds.) Hardware Specification, Verification and Synthesis: Mathematical Aspects. LNCS, vol. 408, pp. 358–379. Springer, Heidelberg (1990)
11. Harrison, J.: A HOL Theory of Euclidean Space. In: Hurd, J., Melham, T. (eds.) TPHOLs 2005. LNCS, vol. 3603, pp. 114–129. Springer, Heidelberg (2005)
12. Hinze, R.: An Algebra of Scans. In: Kozen, D. (ed.) MPC 2004. LNCS, vol. 3125, pp. 186–210. Springer, Heidelberg (2004)
13. Iyoda, J.: Translating HOL functions to hardware. Technical Report UCAM-CL-TR-682, Cambridge Univ., Computer Lab (April 2007)
14. Hunt Jr., W.A., Brock, B.: The Verification of a Bit-slice ALU. In: Leeser, M., Brown, G. (eds.) Hardware Specification, Verification and Synthesis: Mathematical Aspects. LNCS, vol. 408, pp. 282–306. Springer, Heidelberg (1990)
15. Lafont, Y.: Towards an algebraic theory of boolean circuits. Journal of Pure and Applied Algebra 184, 257–310 (2003)
16. Leroy, X.: A formally verified compiler back-end. Journal of Automated Reasoning 43(4), 363–446 (2009)
17. Melham, T.: Higher Order Logic and Hardware Verification. Cambridge Tracts in Theoretical Computer Science, vol. 31. Cambridge University Press (1993)
18. Paulin-Mohring, C.: Circuits as Streams in Coq: Verification of a Sequential Multiplier. In: Berardi, S., Coppo, M. (eds.) TYPES 1995. LNCS, vol. 1158, pp. 216–230. Springer, Heidelberg (1996)
19. Sheeran, M.: μFP, A Language for VLSI Design. In: LISP and Functional Programming, pp. 104–112 (1984)
20. Sheeran, M.: Hardware Design and Functional Programming: a Perfect Match. J. UCS 11(7), 1135–1158 (2005)
21. Slind, K., Owens, S., Iyoda, J., Gordon, M.: Proof producing synthesis of arithmetic and cryptographic hardware. Formal Asp. Comput. 19(3), 343–362 (2007)

First Steps towards the Certification of an ARM Simulator Using Compcert*

Xiaomu Shi[1], Jean-François Monin[1,2], Frédéric Tuong[3], and Frédéric Blanqui[3]

[1] Université de Grenoble 1 - LIAMA
[2] CNRS - LIAMA
[3] INRIA - LIAMA

Abstract. The simulation of Systems-on-Chip (SoC) is nowadays a hot topic because, beyond providing many debugging facilities, it allows the development of dedicated software before the hardware is available. Low-consumption CPUs such as ARM play a central role in SoC. However, the effectiveness of simulation depends on the faithfulness of the simulator. To this effect, we propose here to prove significant parts of such a simulator, SimSoC. Basically, on one hand, we develop a Coq formal model of the ARM architecture while on the other hand, we consider a version of the simulator including components written in Compcert-C. Then we prove that the simulation of ARM operations, according to Compcert-C formal semantics, conforms to the expected formal model of ARM. Size issues are partly dealt with using automatic generation of significant parts of the Coq model and of SimSoC from the official textual definition of ARM. However, this is still a long-term project. We report here the current stage of our efforts and discuss in particular the use of Compcert-C in this framework.

1 Introduction

1.1 Simulation of Systems-on-Chip

Systems-on-Chip (SoC), used in devices such as smart-phones, contain both hardware and software. A part of the software is generic and can be used with any hardware systems, and thus can be developed on any computer. In contrast, developing and testing the SoC-specific code can be done only with this SoC, or with a *software executable model* of the SoC. To reduce the time-to-market, the software development must start before the hardware is ready. Even if the hardware is available, simulating the software on a model provides more debugging capabilities. The fastest simulators use native simulation. The software of the *target* system (i.e., the SoC) is compiled with the normal compiler of the computer running the simulator, but linked with special system libraries. Examples of such simulators are the Android and iOS SDKs.

In order to develop low-level system code, one needs a simulator that can take the real binary code as input. Such a simulator requires a model of the

* Work partly supported by ANR (SIVES, ANR-08-BLAN-0326-01).

J.-P. Jouannaud and Z. Shao (Eds.): CPP 2011, LNCS 7086, pp. 346–361, 2011.

processor and of its peripherals (such as UART, DMA, Ethernet card, etc). When simulating a smart-phone SoC, this kind of functional simulator can be from 1 to 100 times slower than the real chip. These simulators have other uses, for example, as reference models for the hardware verification. An error in the simulator can then mislead both the software and the hardware engineers. QEMU [2] is an open-source processor emulator coming with a set of device models; it can simulate several operating systems. Other open-source simulators include UNISIM [1] (accurately-timed) and SimSoC [8], developed by some colleagues, which is loosely-timed (thus faster). Simics [12] is a commercial alternative. The usual language to develop such simulators is C++, combined with the SystemC [13] and OSCI-TLM [15] libraries.

The work reported here is related to SimSoC.

1.2 The Need for Certification

Altogether, a functional simulator is a complex piece of software. SimSoC, which is able to simulate Linux both on ARM and PowerPC architectures at a realistic speed (over 10 Millions of instructions per second per individual core), includes about 60,000 lines of C++ code. The code uses complex features of the C++ language and of the SystemC library. Moreover, achieving high simulation speeds requires complex optimizations, such as dynamic translation [2].

This complexity is problematic, because beyond speed, *accuracy* is required: all instructions have to be simulated exactly as described in the documentation. There is a strong need to strengthen the confidence that simulations results match the expected accuracy. Intensive tests are a first answer. For instance, as SimSoC is able to run a Linux kernel on top of a simulated ARM, we know that many situations are covered. However it turned out, through further experiments, that it was not sufficient: wrong behaviors coming from rare instructions were observed after several months. Here are the last bugs found and fixed by the SimSoC team while trying to boot Linux on the SPEArPlus600 SoC simulator.

- After the execution of an `LDRBT` instruction, the contents of the base register (`Rn`) was wrong. It was due to a bug in the reference manual itself; the last line of the pseudo-code has to be deleted.
- After a data abort exception, the base register write-back was not canceled.
- Additionally, a half-word access to an odd address while executing some SPEArPlus600 specific code was not properly handled.

Therefore we propose here to certify the simulator, that is, to prove, using formal methods – here: the Coq proof assistant [5,3] – that it conforms to the expected behavior.

This is a long term goal. Before going to the most intricate features of a simulator such as SimSoC, basic components have to be considered first. We then decided to focus our efforts on a sensible and important component of the system: the CPU part of the ARMv6 architecture (used by the ARM11 processor family). This corresponds to a specific component of the SimSoC simulator, which was previously implementing the ARMv5 instruction set only. Rather than certifying this component, it seemed to us more feasible to design a new one directly in C, in such a way that it can be executed alone, or integrated in SimSoC (by including the C code in the existing C++ code). We call this new component `simlight` [4]. Combined with a small `main` function, `simlight` can simulate ARMv6 programs as long as they do not access any peripherals (excepted the physical memory) nor coprocessors. There is no MMU (Memory Management Unit) yet. Integrating it in SimSoC just requires to replace the memory interface and to connect the interrupts (IRQ and FIQ) signals.

The present paper reports our first efforts towards the certification of `simlight`. We currently have a formal description of the ARMv6 architecture, a running version of `simlight`, and we are in the way of performing correctness proofs. The standard way for doing this is to use Hoare logics or a variant thereof. Various tools exist in this area, for example Frama-C [6]. We chose to try a more direct way, based on an operational semantics of C; more precisely the semantics of Compcert-C defined in the Compcert project [10]. One reason is that we look for a tight control on the formulation of proof obligations that we will have to face. Another advantage is that we can consider the use of the certified compiler developed in Compcert, and get a very strong guarantee on the execution of the simulator (but then, sacrificing speed to some extent[1]).

Another interesting feature of our work is that the most tedious (hence error prone) part of the formalization – the specification of instructions – is automatically derived from the reference manual. It is well known that the formal specification of such big applications is the main weak link in the whole chain. Though our generators cannot be proved correct, because the statements and languages used in the reference manual have no formal semantics, we consider this approach as much more reliable than a manual formalization. Indeed, a mistake in a generator will impact several or all operations, hence the chances that it will be detected through a visibly wrong behavior are much higher than with a manual translation, where a mistake will impact only one (eventually rarely used) operation.

Note that after we could handle the full set of ARM instructions, our colleagues of the SimSoC team decided to use the same technology for the SimSoC itself: the code for simulating instructions in `simlight`, i.e., the current component dedicated to the ARM v6 CPU in SimSoC, is automatically derived using a variant of our generator, whereas the previous version for ARM v5 was manually written [4].

Fig. 1 describes the overall architecture. The contributions of the work presented in this paper are the formal specification of the ARMv6 instruction set and the correctness proof of a significant operation. More precise statements on the current achievements are given in the core of the paper.

Related Work. A fully manual formalization of the fm8501 and ARMv7 architectures are reported in [9] and [7]. The formal framework is respectively ACL2 and

[1] According to our first experiments, `simlight` compiled with Compcert is about 50 % to 70 % slower than `simlight` compiled with `gcc -O0`.

HOL4 instead of Coq, and the target is to prove that the hardware or microcode implementation of ARM operations are correct wrt the ARM specification. Our work is at a different level: we want to secure the simulation of programs using ARM operations. Another major difference is the use of automatic generation from the ARM reference manual in our framework, as stated above.

The rest of the paper is organized as follows. Section 2 presents the overall architecture of `simlight` and indicates for which parts of `simlight` formal correctness is currently studied. A informal statement of our current results is also provided there. Sections 3 and 4 present respectively our Coq formal reference model of ARM and the (Coq model of) Compcert-C programs targeted for correctness. A precise statement of our current results and indications on the proofs are given in Section 5. We conclude in Section 6 with some hints on our future research directions. Some familiarity with Coq is assumed in Sections 3, 4 and 5.

2 Main Lines of SimSoC-Cert

2.1 Overall Architecture

The overall architecture of our system, called SimSoC-Cert, is given in Fig. 1. More specifically, we can see the data flow from ARMv6 Reference Manual to the simulation code. Some patches are needed from the textual version of the reference manual because the latter contains some minor bugs. Three kinds of information are extracted for each ARM operation: its binary encoding format, the corresponding assembly syntax and its body, which is an algorithm operating on various data structures representing the state of an ARM: registers, memory, etc., according to the fields of the operation considered. This algorithm may call general purpose functions defined elsewhere in the manual, for which we provide a Compcert-C library to be used by the simulator and a Coq library defining their semantics. The latter relies on Integers.v and Coqlib.v from CompCert library which allows us, for instance, to manipulate 32-bits representations of words. The result is a set of abstract syntax trees (AST) and binary coding tables. These ASTs follow the structure of the (not formally defined) pseudo-code. Then two files are generated: a Coq file specifying the behavior of all operations (using the aforementioned Coq library) and a Compcert-C file to be linked with other components of SimSoC (each instruction can also be executed in standalone mode, for test purposes for instance). More details are provided in [4].

The decoding of ARM operations is not considered in the present paper: this is important and planned for future work, but is less urgent since we were already able to automatize the generation of intensive tests, as reported in [4]. We therefore focus first on the algorithmic body of operations. In order to state their correctness, we need Coq ASTs for the Compcert-C statements of `simlight`. The code generator directly generates such ASTs. Another option would be to work on a generated textual (ASCII) presentation of the Compcert-C code, but we prefer to avoid additional (and possibly unreliable) parsing step as far as

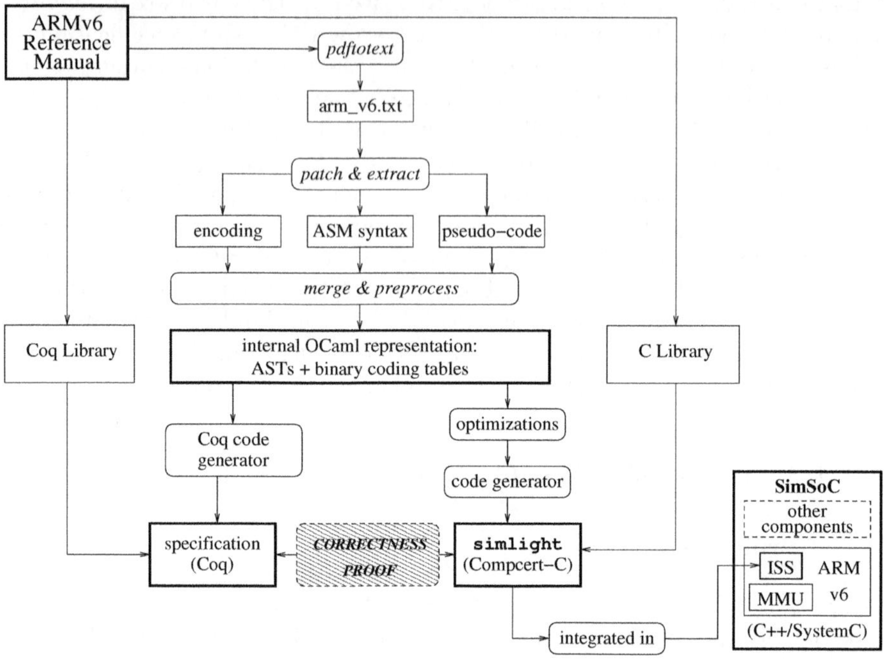

Fig. 1. Overall Architecture

possible. We will see in Section 4 that these ASTs are moreover presented in a readable form using suitable notations and auxiliary definitions.

The whole `simlight` project is currently well-compiled by Compcert (targeting Intel code) and gcc; moreover validation tests succeed completely with both simulators. The version of `simlight` compiled with Compcert can serve as a reference simulator, but for most purposes the version compiled with gcc is prefered for its higher speed.

2.2 Stating Correctness Theorems

Let us now present the purpose of the gray box of Fig. 1, which represents our main target.

The correctness of simulated ARM operations is stated with relation to the formal semantics of ARM as defined by our Coq library and partly automatically produced by the Coq code generator (the box called "specification" in Fig. 1). Note that ARM operations are presented in a readable way using suitable monadic constructs and notations: apart from the security provided by automatic generation, this greatly facilitates the comparison with the original pseudo-code of the reference manual. That said, it should be clear that the reference semantics of ARM is the Coq code provided in these files. Much effort has been spent in order to make them as clear and simple as possible.

In contrast, the Coq description of the behavior of corresponding operations (as simulated by SimSoC – Compcert-C programs) is far more complicated, though the superficial structure is quite similar. This will be detailed in Section 4. In particular, the memory model of the latter version is much more complex. In order to state correctness theorems, we define a relation between an ARM state represented by the Compcert-C memory model and another ARM state, as defined by the Coq formal semantics. Essentially, we have a projection from the former to the latter. Then for each ARM operation, we want a commutative diagram schematized in Fig. 2.

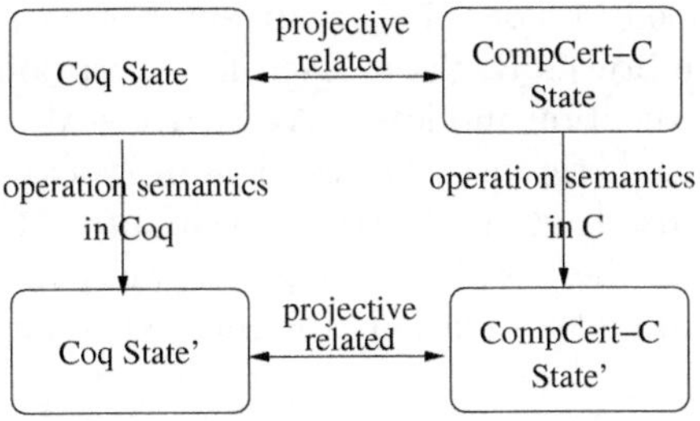

Fig. 2. Correctness of the simulation of an ARM operation

For now, our automatic generation tools operate completely, i.e., we have a Coq formal specification and a Compcert-C simulator for the full instruction set of ARM V6. About proofs, the relationship between the abstract and the concrete memory models is available; we can then state correctness theorems for all ARM operations. The work on correctness proofs themselves started recently. We considered a significant ARM operation called `ADC` (add with carry). Our main theorem (Theorem 1 in Section 5) states intuitevely that the diagram given in Fig. 2 commutes for `ADC`. Its proof is completed up to some axioms on library functions, details are given in Section 5.

3 ARM Model

3.1 Processor Behavior

A processor is essentially a transition system which operates on a state composed of registers (including the program counter) and memory. The semantics of its behavior amounts to repeat the following tasks: fetch the binary code at a given address, decode it as a processor operation and execute it; the last task includes the computation of the address of the next operation. The two main components of a processor simulator are then:

– The decoder, which, given a binary word, retrieves the name of an operation and its potential arguments.

– The precise description of transformations performed by an operation on registers and memory. In the reference manual of ARM, this is defined by an algorithm written in "pseudo-code" which calls low-level primitives for, e.g., setting a range of bits of a register to a given value. Some situations are forbidden or left unspecified. For ARM processors, this results in a so-called "UNPREDICTABLE" state. The best choice, for a simulator, is then to stop with a clear indication of what happens.

Let us illustrate this on a concrete example. Here is the original pseudo-code of the ADC (add with carry) operation of ARMv6. As most operations of ARM, this operation has an argument called cond which indicates whether the operation should be skipped or not. CPSR (Current Program Status Register) and SPSR (Saved Program Status Register, used for exception handling) are special registers related to execution modes of ARM; they also contain flags (N, Z, C and V) relevant to arithmetic instructions. The instruction has four parameters: S is a bit which specifies that the instruction updates CPSR, Rn is a register for the first operand, Rd is the destination register, shifter_operand specifies the second operand according to a (rather complicated) addressing mode.

```
A4.1.2 ADC
    if ConditionPassed(cond) then
        Rd = Rn + shifter_operand + C Flag;
        if S == 1 and d == 15 then
            if CurrentModeHasSPSR() then
                CPSR = SPSR;
            else UNPREDICTABLE
        else if S == 1 then
            N Flag = Rd[31];
            Z Flag = if Rd == 0 then 1 else 0;
            C Flag = CarryFrom(Rn + shifter_operand + C Flag);
            V Flag = OverflowFrom(Rn + shifter_operand + C Flag);
```

In the sequel, this version of ADC is referred to as ADC_pseudocode.

3.2 Coq Semantics of ARM Operations

Each operation O from the reference manual is mechanically translated to a corresponding Coq function named O_Coq. First we define a type state, which is a record with two fields Proc and SCC (System Control Coprocessor) containing respectively the components related to the main processor (status register CPSR, SPSR, other registers...) and the corresponding components related to the coprocessor, as well as the ARM memory model. Then we use a monadic style [14] in order to take the sequentiality of transformations on the state into account. Beyond the state st, two other pieces of information are handled: loc, which represent local variables of the operation and bo, a Boolean indicating whether the program counter should be incremented or not; they are registered in the following record which is used for defining our monad:

```
Record semstate := mk_semstate
  { loc : local
  ; bo : bool
  ; st : state }.

Inductive result {A} : Type :=
| Ok (_ : A) (_ : semstate)
| Ko (m : message)
| Todo (m : message).

Definition semfun A := semstate -> @result A.
```

Note that in general, every *O*_Coq functions terminate with `Ok` as value. However for "UNPREDICTABLE" states for example, errors are implicitely propagated with our monadic constructors for exceptions : `Ko` and `Todo`.

We present now the translation of `ADC_pseudocode`. To this effect, we introduce `_get_st`, a monadic function giving access to the current state `st` in its body, represented by this notation:

```
Notation "'<' st '>' A" := (_get_st (fun st => A))
  (at level 200, A at level 100, st ident).
```

This yields the following code for `ADC_Coq`:

```
(* A4.1.2 ADC *)
Definition ADC_Coq
   (S : bool) (cond : opcode)
   (d : regnum) (n : regnum) (shifter_operand : word) : semfun _ := <s0>
 if_then (ConditionPassed s0 cond)
   ([ <st> set_reg d (add (add (reg_content s0 n) shifter_operand)
                               ((cpsr st)[Cbit]))
   ;If (andb (zeq S 1) (zeq d 15))
     then    (<st> if_CurrentModeHasSPSR (fun em =>
       (<st> set_cpsr (spsr st em))))
     else         (if_then (zeq S 1)
       ([ <st> set_cpsr_bit Nbit ((reg_content st d)[n31])
       ; <st> set_cpsr_bit Zbit (if zeq (reg_content st d) 0 then repr 1
                                 else repr 0)
       ; <st> set_cpsr_bit Cbit
                (CarryFrom_add3 (reg_content s0 n) shifter_operand
                                 ((cpsr st)[Cbit]))
       ; <st> set_cpsr_bit Vbit (OverflowFrom_add3 (reg_content s0 n)
                                shifter_operand ((cpsr st)[Cbit])) ]))]).
```

4 ARM Operations in Simlight

In the right branch of the overall architecture (Fig. 1), we generate `simlight` according to the C syntax given by Compcert. Here, actually we have two presentations of the corresponding code. The first one is in a C source which is integrated into SimSoC (see [4] for more details), its contents is:

```
/* A4.1.2 ADC */
void ADC_simlight(struct SLv6_Processor *proc,
    const bool S,
    const SLv6_Condition cond,
    const uint8_t d,
    const uint8_t n,
    const uint32_t shifter_operand)
{
const uint32_t old_Rn = reg(proc,n);
if (ConditionPassed(&proc->cpsr, cond)) {
 set_reg_or_pc(proc,d,((old_Rn + shifter_operand) + proc->cpsr.C_flag));
 if (((S == 1) && (d == 15))) {
  if (CurrentModeHasSPSR(proc))
   copy_StatusRegister(&proc->cpsr, spsr(proc));
  else
   unpredictable();
  } else {
   if ((S == 1)) {
    proc->cpsr.N_flag = get_bit(reg(proc,d),31);
    proc->cpsr.Z_flag = ((reg(proc,d) == 0)? 1: 0);
    proc->cpsr.C_flag =
     CarryFrom_add3(old_Rn, shifter_operand, proc->cpsr.C_flag);
    proc->cpsr.V_flag =
     OverflowFrom_add3(old_Rn, shifter_operand, proc->cpsr.C_flag);
} } } }
```

This piece of code uses a function called `set_reg_or_pc` instead of `set_reg`: the latter also exists in `simlight` and the function to be used depends on tricky considerations about register 15, which happens to be the PC. More details about this are given in Section 4.1.

The second presentation is an AST according to a Coq inductive type defined in Compcert.

```
Definition ADC_Coq_simlight := (ADC, Internal
    {| fn_return := void; fn_params :=
      [proc -: `*` typ_SLv6_Processor;
       S -: uint8; cond -: int32; d -: uint8; n -: uint8;
       shifter_operand -: uint32];
      fn_vars := [ old_Rn -: uint32];
      fn_body :=
($ old_Rn`:○) `= (call (\reg`:○) E[\proc`:○; \n`:○] ○)`:○;;
`if (• (\ConditionPassed`:○) E[&((`*(\proc`:○)`:○)|cpsr`:○)`:○; \cond`:○] ○)
 then (• (\set_reg_or_pc`:○) E[\proc`:○; \d`:○; ((\old_Rn`:○)+•)+•:○] ○);;
 `if ((($ S`:○)==(#1`:○)`:○)&(($ d`:○)==(#15`:○)`:○)`:○)
  then `if (call (\CurrentModeHasSPSR`:○) E[\proc`:○] ○)
   then (call (\copy_StatusRegister`:○) E[&(•|cpsr`:○)`:○; •] ○)
   else (call ($ unpredictable`:○) E[] ○)
  else `if (($ S`:○)==(#1`:○)`:○)
   then ((($ proc`:○)|cpsr`:○)|N_flag`:○) `=
    (• (\get_bit`:○) E[(• (\reg`:○) E[\proc`:○; \d`:○] ○); #31`:○] ○)`:○;;
```

```
      ((($ proc`:∘)|cpsr`:∘)|Z_flag`:∘) `=
   (((• (\reg`:∘) • ∘)==(#0`:∘)`:∘)?(#1`:∘)`:(#0`:∘)`:∘)`:∘;;
      ((($ proc`:∘)|cpsr`:∘)|C_flag`:∘) `=
   (• (\CarryFrom_add3`:∘) E[•; •; (• (•|C_flag`:∘) ∘)] ∘)`:∘;;
      ((($ proc`:∘)|cpsr`:∘)|V_flag`:∘) `=
   (• (\OverflowFrom_add3`:∘) E[(• (\old_Rn`:∘) ∘); •; •] ∘)`:∘
  else skip
else skip |}).
```

The symbols "∘" and "•" do not belong to the real notations, they stand for types and sub-terms not represented here for the sake of simplicity. Indeed, an important practical issue is that Compcert-C ASTs include types everywhere, hence a naive approach would generates heavy and repetitive expressions at the places where ∘ occurs, thus making the result unreadable (and space consuming). We therefore introduce auxiliary definitions for types and various optimizations for sharing type expressions. We also introduce additional convenient notations, as shown above for `ADC_Coq_simlight`, providing altogether a C-looking presentation of the AST.

We plan to generate the first form from the AST using a pretty-printer. The following discussion is based on the AST presentation.

4.1 Differences with the Coq Model of ARM Operations

Although the encoding of operations in `simlight` and in the Coq semantics of ARM are generated from the same pseudo-code AST, results are rather different because, on one hand, they are based on different data types and, on the other hand, their semantics operates on different memory models. Therefore, the proof that the simulation of an operation in `simlight` behaves as expected according to the Coq semantics is not trivial.

In the Coq model of ARM, everything is kept as simple as possible. ARM Registers are presented by words, the memory is a map from address to contents, the initial value of parameters such as `Rn` is available for free – we are in an functional setting, etc. In contrast, `simlight` uses an imperative setting (hence the need to store the initial value of `Rn` in `Old_Rn`, for instance). More importantly, complex and redundant data structure are involved in order to get fast speed. For example, a 32 bits wide status register is defined as a data structure containing, for every significant bit, a field of Boolean type – internally, this is represented by a byte. A more interesting example is the program counter, which is the register 15 at the same time. As this register is sometimes used as an ordinary register, and sometimes used as the PC, the corresponding data structure implemented in `simlight` includes an array which stores all the registers and a special field `pc`, which is a pointer aliasing register 15. This register plays an important role in ARM architecture. Its value is used in the `may_branch` condition for simulating basic blocks [4]. And during the simulation loop, it is read many times. Note that this special field `pc` is read-only.

Moreover we have to work with the Compcert memory model of such data structures. This, model detailed in [11], introduces unavoidable complications in order to take low-level considerations, such as overlapping memory blocks into account. Another source of complexity is that, in a function call, a local variable takes a meaningful value only after a number of steps representing parameter binding. More details are given in Section 5.

Another important difference is that, in the Coq specification, the semantics is defined by a function wheras, in Compcert-C, the semantics is a relation between the initial memory and the final memory when evaluating statements or expressions.

4.2 Translation from Pseudo-code AST to Compcert-C AST

We describe here the mapping from the pseudo-code AST to Compcert-C AST. This translation is not only to Compcert-C AST, but more specifically to the Compcert-C AST for `simlight`. It makes use of an existing library of functions dedicated to `simlight`. For example in `ADC_pseudocode`, the occurrence of `CPSR` stands for an expression representing the contents of `CPSR` in the current state. But in `simlight`, this corresponds to a call to a library function `StatusRegister_to_uint32`. The translation deals with many similar situations.

Let us now sketch the translation process. Both the definitions of pseudo-code AST and Compcert-C AST include inductive types for expressions, statements and programs. Compcert-C expressions are limited to common programming operations like binary arithmetic operations, type cast, assignments, function calls, etc. For many constructors of pseudo-code AST the mapping is quite natural, but others require a special treatment: the ones which are specific to ARM, for representing registers, memory and coprocessor expressions, invocation of library functions, or bit range expressions. Those special expressions are translated to Compcert-C function calls. For example, the pseudo-code expression `Reg (Var n, Some m)`, designates the contents of register number `n` with the ARM processor mode `m`. In `simlight`, this becomes a call to `reg_m` with parameters `proc`, `n` and `m`.

In summary, the translation of expressions looks as follows:

```
let rec Transf_exp = function
  | Reg (e, m) -> Ecall reg_m ...
  | CPSR -> Ecall StatusRegister_to_uint32 ...
  | Memory (e, n) -> Ecall read_mem ...
  | If_exp (e1, e2, e3) -> Econdition ...
  | BinOp (e1, op, e2) -> Ebinop ...
  | Fun (f, e) -> Ecall f ...
  ...
```

For statements, we have a similar situation. Here, assignments require a special attention. For example in `ADC_pseudocode`, there is an assignment `CPSR=SPSR`. In `simlight`, this assignment is dealt with using a call to the function `copy_StatusRegister`. The corresponding Compcert-C AST embeds this call as an argument of the constructor `Sdo`.

In summary, the translation of statements looks as follows:

```
let rec Transf_stm = function
  | Assign (dst, src) -> Sdo (Ecall funct ...)
  | For (c, min, max, i) -> Sfor ...
  | If (e, i1, i2) -> Sifthenelse ...
  | Case (e, s, default) -> Sswitch ...
  ...
```

In our case, each operation is transformed to a Compcert-C program where there are no global variables, the function list contains only the function corresponding to the considered ARM operation (let us call it f, it is of course an internal function), and with an empty `main`. When the program is called, the global environment built for this program will only contain a pointer to f.

The translation from pseudo-code AST program to Compcert-C AST program has the following shape:

```
let Transformed_program =
  { vars = []
  ; functs = [ Internal (instr_id,
                 { fn_return = Tvoid
                 ; fn_params = ... (* operation parameters *)
                 ; fn_vars = ... (* operation local variables *)
                 ; fn_body = ... (Transf_stm ...) }) ]
  ; main = empty_main }
```

5 Current Proofs

On both sides, the Compcert-C `simlight` model and the Coq ARM model, the state of the processor is expressed by a big Coq term. In the Compcert-C `simlight` model, the processor state information is gathered in a data structure `SLv6_Processor`, which includes the MMU, the status registers `CPSR` and `SPSR`, the system coprocessor and the registers. In the Coq formal model of ARM, the processor state is represented by a value of type `result`, described in Section 3.2.

It is clearly possible to define a projection from a `SLv6_Processor` M to a `result` r. Then we say that M and r are *projective-related*, denoted by *proc_state_related* M r. The evil is in the details of the different type definitions, especially for the memory models. Here are the guiding ideas. Once a function such as `ADC_Coq_simlight` is called, parameters are allocated in memory, and a local environment is built. This local environment contains the mapping from identifiers to a memory block reference. For a variable of type `struct`, such as the ARM processor, the environment only yields an entry pointer to the structure. Here, the type information generated for our Compcert-C AST is needed in order to find fields inside Compcert-C memory, and to retrieve the processor model. The main function used there from Compcert is `load`. Its arguments are a memory M, a block b, an offset *ofs* and the type τ of the value to be loaded from b at *ofs*. Other variables who have a simple type like `int32`, are directly accessed by their identifier from the environment.

Let us now consider a specific instance of Fig. 2, applied to `ADC`. We choose it first because it is a typical ARM operation, which involves various ways of changing the processor state, and arithmetic calculations. Moreover, all data-processing operations have a very similar structure. If we prove the correctness of the `simlight` implementation of `ADC`, we can expect to automate the proofs for the others data-processing operations.

The proof exploits the formal operational semantics of Compcert-C, which is defined as a transition system

$$G, E \vdash \langle \text{piece of code} \rangle, M \stackrel{t}{\Rightarrow} out, M'$$

where G represents the global environment (constants) of the program, E represents the local environment, M and M' represent memory states, t is the trace of input/output events and *out* is the outcome. In our case, the piece of code is `ADC_Coq_simlight`, and the trace of input/output event (t) is empty: all function calls are internal calls. Compcert-C offers two kinds of operational semantics: small-step and big-step semantics. The latter is better suited to our needs because the statement of correctness, along the diagram in Fig. 2, relates states before and after the execution of the body of an operation. The precise statement of our theorem is as follows.

Theorem 1. *Let M and M' be the memory contents respectively before and after the simulation of* `ADC_Coq_simlight`*; similarly, let st and st' be the state of ARM in its formal model. If M and st are projective-related, as well as the arguments of the call to* `ADC`*, then M' and st' are projective-related as well. Formally, if:*

- *proc_state_related M (`Ok` st)*
- *similarly for the arguments of* `ADC`
- $G, E \vdash$ `ADC_Coq_simlight`$, M \stackrel{t}{\Rightarrow} out, M'$

then proc_state_related M' (`ADC_Coq` (arguments, st)).

In the Coq formal model of ARM, transitions are terminating functions returning a result of type `result`, as defined in section 3.

The proof process is driven by the structure of the operation body. Step by step, we observe the memory changes on the Compcert-C side and the state changes on the Coq side, and we check whether the relation still holds between the current Compcert-C memory state and the Coq state. To this effect, we apply theorems on load/store functions from Compcert [11]. Proof by computation does not work because the types involved are complex – they embed logical information – and many definitions are opaque.

In `ADC_Coq`, conditional expressions and function calls for getting values have no side effect on the state. On the Compcert-C side, declaring a local variable in a function has no impact on the memory model of the processor. The state may only change when a function for setting values is called, like `set_reg`, `copy_StatusRegister`, or assignment of bits in register fields. Such calls will return a new memory state on the Compcert-C side and a new `Ok` state on the Coq side. We use small-step semantics for such steps.

Now we need some lemmas for these proof steps. Lemmas can be organized into four kinds. We give an instance of each kind.

Lemma 1. *The conditional expression S==1 has no effect on Compcert-C memory state:*
if $G, E \vdash \texttt{condition}_C ?a1 : a2, M \overset{E0}{\Rightarrow} vres, M'$,
then M=M'.

Lemma 1 is easy to prove by some inversions. All lemmas of this kind have been discharged.

Lemma 2. *The conditional expression S==1 has the same result in the Compcert C model as in the Coq model:*
if $G, E \vdash \texttt{condition}_C ?a1 : a2, M \overset{E0}{\Rightarrow} vres, M'$*:*
- and if $is\_true\ vres$*, then* $condition_{Coq} = true$
- and if $is\_false\ vres$*, then* $condition_{Coq} = false$.

To prove lemma 2, we need to apply small-step semantics, to check the type of `S` and the value of the Boolean result `vres`. Note that in Compcert-C, non-zero integer, non-zero floats and non-null pointer can be interpreted as the Boolean value true, which adds some complexity in the proof.

The proof is by case analysis according to the type of `vres`. As the expression involves a parameter (S), the projective relation about this parameter between Compcert-C memory and the formal model of ARM is required.

All lemmas of this kind have been discharged.

A lemma of the two next kinds is stated for each `simlight` library function which changes the state, e.g., `set_reg`.

Lemma 3. *If proc_state_related M* (`Ok` *st*),
and if $G, E \vdash \texttt{set\_reg}_c(proc, reg\_id, data), M \overset{E0}{\Rightarrow} vres, M'$,
then proc_state_related M' ($\texttt{set\_reg}_{Coq}$ *st*).

For the moment, such lemmas are considered as axioms on the library. In order to be properly stated, we need the Compcert-C ASTs of such library functions, which are not automatically generated. We have 6 lemmas/axioms of this kind for `ADC`.

The next lemma is stated for a given call to `set_reg` in the body of the function `ADC_Coq_simlight` and a parameter P of `ADC_Coq_simlight` which is not used as an argument of `set_reg`.

Lemma 4. *After the call to* `set_reg`*, the value of* P *remains unchanged:*
if $G, E \vdash \texttt{set\_reg}_c(proc, reg\_id, data), M \overset{E0}{\Rightarrow} vres, M'$,
then $P(M) = P(M')$.

Lemma 4 can be proved with the help of theorems of Compcert on “load after store”. A typical proof step looks like:

Table 1. Sizes (in number of lines)

Original ARM ref man (txt)	49655
ARM Parsing to an OCaml AST	1068
Generator (Simgen) for ARM and SH with OCaml and Coq pretty-printers	10675
Generated C code for Simlight ARM operations	6681
General Coq libraries on ARM	1569
Generated Coq code for ARM operations	2068
Generated Coq code for ARM decoding	592
Proof script on `ADC`	1461

If we store a value v on block b ($\texttt{store}(M1, \tau, b, \mathit{ofs}, v) = M2$), then the contents of block b' remains unchanged ($\texttt{load}(\tau', M2, b', \mathit{ofs}') = \texttt{load}(\tau', M1, b', \mathit{ofs}')$) for any type τ' and offset ofs', which makes the accesses disjoint ($b' \neq b$ or $\mathit{ofs} + |\tau| \leq \mathit{ofs}'$ or $\mathit{ofs}' + |\tau'| \leq \mathit{ofs}$).

As for lemmas 3, we need additional axioms on `simlight` library functions.

Our current result is that, with the help of these lemmas, we have a complete correctness proof for `ADC` (Theorem 1).

The whole proof structure of this theorem and all twenty lemmas of kinds 1 and 2 were completed within 2 weeks. The 10 remaining lemmas, of kinds 3 and 4, should require a similar effort. Here, we first need to generate Compcert-C ASTs for the relevant library functions using the C parser available in Compcert.

6 Conclusion

The trust we may have in our result depends on the faithfulness of its statement with relation to the expected behavior of the simulation of `ADC` in `simlight`. It is mainly based on the manually written Coq and C library functions, the translators written in OCaml described in Section 2 (including the pretty-printer for Coq), the final phase of the Compcert compiler, and the formal definition of *proc_state_related*.

The current development is available online[2]. Figures on the size of our current development are given in Table 1.

In the near future, we will extend the work done on `ADC` to all other operations. The first step will be to design relevant suitable tactics, from our experience on `ADC`, in order to shorten a lot the current proof and make it easier to handle and to generalize. We are confident that the corresponding work on the remaining ARM operations will then be done much faster, at least for arithmetical and Boolean operations.

Later on, we will consider similar proofs for the decoder – as for the body of operations, it is already automatically extracted from the ARM reference manual. Then a proven simulation loop (basically, repeat decoding and running operations) will be within reach.

[2] `http://formes.asia/media/simsoc-cert/`

In another direction, we also reuse the methodology based on automatic generation of simulation code and Coq specification for other processors. The next one which is already considered is SH4. In fact, the same approach as the ARMv6 has been followed, and a similar Coq representation can currently be generated from the SH4 manual. Moreover, as the SH pseudo-code is simpler than the ARM, we are hence impatient to work on its equivalence proof.

Acknowledgement. We are grateful to Vania Joloboff and Claude Helmstetter for their many explanations on SimSoC. We also wish to thank the anonymous reviewers for their detailed comments and questions.

References

1. August, D., et al.: Unisim: An open simulation environment and library for complex architecture design and collaborative development. Computer Architecture Letters 6(2), 45–48 (2007)
2. Bellard, F.: QEMU, a fast and portable dynamic translator. In: ATEC 2005: Proceedings of the Annual Conference on USENIX Annual Technical Conference, Berkeley, CA, USA, pages 41. USENIX Association (2005)
3. Bertot, Y., Castéran, P.: Interactive Theorem Proving and Program Development. Coq'Art: The Calculus of Inductive Constructions. Texts in Theoretical Computer Science. Springer, Heidelberg (2004)
4. Blanqui, F., Helmstetter, C., Joloboff, V., Monin, J.-F., Shi, X.: Designing a CPU model: from a pseudo-formal document to fast code. In: Proceedings of the 3rd Workshop on Rapid Simulation and Performance Evaluation: Methods and Tools, Heraklion, Greece (January 2011)
5. Coq Development Team. The Coq Reference Manual, Version 8.2. INRIA Rocquencourt, France (2008), `http://coq.inria.fr/`
6. Correnson, L., Cuoq, P., Puccetti, A., Signoles, J.: Frama-C User Manual, Release Boron-20100401. In: CEA LIST, Software Reliability Laboratory, Saclay, France (2010)
7. Fox, A.C.J., Myreen, M.O.: A Trustworthy Monadic Formalization of the ARMv7 Instruction Set Architecture. In: ITP, pp. 243–258 (2010)
8. Helmstetter, C., Joloboff, V., Xiao, H.: SimSoC: A full system simulation software for embedded systems. In: IEEE (ed.), OSSC 2009 (2009)
9. Hunt Jr., W.A. (ed.): FM8501: A Verified Microprocessor. LNCS (LNAI), vol. 795. Springer, Heidelberg (1994)
10. Leroy, X.: Formal verification of a realistic compiler. Communications of the ACM 52(7), 107–115 (2009)
11. Leroy, X., Blazy, S.: Formal Verification of a C-like Memory Model and Its Uses for Verifying Program Transformations. J. Autom. Reason. 41(1), 1–31 (2008)
12. Magnusson, P.S., et al.: Simics: A full system simulation platform. Computer 35(2), 50–58 (2002)
13. Open SystemC Initiative. SystemC v2.2.0 Language Reference Manual (IEEE Std 1666-2005) (2006), `http://www.systemc.org/`
14. Peyton Jones, S.: Tackling the Awkward Squad: monadic input/output, concurrency, exceptions, and foreign-language calls in Haskell. Online lecture notes (2010)
15. OSCI SystemC TLM 2.0.1 (2007), `http://www.systemc.org/`

Full Reduction at Full Throttle

Mathieu Boespflug[1], Maxime Dénès[2], and Benjamin Grégoire[2]

[1] McGill University
mboes@cs.mcgill.ca
[2] INRIA Sophia Antipolis - Méditerranée
{Maxime.Denes,Benjamin.Gregoire}@inria.fr

Abstract. Emerging trends in proof styles and new applications of interactive proof assistants exploit the computational facilities of the provided proof language, reaping enormous benefits in proof size and convenience to the user. However, the resulting proof objects really put the proof assistant to the test in terms of computational time required to check them. We present a novel translation of the terms of the full Calculus of (Co)Inductive Constructions to OCaml programs. Building on this translation, we further present a new fully featured version of Coq that offloads much of the computation required during proof checking to a vanilla, state of the art and fine tuned compiler. This modular scheme yields substantial performance improvements over existing systems at a reduced implementation cost.

The work presented here builds on previous work described in [11], but we place particular emphasis in this paper on the fact that this scheme is in fact an instance of untyped normalization by evaluation [8, 14, 1, 4].

Introduction

Many proof assistants share with many programming language a common basis in the typed λ-calculus. Systems in the lineage of Church's original higher order logic reuse the typed λ-calculus as both the language for the objects of the discourse (terms) and the language of formulae and propositions about these objects. Following Automath dependently typed systems push the envelope even further to unify objects of discourse, propositions, and proofs, all as terms of the typed λ-calculus. Either way, seeing as the λ-calculus is also a programming language, both strands of proof assistants thus intrinsically support discoursing about programs. In dependently typed programming languages, this is made plain by the following inference rule,

$$\frac{\Gamma \vdash M : A \qquad A =_\beta B}{\Gamma \vdash M : B}\ (\text{conv})$$

which allows replacing part of a proposition for another if an oracle agrees that the two propositions are related to each other through computation (i.e. they are *convertible*).

As a matter of fact, in dependently typed theories, support for reasoning about programs is so good that new proof methodologies have emerged in recent years [5, 16, 9, 12, 10] to establish results in pure and applied mathematics by reducing them to the computation of a program proven to be correct. The point of this approach is to turn many of the deduction steps into computation steps instead, which

J.-P. Jouannaud and Z. Shao (Eds.): CPP 2011, LNCS 7086, pp. 362–377, 2011.

do not appear explicitely in the proof term, so as to yield smaller proofs whose size is independent of the number of computation steps required. However, efficiency of proof checking hinges on the complexity of the algorithm realized by the computation steps and on the performance of the evaluator used to carry them out.

Proof by reflection [5,13] is one such methodology. Consider subclasses $\mathscr{C}$ of the class of all propositions. The idea is that the propositions of somesuch subclasses $\mathscr{C}$ are better established by appeal to some metalogical property shared by these propositions rather than to write independent proofs for each of the properties in $\mathscr{C}$. For instance, the standard library of Coq gives $(<)$ as an inductively defined predicate. But notice that ground instances of $n < m$ are provable only if the computation $f(n,m)$ yields the boolean value true as an answer, where

$$f(x,y) \hat{=} \begin{cases} \mathsf{true} & \text{if } \mathsf{max}(x+1-y,0)=0 \\ \mathsf{false} & \text{otherwise} \end{cases}$$

One can prove $f(x,y) = \mathsf{true}$ by reflexivity in Coq because this proposition is equivalent to $\mathsf{true} = \mathsf{true}$ by the inference rule (conv) above. Now if we have at hand a proof p showing that $\forall xy.\ f(x,y) = \mathsf{true} \Rightarrow x < y$, then the term $p\ x\ y\ \mathsf{refl}$ is a proof of $x < y$. This is a single, generic proof that works for all ground inequalities and that is independent of the sizes of x and y. This style of reflection scales up in the large to complicated semi-decision procedures, as in the proof of the four colour theorem [9], and is promoted in the small by the SSReflect methodology [10]. Proofs by reflection also extend to non-ground propositions via *quoting* (lifting to an object representation) and syntactic transformations (of which a generic proof of correctness is established) [5,12].

Existing functional programming language runtime environments would be most appropriate to decide the convertibility condition which lies at the heart of most every proof by reflection. However, this strategy faces two problems:

- In a conventional functional programming language, one can only analyze and compare values at base type (lists, integers,...). Inhabitants of function types are effectively black boxes that cannot be compared.
- Programs of a functional language are always closed terms, whereas in our context, we may have to compare (and so to evaluate) open terms (with free variables referencing assumptions in the typing context).

These constraints allow runtimes to only deal with substitutions of closed terms into closed terms, which allows for efficient implementation strategies and elides any issues with name capture. Normal forms cannot always be reached purely by closed substitutions. Only weak head normal forms are computed, meaning that they are only *weakly reduced*. Our objective is to implement *full reduction* to normal form of potentially open terms of dependently typed terms at higher order type, with much of the same performance profile and optimizations as might be available in a mature, optimizing compiler for run-of-the-mill programming languages. We conspicuously avoid modifying any existing compiler to do so, let alone writing our own, by reusing as-is the OCaml compiler.

This design choice follows a long history of previous approaches to normalization using off-the-shelf components. Normalization by Evaluation (NbE) is one such appealingly elegant family of approaches. The point there is to obtain the sought normal forms not by the usual iteration of a one-step reduction relation, but rather by constructing a *residualizing* model D of the set Λ of terms, given by a denotation $[\![\_]\!] : \Lambda \to D$, supporting an inverse $\downarrow: D \to \Lambda$ functional [3] (called *reification*) such that

1. if $t \longrightarrow t'$ then $[\![t]\!] = [\![t']\!]$ (soundness) ;
2. if t is a term in normal form, then $\downarrow [\![t]\!] = t$ (reproduction).

Then it is easy to see that if $t \longrightarrow^* t'$ where t' is normal, $\downarrow [\![t]\!] = \downarrow [\![t']\!] = t'$, so composing the interpretation with reification gives a normalization function for terms whose normal form exists. In typed NbE and type directed partial evaluation (TDPE) [7], reification is actually done in a type directed way. But such approaches need to be adapted as the type system changes and scaling them up to powerful type systems such as the Calculus of Inductive Constructions (CIC) [6,17] with a hierarchy of universes as implemented in Coq is non-trivial. Untyped variants of NbE have been proposed [8,14,1,4], but the generality of the untyped approaches have so far come at the cost of adding tags to the interpretation of object syntax to deeply embed it into the host language. While the performance penalty of this tagging can be mitigated in many circumstances [4], some of the interpretive overhead introduced by the tagging invariably remains. Memory allocation and locality of code is negatively impacted and some simple common compiler optimizations (such as uncurrying) need to be redone at the level of the object syntax interpretation.

The implementation of full reduction that we describe here matches the generality of untyped NbE, since it works for all (open) terms of the λ-calculus. It also matches the performance of typed NbE, since the interpretation of terms introduces zero interpretive overhead. One no longer needs to choose between full generality and full performance. The approach used here is inspired by [11], but unlike this earlier work which requires modified versions of the stock OCaml compiler and virtual machine, we achieve full genericity in the underlying evaluator. We do not need to maintain a custom version of this underlying evaluator — meaning a better separation of concern between writing proof assistants[1] and writing compilers.

The structure of this paper is as follows. In Section 1, we offer as a first contribution a unifying view of untyped normalization by evaluation and of the normalization algorithm of [11], showing that the latter can be seen as an instance of the former. We then show how to implement this algorithm by translation of the source language to a functional language, without modifying the underlying compiler or virtual machine, and generalize the algorithm to full reduction of terms of the CIC (Section 2). We proceed to adding coinductive datatypes (Section 3). These encodings are, to the best of our knowledge, novel features in an NbE algorithm. In Section 4, we show through a number of high-level and real world use cases that our solution fares very favourably relative to existing implementations of the conversion rule in Coq, typically attaining a fivefold speedup.

[1] Our implementation is available in a development branch of Coq at http://github.com/maximedenes/native-coq, and in a future release of the proof assistant.

1 Implementations for the λ-Calculus

1.1 Calculus of Symbolic Reduction

Finding the normal form of a term t by evaluation hinges upon distinguishing the head of the value of t, to continue reduction in the body of the abstraction if the value of t is of the form $\lambda x.\ t'$. However, t' is not, in general, a closed term (x can appear free), which evaluators for functional programming languages cater for exclusively.

The symbolic calculus of [11] introduces a new kind of value to represent free variables and terms whose evaluation is "stuck" because of a free variable in head position, as well as a new reduction rule. Hence, the weak reduction of open terms can be simulated by weak reduction of closed symbolic terms. The syntax of the symbolic calculus is as follows:

$$\begin{aligned} \textbf{Term} \ni t &::= x \mid t_1\, t_2 \mid v \\ \textbf{Val} \ni v &::= \lambda x.t \mid [\tilde{x}\, v_1 \ldots v_n] \end{aligned}$$

Where $[\tilde{x}\, v_1 \ldots v_n]$ is a value, called an accumulator, representing the free variable x applied to the arguments $v_1 \ldots v_n$. The reduction rules of the calculus are:

$$\begin{aligned} (\lambda x.t)\, v &\rightarrow t\{x \leftarrow v\} && (\beta_v) \\ [\tilde{x}\, v_1 \ldots v_n]\, v &\rightarrow [\tilde{x}\, v_1 \ldots v_n\, v] && (\beta_s) \\ \Gamma(t) &\rightarrow \Gamma(t') \quad \text{if } t \rightarrow t' \quad (\text{with } \Gamma ::= t\,[\,] \mid [\,]\, v) && \text{context} \end{aligned}$$

The β_v rule is the standard β-reduction rule in call by value[2], the context rule allows reduction in any subterm which is not an abstraction (so called weak reduction). Finally, the β_s rule expresses that free variables behave like boxes accumulating their arguments when applied.

We write $\stackrel{*}{\rightarrow}$ for the reflexive and transitive closure of $\rightarrow$. We define the value $\mathcal{V}(t)$ of a closed symbolic term t as the normal form of t for the relation $\stackrel{*}{\rightarrow}$. Since infinite reduction sequences are possible, this normal form does not necessarily exist. However, if the normal form exists then it must be a value because the reduction cannot get stuck on closed symbolic terms [11].

Given a translation from terms t to symbolic terms $\ulcorner t \urcorner$, we can now express precisely how to get the normal form of t with respect to the β rule, by iteration of weak symbolic reduction and readback: first, compute $\mathcal{V}(\ulcorner t \urcorner)$ by weak symbolic reduction (equation 1); second, inspect the resulting value and recursively normalize the subterms (*readback*).

$$\mathcal{N}(t) = \mathcal{R}(\mathcal{V}(t)) \tag{1}$$

$$\mathcal{R}(\lambda x.t) = \lambda y.\mathcal{N}((\lambda x.t)\,[\tilde{y}]) \text{ where } y \text{ is fresh} \tag{2}$$

$$\mathcal{R}[\tilde{x}\, v_1 \ldots v_n] = x\, \mathcal{R}(v_1) \ldots \mathcal{R}(v_n) \tag{3}$$

[2] We commit to call-by-value in the rest of this paper, but we could equally well have chosen any other standard evaluation strategy, such as call-by-need. Anyway, since CIC is strongly normalizing and confluent, the result is independent of the evaluation strategy.

The normalization algorithm takes a closed symbolic term and returns a λ-term in normal form. If the value is an accumulator the readback function simply injects the symbolic variable $\tilde{x}$ into the source variable x and applies it to the readback of the accumulated arguments (equation 3).

If the value is a function $\lambda x.\ t$, the readback function simply normalizes the application of the function to a fresh[3] symbolic variable $[\tilde{y}]$ (equation 2). Note that the application reduces in one step to $t\{x \leftarrow [\tilde{y}]\}$, so $\mathcal{N}((\lambda x.t)[\tilde{y}])$ will return the normal form of t modulo renaming of x into y. This trick is key to using stock evaluators for symbolic reduction: it means that we do not need to be able to peek inside a function to examine its body. Functions are black boxes whose only observable behaviour is through applying it to some value, as usual.

In [11], the authors use a modification of the OCAML abstract machine to get an efficient reducer for the symbolic calculus. This come at a price: while the implementation effort is reduced, the new abstract machine and associated compiler must be maintained separately, *ad infinitum*. Furthermore, the efficiency is limited to what abstract machines can afford us, which is often much less than compilation to native code. This paper could have been about a new ahead-of-time or just-in-time native code compiler for this abstract machine, but such a specialist task should hardly be the concern of developers of proof assistants and would likely achieve little over reusing as is an existing compiler to native code for an existing language. In the next section, we present a modular interface that we instantiate in two ways, giving two implementations of the above symbolic calculus, both times as vanilla OCAML programs. The first is inspired by higher order abstract syntax (HOAS) and the second uses the reflective capabilities of OCAML.

1.2 Abstract Setting

To perform the normalization of a source term (here a λ-term), we first translate it into a program of our target language (here OCAML) that computes a value. Then, by inspection of the head of the obtained value, we will readback the value into a source term in normal form. To do so we assume that we have a module for values with the following interface:

```
module type Values = sig
  type t
  val app : t -> t -> t
  type atom =
    | Var of var
  type head =
    | Lam of t -> t
    | Accu of atom * t list
  val head : t -> head
  val mkLam : (t -> t) -> t
  val mkAccu : atom -> t
end
```

[3] The freshness condition here can be made precise by using de Bruijn *levels* for symbolic variables.

The first component of the signature is the type representing values. We assume that given two values we are able to compute the value corresponding to the application of one to the other (the app function). Secondly we assume that we are able to discriminate on the head of any value (the head function). In the case of the λ-calculus, a value is either a λ-abstraction (constructor Lam) or an accumulator which is an atom applied to its arguments[4] (constructor Accu). Finally, we assume an injection function for atoms. We assume that term abstractions are represented as OCAML functions, with mkLam injecting functions to terms. The following laws should hold: head (mkAccu a) = Accu(a,[]) and head (mkLam f) = Lam f.

The compilation of a λ-term to an OCAML program is easily given, as follows:

$$\ulcorner x \urcorner^B = \begin{cases} x & \text{if } x \in B \\ \text{mkAccu}(\text{Var } x) & \text{otherwise} \end{cases}$$

$$\ulcorner \lambda x.t \urcorner^B = \text{mkLam (fun } x \rightarrow \ulcorner t \urcorner^{B \cup \{x\}})$$

$$\ulcorner t_1\ t_2 \urcorner^B = \text{app } \ulcorner t_1 \urcorner^B \ulcorner t_2 \urcorner^B$$

The compiler takes as input a λ-term t and a set of bound variables B, returning an OCAML program computing the weak head normal form of t, viewed as a symbolic value. Bound variables in t are compiled to OCAML variable. Otherwise, the code builds the accumulator $[\tilde{x}]$ corresponding to the symbolic variable $\tilde{x}$. The compilation of an abstraction builds an OCAML function (the set B of bound variables is extended with bound variable x). For the application we use the app function to perform application of the functional part to the argument.

The normalization algorithm is thus a straightforward translation of the normalization algorithm presented in the previous section.

$$\mathcal{N}_\Lambda\ t = \mathcal{R}_V\ \ulcorner t \urcorner^\emptyset$$

$$\mathcal{R}_V\ v = \mathcal{R}\ (\text{head } v)$$

$$\mathcal{R}\ (\text{Lam } f) = \lambda y.\mathcal{R}_V\ (f\ (\text{mkAccu (Var } y)) \text{ where } y \text{ is fresh}$$

$$\mathcal{R}\ (\text{Accu}(a, [v_n; \ldots v_1])) = (\mathcal{R}_\Lambda\ a)\ (\mathcal{R}_V\ v_1) \ldots (\mathcal{R}_V\ v_n)$$

$$\mathcal{R}_A\ (\text{Var } x) = x$$

1.3 Tagged Normalization

A natural implementation for the type t of the Values module, suggested in [4, 1], consists in using the type head directly:

```
type t = head
let head v = v
let app t v = match t with
  | Lam f -> f v
  | Accu(a, args) -> Accu(a,v::args)
let mkLam f = Lam f
let mkAccu a = Accu(a,[])
```

[4] Arguments are stored in reverse order to allow efficiently extending the list of arguments when applying an accumulator.

In this case, much of the implementation follows immediately from the laws above. If the first argument models an abstraction, the app function must unbox the first argument to get the representing function and perform the substitution (modelled as application of OCAML values). Otherwise, the first argument is an accumulator. The fact that an accumulator was applied to v is recorded by extending the list of arguments of the accumulator with this new argument.

The representation we described features a succinct implementation. However, explicit tagging of all values to inform about the shape of their heads entails a sizeable performance penalty. Several optimizations are suggested in [4] to mitigate the impact of the costly representation of application, such as uncurrying or specialization of constructors. Although the improvement is significant, we remove the cost of tagging entirely by doing away with tags. We show in the next section how to encode accumulators as special infinite arity functions. Applications then no longer need to dispatch on heads, given this uniform calling convention for both ordinary functions and accumulators.

1.4 Tagless Normalization

Already Grégoire and Leroy [11] remark that an accumulator can be viewed as a primitive function keeping a trace of all arguments fed to it. We show how to write such a function within OCAML, without extending the language with a new primitive.

The OCAML runtime manipulates a variety of different kinds of values: integers, floats, or pointers to arrays, constructed values of a user defined datatype, closures, etc. At a low-level, integers are distinguished from pointers to heap allocated blocks by the content of their least significant bit, which is always set to 1 in the case of integers[5]. A memory block, written $[T : v_0;...;v_n]$, is composed by its tag T (a small integer) and its fields $v_0 \ldots v_n$. Closures are encoded by a block $[T_\lambda : C; v_1; \ldots; v_n]$ where the first field C is a code pointer and $v_1, \ldots; v_n$ are the values associated to the free variables of the function (i.e. the environment of the closure).

In [11], accumulators are represented using the same layout as closures: $[0 : \mathrm{ACCU}; k]$ where k is the memory representation of the accumulator and ACCU is a code pointer to a single instruction. When applied to an argument this instruction builds a fresh accumulator block containing the representation of k applied to the new argument. The major advantage of this technique is that the compilation scheme of the application is unchanged (accumulators can be seen as closures), so there is no penalty to the evaluation of an application. In particular, there is no penalty to the evaluation of closed terms. A second advantage is that the tag used for accumulator blocks is 0 which allows to distinguish the kind of block we obtain by a simple inspection of the tag (the tag T_λ used for closure is not 0).

Our idea is to use the same trick but directly in OCAML. Remember that an accumulator is a function expecting one argument, accumulating this argument and recursively returning an accumulator. Such a function, can be defined as follows[6]:

[5] This information is used by the garbage collector, and it is the reason why Ocaml integers are limited to 31 (resp. 63) bits on a 32 (resp. 64) bits machine.

[6] The -rectypes option is needed.

```
type t = t -> t
let rec accu atom args = fun v -> accu atom (v::args)
let mkAccu atom = accu atom []
```

Given an atom a the value of mkAccu a is a function expecting one argument v. This argument is stored in the list args, and the result is itself an accumulator. This is what we expect, but the tag of the object is T_λ and not 0. Fortunately, the tag of objects can be changed using the Obj module of Ocaml. This leads to the following code for accu:

```
let rec accu atom args =
  let res = fun v -> accu atom (v::args) in
  Obj.set_tag (Obj.repr res) 0; (res : t)
```

The result is an object of type t and its tag is now 0. Finally we need to write the head function. For this we inspect the tag of the value: if the tag is not 0 then it is a closure so we can return the value itself; if the tag is 0 we need to get the atom and the accumulating arguments. The atom is stored a position 3 in the closure and the list of arguments at position 4. The Obj.field function allows to get them. This leads to the following code:

```
type t = t -> t
let app f v = f v
let mkLam f = f
let getAtom o = (Obj.magic (Obj.field o 3)) : atom
let getArgs o = (Obj.magic (Obj.field o 4)) : t list
let rec head (v:t) =
  let o = Obj.repr v in
  if Obj.tag o = 0 then Accu(getAtom o, getArgs o) else Lam(v)
```

Note that the app function simply performs the application (without matching the functional part) and that the mkLam function is the identity. In practice, these operators are inlined by the compiler and hence effectively disappear in the output code. The tags in the previous implementation played two roles: they allowed App f t to do the right thing depending on whether f was a function or an accumulator, and guide the readback. With our tagless implementation, we have made the calling convention for functions and accumulators uniform, and rely on the runtime of the target language to inform readback. We do not need tags during readback because the runtime can already distinguish between different kinds of values. Finally, the presence of unsafe operations, like Obj.repr and Obj.magic (which are untyped identity functions), is not a matter of concern in the sense that our *source* language is typed so we have no particular safety requirement in our *target* language.

2 Extension to the Calculus of Inductive Constructions

In this section, we extend our approach to the Calculus of Inductive Constructions. Decidability of type checking only holds for *domain-full* terms, where variables are all explicitly annotated with their type at their binding site. However, one can still safely erase annotations on terms of the CIC when testing convertibility [2]. Furthermore, given a term with erased annotations it is possible to recover the type annotations from its type. We hence only consider a domain-free variant of the CIC.

2.1 The Symbolic CIC

The syntax of the symbolic calculus is extended with sorts, dependent products, inductive types, constructors, pattern matching and fixpoints:

$$\begin{aligned}
\textbf{Term} \ni t, P &::= x \mid t_1\, t_2 \mid v \mid C_i(\boldsymbol{t}) \mid \textbf{case}_{\langle P\rangle} \textbf{of} (C_i(\boldsymbol{x_i}) \to t_i)_{i\in I} \mid \textbf{fix}_m (f : T := t) \\
\textbf{Val} \ni v &::= \lambda x.t \mid [k\, v] \mid C_i(\boldsymbol{v}) \\
\textbf{Atom} \ni k &::= \tilde{x} \mid s \mid \Pi x : t.t \mid \textbf{case}_{\langle P\rangle}\, k\, \textbf{of} (C_i(\boldsymbol{x_i}) \to t_i)_{i\in I} \mid \textbf{fix}_m (f : T := t)
\end{aligned}$$

It is worth noting that we only represent here fully applied constructors. Indeed, since the η rule is admissible for the Calculus of Inductive Construction (and available in recent implementations of Coq), we may perform η-expansions where needed to preserve this invariant.

$$\begin{aligned}
\textbf{case}_{\langle P\rangle}\, C_i(\boldsymbol{v})\, \textbf{of} (C_i(\boldsymbol{x_i}) \to t_i)_{i\in I} &\to_\iota t_i\{\boldsymbol{x_i} \leftarrow \boldsymbol{v}\} && (\iota_v^1) \\
\textbf{case}_{\langle P\rangle}\, [k]\, \textbf{of} (C_i(\boldsymbol{x_i}) \to t_i)_{i\in I} &\to_\iota [\textbf{case}_{\langle P\rangle}\, k\, \textbf{of} (C_i(\boldsymbol{x_i}) \to t_i)_{i\in I}] && (\iota_s^1) \\
\textbf{fix}_m (f : T := t)\, v_1 \ldots v_{m-1}\, C_i(\boldsymbol{v}) &\to_\iota t\{f \leftarrow \textbf{fix}_m (f : T := t)\}\, v_1 \ldots v_{m-1}\, C_i(\boldsymbol{v}) && (\iota_v^2) \\
\textbf{fix}_m (f : T := t)\, v_1 \ldots v_{m-1}\, [k] &\to_\iota [\textbf{fix}_m (f : T := t)\, v_1 \ldots v_{m-1}\, k] && (\iota_s^2)
\end{aligned}$$

The rules ι_v^1 and ι_v^2 are the usual reduction rules for case analysis and fixpoints in the CIC. Fixpoints reduce only if their recursive argument (with index denoted by m) is a constructor. This prevents infinite unrolling of fixpoints during normalization. ι_s^1 and ι_s^2 are the symbolic counterparts, and handle the case when the argument is an accumulator. A new accumulator is created to represent the application of a fixpoint or a case analysis that cannot be reduced any further.

$$\begin{aligned}
\mathcal{R}_A(\textbf{case}_{\langle P\rangle}\, k\, \textbf{of} (C_i(\boldsymbol{x_i}) \to t_i)_{i\in I}) &= \textbf{case}_{\langle P\rangle}\, \mathcal{R}(k)\, \textbf{of} (C_i(\boldsymbol{x_i}) \to \mathcal{N}(f(C_i(\overrightarrow{[\tilde{x}_i]}))))_{i\in I} \\
&\quad \text{where } f = \lambda x.\textbf{case}_{\langle P\rangle}\, x\, \textbf{of} (C_i(\boldsymbol{x_i}) \to t_i)_{i\in I} \\
\mathcal{R}_A(\textbf{fix}_m (f : T := t)) &= \textbf{fix}_m (f : T := \mathcal{N}((\lambda f.t)\, [\tilde{f}]))
\end{aligned}$$

Fig. 1. Readback algorithm for pattern matching and fixpoints

Fig. 1 describes the readback algorithm. Readback of an accumulator representing a case analysis requires to normalize branches. As is the case for abstractions, bodies cannot be accessed directly, hence the need to apply the expression to trigger reduction. More precisely, if $\textbf{case}_{\langle P\rangle}\, t\, \textbf{of} (C_i(\boldsymbol{x_i}) \to t_i)_{i\in I}$ is in weak head normal form (t evaluates to an accumulator), we apply $\lambda x.\textbf{case}_{\langle P\rangle}\, x\, \textbf{of} (C_i(\boldsymbol{x_i}) \to t_i)_{i\in I}$ successively to the constructors $C_i(\boldsymbol{x_i})$ where the $\boldsymbol{x_i}$ are accumulators representing free variables, recursively applying readback on the result.

$$\ulcorner s \urcorner^B = \text{mkAccu (Sort } s)$$
$$\ulcorner \Pi x : T.U \urcorner^B = \text{mkAccu (Prod}(\ulcorner T \urcorner^B, \ulcorner \lambda x.U \urcorner^B))$$
$$\ulcorner C_i(\boldsymbol{t}) \urcorner^B = \text{mkConstruct i } [|\ulcorner \boldsymbol{t} \urcorner^B|]$$
$$\ulcorner \mathbf{case}_{\langle P \rangle}\ t\ \mathbf{of}(C_i(\boldsymbol{x}_i) \to t_i)_{i \in I} \urcorner^B =$$

```
let rec case c =
  match c with
  | ⌜C_1(x_1)⌝^{B∪{x_1}} -> ⌜t_1⌝^{B∪{x_1}}
  | ...
  | ⌜C_n(x_n)⌝^{B∪{x_n}} -> ⌜t_n⌝^{B∪{x_n}}
  | _ -> mkAccu Match(Ĩ,c,⌜P⌝^B,case)
in case ⌜t⌝^B
```

$$\ulcorner \mathbf{fix}_m (f : T := t) \urcorner^B =$$

```
let fnorm f = ⌜t⌝^{B∪{f}} in
let rec f =
  mkLam (fun x_1 -> ... -> mkLam (fun x_m ->
  if is_accu x_m then
    mkAccu (Fix (fnorm,⌜T⌝^B,m)) x_1 ... x_m
  else fnorm f x_1 ... x_m) ... )
in f
```

Fig. 2. Compilation scheme of the extended calculus of symbolic reduction

2.2 The Translation

The signature of our module Values needs to be extended accordingly to represent all possible heads and shapes of accumulators:

```
module type Values = sig
  type head =
    | ...
    | Construct of int * t array
  type atom =
    | Var of var
    | Sort of sort
    | Prod of t * t
    | Match of annot * t * t * (t -> t)
    | Fix of (t -> t) * t * int
  ...
  val mkConstruct : int -> t array -> t
end
```

Here again, we keep track only of the informations that are relevant for the conversion test. Constructors are identified by an index into the list of constructors of the inductive type it belongs to, and carry a vector of arguments. Case analyses are characterized by the term being matched, the predicate which expresses the (possibly dependent) return type, and the branches.

The compilation scheme is extended accordingly, as shown in Fig. 2, where is_accu is a simple auxiliary function defined as:

```
let is_accu v = match head v with
  | Accu _ -> true
  | _ -> false
```

Compilation of pattern matching builds a recursive closure which can reduce to a branch if applied to a constructor or otherwise to an accumulator storing information that is necessary for the reification. In particular, the recursive closure case is stored in the accumulator. This function plays the role of $\lambda x.\mathbf{case}_{\langle P\rangle}\, x\, \mathbf{of}(C_i(\mathbf{x}_i) \to t_i)_{i\in I}$ in the readback algorithm (cf Fig. 1). The use of such a recursive closure prevents exponential duplication of code.

The same kind of trick is used for fixpoints: the fnorm function parametrized by f encapsulates the body and the fixpoint itself is represented by a recursive closure expecting m arguments. If the m-th argument is an accumulator, then rule ι_s^2 applies, otherwise the fixpoint is reduced, as in ι_v^2.

Following our first approach (with explicit tagging), a concrete implementation of constructors could be:

```
mkConstruct i args = Construct(i,args)
```

Instead, we map inductive types and constructors of the CIC to datatypes and constructors of the host language, thus avoiding any allocation overhead, superfluous indirections and benefiting from the host language's efficient implementation of pattern matching. Thus, the following inductive type

```
Inductive I := C1 : T1 | ... | Cn : Tn
```

is translated to:

```
type I = Accu_I of t | C1 of t * ··· * t | ... | Cn of t * ··· * t
```

where the signatures match the arity of each constructor. This allows us to interpret a constructor in the source term by a constructor of the host language.

```
mkConstruct i v = Obj.magic C_i(v)
```

OCAML represents non-constant constructors by a memory block and distinguishes them according to their tag. Since Accu_I is the first non-constant constructor of the generated type, it will be attributed tag 0. This is compatible with our function head, which relies on the fact that we reserved tag 0 for accumulators.

2.3 Optimizations

To make the compilation scheme presented above more explicit, let us consider the addition over naturals defined as:

```
Fixpoint add (m n:nat):= match m with O => n | S p => S(add p n) end.
```

Strict application of our compilation scheme would yield:

```
let norm_add f m n =
  let rec case_add m = match m with
    | Accu_nat _ ->
      mk_sw_accu [...] (cast_accu m) pred_add (case_add f n)
```

```
    | Construct_nat_0 -> n
    | Construct_nat_1 p -> Construct_nat_1 (f p n)
in
let rec add m n = if is_accu m then
    mk_fix_accu [...] fixtype_add normtbl_add m n
  else norm_add add m n
in add
```

In the code above, some type information and annotations necessary to the reification have been elided, while some others are referred to by pred_add, fixtype_add and normtbl_add.

However, in our real implementation, several optimizations are performed. First, in order to avoid building at each recursive call a closure representing the pattern matching, we inline one level of pattern matching. Then, if a fixpoint starts a case analysis on the recursive argument, as it is often the case, we can avoid the call to head since the pattern matching will capture the case when the argument is an accumulator. On our function add, the final code looks like:

```
let rec case_add f n m = match m with
  | Accu_nat _ ->
    mk_sw_accu [...] (cast_accu m) pred_add (case_add f n)
  | Construct_nat_0 -> n
  | Construct_nat_1 p -> Construct_nat_1 (f p n)

let norm_add f m n = case_add f n m

let rec add m n = match m with
  | Accu_nat _ -> mk_fix_accu [...] fixtype_add normtbl_add m n
  | Construct_nat_0 > n
  | Construct_nat_1 p -> Construct_nat_1 (f p n)
```

3 Coinductive Types

The Calculus of (Co)Inductive Constructions supports the definition of co-recursive data, which can be built using constructors of coinductive datatypes or using cofixpoints. Co-recursive data can be infinite objects like the elements of type stream:

```
CoInductive stream := Cons : nat -> stream -> stream.
CoFixpoint sone := Cons 1 sone.
CoFixpoint snat x := Cons x (snat (1+x)).
```

sone represents the infinite list of 1 and snat x the infinite list of naturals starting from x. To prevent infinite unrolling, the evaluation of cofixpoint is lazy. This means that sone is in normal form, and so is snat 0. Only pattern matching can force the evaluation of a cofixpoint, the reduction rule is the following[7]:

$$\begin{array}{lcl}\textbf{case}\ c\ a\ \textbf{with}\ \ldots & \longrightarrow & \textbf{case}\ (t\{f \leftarrow c\})\ a\ \textbf{with}\ \ldots \\ \text{where}\ c = \textbf{cofix}\ f\ := t & & \end{array}$$

[7] The guard condition ensures that the reduction of a cofixpoint always produces a constructor.

Straightforward implementation of the reduction rule would lead to an inefficient evaluation strategy, since there is no sharing between multiple evaluations of $c\ \boldsymbol{a}$. To get an efficient strategy for cofixpoints, we use the same idea as OCAML, which roughly consists in representing a term of type ‘a Lazy.t by a reference either to a value of type ‘a (when the term has been forced) or to a function of type unit -> ‘a. When forcing a lazy value, two cases appear: the reference points to the result of a previous evaluation, which can be directly returned, or to a function, in which case it is evaluated and the result is stored in the reference.

However, reduction rules of cofixpoints require that we keep track of the original term which has been forced. This leads to the following implementation:

```
type atom =
  | ...
  | Acofix_e of t * (t -> t) * t
  | Acofix of t * (t -> t) * (unit -> t)

let update_atom v a =
    Obj.set_field (Obj.magic v) 3 (Obj.magic a)

let force v =
  if is_accu v then match get_atom v with
      | Acofix_e(_,_,v') -> v'
      | Acofix (t,norm, f)  ->
          let v' = app_list f (args_accu v) () in
          update_atom v (Acofix_e(t,norm v')); v'
      | _ -> v
  else v
```

To force a value, we first check if it is an accumulator. If not, it means that it is a constructor of a coinductive type which is returned unchanged. Otherwise, if the atom is an already evaluated cofixpoint, we return the stored result. If it is a cofixpoint which has not been evaluated, the function is applied to its accumulated arguments (through the app_list routine) and the accumulator is updated with a new atom. In the last case, the accumulator is a neutral term.

Coinductive types have strictly the same compilation scheme as inductive types (c.f. Section 2.2). For cofixpoints, we use the following scheme:

$\ulcorner \mathbf{cofix}\, f\ :\ T\ :=\ t \urcorner^{B} =$

```
let fnorm f = ⌜t⌝^(B∪{f}) in
let f = mk_accu dummy_atom in
update_atom f (Acofix(⌜T⌝^B, fnorm, fun _ -> fnorm f));
f
```

This is directly adapted from the compilation of fixpoints (c.f. Section 2.2). It is worth mentioning that the accumulator f is created first with a dummy atom and then updated with the real one whose definition depends on f (under a lambda abstraction). We use this construction to circumvent a limitation on the right hand side of **let rec** in OCAML. Finally, the compilation of pattern matching adds a force if the matched term has a coinductive type.

Table 1. Benchmarks run on a (a) 64 bits architecture and (b) 32 bits architecture, both with 4GB memory

	Standard reduction	Bytecode interpreter	Native compilation
BDD	4min53s (100%)	21,98s (7,5%)	11,36s (3,9%)
Four colour	not tested	3h7min (100%)	34min47s (18,6%)
Lucas-Lehmer	10min10s (100%)	29,80s (4,9%)	8,47s (1,4%)
Mini-Rubik	Out of memory	15,62s (100%)	4,48s (28,7%)
Cooper	not tested	48,20s (100%)	9,38s (19,5%)
RecNoAlloc	2min27s (100%)	14,32s (9,7%)	1,05s (0,7%)

	Standard reduction	Bytecode interpreter	Native compilation
BDD	6min1s (100%)	18,08s (5,0%)	11,28s (3,1%)
Four colour	not tested	2h24min (100%)	46min34s (32,3%)
Lucas-Lehmer	15min56s (100%)	21,59s (2,3%)	10,04s (1,1%)
Mini-Rubik	Out of memory	14,06s (100%)	3,99s (28,4%)
Cooper	not tested	37,88 (100%)s	10,18s (26,9%)
RecNoAlloc	4min3s (100%)	11,21s (4,6%)	1,47s (0,6%)

4 Benchmarks

In order to assess the performance of our low-level approach, we compared our implementation (**Native compilation**) with two preexisting implementations of term conversion in the Coq proof assistant. The first one (**Standard reduction**) is based on an abstract machine that manipulates syntactic representations of terms using a lazy evaluation strategy. The second one (**Bytecode interpreter**), is the bytecode based virtual machine using a call-by-value evaluation strategy described in [11].

To ensure a meaningful comparison, we extracted most of our benchmarks from real-world use cases in a variety of settings:

BDD is an implementation of binary decisions diagrams [16], which checks if a given proposition is a tautology. In our example, we ran it on an expression of the pigeonhole principle: if n pigeons are put in $n-1$ holes, there cannot be only one pigeon in each hole.

Four colour is the reducibility check of configurations in the formal proof of the four colour theorem by Gonthier and Werner [9], which represents most of the computation time of the whole proof.

Lucas-Lehmer is an implementation of Lucas-Lehmer primality test which decides if a given Mersenne number is prime or not.

Mini-Rubik checks that any position of the 2x2x2 Rubik's is solvable in at most 11 moves, using machine integers and arrays which we were able to port to our approach without extra cost, because the whole OCaml language is accessible to our compiler. The original formalization was described in [15].

Cooper implements Cooper's quantifier elimination on a formula with 5 variables.

RecNoAlloc triggers 2^{27} trivial recursive calls (i.e. without memory allocation to store the result). This aims at measuring pure performance impact, when garbage collection is not significantly involved.

We ran the benchmarks on two different architectures (32 and 64 bits), because some optimizations of Coq's bytecode interpreter like the use of threaded code are available only on 32-bits environments. This accounts for the different ratios between **Bytecode interpreter** and **Native compilation** since the latter is not impacted by such limitations.

Most of the results show a speed-up factor ranging from 2 to 5, which is typical of the expected speed-ups when going from bytecode interpretation to native-code compilation. It is worth noting that the performance improvement is particularly significant on examples involving less garbage collection. This is highlighted by **RecNoAlloc** where the speed-up factor lies between 7 and 14, depending on the architecture.

We also used **Cooper** and **RecNoAlloc** to assess the gap between our reference implementation of tagless normalization with a preliminary tagged version, achieving respectively 1.5 and 2.5 speed-up factors in favour of the former. Also, the program obtained through extraction on **RecNoAlloc** produces an OCaml program running at 76% of the time spent by its **Native compilation** equivalent. The performance penalty over the extracted version can be attributed to the overhead of having to check the guard condition at every recursive call, showing that our compilation scheme achieves close to the best performance that could possibly be achieved on a call-by-value evaluator.

Conclusion

The move towards greater automation and reduced proof sizes shifts away some of the burden of formalizing mathematics from the user and the tactics. The flip side is a greater pressure on the implementation of the proof checker, with the checking of the proof of the four colour theorem quite simply unfeasible without introducing some form of compilation of proof terms. Even checking based on a bytecode compilation scheme takes upward of 3 hours on a desktop machine. We have presented in this paper a more than fivefold improvement on this total checking time, enabling the widespread use of proof by reflection techniques in ever larger and greater developments in mathematics and software verification.

The approach we propose is cheap enough that it can readily be implemented in most any interactive proof environment where computing with proofs is pervasive, without expert knowledge on compilation technology. The correctness of our approach is in part contingent upon the correctness of the compiler, whose entire code enters the trusted base. However, the chosen compiler is already in the trusted base of the proof assistant if the target language of the translation described here and the implementation language of the proof assistant coincide. A certified compiler for the target language would certainly be of interest here to reduce the trusted base.

Since the whole target language is available to our translation routine, we have successfully implemented new features such as machine-native integers and persistent arrays nearly for free, whereas approaches based on ad-hoc compilers or runtimes require extra work each time an extension is needed.

Our tagged implementation is portable across any functional programming language, while our tagless implementation makes use of OCaml specific extensions that are already partially implemented in other typed languages. In particular, the unpackClosure# primitive of the GHC compiler for Haskell might well be sufficient for our purposes. Further investigation into the feasibility and trade-offs of this optimization for other languages is left as future work.

References

1. Aehlig, K., Haftmann, F., Nipkow, T.: A Compiled Implementation of Normalization by Evaluation. In: Mohamed, O.A., Muñoz, C., Tahar, S. (eds.) TPHOLs 2008. LNCS, vol. 5170, pp. 39–54. Springer, Heidelberg (2008)
2. Barras, B., Grégoire, B.: On the Role of Type Decorations in the Calculus of Inductive Constructions. In: Ong, L. (ed.) CSL 2005. LNCS, vol. 3634, pp. 151–166. Springer, Heidelberg (2005)
3. Berger, U., Schwichtenberg, H.: An inverse of the evaluation functional for typed λ-calculus. In: LICS 1991, pp. 203–211 (1991)
4. Boespflug, M.: Conversion by evaluation. In: Proceedings of the Twelfth Internation Symposium on Practical Aspects of Declarative Languages, Madrid, Spain (2010)
5. Boutin, S.: Using Reflection to Build Efficient and Certified Decision Procedures. In: Ito, T., Abadi, M. (eds.) TACS 1997. LNCS, vol. 1281, pp. 515–529. Springer, Heidelberg (1997)
6. Coquand, T., Paulin, C.: Inductively Defined Types. In: Martin-Löf, P., Mints, G. (eds.) COLOG 1988. LNCS, vol. 417, pp. 50–66. Springer, Heidelberg (1990)
7. Danvy, O.: Type-directed partial evaluation. In: Proceedings of the 23rd ACM SIGPLAN-SIGACT Symposium on Principles of Programming Languages, POPL 1996, pp. 242–257. ACM, St. Petersburg Beach (1996)
8. Filinski, A., Korsholm Rohde, H.: A Denotational Account of Untyped Normalization by Evaluation. In: Walukiewicz, I. (ed.) FOSSACS 2004. LNCS, vol. 2987, pp. 167–181. Springer, Heidelberg (2004)
9. Gonthier, G.: The Four Colour Theorem: Engineering of a Formal Proof. In: Kapur, D. (ed.) ASCM 2007. LNCS (LNAI), vol. 5081, p. 333. Springer, Heidelberg (2008)
10. Gonthier, G., Mahboubi, A.: A Small Scale Reflection Extension for the Coq system. Research Report RR-6455, INRIA (2008)
11. Grégoire, B., Leroy, X.: A compiled implementation of strong reduction. In: Proceedings of the Seventh ACM SIGPLAN International Conference on Functional Programming, pp. 235–246. ACM (2002)
12. Grégoire, B., Mahboubi, A.: Proving Equalities in a Commutative Ring Done Right in Coq. In: Hurd, J., Melham, T. (eds.) TPHOLs 2005. LNCS, vol. 3603, pp. 98–113. Springer, Heidelberg (2005)
13. Harrison, J.: Metatheory and reflection in theorem proving: A survey and critique. Technical Report CRC-053, SRI Cambridge, Millers Yard, Cambridge, UK (1995)
14. Lindley, S.: Normalisation by evaluation in the compilation of typed functional programming languages. Ph.D. thesis, University of Edinburgh (2005)
15. Théry, L.: Proof Pearl: Revisiting the Mini-Rubik in Coq. In: Mohamed, O.A., Muñoz, C., Tahar, S. (eds.) TPHOLs 2008. LNCS, vol. 5170, pp. 310–319. Springer, Heidelberg (2008)
16. Verma, K.N., Goubault-Larrecq, J., Prasad, S., Arun-Kumar, S.: Reflecting Bdds in Coq. In: He, J., Sato, M. (eds.) ASIAN 2000. LNCS, vol. 1961, pp. 162–181. Springer, Heidelberg (2000)
17. Werner, B.: Une Théorie des Constructions Inductives. Ph.D. thesis, Université Paris-Diderot - Paris VII (May 1994)

Certified Security Proofs of Cryptographic Protocols in the Computational Model: An Application to Intrusion Resilience*

Pierre Corbineau, Mathilde Duclos, and Yassine Lakhnech

Université de Grenoble, CNRS – Verimag, Grenoble, France

Abstract. Security proofs for cryptographic systems can be carried out in different models which reflect different kinds of security assumptions. In the symbolic model, an attacker cannot guess a secret at all and can only apply a pre-defined set of operations, whereas in the computational model, he can hope to guess secrets and apply any polynomial-time operation. Security properties in the computational model are more difficult to establish and to check.

In this paper we present a framework for certified proofs of computational indistinguishability, written using the Coq proof assistant, and based on CIL, a specialized logic for computational frames that can be applied to primitives and protocols. We demonstrate how CIL and its Coq-formalization allow proofs beyond the black-box security framework, where an attacker only uses the input/output relation of the system by executing on chosen inputs without having additional information on the state. More specifically, we use it to prove the security of a protocol against a particular kind of side-channel attack which aims at modeling leakage of information caused by an intrusion into Alice and Bob's computers.

Keywords: Provable Cryptography, Formal Verification, Computational Model, Security Protocol, Intrusion Resilience.

1 Introduction

Context. Nowadays, security properties, like most proofs of programs, are more and more often checked by computer programs. This is especially the case for proofs carried out in the symbolic model, where an adversary's possible behaviours are usually modelled by a set of Horn clauses modulo some equational theory or using a process algebra such as the applied π-calculus. Those approaches rely on the perfect security assumption and do not take into account the possibility of guessing a secret value by chance at random. The description of the cryptographic scheme or protocol is fed to an adhoc verification tool [Bla01] or sometimes simply a resolution theorem prover, which checks that the adversary is unable to compute the secret value. Most attempts at formalising security proofs so far have also been carried out in this model.

To work under a more realistic assumption, the so-called computational model deals with probabilities and takes into account the possibility of random guessing: to successfully attack a cryptographic scheme, an adversary must be able to win significantly

* This work has been partially supported by the ANR SCALP project.

J.-P. Jouannaud and Z. Shao (Eds.): CPP 2011, LNCS 7086, pp. 378–393, 2011.

more often than what is possible through random guessing. Proofs in the computational model rely on a specialized notion of probabilistic games called computational frames. A frame pits an adversary against a set of blackbox primitives called oracles. In practice the adversary is implicit in the description of frames and proofs rely on sequences of oracle transformations: we show that the adversary cannot win significantly more often against the real oracles than against ideal ones which do not reveal any information. Of course, this requires limiting the resources used by the adversary. The usual requirement uses the Probabilistic Polynomial Time model.

Security properties in the computational model thus have complex definitions, and for that reason, most security proofs are only done on paper. Furthermore, these proofs contain a lot of oracle descriptions which are similar but for localised modifications. This make hand-checking those proofs nearly intractable and as a result, several flawed proofs having been published [RB95].

This complexity can be addressed by two methodological means: the move from *ad hoc* proofs to proofs based on a specialized logic — which CIL (Computational indistinguishability logic [BDKL10]) aims to be — and the certifications of proofs by a formal proof assistant — we used Coq to establish our theorems [Coq]. In this paper we illustrate how these two approaches can be combined by first formalizing computational frames and CIL rules in Coq, and then by using this development as a foundation for our case study: we prove the leakage resilience property of a concrete protocol against side-channel attacks.

The Coq-CIL Library. The formalization[1] of CIL in Coq is composed of several layers. We use the probability distributions from the existing ALEA library [APM09] which implements probability distributions and their properties. The model we use to prove the correctness of CIL rules is based on probabilistic games between a set of cryptographic oracles and an attacker. First we define the interface between the two players — the set of possible moves — which we call an *oracle signature*. Then we define oracle families implementing the chosen signature: they will behave as black boxes sharing a common secret state. Finally we model adversaries as probabilistic processes able to interact with the oracles of the chosen signature, and we define inductively the outcome — a distribution of pairs of adversary outputs and interaction traces — of such a game.

In the paper we present several approaches for the implementation of adversaries and explain why we chose the probabilistic input/output automaton view of adversaries. We then explain how such adversaries can be constructed using four basic *monadic operators* for constant output, sequential composition, random sampling and oracle call. These operators provide proto-language for adversaries that can be easily mixed with Coq functions. We also put exact resource usage restrictions on adversaries that can be linked to exact information leakage.

Using these base definitions we can express the negligibility and indistinguishability notions which are the two main judgement forms of CIL. We then proceed to prove a theorem corresponding to each CIL rule. The whole CIL logic has been formalized, this

[1] The Coq-CIL library can be downloaded at `http://www-verimag.imag.fr/corbinea/ftp/programs/coq-cil.zip`

is what we call the COQ-CIL library. However in this paper, we will mostly describe the indistinguishability judgement and restrict ourselves to the CIL fragments that has been actually used in our case study: reflexivity, symmetry, transitivity (triangular inequality) and substitution (congruence) rules.

The SUB rule is a very powerful tool in CIL proofs: it expresses a congruence property for indistinguishability. In order to state it properly, we need to define a notion of context for which indistinguishability is indeed congruent. The contexts we use are pieces of probabilistic processes that can be either composed with the oracle family or simulated by the adversary, yielding the same result distribution in both cases.

Case Study : An Intrusion Resilient Security Protocol. Using the *Coq-CIL* framework enables us to go beyond black-box security: in the following case study we illustrate a specific white-box property, namely the ability by the adversary to obtain a limited amount of information about the internal state of the protocol participant, which encompasses a certain class of side-channel attacks.

The study of side-channel attacks on cryptographic schemes has allowed to reveal design flaws that had previously gone undetected: although RSA had been proved secure in weaker attack models, cryptanalysts were able to find a serious timing attack [Koc96]. Despite a fair amount of work on the subject, the usual models by themselves are not sufficient to reason upon side-channel attacks.

Leakage resilient cryptography fills the gap in extending models to achieve provable cryptography considering the possible threat of side-channel. Several models try to take into account leak from the scheme. Some try to make the scheme independent from the hardware, as physically observable cryptography [MR04b], some other focus in a specific form of covert channel.

Maurer et al. [Mau92] have designed the Bounded Storage Model (BSM) to address the issue of viruses. In this model, the adversary is able to retrieve secret data from an honest infected machine via a malware that he or she has installed on these machines and that can send back a limited amount of private data to the adversary: for example a share of a long-term secret key. Using the COQ-CIL library, we formalize and certify secure an existing key exchange protocol in the BSM, first presented in [Dzi06]. Our work provides the first certified proof about a leakage resilient cryptographic scheme.

Related Work. Providing an extended account for all the recent work on leakage-resilient cryptography would cover a fair amount of pages and is beyond the scope of this article. We just mention some natural extensions to the BSM model, and refer to [ADW09] for a survey. Models for both intrusion-resilience and leakage-resilience were proposed in [MR04a] by Micali *et al.*. Dziembowski *et al.* in [DP08] followed with a leakage resilient stream cipher. One of the latest achievement was a leakage resilient variant of ElGamal encryption, developed by Kiltz and Pietrzak in [KP10].

The reader may want to read [CKW10] for a recent survey on symbolic methods and computational soundness results. In the computational model, recent work has allowed to design indistinguishability logics (among others), initiated by Impaglazzio and Kapron [IK06] and further developed by Zhang [Zha08] and Datta et al. [DDMW06]. We chose to use CIL rather than these logics because its semantics could be extended naturally to account for the bounded storage model, and its high level rules allowed a more natural presentation of proof steps.

Tools have also been developed such as CryptoVerif [Bla06], which is an automated verifier in the computational model, and CertiCrypt [BGZ09], a framework to construct machine-checked proofs in Coq. CryptoVerif is not a certifying tool and thus offers no formal guarantee. Certicrypt uses Coq for certification but it uses a language based approach rather than the more flexible model-based approach we have developed with COQ-CIL.

2 Formalising Computational Frames for *Coq-CIL*

Cryptographic frames are composed of three parts :

- the oracles, which share a common *secret* internal state and expose a public interface described by an oracle signature.
- the adversary, which is our *malicious opponent* and can interact with the oracles using limited resources.
- the event, which is a property of the execution trace (including oracle states) and of the output of the adversary. It can either express a winning condition for the adversary in a negligibility frame or simply observe the boolean answer in an indistinguishability challenge.

2.1 Oracles and Signatures

We have chosen to model oracles as a family of probabilistic functions which take an input state and input data and return a distribution of pairs of output states and output data. The Coq modeling of the family starts with a set of indices which we call *oracle names* and two oracle *typing functions* which return an input and an output type corresponding to each name. Together they form an *oracle signature* which constitutes the public interface of the frame against which an adversarial process can be executed.

Then, given a certain type of private *oracles states*, we can define the type of oracle families carrying the chosen signature while using this particular type of states.

```
Record oracle_signature := mkOS
{ oracle_name : Type;
  oracle_input: oracle_name -> Type;
  oracle_output: oracle_name -> Type }.

Variable os : oracle_signature.
Variable State:Type.

Definition oracle_fun (input:Type) (output:Type) :=
  input * State -> distr (output * State).

Definition oracle_functions := forall name,
  oracle_fun (oracle_input os name) (oracle_output os name).
```

2.2 About Probabilities

In order to model probabilistic processes, we have used the ALEA library for probabilistic distributions, developed by Christine Paulin *et al* [APM09]. It is based on an axiomatisations of the real interval $[0;1]$ denoted by the Coq type `U`. The type `U` is equipped with a total ordering, addition and multiplication, distance function, and a lowest upper bound operator $\bigsqcup$ for monotonic sequences.

Given an arbitrary type `A`, *events over* `A` are arbitrary functions of type `A`→`U`. The classical view of events is that these functions should be $\{0;1\}$ valued but any function is actually allowed in this model in order to enable the definition of the monadic `Mlet` operator (see below). The ordering over $[0;1]$ can be extended to events using pointwise ordering.

Using the event ordering, one can define a (partial) *distribution over* `A` as a monotonic and continuous (*i.e.* lub-preserving) function from *events over* `A` to $[0;1]$ with additional properties about addition, scaling and complementation of events. A distribution over type `A` (type `distr A`) thus maps any event over A to its probability value.

Distributions can be combined into probabilistic functions using two monadic operators. First, `Munit` is a constant distribution of weight 1 for a given element in `A` : $\mathcal{P}_{\texttt{Munit}(x)}(ev) = ev(x)$. Then, we can compose two probabilistic expressions using the `Mlet` operator. `Mlet` acts as a probabilistic binding operator by combining a distribution over `A` and a function of type `A` $\rightarrow$ `distr B` into a distribution over type `B` which satisfies the following equation: $\mathcal{P}_{\texttt{Mlet}(d,f)}(ev) = \mathcal{P}_{\texttt{d}}(x \mapsto \mathcal{P}_{f(x)}(ev))$. Please notice that in general, the event $x \mapsto \mathcal{P}_{f(x)}(ev)$ can have values strictly between 0 and 1.

One last point we need to build our probabilistic processes is probabilistic recursion, in which the function may chose probabilistically to terminate or make a another recursive call. This can be achieved by using the classical view of recursive functions as least fixpoints of monotonic functions transformers. Given a monotonic function transformer F of type (`A` $\rightarrow$ `distr B`) $\rightarrow$ (`A` $\rightarrow$ `distr B`), one can simulate the recursive function definition `let rec` $f(a) := F(f)(a)$ by the following fixpoint definition: `let`$f = \bigsqcup_{n \in \mathbb{N}} F^n(\_ \mapsto \bot)$, where $\bot$ stands for the null partial distribution: $\mathcal{P}_{\bot}(ev) = 0$.

2.3 Adversarial Process

Given an oracle signature, we define an adversary as a player which can probabilistically take the following actions in its turn: either return an answer and end the game (successfully or not) or request an oracle query for a specific name and input data, and continue with its output.

There are three approaches to how we can give the adversary access to oracles. The *extensional* approach which where we define the adversary as a *higher-order function* which takes the oracles functions as arguments and calls them himself. The *intentional* or syntactic approach, consist in defining a fixed language for the code of the adversary, and write an interpretation function or relation for this language [BGZ09]. The *operational* approach consists in describing the adversary as a probabilistic transition system whose transitions consist in oracle interactions.

The *extensional* approach gives enormous power to the adversary because even if he doesn't cheat by looking at the oracle state, which we can prevent by requesting he be polymorphic with respect to to the said oracle state, he can still cheat by e.g. calling several times an oracle with the same input state, waiting for a favorable output, and then carry on, thereby unbalancing the odds of getting to that favorable configuration. It is our opinion that any attempt at correcting the problem by specifying adversarial honesty would amount to assert behavioural equivalence to an operational adversary. We therefore view this approaoch as redundant.

The *intentional* approach lacks flexibility and introduces combinatory explosion w.r.t. the number of languages constructs (a common problem with analog deep reflection approaches), and in the end the interpretation of programs leads us to look at their operational behaviour (i.e. the call sequences).

The *operational* approach ensures adversarial honesty by isolating the oracle state data-flow from the adversary data-flow, using a third party, the interaction function, which can also record a trace of the exchanges. The trace can then be used to express properties. However, since it is often awkward to express processes as probabilistic transition systems, we constructed a monad structure with combinators acting as language constructs to combine sub-processes.

The Monad of Adversarial Processes. An adversary is encoded as a type of states and a probabilistic transition function. The transition function returns a distribution of *adversary moves*. The two kinds of moves an adversary can make are: yield a result and end the game which we called a *Return* move, or ask for an oracle call, which we call a *Request* move. When calling an oracle, the adversary must provide the name of the called oracle, its input data, and a *continuation* function which turns the oracle output data into a new state of the transition system.

```
Inductive Response state A :=
  Request : forall (name : oracle_name os),
            (oracle_input os name) ->
            (oracle_output os name -> state) ->
            Response state A
| Return : A -> Response state A.

Definition run_function state A :=
  state -> distr (Response state A).

Record Computation (A:Type) := mkC
{ c_state: Type;
  c_init: c_state;
  c_run:> run_function c_state A }.
```

The monad of adversarial computations is composed of two generic monadic operators and two specific ones :

Cunit(a) builds a process returning a constant value a.
Clet(v, f) allows the sequential composition of two processes: it computes the output x of v before computing the output of $f(x)$.
Cdraw(d) returns a value sampled from a distribution d.
Ccall(o, x) calls the oracle o with input x and returns the output data.

2.4 Resource Constraints and Indistinguishability

The formalism we use deals with exact security measurement rather than the usual asymptotic one, that is to say that indistinguishability is defined relatively to a certain probability distance ϵ in U and a maximum allowed number $k_o \in \mathbb{N}$ of calls to every oracle o. We say an adversarial process is a valid adversary if and only if it satisfies all following conditions:

Totality An adversary must terminate with probability 1 against any total oracle family, i.e. whenever each oracle call also returns with probability 1: $\forall \mathbb{O}.\mathcal{P}_{(\mathcal{A}|\mathbb{O})}(\top) = 1$

Call bound An adversary must have a 0 probability of exceeding the number of allowed oracle calls of any given name, i.e. of presenting more than k_o calls to a given oracle o in the final execution trace: $\forall \mathbb{O}, o.\mathcal{P}_{(\mathcal{A}|\mathbb{O})}(\#_o > k_o) = 0)$

Runtime A complete formalisation of resource constraints would require to specify precisely the adversary's `c_run` function runtime, and how it is modified by the reduction steps of CIL proofs. At this time, we have not yet formalised this part of the model. We plan to address it in the near future.

Given those definitions, we say that two oracle families $\mathbb{O}_1$ and $\mathbb{O}_2$ against the same signature are $(\boldsymbol{k}, \epsilon)$ indistinguishable if, and only if, for all valid boolean adversaries, the probabilities of returning true against $\mathbb{O}_1$ and $\mathbb{O}_2$ differ by at most ϵ:

$$\mathbb{O}_1 \sim_{\boldsymbol{k},\epsilon} \mathbb{O}_2 \Leftrightarrow \begin{array}{l} \forall \mathcal{A}, (\forall \mathbb{O}.\mathcal{P}_{(\mathcal{A}|\mathbb{O})}(\top) = 1) \Rightarrow \\ \quad (\forall \mathbb{O}, o.\mathcal{P}_{(\mathcal{A}|\mathbb{O})}(\#_o > k_o) = 0) \Rightarrow \\ \quad \left|P_{(\mathcal{A}|\mathbb{O}_1)}(\mathtt{answer} = \mathtt{true}) - P_{(\mathcal{A}|\mathbb{O}_2)}(\mathtt{answer} = \mathtt{true})\right| \le \epsilon \end{array}$$

3 *Coq-CIL*: Computational Indistinguishability Logic

3.1 Indistinguishability as a Distance Relation on Oracles

The three basic rules of CIL assert that computational indistinguishability behaves as a distance relation :

$$\frac{}{\mathbb{O} \sim_{\boldsymbol{k},0} \mathbb{O}}(\textsc{Refl}) \quad \frac{\mathbb{O}_1 \sim_{\boldsymbol{k},\epsilon} \mathbb{O}_2}{\mathbb{O}_2 \sim_{\boldsymbol{k},\epsilon} \mathbb{O}_1}(\textsc{Sym}) \quad \frac{\mathbb{O}_1 \sim_{\boldsymbol{k},\epsilon_1} \mathbb{O}_2 \quad \mathbb{O}_2 \sim_{\boldsymbol{k},\epsilon_2} \mathbb{O}_3}{\mathbb{O}_1 \sim_{\boldsymbol{k},\epsilon_1+\epsilon_2} \mathbb{O}_3}(\textsc{Trans})$$

The reflexivity rule does not only deal with equality in the sens of Coq's $\beta\iota$-conversion: it is also compatible with observational equivalence. The CIL logic also contains rules based on state bisimulation and failure events that have been been proved sound in Coq but their description is beyond the scope of this paper.

3.2 Contexts and the Sub rule

One of the most powerful forms of reasoning allowed by CIL and used in this paper is the Sub rule. It states that ϵ-indistinguishable frames $\mathbb{O}_1$ and $\mathbb{O}_2$ cannot be made more distinguishable by putting them inside a *context* that an adversary can simulate.

```
Variables sigma sigma' : oracle_signature.
Variable complement : Type.

Definition base_context := forall name',
  (oracle_input sigma' name' * complement) ->
  Computation sigma (oracle_output sigma' name' * complement).
```

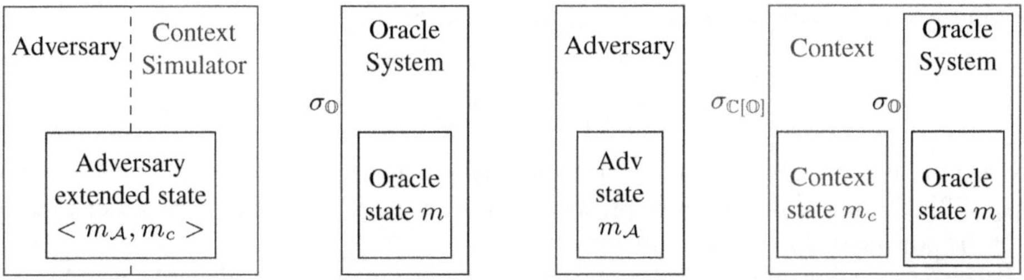

(a) Context Composed with an Adversary (b) Context Applied to an Oracle System

Fig. 1. Fundamental Property of Contexts

Contexts. The aim of a context is to simulate a signature σ' against any frame of signature σ, using some persistent information m_c to extend the internal oracle state. A context $\mathbb{C}$ is made of a family $\mathcal{C}_{o', o' \in \sigma'}$ of adversaries playing against the oracle signature σ to simulate an oracle in σ'. Given a σ-oracle family $\mathbb{O}$, we define the combined σ'-oracle family $\mathbb{C}[\mathbb{O}]$ by the following equation where $\mathbb{O}(m)$ means the oracle family $\mathbb{O}$ starting at state m :

$$\mathbb{C}[\mathbb{O}]_{o'}(in, (m_c, m)) \equiv (\mathcal{C}_{o'}(in, m_c)|\mathbb{O}(m))$$

Please note that the internal state of $\mathbb{C}[\mathbb{O}]$ consists of a pair of a context state m_c and an $\mathbb{O}$-state m.

This definition allows us to turn any adversary $\mathcal{A}$ against the $\mathbb{C}[\mathbb{O}]$ oracles into an adversary $\mathcal{A} \circ \mathbb{C}$ against the σ-oracle family $\mathbb{O}$ *with the same output distribution*. The combined adversary is simply obtained by *inlining* the code of the simulators from $\mathbb{C}$ into $\mathcal{A}$.

Lemma 1 (Fundamental property of contexts). *For any given adversary $\mathcal{A}$ against σ', any context $\mathbb{C}$ from σ to σ', any oracle family $\mathbb{O}$ implementing σ, and any event ev, we have:*

$$\mathcal{P}_{\mathcal{A}|\mathbb{C}[\mathbb{O}]}(ev) = P_{\mathcal{A} \circ \mathbb{C}|\mathbb{O}}(ev)$$

The Coq proof of this property can be done by showing that distributions on both sides over- and under-approximate each other when seen as a limit of a monotonic sequence of distributions (term by term equality cannot work here).

The Sub rule. In order to state the Sub rule properly, we have to address the issue of resources: each call to a simulator of the σ'-oracle o' in an instance $\mathbb{C}[\mathbb{O}]$ of the context may result in a certain number of calls made by $\mathbb{C}$ to the σ-oracle o in $\mathbb{O}$. We name $m_{o,o'}$ a bound on calls to σ-oracle o by the simulator for the σ'-oracle o', independent of the implementation of $\mathbb{O}$, and we call M the matrix made of all those bounds.

This matrix can then be used to compute the call bound for $\mathbb{O}$ given a call bound for $\mathbb{C}[\mathbb{O}]$. In most cases, the matrix has at most one non null coefficient per column and this coefficient is 1. For any given adversary $\mathcal{A}$ against σ' with call bound $\boldsymbol{k}$, the adversary $\mathcal{A} \circ \mathbb{C}$ has call bound $M.\boldsymbol{k}$. This fact together with the fundamental property of contexts allows use to state the Sub rule:

Lemma 2 (SUB rule). *The following CIL rule is correct:*

$$\frac{\mathbb{O}_1 \sim_{M.\boldsymbol{k},\epsilon} \mathbb{O}_2}{\mathbb{C}[\mathbb{O}_1] \sim_{\boldsymbol{k},\epsilon} \mathbb{C}[\mathbb{O}_2]} (\text{SUB})$$

The proof is a straightforward use of the fundamental property for both $\mathbb{O}_1$ and $\mathbb{O}_2$, the big difficulty being the proof of preservation of resource usage in the proof of totality for the adversary composed with contexts.

The SUB rule is mostly used backwards in a proof in order to reduce the problem by removing identical parts in both frames; the key being how to identify a proper *memory separation* between context and oracles. Now that we are properly equipped, we can move on to the the example of security proof we are presenting in this paper.

4 Intrusion Resilience and Bounded Storage Model

A cryptographic scheme is said to be intrusion resilient secure if it ensures some security property (as secrecy, for example), even if its adversaries can send a virus to the honest parties. We verify this property in the Bounded Storage Model, deeply explained by Dziembowski [DM04]: in this model, the virus has complete view on honest parties hardware, but can only send back short information on it to the adversary. For compatibility, we adopt the notations from [DM04], in particular:

- k is the security parameter,
- $\sigma(k)$ is the size of the output of the virus,
- s is the size of the adversary's memory,
- K is a (huge) common random input to the honest participant,
- $\alpha(k)$ is the size of K.

4.1 Bounded Storage Model

Dziembowski [Dzi06] defines in details the underlying model specialized in key exchange protocols in the BSM. As our point here is to give a rough idea of the model and protocol, we only informally introduce them, and refer to [Dzi06] for more details.

For simplicity, we assume that there are two honest parties Alice and Bob, whose goal is to set up a common session key. Alice and Bob share a long random $\alpha(k)$-bit string K, from a randomizer space $\mathcal{R} \stackrel{\text{def}}{=} \{0,1\}^{\tau(k)}$, either temporarily accessible by anyone, or broadcasted by Alice or Bob (for all sessions). Moreover, they have fresh (independent) random inputs at the beginning of each session. At the end of the session, they both obtain a session key κ_A and κ_B respectively.

Additionally, we assume the presence of an adversary. As usual, the adversary has complete control over the network: she can delay, change or even stop messages between the parties, and send her own messages. Moreover, the adversary can corrupt Alice and Bob's machines: at the beginning of each session she can install a virus, i.e. get back the result of a function (Γ) over Alice's and Bob's internal states. The BSM mandates that the result is of bounded size, but allows the size of the function to be unbounded (so that Γ can be any arbitrary function). For now, the BSM does not take into account active viruses which could alter the memory or change the behavior of the honest participant.

The goal of the adversary is to compromise the session key in an uncorrupted session — where no virus was sent to infect Alice or Bob's systems — i.e.:

- Either make Alice and Bob agree on different session keys,
- Or obtain significant information about the key Alice or Bob agree on.

The protocol studied in this paper uses an intrusion resilient key expansion function. Such a function is able to expend a key Y into V using a huge random string R such that even if the adversary has had a temporary access to R and that she knows Y, the adversary is not able to get significant information on V (once she has lost her access to R).

We assume that the adversary has a memory of size $\sigma(k)$ considerably smaller than $\tau(k) \stackrel{\text{def}}{=} |R|$. Thus she can only store partial information of size $\sigma(k)$ on R, via any storage function $h : \mathcal{R} \rightarrow \{0,1\}^{\sigma(k)}$ she wants to use. Once she retrieves this information, she looses her access to R. As R is truly random, the adversary cannot recompute R from $h(R)$. It follows that knowing $(h(R), Y)$, the adversary has no meaningful information about $f(R, Y)$.

Such a function is called a (σ, τ, ν, μ)-*secure intrusion resilient expansion key function*. Such a function has been designed and proved intrusion-resilient in [DM04].

4.2 Intrusion Resilient Session Key Generation Protocol

The original protocol described in [Dzi06] works in two phases: first, both parties compute a common authentication intermediate key S' from their long shared key. Then, using this key with a MAC function, they exchange the session key, using an asymmetric encryption scheme. The protocol is described in Fig. 2.

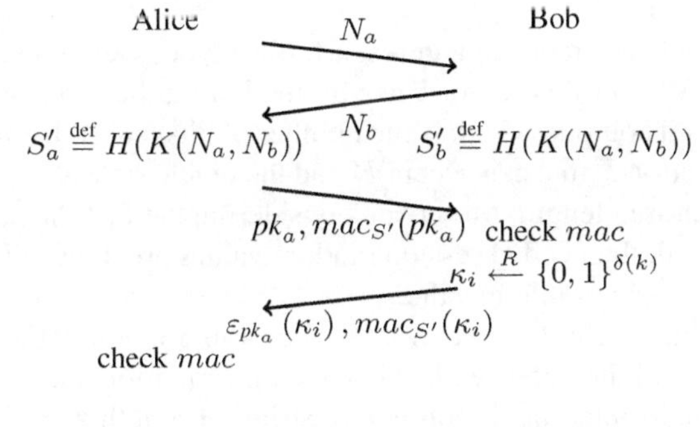

Fig. 2. Description of the Protocol [Dzi06]

For some fixed security parameter k, let $\mathcal{R} = \{0,1\}^{\tau(k)}$, $\mathcal{Y} = \{0,1\}^{\mu(k)}$, $K = (R_A, R_B) \in \mathcal{R}^2$, $\alpha(k) \stackrel{\text{def}}{=} 2\tau(k)$, $f : \mathcal{R} \times \mathcal{Y} \rightarrow \{0,1\}^{\nu(k)}$ is a (σ, τ, ν, μ)-secure intrusion resilient expansion key function, and $H : \{0,1\}^{\nu(k)} \to \{0,1\}^{\lambda(k)}$ is a hash function (modeled as a random oracle function), MAC uses a key of length $\lambda(k)$ and the encryption scheme is an asymmetric encryption scheme semantically secure.

The protocol takes place in seven steps:

1. A picks up randomly Y_A from $\mathcal{Y}$ and sends it to Bob.
2. B picks up randomly Y_B from $\mathcal{Y}$ and sends it to Alice.
3. A and B compute $S_A = S_B = S \stackrel{\text{def}}{=} f(R_A, Y_A) \oplus f(R_B, Y_B)$ and $S'_A = S'_B = S' \stackrel{\text{def}}{=} H(S)$. This terminates the first part of the protocol.
4. A generates a public and private key pair[2], and sends the public key along with a MAC using S' of it (labelled with her identity) to B.
5. B checks the MAC, if it is correct then it generates randomly a session key $\kappa_i \in \{0,1\}^{\delta(k)}$, encrypts it, sends it along with its MAC (labelled once again with B identity) to A and outputs κ_i.
6. A checks the MAC, decrypts the message and outputs κ_i.
7. At the end of the session, A and B erase all their internal data, except K.

If one of the checks fails, the party aborts the session.

This protocol has been proved secure by hand in [Dzi06]. The security proof is done by reducing to the security of the MAC, encryption, and to the intrusion-resilience of f.

5 Intrusion Resilient Session Key Generation Protocol Certified Correct in CIL

As the aim of f is to extract a random output from K, to simplify notations (and proofs), we can consider $f(K, \cdot)$ as a randomizer. As such, we can write $K(n_a, n_b)$ as a shorthand for $f(K, n_a) \oplus f(K, n_b)$. In the Coq proof we simply assume K is an arbitrary function that turns a pair of nonces into a session key, and this is the only way to access the long-term key. However, the attacker can call the K function an arbitrary number of times.

In the Coq model for oracle signatures, arbitrary types can be used for oracle input and output types. We put this to good use by modelling the virus as a dependent pair made of the size of its output, which is the number of bits leaked from the malware, and a higher-order function Γ that can accept K and the oracle state as arguments and return a bitstring of the chosen length. Moreover, considering the fact that an oracle system is deterministic once all the needed session random values are drawn, Γ will actually just need K and these session random values.

The initialization oracle thus receives Γ, draws its random values and executes Γ over K and them, and then and sends the leaked information back to the adversary. A session is then uncorrupted if Γ returns a bitstring of length zero. The oracle system checks that the adversary does not retrieve more information than he is allowed to by incrementing an internal counter of information leakage, and censoring Γ when the quota is reached.

[2] Note that this particular key pair is only valid in this session.

The theorem that we want to prove is the indistinguishability of the original protocol with an idealized one. For that, we make a few (reasonable) assumptions on probability bounds for successful attacks against MAC, encryption schemes and the intrusion-resistant function f.

The idealized protocol is essentially the same as the original one, except for sensible data: in no case they depend on the exchange data. Then if the adversary knows or modifies the exchanged data, she cannot have any valuable information over sensible data.

In order to have the same behavior in both protocols, we adapt the original protocol at two points:

- the computation of S' is no longer $H(K(N_a, N_b))$ but is instead drawn from the output distribution of H. If N_a and N_b are exchanged without any intervention from the adversary, then we draw the same value for Alice and Bob. Otherwise we draw one S'_a for Alice and another S'_b for Bob.
- Bob does not send the encryption of the session key κ_i to Alice. Instead he sends the encryption of zero. If this encryption reaches Alice unmodified by the adversary, then Alice retrieves "magically" κ_i.

By modifying the protocol this way, we are sure that the adversary cannot compute S' from the exchanged nonces, thus she cannot impersonate Alice or Bob later on. She cannot retrieve κ_i either, or modify the cipher to change the value of Alice's κ_i.

Theorem 1. *Assuming the following hypotheses:*

1. *MAC is an Message Authentication Code scheme which cannot be distinguished from an ideal one with a probability more than* ϵ_{mac}
2. *Encr is an encryption scheme, for which the probability to distinguish between an encryption of the zero string and any other encryption is at most* ϵ_{encr},
3. *let f be an (σ, τ, ν, μ)-secure intrusion-resilient function with probability ϵ_f to be broken.*

The protocol presented in Sect. 4.2 can only be distinguished from the idealized protocol with probability at most $\epsilon_{mac} + \epsilon_{encr} + \epsilon_f$, which in CIL means: $\pi \sim_{\epsilon_{mac}+\epsilon_{encr}+\epsilon_f} \pi_{id}$

The next step is to turn both protocols into oracle systems on which we can work using CIL. The mandatory changes to be able to study Dziembowski's protocol are explained in Fig. 3. We recall that the protocol is presented in Fig. 2. As the adversary is both the scheduler and the network of the protocol, we can describe it as in Fig. 3(a): the adversary is the only way for the two honest parties to communicate. From this view, we can design our oracle system (Fig. 3(b)): the adversary can now request a set of oracles, standing for each action of each party. These oracles share a common memory. This common memory is used to store computed values from one action to another. It also stores information to enforce a proper sequencing of requests to Alice and to Bob (*i.e.* avoid duplicate requests). This way, we can transform the protocol π into an oracle system (game) $\mathbb{O}_\pi$.

We want to reduce the security of the protocol (defined by the indistinguishability from the idealized version of the protocol) to the intrusion-resilience of f, and to the security of the MAC and encryption schemes.

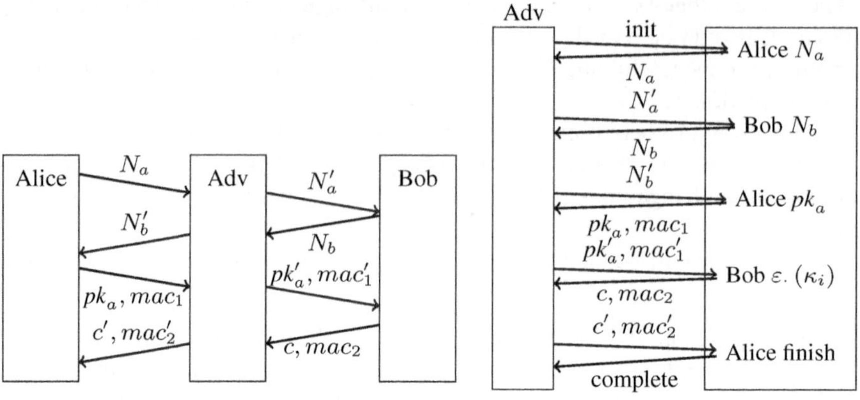

(a) First Step towards Oracle System (b) Oracle System

Fig. 3. Describing Dziembowski: from Protocol to Oracle System

To do so, we will use contexts to divide the protocol (π) in two frames:

- the first part of the protocol (π_0) which aims at establishing S',
- the second part of the protocol (π_1) which uses S' to authenticate and securely exchange the session key.

To end up with a oracle systems having the same behaviour, we write two contexts (Fig. 4): $\mathbb{C}_{\pi_1}$ (to complete $\mathbb{O}_{\pi_0}$) and $\mathbb{C}_{\pi_0}$ (to complete $\mathbb{O}_{\pi_1}$) which simulate respectively the second and first part of $\mathbb{O}_\pi$, and pass the queries from the adversary to the concerned oracles. Since we ultimately want to prove $\mathbb{O}_\pi \sim \mathbb{O}_{\pi^{id}}$, we write them as this combination: $\mathbb{O}_\pi = \mathbb{C}_{\pi_0}\left[\mathbb{O}_{\pi_1}\right], \mathbb{O}_{\pi^{id}} = \mathbb{C}_{\pi_1^{id}}\left[\mathbb{O}_{\pi_0^{id}}\right]$ (equality is here equality on message distributions).

We come to the sensitive part of the proof: we are to keep the value of S' secret from the adversary, and available for the oracles (and the contexts) that need this value to compute MACs to authenticate every messages in $\mathbb{O}_{\pi_1}$.

In the oracle system $\mathbb{C}_{\pi_1}\left[\mathbb{O}_{\pi_0}\right]$ (described in Fig. 4(a)), S' is computed as seen by Alice and Bob in the protocol by the values S'_a and S'_b, which are respectively passed to the context by `Alice_MAC_key_sender` and `Bob_nonce_sender`. In this way the oracles `Alice_MAC_key_sender` and `Bob_sessionkey_sender` are able to authenticate themselves, and the adversary has no access to the value of S'.

In the same manner, in the oracle system $\mathbb{C}_{\pi_0}\left[\mathbb{O}_{\pi_1}\right]$, described in Fig. 4(b), `Alice_MAC_key_sender` passes out to the oracle system $\mathbb{O}_{\pi_1}$ a boolean stating whether or not Alice and Bob computed the same value for S'. From that information, $\mathbb{O}_{\pi_1}$ either draw one (same) S', or two independent for each party.

In Fig. 5, we detail the proof. The first line is our conclusion. To reach it, we swap oracle systems from an idealized version to the original one and vice versa. Thus we can focus on two subgoals: $\mathbb{O}_{\pi_1} \sim \mathbb{O}_{\pi_1^{id}}$ and $\mathbb{O}_{\pi_0} \sim \mathbb{O}_{\pi_0^{id}}$ (third line).

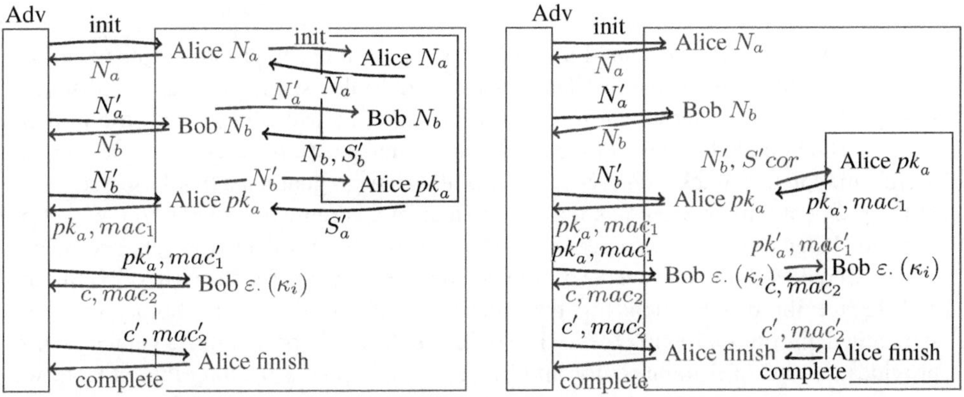

(a) $\mathbb{C}_{\pi_1^{id}}(\mathbb{O}_{\pi_0^{id}})$ (b) $\mathbb{C}_{\pi_0}(\mathbb{O}_{\pi_1})$

Fig. 4. Contexts: Focus on first and second part of the protocol

For both subgoals, we again focus on the key parts of the protocol, using context. Regarding $\mathbb{O}_{\pi_1} \sim \mathbb{O}_{\pi_1^{id}}$, we rely on the encryption and mac scheme. The contexts used are then a simulation of π_1 and of π_1^{id} calling respectively the idealized or usual MAC scheme, and the second the idealized encryption (which always returns an encryption of zero) or the usual encryption scheme. Then this part of the proof relies on:

- the encryption of zero is indistinguishable from any other encryption,
- the MAC scheme is indistinguishable from an idealized MAC scheme.

For $\mathbb{O}_{\pi_0} \sim \mathbb{O}_{\pi_0^{id}}$, we rely on K. The context is then a simulation of π_0 essentially using the oracle K. The basic assumption is that K's result is indistinguishable from the uniform distribution: that is to say that the adversary has no information on K.

$$
\begin{array}{ccc}
\mathbb{O}_{\pi_1} & \sim & \mathbb{O}_{\pi_1^{id}} \\
\| & & \| \\
\mathbb{C}_{\varepsilon_{pk}(\cdot)}[\mathbb{O}_{MAC}] & & \mathbb{C}_{MAC^{id}}[\mathbb{O}_{\varepsilon_{pk}(0)}] \\
\sim & & \sim \\
\mathbb{C}_{\varepsilon_{pk}(\cdot)}[\mathbb{O}_{MAC^{id}}] & = & \mathbb{C}_{MAC^{id}}[\mathbb{O}_{\varepsilon_{pk}(\cdot)}] \\
\Uparrow & & \Uparrow \\
\mathbb{O}_{MAC^{id}} \sim \mathbb{O}_{MAC} & & \mathbb{O}_{\varepsilon_{pk}(\cdot)} \sim \mathbb{O}_{\varepsilon_{pk}(0)}
\end{array}
\Longrightarrow
\begin{array}{ccc}
\mathbb{O}_{\pi} & \sim & \mathbb{O}_{\pi^{id}} \\
\| & & \| \\
\mathbb{C}_{\pi_0}[\mathbb{O}_{\pi_1}] & & \mathbb{C}_{\pi_0^{id}}[\mathbb{O}_{\pi_1^{id}}] \\
\sim & & \sim \\
\mathbb{C}_{\pi_0}[\pi_1^{id}] & = & \mathbb{C}_{\pi_1^{id}}[\mathbb{O}_{\pi_0}]
\end{array}
\Longleftarrow
\begin{array}{ccc}
\mathbb{O}_{\pi_0} & \sim & \mathbb{O}_{\pi_0^{id}} \\
\| & & \| \\
\mathbb{C}_H[\mathbb{O}_K] & \sim & \mathbb{C}_H[\mathbb{O}_U] \\
& \Uparrow & \\
\mathbb{O}_K & \sim & \mathbb{O}_U
\end{array}
$$

Fig. 5. Sketch of the CIL proof

6 Conclusion

In this paper, we have presented a security proof for a cryptographic protocol in the intrusion resilient attack model. This attack model allows the adversary to discover some of the protocol participants' secret information by executing a malware in their system. The bounded-storage model allows us to make assumptions about the amount of information the attacker can retrieve. This allows us to guarantee that future sessions of the protocol remain secure as long as the malware is removed first. To the extent of our knowledge, this is the first certified security proof for a leakage resilient protocol.

The Coq development we have presented consists of two parts: the COQ-CIL library and the specific development for this security proof. The COQ-CIL library provides semantic definitions allowing the user to define his/her own probabilistic games. It also provides a proof of soundness for all CIL rules. As a result of doing this in Coq, we now have a more precise semantics for the judgements and rules in CIL — especially the definition for contexts — and a better assurance as to the soundness of CIL itself. On top of that, the COQ-CIL library now provides us a toolbox for defining and proving security properties in the computational model.

The first article describing the CIL proof [BDL11], we had to provide an extension of CIL to adapt it to the BSM. In the Coq proof this was not the case because the COQ-CIL is general enough to allow us powerful definitions, e.g. the use of higher-order functions as the input type for an oracle in order to model the actual intrusion. This gives some confidence as to the generality and the faithfulness of the formalisation.

Further work. First, since we only used a very limited fragment of the COQ-CIL rules for this development, it would be interesting to do more complex CIL proofs in Coq. There are plenty of cryptographic schemes and protocols out there that would be interesting to formalise starting with those for which we already have CIL proofs. These could include more diverse applications such as signing and e-voting protocols.

A continuation of this work will be the development of a tool to automate the generation of the Coq code for oracles and contexts. The automated code generation will help avoid the error prone cutting and pasting of code, simplify the handling of multiple state types and enable some automation in the definition of contexts by using dependency analysis.

References

[ADW09] Alwen, J., Dodis, Y., Wichs, D.: Survey: Leakage Resilience and the Bounded Retrieval Model. In: Kurosawa, K. (ed.) Information Theoretic Security. LNCS, vol. 5973, pp. 1–18. Springer, Heidelberg (2010)

[APM09] Audebaud, P., Paulin-Mohring, C.: Proofs of randomized algorithms in Coq. Science of Computer Programming 74(8), 568–589 (2009)

[BDKL10] Barthe, G., Daubignard, M., Kapron, B., Lakhnech, Y.: Computational indistinguishability logic. In: Proceedings of the 17th ACM Conference on Computer and Communications Security. ACM, New York (2010)

[BDL11] Barthes, G., Duclos, M., Lakhnech, Y.: A computational indistinguishability logic for the bounded storage model. In: FPS 2011 (2011)

[BGZ09] Barthe, G., Grégoire, B., Zanella Béguelin, S.: Formal certification of code-based cryptographic proofs. In: Proceedings of POPL 2009, pp. 90–101 (2009)

[Bla01] Blanchet, B.: An Efficient Cryptographic Protocol Verifier Based on Prolog Rules. In: 14th IEEE Computer Security Foundations Workshop (CSFW-14), Cape Breton, Nova Scotia, Canada, pp. 82–96. IEEE Computer Society (June 2001)

[Bla06] Blanchet, B.: A computationally sound mechanized prover for security protocols. In: IEEE Symposium on Security and Privacy, pp. 140–154 (2006)

[CKW10] Cortier, V., Kremer, S., Warinschi, B.: A survey of symbolic methods in computational analysis of cryptographic systems. J. Autom. Reasoning, 1–35 (2010)

[Coq] The Coq Proof Assistant, `http://coq.inria.fr/`

[DDMW06] Datta, A., Derek, A., Mitchell, J.C., Warinschi, B.: Computationally sound compositional logic for key exchange protocols. In: Proceedings of CSFW 2006, pp. 321–334 (2006)

[DM04] Dziembowski, S., Maurer, U.: Optimal randomizer efficiency in the bounded-storage model. Journal of Cryptology 17(1), 5–26 (2004)

[DP08] Dziembowski, S., Pietrzak, K.: Leakage-resilient cryptography. In: IEEE 49th Annual IEEE Symposium on Foundations of Computer Science, FOCS 2008, pp. 293–302 (2008)

[Dzi06] Dziembowski, S.: Intrusion-Resilience Via the Bounded-Storage Model. In: Halevi, S., Rabin, T. (eds.) TCC 2006. LNCS, vol. 3876, pp. 207–224. Springer, Heidelberg (2006)

[IK06] Impagliazzo, R., Kapron, B.: Logics for reasoning about cryptographic constructions. Journal of Computer and Systems Sciences 72(2), 286–320 (2006)

[Koc96] Kocher, P.C.: Timing Attacks on Implementations of Diffie-Hellman, RSA, DSS, and Other Systems. In: Koblitz, N. (ed.) CRYPTO 1996. LNCS, vol. 1109, pp. 104–113. Springer, Heidelberg (1996)

[KP10] Kiltz, E., Pietrzak, K.: Leakage Resilient ElGamal Encryption. In: Abe, M. (ed.) ASIACRYPT 2010. LNCS, vol. 6477, pp. 595–612. Springer, Heidelberg (2010)

[Mau92] Maurer, U.M.: Conditionally-perfect secrecy and a provably-secure randomized cipher. Journal of Cryptology 5(1), 53–66 (1992)

[MR04a] Micali, S., Reyzin, L.: Physically Observable Cryptography. In: Naor, M. (ed.) TCC 2004. LNCS, vol. 2951, pp. 278–296. Springer, Heidelberg (2004)

[MR04b] Micali, S., Reyzin, L.: Physically Observable Cryptography (extended abstract). In: Naor, M. (ed.) TCC 2004. LNCS, vol. 2951, pp. 278–296. Springer, Heidelberg (2004)

[RB95] Rogaway, P., Bellare, M.: Optimal asymmetric encryption how to encrypt with rsa (1995)

[Zha08] Zhang, Y.: The computational SLR: a logic for reasoning about computational indistinguishability. IACR ePrint Archive 2008/434, 2008. Also in Proc. of Typed Lambda Calculi and Applications (2009)

Proof Pearl: The Marriage Theorem

Dongchen Jiang[1,2] and Tobias Nipkow[2]

[1] State Key Laboratory of Software Development Environment, Beihang University
[2] Institut für Informatik, Technische Universität München

Abstract. We describe two formal proofs of the finite version of Hall's Marriage Theorem performed with the proof assistant Isabelle/HOL, one by Halmos and Vaughan and one by Rado. The distinctive feature of our formalisation is that instead of sequences (often found in statements of this theorem) we employ indexed families, thus avoiding tedious reindexing of sequences.

1 Introduction

This paper describes two machine-checked proofs [6] of the marriage theorem, also known as Hall's theorem. The theorem was first proved by Hall in 1935 [4]. It provides a necessary and sufficient condition for the ability of selecting distinct elements from a collection of sets.

The standard statement of the theorem is phrased in terms of a finite sequence of sets $A_1, \ldots, A_n$. A sequence of elements $x_1, \ldots, x_n$ is called a *system of distinct representatives* (or *SDR* for short) for $A_1, \ldots, A_n$ iff

1. $x_i \in A_i$ for all $1 \leq i \leq n$, and
2. $x_i \neq x_j$ for all $1 \leq i, j \leq n$ such that $i \neq j$.

Now we can formulate the *Marriage Theorem*:

> A sequence of finite sets $A_1, \ldots, A_n$ (which need not be distinct) has an SDR iff the union of any $m \leq n$ of the A_i contains at least m elements.

The condition for the existence of an SDR is called the *marriage condition*. Note that we restrict ourselves to the finite version. Hall proved it for arbitrary sets A_i. Later work also relaxed the finiteness of the number of sets A_i — see Rado [8] for details.

We started our formalisation with "Proofs from the Book" [1], surely the ultimate reference for beautiful proofs, which also treats the finite version only. This lead to a lengthy proof that required additional concepts and lemmas about sequences. Then it dawned on us that sequences are a complication: the order of the sets is irrelevant, it only matters that there are only finitely many. Hence we replaced sequences by functions with a finite domain. This reduced the length of the proof by a factor of more than 2 to 140 lines. When we went back to the literature we discovered that, as far as the representation is concerned, we had ended up with the indexed families model by Everett and Whaples [3] (although they

J.-P. Jouannaud and Z. Shao (Eds.): CPP 2011, LNCS 7086, pp. 394–399, 2011.

call them sequences). Sequences are easy to understand, even for laymen, which may be why Aigner and Ziegler chose them. Hall himself had used sequences, too. However, sequences are inconvenient as a formal model, as we discovered in our first formalisation and as we shall detail later.

Alerted by a referee, we also formalised the proof by Rado [8]. This proof is already phrased in terms of a family of sets. Its formalisation was shorter again and required only 80 lines. Neither of our two proofs require additional definitions or lemmas beyond what is in the library.

After a review of related work in the next section, Section 3 gives a brief introduction to our logical language, explains the model and states the theorem formally. Section 4 presents our formalisation of the proof by Halmos and Vaughan and we compare our model with the one based on sequences. Section 5 presents our formalisation of Rado's proof, which is very close to Rado's text, and we compare it with the Mizar formalisation by Romanowicz and Grabowski.

2 Related Work

A number of different proofs of Hall's theorem have appeared in the literature [4,3,8], usually by some form of induction (for the finite case) that corresponds to an algorithm constructing an SDR. Aigner and Ziegler [1] follow the proof by Halmos and Vaughan [5], which is beautifully written, avoids almost all technical terminology (not just sequences), and takes a mere 13 lines.[1]

Romanowicz and Grabowski [9] formalised Rado's proof [8] (which is of the order of 15 lines) in the Mizar theorem prover. Their formalisation requires more than 1000 lines; for more details about their proof see Section 5.1. Initially we had ignored Rado's proof because of the length of its Mizar formalisation. But a referee pointed out that this was not Rado's fault: the referee had formalised it in his favourite theorem prover in 40 lines. This was the motivation for our own formalisation of this proof.

The proof on Wikipedia [10] also follows Halmos and Vaughan, but is phrased in terms of "collection of sets" (rather than sequences), and employs set-theoretic notation. If collections are interpreted as sets, the proof does not quite work and the statement is weaker than Hall's. If collections are interpreted as multisets, it works, but drags in multisets gratuitously.

3 Language and Formalisation

Our work was performed with the help of the theorem prover Isabelle/HOL [7], whose set theoretic language is close to that of standard mathematics, with a few minor exceptions. Set difference is written as $X - Y$, and the image of a function f over a set X, i.e. $\{f\ x \mid x \in X\}$, is written as $f\ `\ X$.

[1] Aigner and Ziegler state that Halmos and Vaughan merely rediscovered the proof by Easterfield [2]. This is technically correct, but because Easterfield did not even realise himself that he had proved Hall's theorem, and did not phrase the lemma as abstractly as Hall did, we refer to Halmos and Vaughan for the proof.

HOL is a typed logic with type variables (α, β, etc), function types ($\alpha \Rightarrow \beta$) and set types (α *set*). To express that x is of some type τ we write $x :: \tau$. Predicate *finite* expresses that a given set is finite. The fact that some function f is injective on some set X is written as f *inj-on* X. Updating a function f at argument x with new result value y is written $f(x := y)$.

Earlier on we stated that we would model the collection of sets A_i as a function with a finite domain, thus avoiding sequences with their irrelevant order. In HOL, we express this as a function $A :: \alpha \Rightarrow \beta\ set$ together with a finite set $I :: \alpha\ set$. We call I the *index set.* This model subsumes sequences (let I by the set $\{1, \ldots, n\}$) but is more flexible: we can remove arbitrary subsets from I without the need to renumber the result. Of course, mathematically speaking, renumbering is trivial, but in formal proofs it requires additional machinery and proof steps (see Section 4.1). The marriage condition (for A and I) can now be expressed as follows:

$$\forall J \subseteq I.\ |J| \leq |\bigcup_{i \in J} A\ i|$$

To avoid unnecessary index variables we will write $\bigcup_J A$ instead of $\bigcup_{i \in J} A\ i$.

An SDR (for A and I) is formalised as a function $R :: \alpha \Rightarrow \beta$ that returns the representative for each index and satisfies the following conditions:

1. $\forall i \in I.\ R\ i \in A\ i$, and
2. R *inj-on* I.

Thus the marriage theorem can be stated as follows in Isabelle:

assumes *finite* I **and** $\forall i \in I.\ \mathit{finite}(A\ i)$
shows $(\exists R.\ \forall i \in I.\ R\ i \in A\ i \wedge R\ \mathit{inj\text{-}on}\ I)$
$\longleftrightarrow (\forall J \subseteq I.\ |J| \leq |\bigcup_J A|)$

Necessity of the marriage condition is easy (and takes us 13 lines to formalise). Let R be an SDR for A and I, and let $J \subseteq I$. Hence $R\ `\ J \subseteq \bigcup_J A$ because $\forall i \in I.\ R\ i \in A\ i$. Thus $|R\ `\ J| \leq |\bigcup_J A|$. Because R is injective on $I \subseteq J$, we also have $|J| = |R\ `\ J|$. Combining the two cardinality facts yields the desired $|J| \leq |\bigcup_J A|$.

We will now present the formalisation of two proofs of sufficiency of the marriage condition: we assume the marriage condition and construct an SDR for A and I.

4 The Proof by Halmos and Vaughan

The proof is by induction on the finiteness of I: we may assume that the proposition holds for all proper subsets of I and we have to show it for I. This is slightly more convenient than a proof by induction on the cardinality of I. The proposition to be proved must now include all assumptions of the theorem about I, including the marriage condition, and becomes

$$\begin{aligned} \mathit{finite}\ I \longrightarrow (\forall i \in I.\ \mathit{finite}(A\ i)) \longrightarrow (\forall J \subseteq I.\ |J| \leq |\textstyle\bigcup_J A|) \\ \longrightarrow (\exists R.\ \forall i \in I.\ R\ i \in A\ i \wedge R\ \mathit{inj\text{-}on}\ I) \end{aligned} \quad (1)$$

This is what we prove by induction on *finite* I. The case $I = \emptyset$ is trivial. Otherwise assume $I \neq \emptyset$ and make a case distinction on whether there is a *critical family* (as Aigner and Ziegler call it), i.e. a nonempty $K \subset I$ such that $|K| = |\bigcup_K A|$.

First we assume there is no critical family, i.e.

$$\forall K \subset I.\ K \neq \emptyset \longrightarrow |\bigcup_K A| \geq |K| + 1 \tag{2}$$

Because I is nonempty, we obtain an index $n \in I$. We also have $\forall i \in I.\ A\ i \neq \emptyset$ because an empty $A\ i$, $i \in I$, would imply, by the marriage condition, that $1 = |\{i\}| \leq |\bigcup_{\{i\}} A| = 0$, a contradiction. Thus we obtain some $x \in A\ n$, which we take as the representative for $A\ n$. Then we apply the induction hypothesis to the reduced problem A' and I':

$$A' = \lambda i.\ A\ i - \{x\} \qquad I' = I - \{n\}$$

From the assumption that A and I satisfy the marriage condition, it is easy to show that A' and I' still satisfy the marriage condition. Let J be an arbitrary subset of I'. Because we delete the same element x from each $A\ i$, $|\bigcup_J A'|$ can only be 1 smaller than $|\bigcup_J A|$, which, by (2), means that still $|J| \leq |\bigcup_J A'|$. Thus the induction hypothesis actually applies and yields an SDR R' for A' and I'. Because $x \notin A'\ i$ for $i \in I'$, it is easy to prove that the following R is indeed an SDR for A and I:

$$R = R'(n := x)$$

If there is a critical family, i.e. in the negation of case (2), we obtain a nonempty index set $K \subset I$ such that $|\bigcup_K A| < |K| + 1$. By the marriage condition we have $|K| \leq |\bigcup_K A|$. Together this implies that K is indeed a critical family:

$$|K| = |\bigcup_K A|$$

Because $K \subset I$, the induction hypothesis applies and we obtain an SDR R_1 for A and K. It remains to find an SDR for $I - K$. We simply remove $\bigcup_K A$ from each $A\ j$:

$$A' = \lambda j.\ A\ j - \bigcup_K A \qquad I' = I - K$$

As the cardinality of K equals the cardinality of $\bigcup_K A$, the marriage condition still holds for A' and I'. As also $I' \subset I$, the induction hypothesis applies and we obtain an SDR R_2 for A' and I'. Let

$$R = \lambda i.\ \text{if } i \in K \text{ then } R_1\ i \text{ else } R_2\ i$$

Because we excluded $\bigcup_K A$ from A' ' I', it is clear that $A\ i \neq A'\ j$ for any $i \in K$ and $j \in I'$. Therefore R is an SDR for A and I.

This concludes the inductive proof of (1). The sufficiency of the marriage condition for the existence of an SDR follows trivially.

4.1 Sequences versus Indexed Families

Both our proof and the one by Aigner and Ziegler are more detailed expositions of the proof by Halmos and Vaughan. The only difference is the underlying model: sequence versus indexed family of sets. Sequences are familiar to everybody and their finiteness is built in. The renumbering necessary in the critical-family case is easy for a human, but tedious for the machine. Not only do we need a function to remove a set of indices from a sequence (to allow us to apply the induction hypothesis to a subsequence) but we also need to compute the function that maps indices of the subsequence back to indices of the original sequence (to allow us to turn the SDR obtained for the subsequence into an SDR for the original sequence). And then we need to prove a number of tedious lemmas about how SDRs stay SDRs when they are lifted from the subsequence to the original sequence. All of this because we have introduced an irrelevant order.

5 Rado's Proof

The proof is by induction on the number of $A\ i$ that contain two or more elements. If $|A\ i| \geq 2$ for some i, then Rado shows that there is an $x \in A\ i$ such that $A(i := A\ i - \{x\})$ still satisfies the marriage condition. If all $A\ i$ are singletons, the marriage condition implies that the $A\ i$ directly yield the desired SDR. We merely present the key step of the induction: if $x_1, x_2 \in A\ i$ and $x_1 \neq x_2$, then $A(i := A\ i - \{x_1\})$ or $A(i := A\ i - \{x_2\})$ must satisfy the marriage condition.

Let A satisfy the marriage condition and let $x_1, x_2 \in A\ i$ be such that $x_1 \neq x_2$. For a contradiction, let $A_k = A(i := A\ i - \{x_k\})$ and assume that neither A_1 nor A_2 satisfy the marriage condition. Hence, for both k there is a $J'_k \subseteq I$ such that $|J'_k| > |\bigcup_{J'_k} A_k|$. Because A satisfies the marriage condition, $i \in J'_k$. Let $J_k = J'_k - \{i\}$. Hence $|J_k| \geq |(\bigcup_{J_k} A) \cup (A\ i - \{x_k\})|$. Let $U_k = \bigcup_{J_k} A$ and $U'_k = U_k \cup (A\ i - \{x_k\})$. This leads to the following contradiction:

$$
\begin{aligned}
|J_1| + |J_2| &\geq |U'_1| + |U'_2| \\
&= |U'_1 \cup U'_2| + |U'_1 \cap U'_2| \\
&= |U_1 \cup U_2 \cup A\ i| + |U'_1 \cap U'_2| \\
&\geq |U_1 \cup U_2 \cup A\ i| + |U_1 \cap U_2| \\
&\geq |\bigcup_{J_1 \cup J_2 \cup \{i\}} A| + |\bigcup_{J_1 \cap J_2} A| \\
&\geq |J_1 \cup J_2 \cup \{i\}| + |J_1 \cap J_2| \qquad (3) \\
&= |J_1 \cup J_2| + 1 + |J_1 \cap J_2| = |J_1| + |J_2| + 1
\end{aligned}
$$

Every step can be justified by set theory and side conditions like $i \notin J_1 \cup J_2$ and $x_1 \neq x_2$. Step (3) holds because A satisfies the marriage condition.

5.1 The Mizar Formalisation

Following Rado, Romanowicz and Grabowski [9] provided the first formal proof of the Marriage Theorem, in the Mizar prover. They used sequences, although

Rado used indexed families. However, this should not have a large impact on their proof because there is no need for reindexing, the set I remains fixed throughout the proof. Nevertheless the Mizar proof is much longer than ours. Of course a comparison is difficult because the two provers differ, although at least both computer proofs are declarative and not unreadable proof scripts. The Mizar proof comes to 1600 lines, consisting of 8 definitions and 32 lemmas. Even if we exclude the definitions and lemmas in the Preliminaries and Union of Finite Sequences sections (which can be seen as general background knowledge), there still are more than 1200 lines and 22 lemmas left. But line counts are misleading: after sending those 1200 lines through gzip, 7.3 kB remain, as compared with 1.9 kB for our corresponding proof. This factor of 3.8 probably reflects differences such as automation, the library and sequences vs families.

Acknowledgement. We are grateful to the anonymous referee for motivating us to formalise Rado's proof, too.

References

1. Aigner, M., Ziegler, G.M.: Proofs from the Book. Springer, Heidelberg (2001)
2. Easterfield, T.E.: A combinatorial algorithm. Journal London Mathematical Society 21, 219–226 (1946)
3. Everett, C.J., Whaples, G.: Represetations of sequences of sets. American Journal of Mathematics 71, 287–293 (1949)
4. Hall, P.: On representatives of subsets. Journal London Mathematical Society 10, 26–30 (1935)
5. Halmos, P.R., Vaughan, H.E.: The marriage problem. American Journal of Mathematics 72, 214–215 (1950)
6. Jiang, D., Nipkow, T.: Hall's marriage theorem. In: Klein, G., Nipkow, T., Paulson, L. (eds.) The Archive of Formal Proofs (December 2010), `http://afp.sf.net/entries/Marriage.shtml`; formal proof development
7. Nipkow, T., Paulson, L.C., Wenzel, M.T.: Isabelle/HOL. LNCS, vol. 2283. Springer, Heidelberg (2002)
8. Rado, R.: Note on the transfinite case of Hall's Theorem on representatives. Journal London Mathematical Society 42, 321–324 (1967)
9. Romanowicz, E., Grabowski, A.: The Hall marriage theorem. Formalized Mathematics 12(3), 315–320 (2004)
10. Wikipedia: Hall's marriage theorem — wikipedia, the free encyclopedia (2011), `en.wikipedia.org/w/index.php?title=Hall%27s_marriage_theorem&oldid=419179777` (accessed September 8, 2011]

Author Index

Batch number: 10844619

Printed by Printforce, the Netherlands